Science through Biology

J. J. Head

Team leader responsible for developing biology examinations
in the Nuffield Science Teaching Project.

Head of Science Division, School of Education,
Polytechnic of North London.

Illustrated by Derek Whiteley

Edward Arnold

© J. J. Head 1976

First published 1976
by Edward Arnold (Publishers) Ltd
25, Hill Street, London WIX 8LL

ISBN   0 7131 1972 1

We shall not cease from exploration
And the end of all our exploring
Will be to arrive where we started
And know the place for the first time.

(T. S. Eliot, *Little Gidding*)

Photoset and printed by Interprint (Malta) Ltd

Preface

Science Through Biology contains all the material required for a course in introductory secondary (high) school biology. The book is given coherence by constant reference to the two great biological paradigms on which it is based: the cell, and evolution by natural selection. Some classical experiments are described, and a few 'open-ended' questions are asked, but there is no attempt to expound, in terms that must inevitably be unsatisfactory at this level, what 'scientific method' is. That understanding develops from a reading of the whole book. A great deal can be learned from a consideration of Mendel's experiments alone. Teachers may like to supplement *Science Through Biology* with the open-ended questions in my *Discovering Biology* (Oxford University Press, 1972).

It is a strange fact that although the electron microscope has contributed as much to our biological understanding since the late 1950s as the light microscope did in the whole of the nineteenth century, few introductory school biology textbooks make use of electron micrographs. Yet they are first-hand evidence, their clarity is enormously greater than that of light micrographs, and they are consequently easier to understand. *Science Through Biology* contains many electron micrographs, and is very well illustrated in other ways.

I dedicate this book to my daughter Ellen.

Hampton, June 1976

Contents

Acknowledgements

I should like to thank the following for their permission to reproduce photographs and diagrams.

1.2 Dr. J. A. G. Rhodin, 1.5 After Fawcett, 1.6 Professor A. S. Breathnach, 1.7 After Fawcett, 1.8 Dr. A. M. Glauert, 1.9(a) Professor D. W. Fawcett, 1.9(b) Professor D. H. Northcote, 1.10(a) Dr. E. Munn, 1.10(c) Dr. E. Munn, 1.11(b) Professor D. W. Fawcett, 1.12(a) Dr. Gordon F. Leedale, 1.13 S. D. Douglas and L. N. Chessin, 1.14 Dr. A. C. Allison, 1.16 Professor R. D. Preston, 1.17(a) E. Kellenberger and W. H. Schreil, 1.17(b) T. E. Jensen and C. C. Bowen, 2.1 Professor W. Montagna, 2.2 Professor W. Montagna, 2.4(a) Hinsdale Health Museum, Illinois, 2.4(b) Hinsdale Health Museum, Illinois, 2.5 After A. A. Maximow, 2.6 From *Introduction to Biology* by D. G. Mackean, John Murray, 4.2 Dr. M. B. V. Roberts, 4.5 Dr. E. Munn, 5.5 Mr. A. D. Greenwood, 5.6 Mr. A. D. Greenwood, 5.7 Mr. A. D. Greenwood, 5.8 Mr. A. D. Greenwood, 5.11 Dr. T. A. Mansfield, 5.12 Dr. C. M. Willmer, 5.13 K. Oates, 6.1 Professor D. W. Fawcett, 6.8 From *Introduction to Biology* by D. G. Mackean, John Murray, 6.12 Royal Free Hospital School of Medicine, 6.18(a) After Braus, 6.18(b) After Braus, 6.19 Professor R. M. H. McMinn, 6.20 Professor R. M. H. McMinn, 6.21(a) R. A. G. Stokes, 6.21(c) Dr. P. G. Toner, 6.21(d) Professor D. W. Fawcett, 6.21(e) Professor R. M. H. McMinn, 6.22 Dr. E. Munn, 6.23 C. H. and E. Hewer, 6.25(a) Wellcome Foundation, 6.25(b) after D. G. Mackean 7.8 Royal College of Surgeons of England, 7.16(a) Royal College of Surgeons of England, 7.16(b) Anatomical Institute, Bern, 7.19 Dr. J. A. G. Rhodin, 7.20 (a),(b) Professor E. R. Weibel, 7.21 Royal College of Surgeons of England, 7.23 Professor A. S. Breathnach, 8.1 Professor D. W. Fawcett, 8.2(b) Professor H. Z. Movat, 8.5 T. Kuwabara, 8.13 Royal College of Surgeons of England, 8.19 Royal College of Surgeons of England, 8.21 After Schaffer, 8.23 Dr. G. D. Pappas, 10.1(b) Professor D. H. Northcote, 10.4 Dr. F. A. L. Clowes, 10.6 After W. M. M. Baron, 10.10 D. J. Avery, 10.15 Dr. J. Cronshaw, 10.16 F. B. P. Wooding, 10.18 Dr. G. T. O'Loughlin, 11.3 Royal College of Surgeons of England, 11.7 Professor D. B. Moffat, 11.8 Professor D. B. Moffat, 12.6 After Maximow, 12.7 After Cynthia Clarke 12.11(a) Professor D. W. Fawcett, 12.11(b) Dr. A. C. Allison, 12.12 Central Press Photos, 12.13(a) Carnegie Institute, 12.13(b) Carnegie Institute, 12.14(a) Carnegie Institute, 12.14(b) Carnegie Institute, 12.15 Carnegie Institute, 12.16 Carnegie Institute, 12.17 Professor W. J. Hamilton, 12.18 Professor W. Montagna, 12.21 Professor W. J. Hamilton, 12.23 San Francisco Medical Center, University of California, 12.26 (whole sequence) From *'You and your baby'* Family Doctor Publications, 12.29 Syndication International and *Daily Mirror*, 13.17(a) Dr. M. Black, 13.17(b) Dr. M. Black, 14.11 Dr. K. R. Lewis and Professor B. John, 14.12 Professor C. D. Darlington, 14.13(a) Dr. A Bajer, 14.13(b) Dr. A. Bajer, 14.14 Dr. K. R. Lewis and Professor B. John, 14.15 Dr. S. A. Henderson, 15.5 Professor S. Polyak, 15.6 Professor H. de F. Webster, 15.7 Professor E. G. Gray, 15.9(a) Professor J. Z. Young, 15.10(a) Professor E. G. Gray, 15.11(a) Professor E. G. Gray, 15.11(b) Professor E. G. Gray, 15.12(a) Professor E. G. Gray, 15.12(b) Professor E. G. Gray, 15.15 C. H. and E. Hewer, 15.22(a),(b) Professor E. G. Gray, 15.23 Professor D. M. MacKay, 15.25 Department of Human Anatomy, Oxford, 15.27 Professor L. Orci, 16.1 Dr. C. K. Brain, 16.2 After T. N. Cornsweet, 16.3 Professor H. Mizoguti, 16.6 After D. G. Mackean, 16.9 E. F. Fincham, 16.11 J. Marshall and P. L. Ansell, 16.13 Professor R. L. Gregory, 16.15 Professor I. Friedmann, 16.16(b) Royal College of Surgeons of England, 16.17 Royal College of Surgeons of England, 16.19(a) (b) (c) Ferens Institute of Otolaryngology, 17.2 Professor J. J. Pritchard, 17.3 Professor J. J. Pritchard, 17.4 From Boyde and Hobdell, 1969, 17.5 Professor J. J. Pritchard, 17.20(b) Professor J. J. Pritchard, 17.23 Professor D. W. Fawcett, 17.24 Professor H. Mizoguti, 17.25 Professor A. W. Ham, 17.26 Professor J. J. Pritchard, 18.2 Reproduced from St. Mary's Hospital Medical School Pamphlet, 18.3(a) Dr. Gordon F. Leedale, 18.3(b) After Leedale, 18.4(a) Dr. Gordon F. Leedale, 18.4(b) Dr. Gordon F. Leedale, 18.5 Dr. Gordon F. Leedale, 18.7 Dr. E. G. Jordan, 18.19(a) Professor V. B. Wigglesworth, 18.19(b) After V. B. Wigglesworth, 18.19(c) R. A. Haynes, 18.20(b) Professor J. W. L. Beament, 18.22 Dr. R. F. Chapman and the Centre for Overseas Pest Research, 18.23(a) M. D. Kendall, Centre for Overseas Pest Research, 18.23(b) M. D. Kendall, Centre for Overseas Pest Research, 18.24(a) M. D. Kendall, Centre for Overseas Pest Research, 18.24(b) Cambridge Instruments Ltd., 18.31(a) Zoological Society of London, 18.31(b) Zoological Society of London, 18.31(c) Zoological Society of London, 18.32 after D. G. Mackean, 18.34 Professor E. J. Denton, 18.35 After G. M. Hughes, 19.1 O. N. Allen, 19.2 Professor C. E. Fogg, 19.5 J. F. Farrar, 19.6 J. F. Farrar, 19.7(a) Wellcome Museum of Medical Science, 19.7(b) K. O. Wood and R. G. Bird, 19.8 British Museum (Natural History), 19.16 Based on Donald J. Bogue, Principles of Demography (New York: John Wiley and Sons, 1969), 19.20 Based on Stig H. Fonselius, *Stagnant Sea, Environment* July 1970, 20.2 Hale Observatories, 20.3 Hale Observatories, 20.4 Professor R. W. Horne, 20.6(f) British Museum (Natural History), 20.7 Dr. M. Gravelle, 20.8 Professor E. S. Barghoorn, 20.13(a) British Museum (Natural History), (b) Zoological Society of London, 20.17 Dr. H. D. B. Kettlewell, 20.19(a) Professor E. B. Ford and J. S. Haywood, 20.19(b) Professor E. B. Ford and J. S. Haywood, 20.19(c) Professor E. B. Ford and J. S. Haywood, 20.20 Dr. A. C. Allison.

Cover photograph: Vertical section through a capillary from the lung of a dog, Courtesy of E. R. Weibel (electron micrograph).

The remaining photographs are by the author.

1 Cells, Units of Plant and Animal Structure

The finest detail that the human eye can see or *resolve* is about 0.1 mm. That is, most people could see two small dots which were 0.1 mm apart as separate dots. If they were closer they would appear to be one dot.

The microscope was invented about 1590, but it was a crude instrument by modern standards, two lenses at the opposite ends of a tube. Some isolated observations were made with it, but it was not until the nineteenth century, over 200 years later, that microscopy led to major advances in man's understanding of living organisms. Some of the observations were nevertheless thorough. In 1660, Malpighi discovered the existence of blood capillaries in the lung of a frog. He inflated the lung, dried it and examined it with his microscope. Capillaries are the smallest blood vessels in the body and cannot be seen with the naked eye. Fig. 8.5 is a photograph of capillaries in the retina (part of the eye). Malpighi also described the cell structure of parts of plants, including pith, buds, and leaves.

Nehremiah Grew (1641–1712), an English doctor, made many observations on the structure of plant tissues. Although Malpighi and Grew both observed plant cells, the idea that all plants (and animals) are composed of cells was not recognized as important concept until 100 years after Grew was dead.

Robert Hooke examined a thin slice of cork in 1665 and described it as being composed of small air-filled compartments which he called cells. He did not attach any biological importance to his discovery, and we now know that cork is dead tissue and that Hooke was looking only at non-living cell walls.

The microscope was only used as a toy in the eighteenth century. In 1827 a big improvement was made in the quality of microscope lenses. Until then, colourless objects examined under the microscope shone with the colours of the rainbow, making accurate observation difficult. This fault was overcome by Dolland, and thereafter there was a rapid spread of interest in microscopy and some nineteenth century scientists devoted their working lives to improving methods of preparing and staining thin slices of plant and animal tissues and discovering what they contained.

The cell theory

By 1839 enough observations had been made for Theodore Schwann to state that:

'All organized bodies (that is, animals and plants) are composed of essentially similar parts, namely cells.'

This idea, although based on limited observations, very rapidly became accepted, and is called the **cell theory**. Later, in 1858, a German called Virchow added a further idea:

'All cells arise from previously-existing cells.'

We now know that Schwann was wrong in many of the things he thought about cells, but even so the cell theory is one of the greatest ideas that has been formulated in the history of biology, and has probably led to more advances in the science than any other idea.

Looking at cells

Using a good microscope, objects which are as close together as 0.2 μm can be seen separately. One μm is 0.001 mm, and this means that the 'seeing power' of the unaided human eye has been increased some 200 times. Your school microscope will probably not be as good as this, but it will show a lot of interesting details of cells. It probably has two lenses at the base of the tube marked $\times 10$ and $\times 40$. These are called the objective lenses. If the eyepiece lens is marked $\times 10$ you have two magnifications available, $10 \times 10 = 100$ and $10 \times 40 = 400$. To get good results it is important to know how to adjust the light source properly and your teacher will show you how to do this.

Appearances of cells in sections

Blood cells lie freely suspended in liquid; cells from the lining of the cheek are easily detached from the cells which lie beneath them; both these cell types are therefore easily prepared for microscopic examination. But most cells, for example, animal muscle cells or plant stem cells, are joined to other cells. It is sometimes possible to pull such cells apart to examine them, but the usual method is to examine a thin slice of the tissue in which the cells lie. Such sections contain only small parts of cells, and different sections of the same cell will show different features. For example, suppose we were investigat-

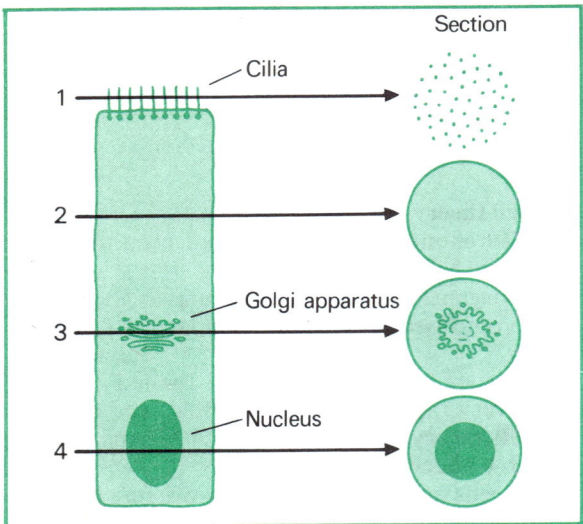

Fig. 1.1 Transverse sections in different planes contain different parts of a cell.

ing a cell shaped as shown in Fig. 1.1. Thin transverse sections made in the planes marked 1, 2, 3, and 4 would contain different parts of the cell. Section 1 would show only the cilia; section 2 would show some of the cytoplasm and we would see little more than the edge of the cell and some apparently structureless contents. Section 3 would show a structure called the Golgi apparatus; section 4 (and none of the others) would show the nucleus. The fact that the first three sections did not show a

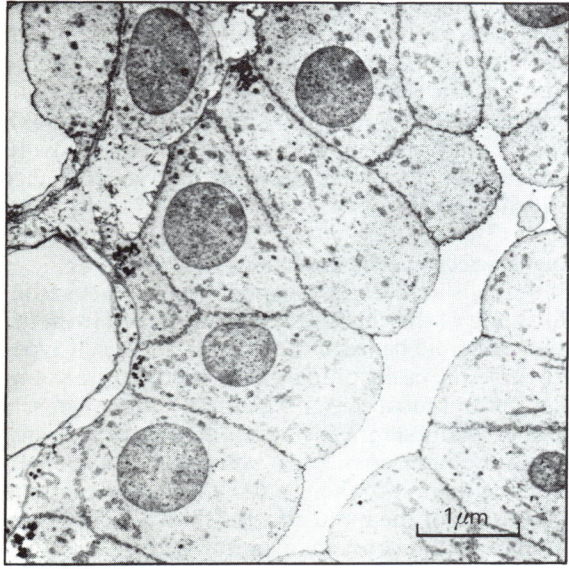

Fig. 1.2 Electron micrograph showing sections of cells in the human kidney.

nucleus does not mean that the cell did not contain one.

Look at Fig. 1.2, a section of some cells in the human kidney. Why do you think that the nuclei appear to be of different sizes? Why do some of the cells in the section show no nuclei? What do you think would be the three-dimensional shape of these cells?

One section can never tell us the true shape of a cell. Figure 1.3(b) is a stereogram of part of a buttercup root and some of the cells are seen both in transverse section and longitudinal section. Look at the cells marked X and Y. By examining only a transverse section we might conclude that both these cells were spherical. However, the longitudinal face of the drawing shows that the cells X are more or less spherical because their shape is similar when they are cut transversely and longitudinally; but cells Y are quite different. Although they have a circular outline when cut transversely, they are long and thin when cut longitudinally.

Now suppose that the section of cell Y is not perfectly transverse but has been cut on a slant. Then, although Y is really a tube of circular cross section, it will appear to be oval (Fig. 1.4). And if the plane of the longitudinal section did not pass through the centre of the hole in cell Y, it may appear

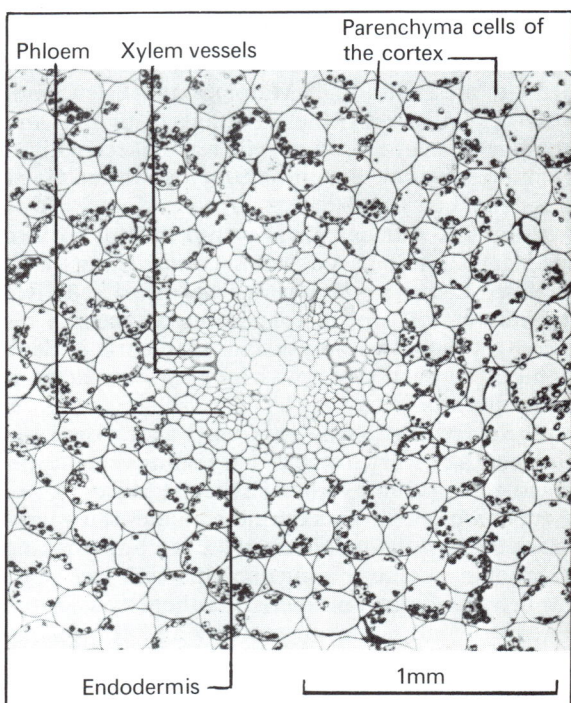

Fig. 1.3(a) Transverse section of part of the centre of a young buttercup root (Courtesy Carolina Biological Supply Co.).

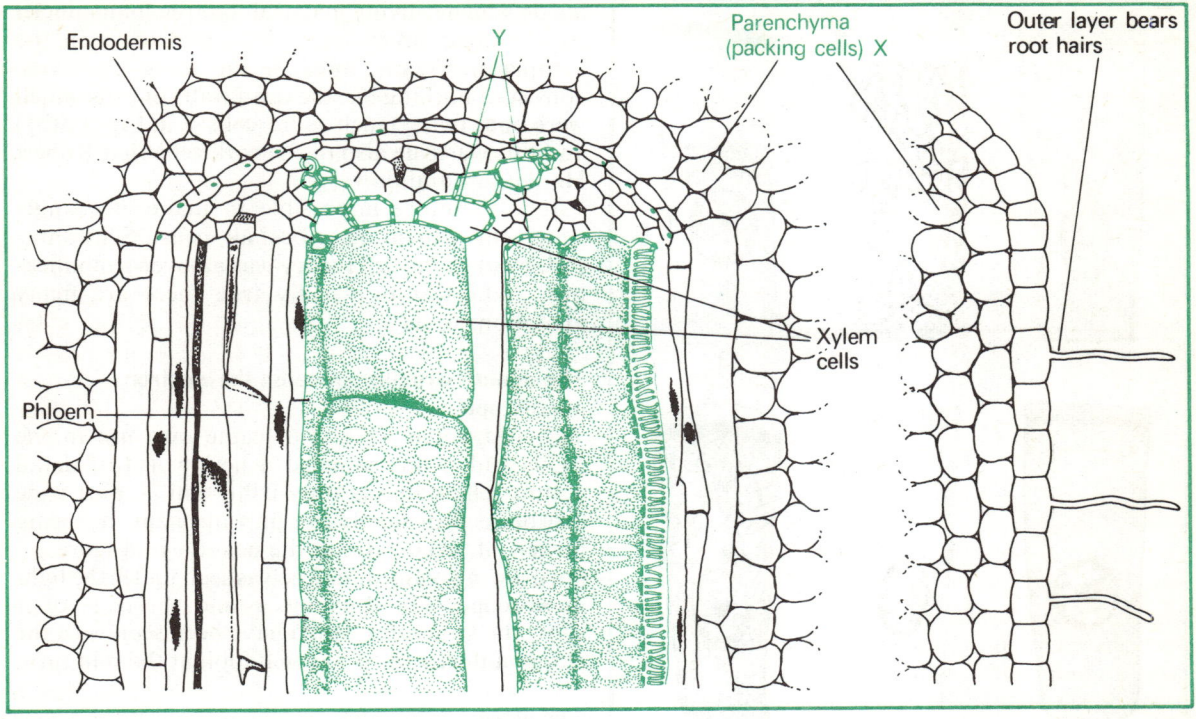

1.3(b) Stereogram of part of the centre of a young buttercup root.

either to be narrower than it really is or to be a solid object with no hole in the centre (Fig. 1.4).

Fixing and staining cells

Very little can be seen if a *living* cell is examined under the microscope. This is because most of the parts of the cell are equally transparent. To see detail, the sections containing cells have to be treated with dyes or stains, which make the different parts visible when light is shone through them. But a living cell soon decays, and before staining it must be treated with chemicals to prevent it disintegrating. This is called **fixation**. The chemicals that are used for this purpose act by altering the chemical structure of the proteins in the cell, making them insoluble. Formaldehyde, mercury (II) chloride, and osmic acid are three commonly used fixatives, and of course they are all strong poisons.

During fixation, insoluble materials may be formed within the cell which do not exist in the living state.

Summarizing: 1. Sections of cells only contain parts of the cell. 2. Fixation may produce 'objects' (artefacts) which may be mistaken for true components of the living cell. Only research can deter-

mine whether such objects are true cell components or are artefacts. 3. Different stains stain different parts of the cell and may give quite different impressions of what is really there. This is well

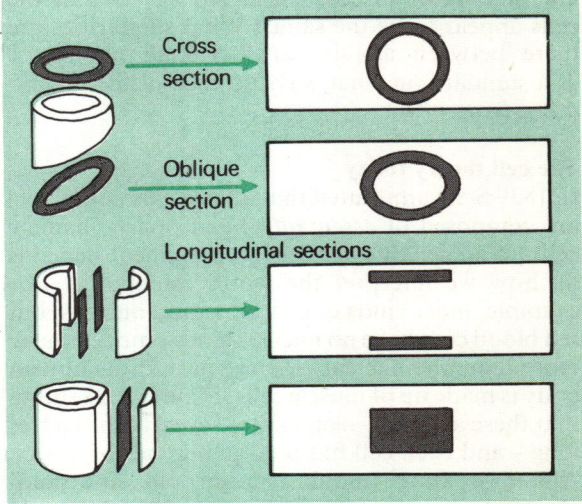

Fig. 1.4 Diagram showing how the plane of a section can alter the microscopic appearance of a tube-shaped cell.

3

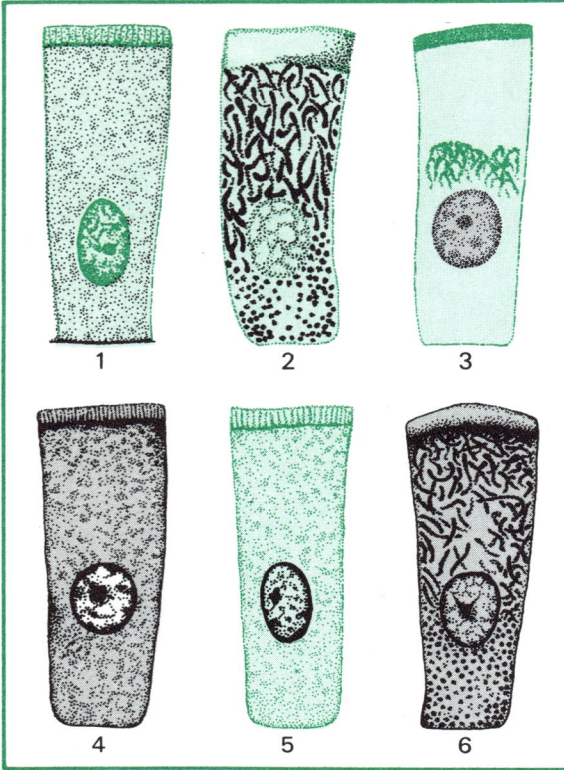

Fig. 1.5 Cells from the lining of the small intestine of a guinea pig. All the cells shown are identical in structure, but the different fixing and staining methods used give them very different appearances.

illustrated in Fig. 1.5 which shows drawings of identical cells from the intestine of a guinea pig fixed and stained in various ways. Do any two of the cells appear to be the same? What similarities are there between all six cells? Taking number 1 as a standard, in what ways do each of the other 5 differ from it?

The cell theory today
In 1839 Schwann stated that all animals and plants are composed of *essentially similar* parts, namely cells. Whether this is a correct statement depends on how we interpret the word 'essentially'. For example, most kinds of cells possess a nucleus, but red blood cells have no nucleus, and some cells have more than one nucleus. A large part of the human body is made up of muscle cells. Figure 17.22 shows that these are long—sometimes several centimetres long—and each cell has a large number of nuclei. Also most fungi (moulds, mushrooms and toadstools) do not consist of cells each with a separate nucleus. Their 'cells' have a distinct outer wall

inside which is living material, but the many nuclei they contain are situated here and there in the cytoplasm. Again, most of the 'cells' in a tree consist of nothing but the dead walls of cells which were once living (such as the cells Y in Fig. 1.3(b)) and this was true also of the cork cells that Robert Hooke saw and drew.

It is therefore impossible to make any single statement which describes all cells. Schwann's statement of the cell theory was an over-simplification. Whilst it is broadly true there are many exceptions to it.

The minute structure of cells: the electron microscope
This instrument (Fig. 1.6) came into use in the 1950s and it has revealed a new world of detail within cells. Objects only 0.001 μm (1 nm) wide can be seen distinctly, an improvement in 'seeing power' of 100 000 times the naked eye. Figure 1.7 is a drawing of an animal cell as seen under the light microscope. The drawings round the edge of it show the structures which have been seen with the electron microscope. We will look at them in turn.

The plasma membrane
This bounds the outside of the cell and separates it from other cells and from the liquid outside the cell. It is not an absolute barrier like the steel hull of a submarine; it allows water and dissolved chemicals to pass through it, but in an orderly and regulated way. It consists of a sandwich of three layers. These

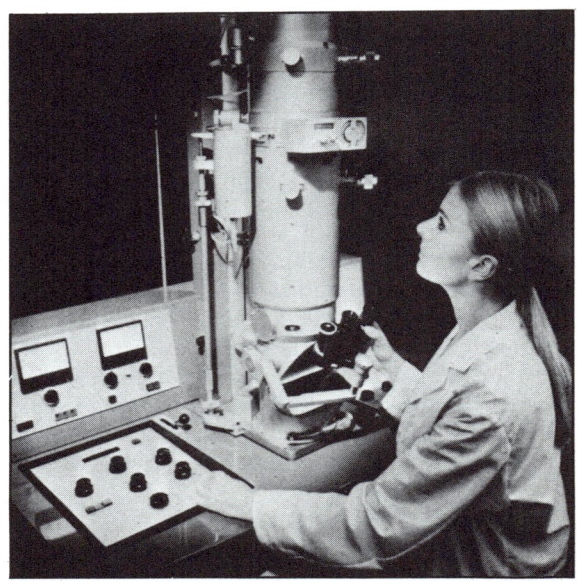

Fig. 1.6 An electron microscope.

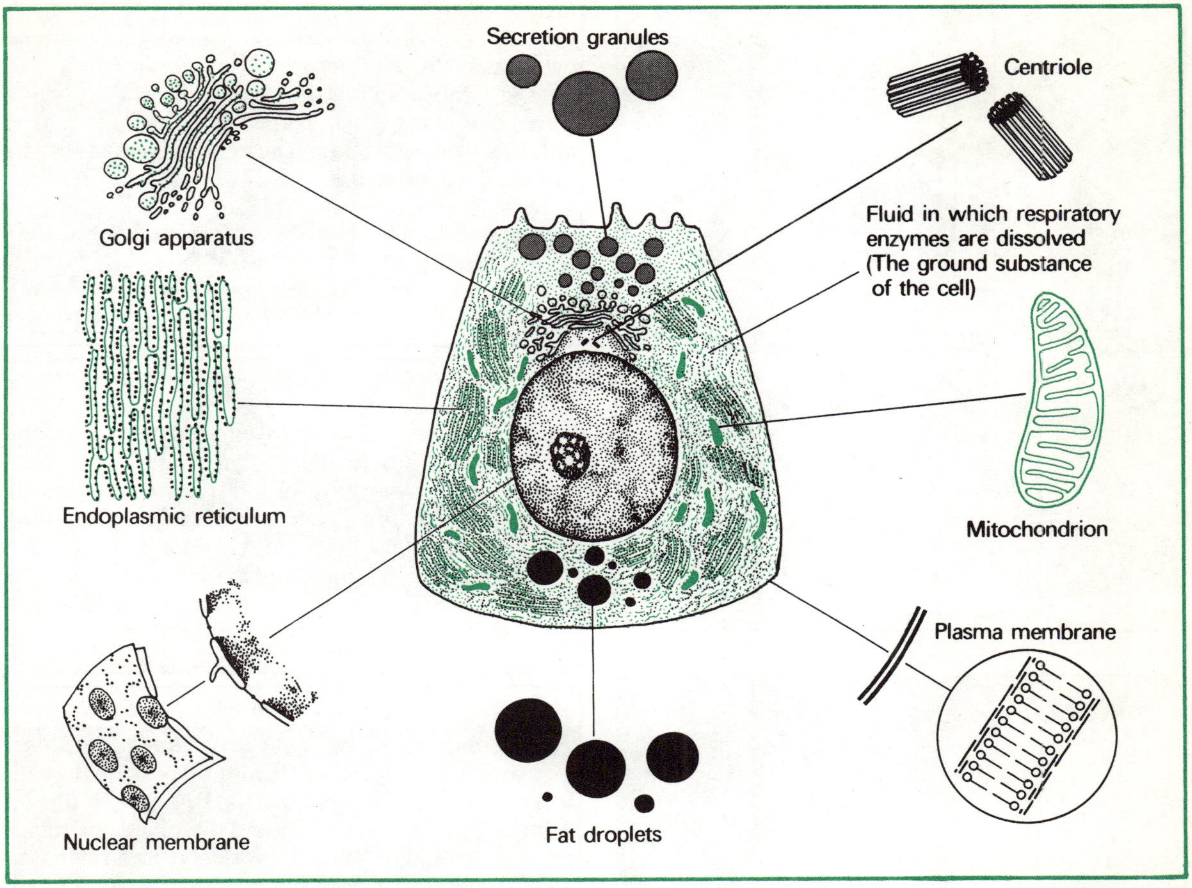

Secretion granules

Centriole

Golgi apparatus

Fluid in which respiratory
enzymes are dissolved
(The ground substance
of the cell)

Endoplasmic reticulum

Mitochondrion

Nuclear membrane

Fat droplets

Plasma membrane

Fig. 1.7 Diagram of an animal secretory cell showing the organelles enlarged as shown by the electron microscope.

can be seen in Fig. 1.8, which is an electron micrograph of a cell membrane. The outer layers (black in the electron micrograph) are made of protein and it is thought that they control the passage into the cell of ions and water-soluble chemicals. In between is a layer of fatty material (grey in the electron micrograph). It is thought that this controls the entry of fat-soluble chemicals. The whole sandwich is about 0.008 μm (8 nm) thick.

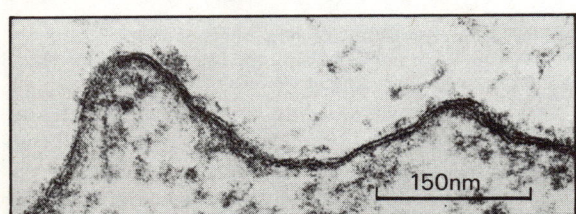

Fig. 1.8 Electron micrograph of a plasma membrane in transverse section.

150nm

The nucleus

This will be considered in more detail in chapter 14. Figure 1.9(a) is an electron micrograph of a section through a nucleus. It is bounded by a membrane essentially similar to the plasma membrane except that there are many holes or pores in it. These can be seen in surface view in Fig. 1.9(b), and they account for about 20 % of the surface area of the nuclear membrane. The nucleus controls the activity of the cell and chemical messages from it pass out through these pores to the remainder of the living cell material, the **cytoplasm**, telling it what to do.

Mitochondria

These can be seen with the light microscope, and some of the 6 pictures in Fig. 1.5 show external views of them. Under the electron microscope they can be examined in section. Figure 1.10(a) shows

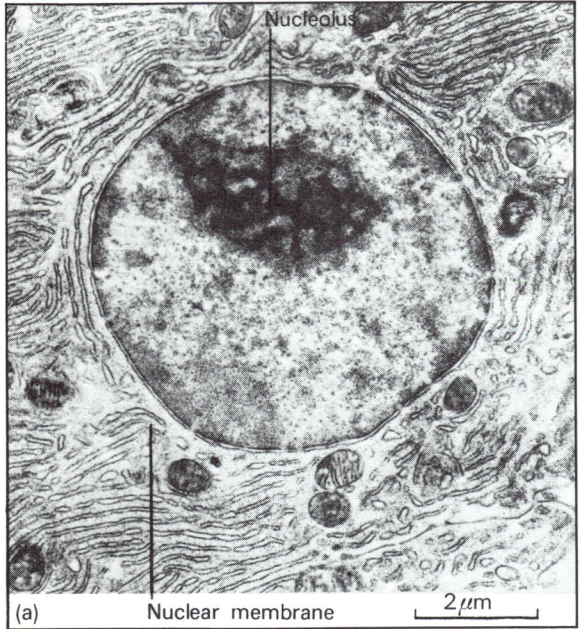

Nucleolus

Nuclear membrane

$2\,\mu m$

(a)

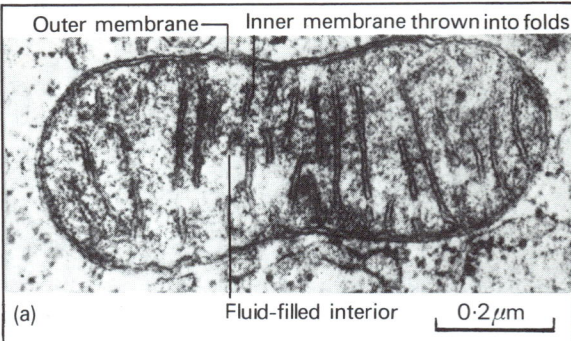

Outer membrane — Inner membrane thrown into folds

Fluid-filled interior

$0\cdot2\,\mu m$

(a)

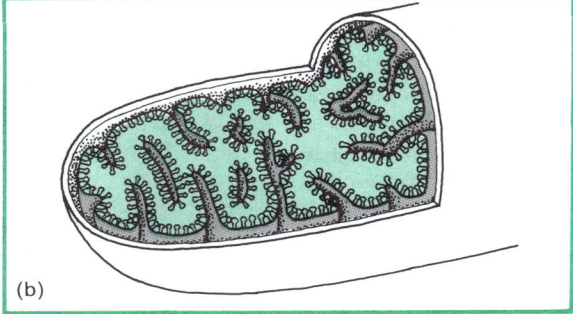

(b)

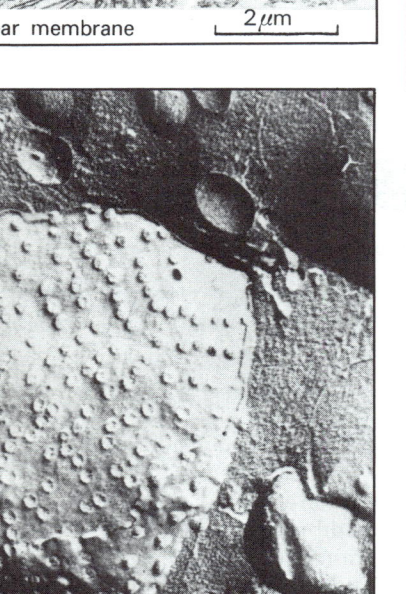

Pores in the nuclear membrane

$2\,\mu m$

(b)

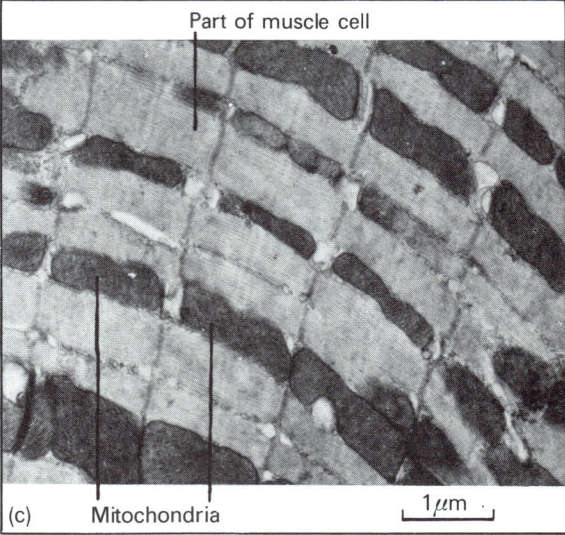

Part of muscle cell

Mitochondria

$1\,\mu m$

(c)

Fig. 1.9(a) Electron micrograph of a section through part of one cell from the pancreas, showing the nucleus. The pores in the nuclear membrane are indicated by arrows.
(b) Scanning electron micrograph showing a surface view of a nucleus. Note the many pores in the nuclear membrane.

Fig. 1.10(a) Longitudinal section of a mitochondrion (electron micrograph).
(b) Stereogram reconstruction of a mitochondrion.
(c) Electron micrograph showing the large numbers of mitochondria situated between muscle cells in heart tissue.

one cut lengthways. They are 2–4 μm long and about 0.5 μm in diameter. They consist of two membranes, the inner one being thrown into a series of folds covered with granules. Between the folds is a dense fluid. Figure 1.10(b) is a stereogram reconstruction of a mitochondrion.

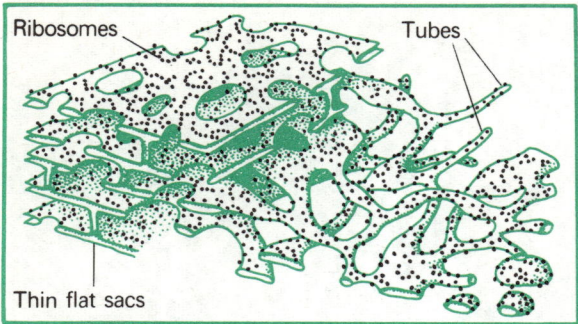

Fig. 1.11(a) Diagram of part of the endoplasmic reticulum.

Their function is to supply chemical energy for the cell. They are the cell's power stations. The number of mitochondria per cell varies from about 50 to about 5000, and active cells, such as muscle cells, always contain a lot of them. Figure 1.10(c) shows the large number of mitochondria present between muscle cells in mouse heart tissue.

Endoplasmic reticulum
The structures, or **organelles**, that we have examined are surrounded by the **ground substance** of the cytoplasm. This appears structureless under the light microscope, but the electron microscope shows that embedded in it are large numbers of thin flat sacs which occasionally narrow down into fine tubes (Fig. 1.11(a)) This network is called the **endoplasmic reticulum**. Both sacs and tubes are made of membranes and are covered on the outside with thousands of minute granules called **ribosomes.** Inside the sacs and tubes is a fluid. In section the endoplasmic reticulum appears as a series of parallel lines (Fig. 1.11(b)). It forms a series of complicated pathways through which materials are moved from one part of the cell to another.

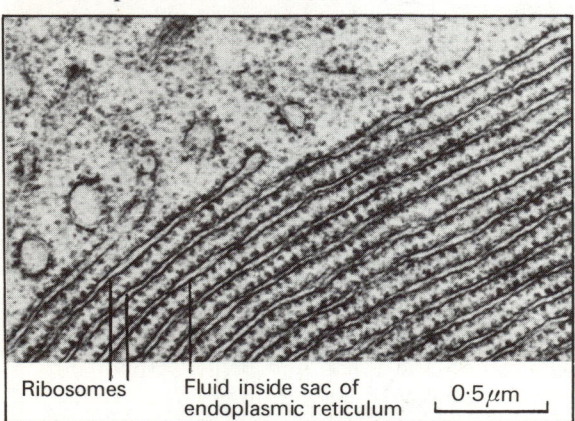

Fig. 1.11(b) Electron micrograph of a section through several sacs of the endoplasmic reticulum.

Golgi apparatus
This organelle was first described by Camillo Golgi in 1898, but for many years there was much controversy as to whether it was a structure that really exists in the living cell or whether it was an artefact produced by the method of fixing and staining cells. However, the electron microscope shows it clearly. It consists of a series of membranes, arranged something like a pile of saucers and attached to the endoplasmic reticulum (Fig. 1.12(b)). When sectioned the organelle looks like a

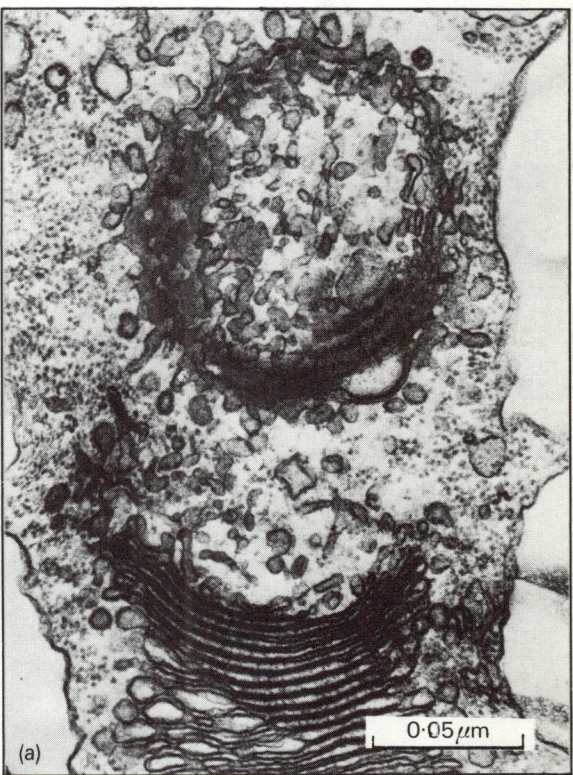

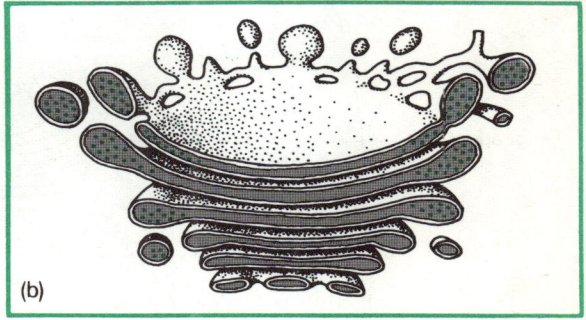

Fig. 1.12(a) Two Golgi apparatuses, upper one seen from above, lower one seen in section.
(b) Stereogram reconstruction of a Golgi apparatus.

series of curved double lines. Fig. 1.12(a) is an electron micrograph showing one Golgi apparatus in section (the lower one) and another viewed from above (the upper one). One of the functions of this organelle is to gather up chemicals that have been made in the cell and parcel them inside a membrane.

Fig. 1.13 Section of a human white blood cell showing lysosomes (electron micrograph).

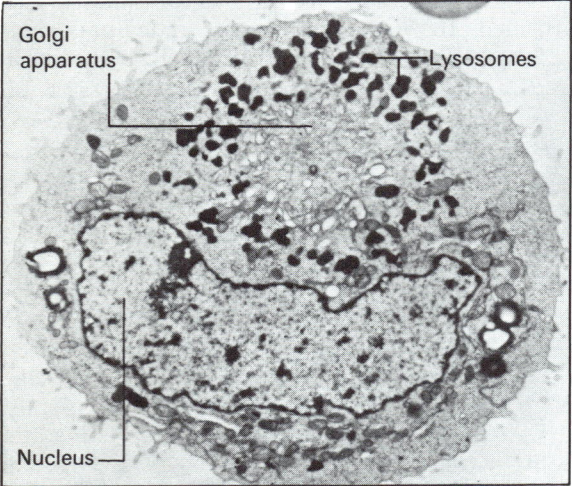

Golgi apparatus

Lysosomes

Nucleus

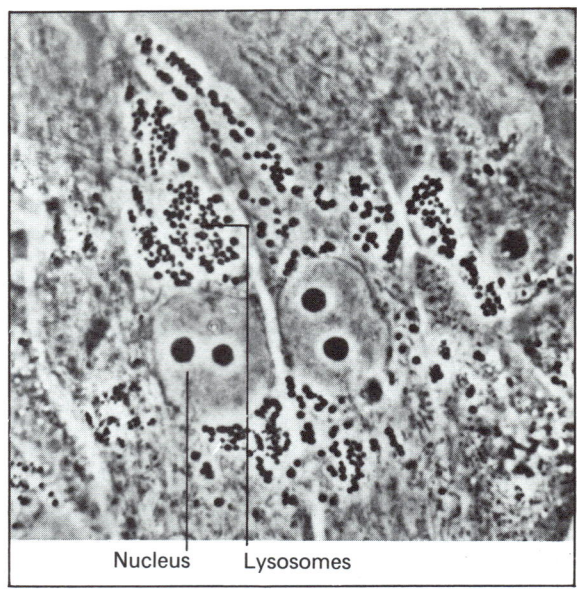

Nucleus   Lysosomes

Fig. 1.14 Cells of monkey kidney, showing large numbers of lysosomes (light microscope).

The material can be exported from the cell or taken to parts of the cell where the chemical is needed. For example, salivary glands secrete saliva, and the pancreas and intestine secrete digestive juices. These are solutions of proteins. The nucleus sends

Fig. 1.15 Diagram showing the activities of lysosomes.

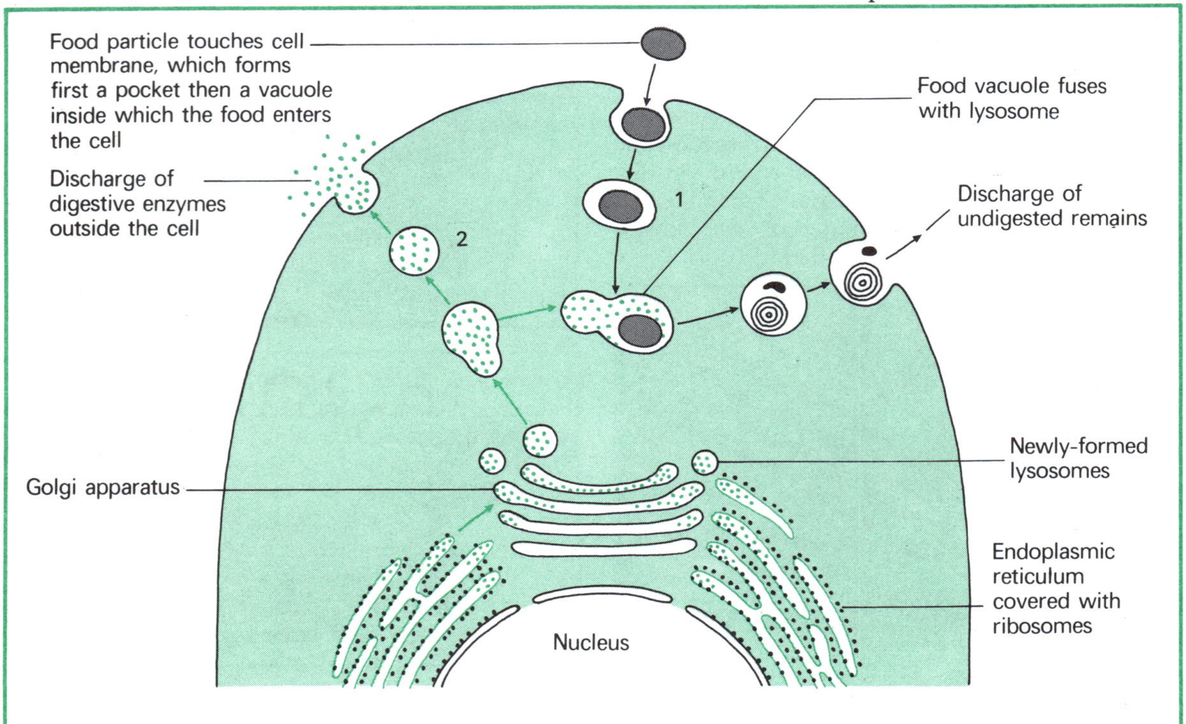

Food particle touches cell membrane, which forms first a pocket then a vacuole inside which the food enters the cell

Discharge of digestive enzymes outside the cell

Golgi apparatus

Food vacuole fuses with lysosome

Discharge of undigested remains

Newly-formed lysosomes

Endoplasmic reticulum covered with ribosomes

Nucleus

1

2

chemical instructions to the ribosomes, which make the appropriate protein. This is passed along the network of the endoplasmic reticulum until it reaches the Golgi apparatus, where it is parcelled into droplets ready for secretion from the cell. In Fig. 1.12(b) you can see spherical droplets enclosed by membranes that have been nipped off from the edge of the saucer-like membranes of the Golgi apparatus.

Lysosomes

These were discovered in 1955 in animal cells. Recently they have been found also in plant cells. They consist of small spherical packets of digestive enzymes each surrounded by a membrane, and they are made by the Golgi apparatus. In Fig. 1.13 they have been stained black and can be seen surrounding the Golgi apparatus. Fig. 1.14 shows their appearance under the light microscope in mouse kidney cells.

Lysosomes are known to have at least two functions. (1) They digest material that has been taken into the cell. (2) They release the digestive enzymes they contain outside the cell. This causes the breakdown of material outside the cell, and an example is in the shortening of tadpole tails when a tadpole is turning into a frog.

Figure 1.15 is a drawing showing some of the activities of lysosomes in diagrammatic form.

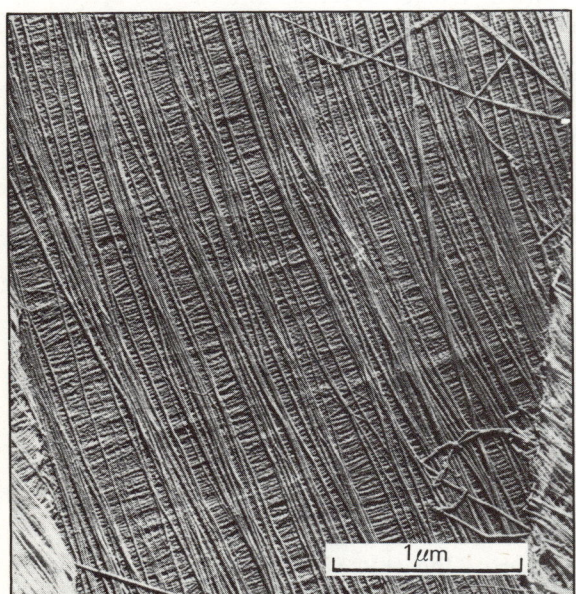

Fig. 1.16 Electron micrograph of a surface view of part of a plant cell wall. The individual strands of cellulose can be clearly seen.

Digestion of materials taken into the cell Some cells take into their cytoplasm solid material from outside themselves. For example white blood cells engulf and digest bacteria, thus helping to defend the body against disease. Also unicellular animals, the Protozoa, of which *Amoeba* is perhaps the best-known example, engulf food particles and digest them. When the food particle touches the plasma membrane, the membrane becomes pinched in, forming a space or **vacuole** into which the food particle is taken. This moves towards the Golgi apparatus and fuses with lysosomes, whose digestive enzymes dissolve the food. The vacuole now moves to the surface of the cell, where undigested remains are discharged (Fig. 1.15, route 1).

Plant cells

All the organelles we have seen so far occur in both animal and plant cells. Three other important structures are found in plant cells but not in animal cells; the **central permanent vacuole**, the **cellulose cell wall**, and the **chloroplasts**. These can all be seen in Fig. 5.6, a section of leaf cells seen under the electron microscope. The bulk of the cell consists of the large central vacuole, which is non-living and contains water with various chemicals dissolved in it. It presses against the cytoplasm and the cell wall, keeping the cell stiff or turgid. The cytoplasm forms a thin layer at the edge of the cell. It is bounded on either side by a plasma membrane and embedded in it are the numerous chloroplasts. These are organelles whose special function is to manufacture sugar from carbon dioxide and water using light energy (photosynthesis). The cell wall, which lies outside the outer plasma membrane, is also non-living and consists of fibres of cellulose laid down in criss-cross layers (Fig. 1.16). Unlike the plasma membranes, the cellulose wall offers no resistance to the passage of dissolved substances. Its function is as part of the plant's skeleton.

Are all cells basically similar?

The cell theory states that all living things are composed of essentially similar parts, namely cells. This is only true in a very broad sense, depending on what we mean by 'cell' and by 'essentially similar'.

Studies with the electron microscope have shown that living organisms can be divided into two groups with regard to their cell structure, the **Prokaryota** (Greek, first cells) and the **Eukaryota** (Greek, proper cells). The first group, the Prokaryota, consists of the bacteria and the blue-green algae, a group of very simple plants some single-

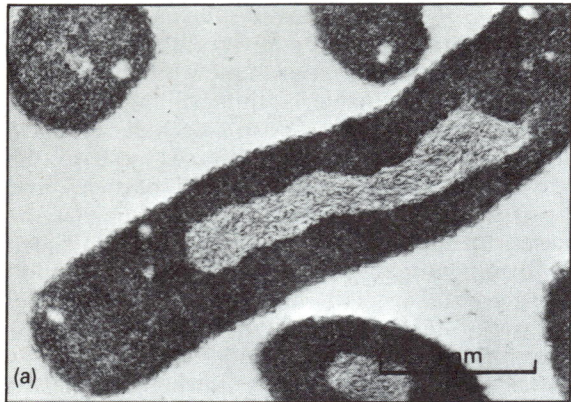

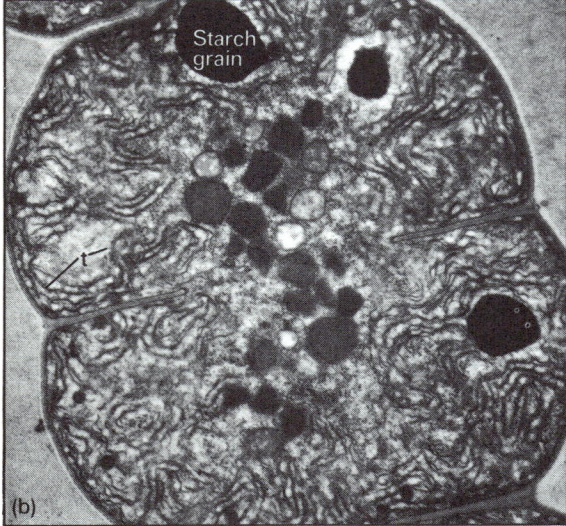

Fig. 1.17(a) Electron micrograph of a section through a bacterium. Note the absence of organelles.
(b) Electron micrograph of a transverse section of a blue-green algal cell. Note the absence of all organelles seen in Fig. 1.7.

The cells of these organisms are much less **differentiated** than those we have previously examined. This means that they are not subdivided to so great an extent into small, specialized structures (organelles), each of which has a particular function.

The second group, the Eukaryota, contains all other organisms, all animals and all other plants. Animal cells differ from living plant cells in not possessing a central permanent vacuole, a cellulose cell wall, or chloroplasts. But they are similar in that they do possess a plasma membrane, a nucleus, mitochondria, endoplasmic reticulum, Golgi apparatus, and lysosomes. These eukaryotic cells are therefore more highly organized and differentiated than prokaryotic cells and it would seem reasonable to suggest that the Prokaryota were the first living organisms to exist on the earth and that the Eukaryota have been derived from them by evolutionary development. There is fossil evidence that life did evolve in this order (Chapter 20).

Horizons revealed by the cell theory
The cell theory, as proposed by Schwann, was based on observations made with primitive light microscopes on a limited range of animal material. We could not accept it today in the words in which Schwann formulated it, but its importance is not that it is a rigid rule that is right in every instance, a sort of master key that unlocks the secrets of nature at one turn, but that it is sufficiently correct to have stimulated thought and work amongst other biologists. Let us look at some of the advances that could not have been made without the concept of the cell.

Different sorts of cell: division of labour
Accepting that organisms are made of cells, we want to know what cells from different parts of the animal and plant body look like. How do they resemble each other and differ from each other? This is the science of **cytology**. It leads to discovering how their structure is suited to the special functions they perform in the body. Some of the cell types found in the animal body are:
a Muscle cells, whose job is to contract and produce movement (Fig. 17.23).
b Surface-layer cells (epithelia). Some of these are protective, as in the lining of the cheek and the outside of the skin (Fig. 15.3). Others, such as those which line the small intestine, absorb dissolved foods (Fig. 6.22).
c Nerve cells which conduct electrical messages and coordinate the body's activity (Fig. 15.4).

celled, others composed of long thin threads consisting of cells joined together end to end. The Prokaryota possess none of the organelles described above. Figure 1.17 shows electron micrographs of sections through a bacterial cell (a) and the cell of a blue-green alga (b). Note:
1 there is no clearly-defined nucleus surrounded by a membrane;
2 there are no mitochondria;
3 the blue-green alga carries out photosynthesis, but there is no chloroplast (instead photosynthesis takes place in membranes (*t*) which are scattered throughout the cell);
4 neither cell contains any endoplasmic reticulum, Golgi apparatus, or lysosomes.

d Bone cells (Fig. 17.5), which make bone from chemical materials delivered by the blood.

e Blood cells, both the red cells which carry oxygen from the lungs to all cells of the body, and the white cells which protect the body against infection in various ways (Fig. 8.1).

f Reproductive cells, sperms and ova (Figs. 12.11(a), 12.12).

Some of the cell types found in the plant body are:

a Packing cells (**parenchyma**, Fig. 1.3(a)).

b Dividing cells (**cambium**), which cause growth (Fig. 10.2).

c Xylem 'cells', that is the non-living walls of once-living cells, which make pipes for the conduction of water around the plant (Fig. 10.1).

d **Phloem** cells, which move the food material made by the leaves to other parts of the plant (Fig. 10.15).

e **Cork** cells, such as those seen by Robert Hooke; these form a protective outer layer in many plants.

Cell division

Virchow said that 'all cells arise from previously-existing cells'. If we accept this, we must ask: how does one cell produce another? Investigations have shown that when a cell divides to form two cells, the nucleus behaves in a complicated and precise manner. Dark-staining rods called **chromosomes** appear within it (Fig. 14.11) and are shared out to form two nuclei. The study of this behaviour has led to a whole new branch of science called genetics.

Cell differentiation: embryology

Sexual reproduction involves two special cells, the **sperm** and the **ovum**. These fuse producing a single cell, the **zygote**, from which the entire animal or plant develops. When the zygote divides it produces two cells. Each of these divides, giving four cells. In a short time millions of cells are produced, but they do not form a shapeless lump. Taking human development as an example, the beginnings of a backbone soon appear, then a heart, a brain, eyes and limbs, and each of these organs contains different kinds of cell. The study of this complex process is the science of embryology.

Biochemistry of the cell: the nature of life

The study of the chemicals, such as proteins, fats and carbohydrates, which occur in cells, and the ways in which they are made and destroyed in the cell, makes it possible to say that there is *no such thing as living matter*, though there are *living systems composed of chemicals*. The study of the biochemical workings of the cell not only helps us to understand the nature of life but has led to most of the progress made in medicine.

To close this chapter with an example: cancer is a prevalent and distressing human disease. It is caused by certain cells dividing at times and in places where they would not normally divide, producing tumours. What causes this, and how can it be prevented or cured? To obtain an answer to this question we must first accept the idea of the cancer cell; then try to understand how it works and behaves in a normal body; and then try to determine why and how its behaviour changes to produce a tumour.

2 The Vertebrate Body

External features of the human body

The body is covered with skin, whose thickness varies. Over the front of the eye it is very thin and transparent; on the soles of the feet and the palms of the hands, it is thick; on the lips it is thin and very sensitive. Some parts of the skin, notably the head and the pubic region, are very hairy (Figs. 2.1, 2.2). Other parts, such as the soles of the feet, the palms of the hands, and in the man the penis, are hairless, yet others bear fine hairs. In places where the skin is thin, parts of the underlying bones can be seen, especially in a man. Find the **clavicle** (collar bone), the hip bone, the knee cap, and the ankle bones in Fig. 2.1. Muscles are attached to bones by **tendons**. Some of these are so hard and thick that you might mistake them for bones. Look at the back of your hand and move your fingers. Notice the tendons which connect the arm muscles to the fingers moving under the skin. Feel the thick tendon behind your knee.

In general shape the man is more angular, the woman more rounded. His shoulders are broader than hers, her hips much broader than his. Her large breasts, which produce milk for feeding her babies, are the most prominent feature of her body. Both man and woman have **nipples**. It is through these that the young baby draws its milk from the mother's breast. In the man there is no breast, and the nipples are small and functionless.

The man's muscles are more strongly developed and less obscured by the thicker, fatty layer of the skin than the woman's, and several of them can be seen in Fig. 2.1; but both man and woman have the same muscles.

The woman's reproductive organs lie inside her body. The birth canal (**vagina**) opens to the outside at the slit-like **vulva**, though this is obscured by **pubic hair** in Fig. 2.2. In contrast, some of the man's reproductive organs are on the outside of his body, his **penis**, and hanging behind it his scrotum, a sac of skin which contains his two **testes**.

Other obvious external features of the body are the eyes, the nose, which has a sensory and a respiratory function, and the external part of the ear. The **navel** marks the point where the **umbilical cord** from the mother was attached to the body

when these adults were developing as babies in their mother's womb.

Beneath the surface

If the skin is removed from the body the underlying muscles, tendons, and **ligaments** are revealed. Ligaments (Latin ligamentum, a bandage) are bands of very tough fibrous tissue which join bones together. Figs. 2.3(a) and (b) show some of the muscles, though many others lie beneath these. Some of the muscles which show through the skin have been numbered in Fig. 2.1. Try to identify them by referring to Fig. 2.3.

Internal organs

The model shown in Figs. 2.4(a) and (b) enable us to see further into the body for in it the muscles and tendons have been replaced by transparent plastic. Now we can see more of the skeleton.

Look for the following bones: the upper arm bone (humerus); shoulder blades; collar bone; backbone; ribs; breastbone (sternum); skull; upper leg bone (femur); hip bones; base of backbone (sacrum) attached to the hip bones. Feel these bones in your own body.

The only other things visible in the arms and legs of the model are arteries, veins, and nerves. But in the torso many more organs can be seen. Outside the rib cage are the breasts. For most of a woman's life these are largely made of fatty tissue. When she is suckling a baby they enlarge and make milk. Beneath the ribs lie the lungs, and between them is the heart, hidden by the breastbone.

In the bottom half of the torso are the digestive, reproductive, and excretory organs. Of these, only parts of the digestive system can be seen in Fig. 2.4. These are part of the stomach, part of the small intestine, and part of the large intestine. The liver is the largest organ in the body and one of its functions is to provide bile for the digestive system. Part of the liver can be seen in Fig. 2.4(a).

The ovaries, womb, and kidneys cannot be seen in Fig. 2.4 because they are hidden by the gut. Neither can we see the brain and spinal cord because they lie inside the skull and backbone respectively.

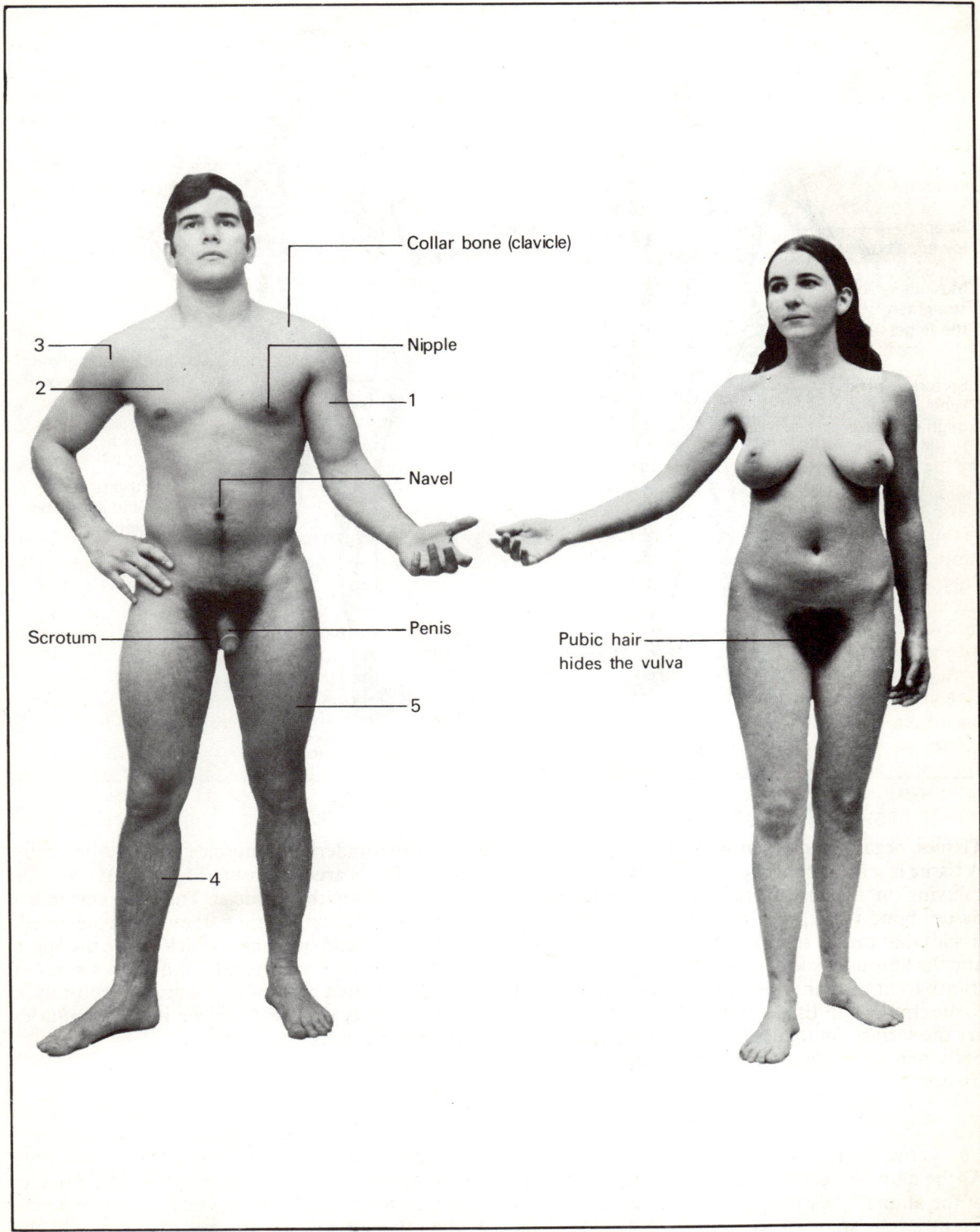

Fig. 2.1 Man   Fig. 2.2 Woman

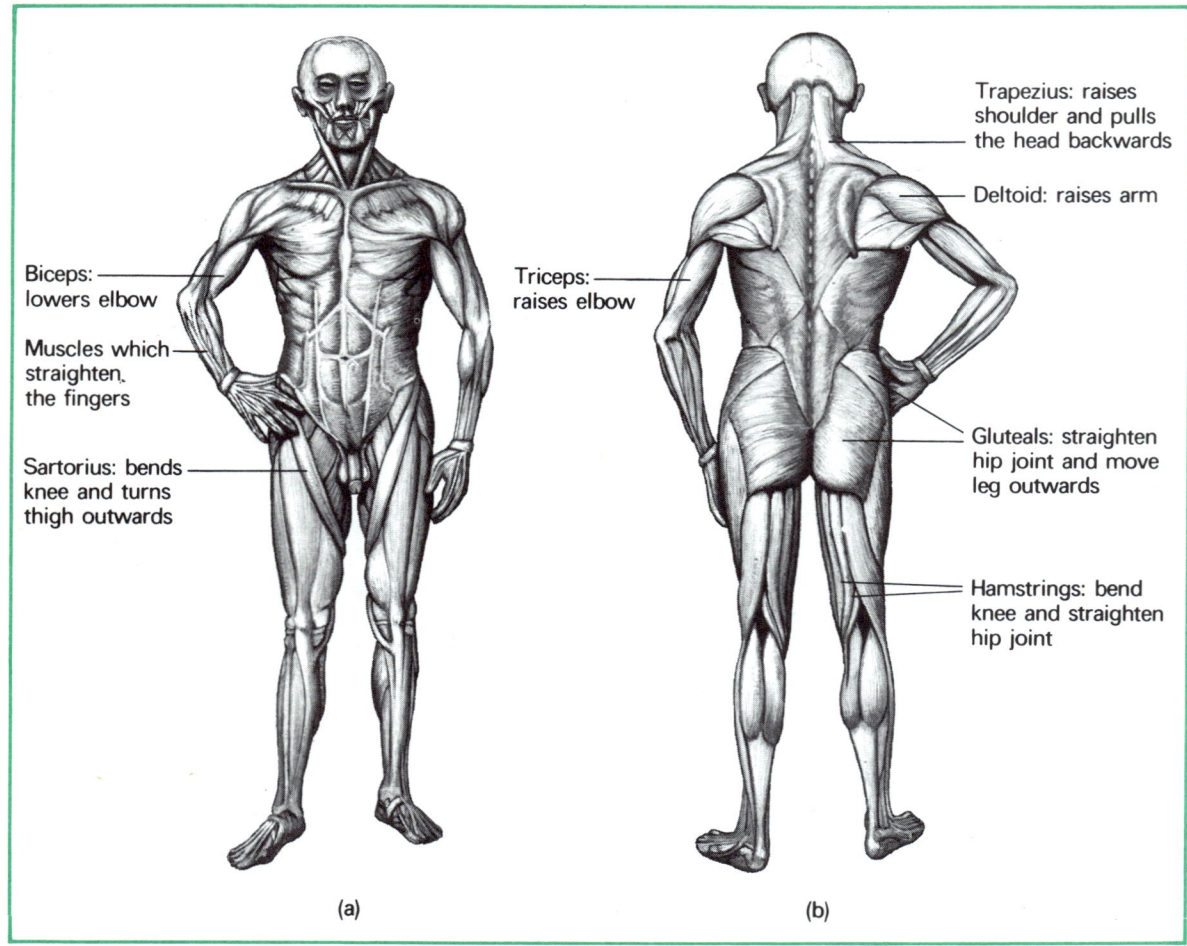

Fig. 2.3(a), (b) Drawing of a flayed man, showing the muscles and tendons which lie beneath the skin.

Tissues, organs, and organ systems

A **tissue** is a large collection of cells of similar kind carrying out similar functions, for example muscle tissue, bone tissue, and nervous tissue. The only tissues that can be seen in Fig. 2.3 are muscle tissue and the fibrous tissue of tendons and ligaments. The transparent woman shows bone tissue and nervous tissue. Important tissues not visible in these figures are the surface-lining tissues—on the outside of the body and on its internal organs—and connective tissues.

Connective tissue

As the name suggests, connective tissues join parts of the animal body together. There are several different kinds, some of which (blood, tendon, ligament) are familiar. Another type found joining the skin to the underlying muscles and also in many other places is **areolar** tissue (Fig. 2.5). It contains large, flat cells with big nuclei. These are embedded in various materials that they have secreted, namely (i) **ground substance**, which fills in the space between the other constituents, and (ii) two kinds of fibre. **Tissue fluid** permeates the ground substance. It is constantly exuded from the blood capillaries lying in the connective tissue.

Organs

The body contains various organs each with its special job, and each made up of several different kinds of tissue. The heart is an organ whose only job is to pump blood. It is made up of muscle tissue, blood tissue, nervous tissue, and connective tissue. Some other organs are the eye, the stomach, the lungs, the liver, the kidney, and the brain.

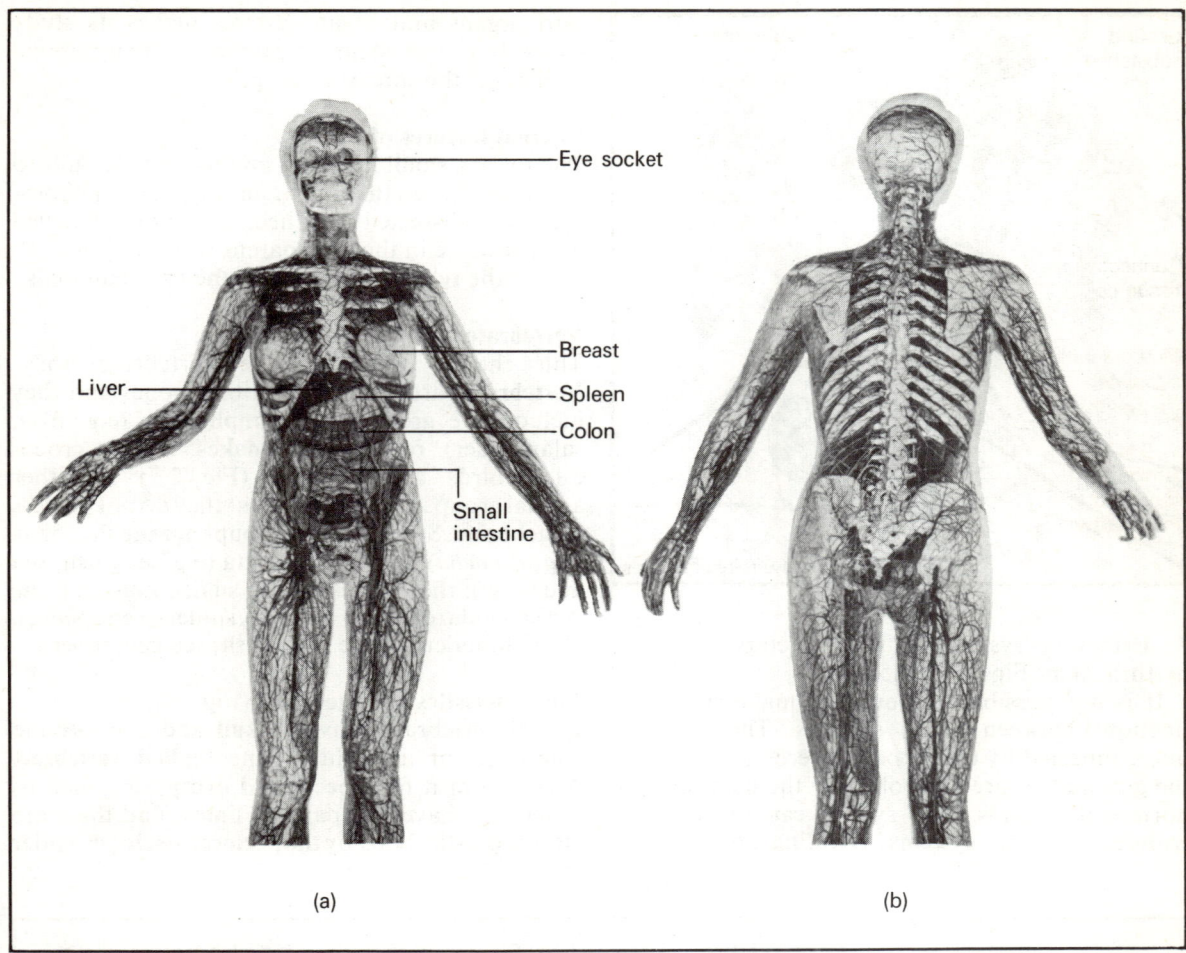

Eye socket

Liver

Breast

Spleen

Colon

Small intestine

(a)

(b)

Fig. 2.4(a), (b) Model showing the internal organs of a woman.

Organ systems

The body depends on all its parts for its existence, but some organs work together in more obviously connected ways than others, forming a system. The excretory system consists of the kidney which makes urine, the ureter which carries urine to the bladder, which stores it, and the urethra which conveys urine from the bladder to the exterior (Fig. 11.2).

Division of labour

In human society the total work to be done is shared out among postmen, shopkeepers, farmers, dustmen, doctors, teachers, and many others. This is called *division of labour*. In the cell, division of labour operates on a minute scale and each organelle has a particular job to do. In man's whole body

it operates at other levels. Special cells, such as muscle cells, bone cells, and nerve cells carry out specific tasks. These cells are collected into organs and organ systems. The main organ systems of the human body are:

1   Circulatory system (heart, blood vessels, Fig. 8.12).

2   Respiratory system (larynx, trachea, lungs, Fig. 7.15).

3   Coordinating system (brain, spinal cord, spinal and other nerves, endocrine glands, Figs. 15.8, 15.28).

4   Sensory system (eye, ear, nose, tongue, skin, Figs. 16.3, 16.14).

5   Digestive system (gut, liver, pancreas, Fig. 6.9).

6   Reproductive system (testes, vasa deferentia, penis; ovaries, Fallopian tubes, womb, vagina, Figs. 12.4, 12.9(b)).

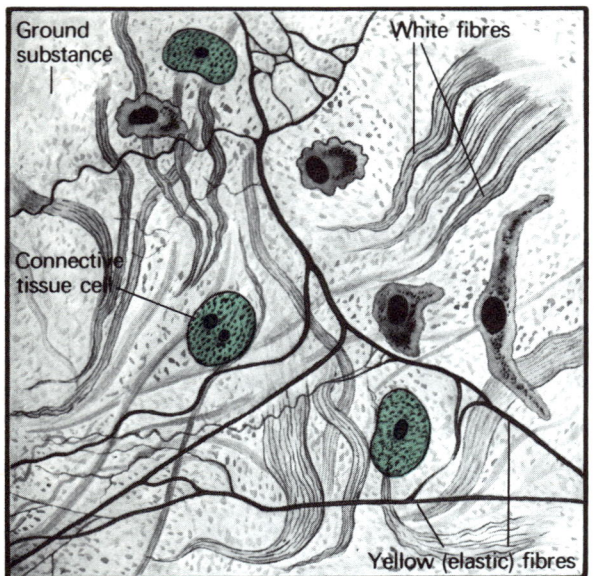

Fig. 2.5 Drawing of areolar tissue.

7 Excretory system (kidneys, ureters, bladder, urethra, skin, Figs. 11.2, 15.3).

It is not possible, however, to make rigid distinctions between organ systems. They are all interconnected by the blood, all receive food from the gut, and all are controlled by the nervous and hormonal systems. One system cannot function without the other systems. Dividing up the body into organs and organ systems makes its study easier, but we must not forget that the living animal or plant is the sum of all its parts.

Internal features of the rat

The rat is a small mammal and its body is built to the same plan as that of a man. Fig. 2.6 is a photograph of a dissected rat. Check the features labelled against those in the transparent woman (Fig. 2.4), noting the relative positions in the two mammals.

Vertebrates and invertebrates

This chapter is called The Vertebrate Body. **Vertebrates** are animals with backbones and they include five groups: fish, amphibians (e.g. frog, salamander), reptiles (e.g. snakes, lizards, crocodiles), birds, and mammals (Fig. 2.7). All other animal groups are **invertebrates**; they do not possess a backbone. Some of these groups are the Protozoa (e.g. *Amoeba*), the Coelenterata (e.g. jelly fish, sea anemone), the Mollusca (e.g. snails, mussels), the Arthropoda (e.g. crabs, insects, spiders, centipedes), the Echinodermata (e.g. starfish, sea cucumber).

Characteristics of vertebrates (Fig. 2.8)

1 All vertebrates have a **skull** and a backbone composed of individual bones called **vertebrae**. Apart from a few specialized exceptions such as snakes, all have two pairs of limbs, and these are attached to the body by the **pectoral girdle** (shoulder

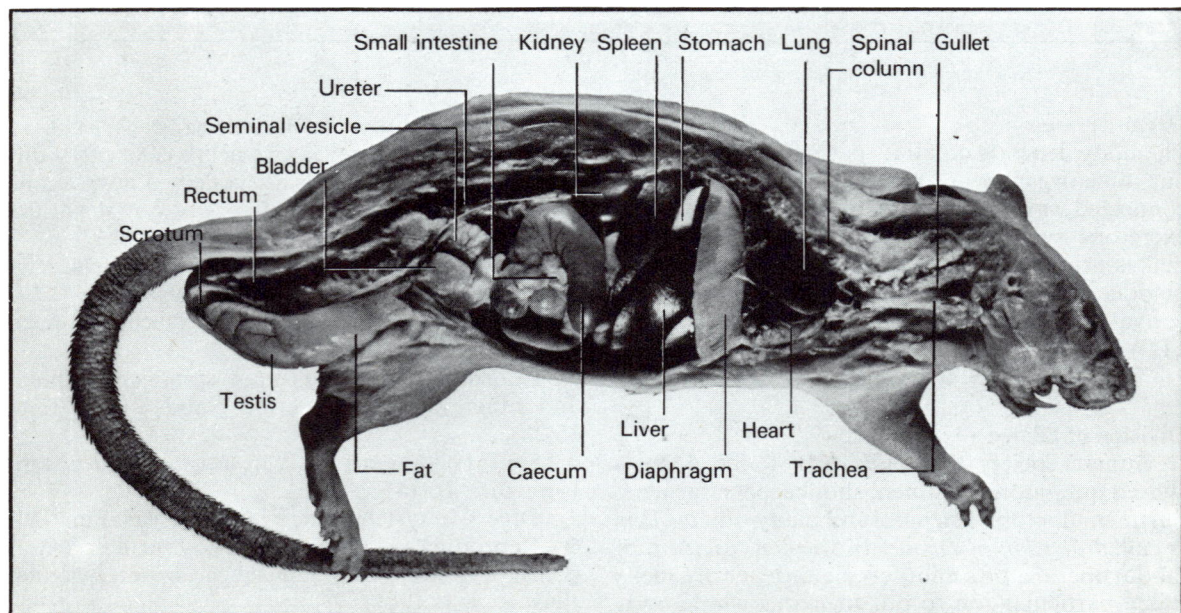

Fig. 2.6 Internal organs of a rat.

16

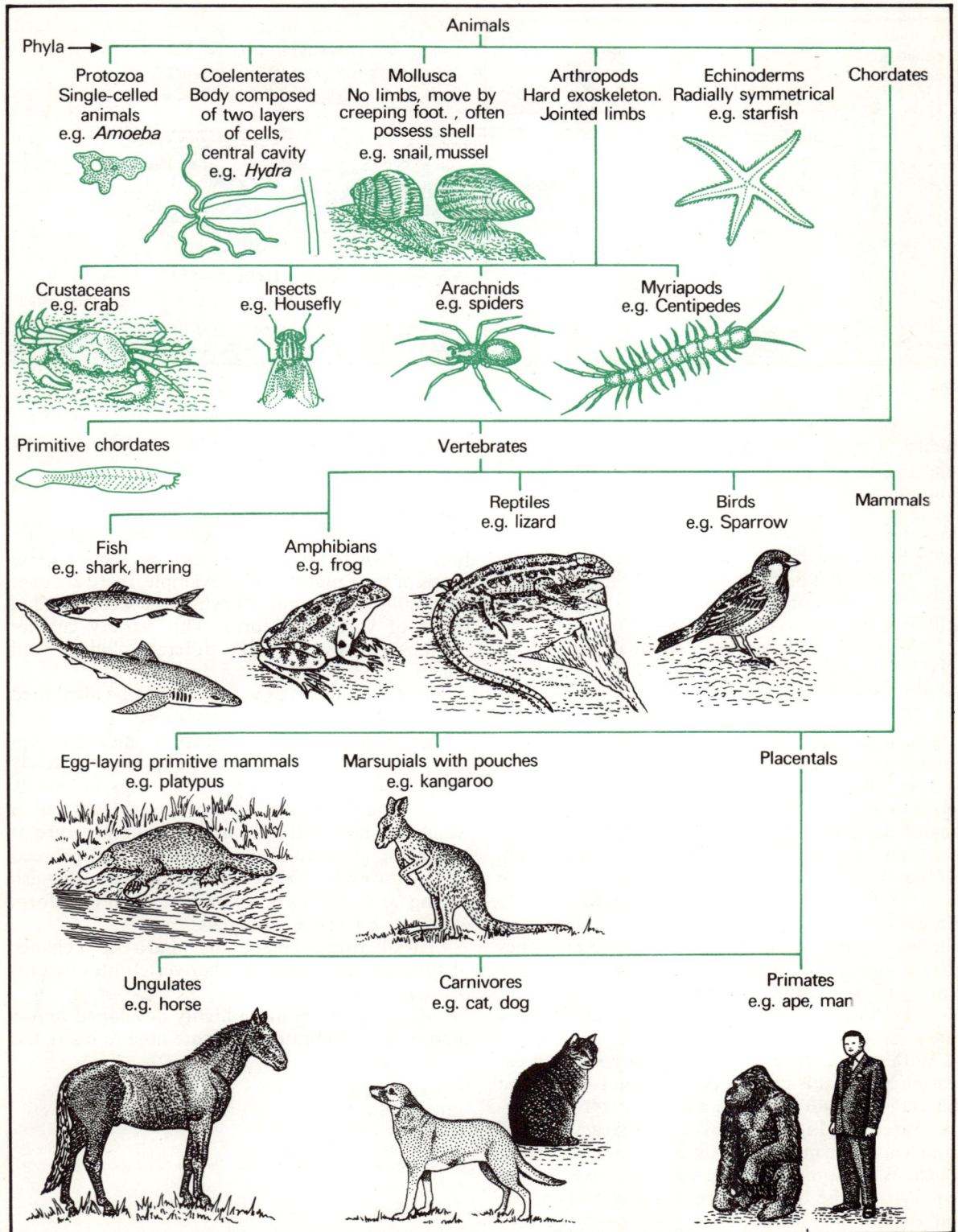

Animals

Phyla →

Protozoa
Single-celled
animals
e.g. *Amoeba*

Coelenterates
Body composed
of two layers
of cells,
central cavity
e.g. *Hydra*

Mollusca
No limbs, move by
creeping foot. , often
possess shell
e.g. snail, mussel

Arthropods
Hard exoskeleton.
Jointed limbs

Echinoderms
Radially symmetrical
e.g. starfish

Chordates

Crustaceans
e.g. crab

Insects
e.g. Housefly

Arachnids
e.g. spiders

Myriapods
e.g. Centipedes

Primitive chordates

Vertebrates

Reptiles
e.g. lizard

Birds
e.g. Sparrow

Mammals

Fish
e.g. shark, herring

Amphibians
e.g. frog

Egg-laying primitive mammals
e.g. platypus

Marsupials with pouches
e.g. kangaroo

Placentals

Ungulates
e.g. horse

Carnivores
e.g. cat, dog

Primates
e.g. ape, man

Fig. 2.7 A classification of the major animal groups.

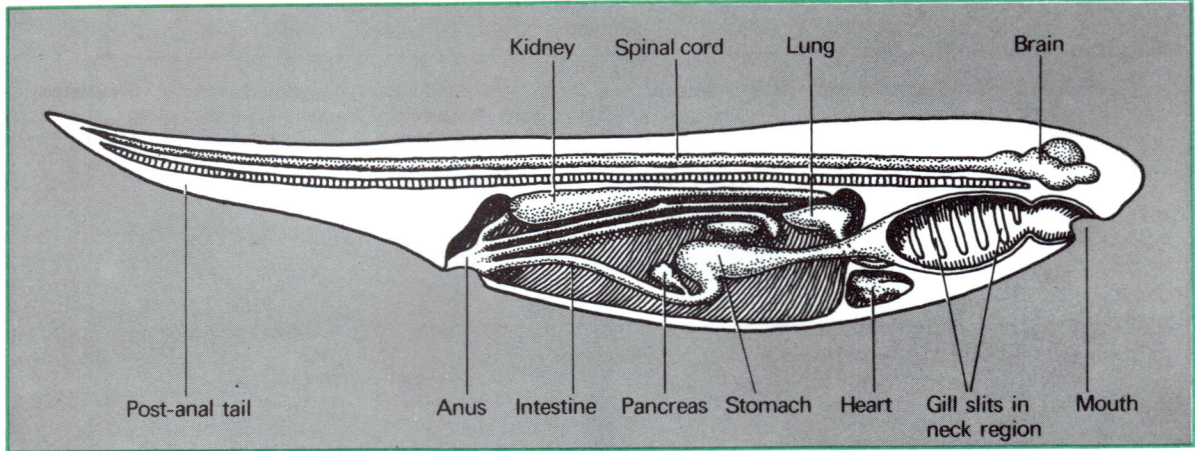

Fig. 2.8 The general vertebrate body plan (girdles and limbs not shown).

blades in man) and the **pelvic girdle** (hip bones in man). In the amphibians, reptiles, birds, and mammals, the skeleton of each limb has the same basic pattern. There is upper limb bone, two lower limb bones, a group of wrist or ankle bones, five palm bones, and five sets of finger or toe bones (Fig. 2.9). This is the pentadactyl (Greek, five fingered) limb, and many modifications of it have evolved in different vertebrate groups (Fig. 2.10).

2   The brain and spinal cord are on the upper side of the body. Both possess a central canal and so are hollow (Fig. 15.15).

3   The heart is on the lower side of the body. (N.B. Man is an unusual vertebrate in that he stands upright.)

4   At some stage of their life history, vertebrates possess a series of paired **gill pouches** in the neck region. In fish and the tadpole stage of amphibians, these develop into **gill slits** leading from the inside of the pharynx to the exterior. (The pharynx is the part of the throat between the mouth and the oesophagus, see Fig. 18.36.) Water passes through them over the gills. In other vertebrates the pouches do not pierce the neck and can only be seen in the embryo (Fig. 2.11).

5   Vertebrates possess a tail which extends beyond the anus, the rear opening of the gut.

Within the vertebrates there are very great variations on this ground plan, but nobody would have any difficulty in telling a vertebrate from an invertebrate. As Fig. 2.11 shows, basic structures and similarities are more obvious in the embryo, before birth. What similarities can you see between these embryos in the head region, the back region, the tail region, the limb region, and the belly region?

What features of the body of a man show that he is a vertebrate?

Characteristics of mammals

Mammals are the most successful group of vertebrates alive today. Success can be measured in terms of their numbers (for example, there are over 4000 million men and women in the world) and in terms of variety of form. This variety enables mammals to live in many different climates and places.

The great success of mammals can be attributed to three outstanding characteristics:

1   More than any other group of animals except birds, they are able to keep their bodies at a steady temperature of about 35°–40°C. This is usually higher than the temperature of the environment in which the mammal lives and is a temperature at which the chemical reactions in the cell proceed most efficiently. Other vertebrates are sluggish during cold seasons and at night, when they form easy prey for active mammals.

2   The offspring of mammals stand a better chance of survival than those of other vertebrates (see (ii) below).

3   Mammals have more highly developed brains than other vertebrates. They are able to learn and to remember.

Fig. 2.9(a) Skeleton of a perch. Note skull, backbone, pectoral girdle and pectoral fins, pelvic girdle and pelvic fins.
(b) Skeleton of a frog (an amphibian). Note skull, backbone, girdles, and pentadactyl limbs.
(c) Skeleton of an alligator (a reptile). Note skull, backbone, girdles, and pentadactyl limbs.
(d) Skeleton of a pigeon (a bird). Note the modified pelvic girdle.

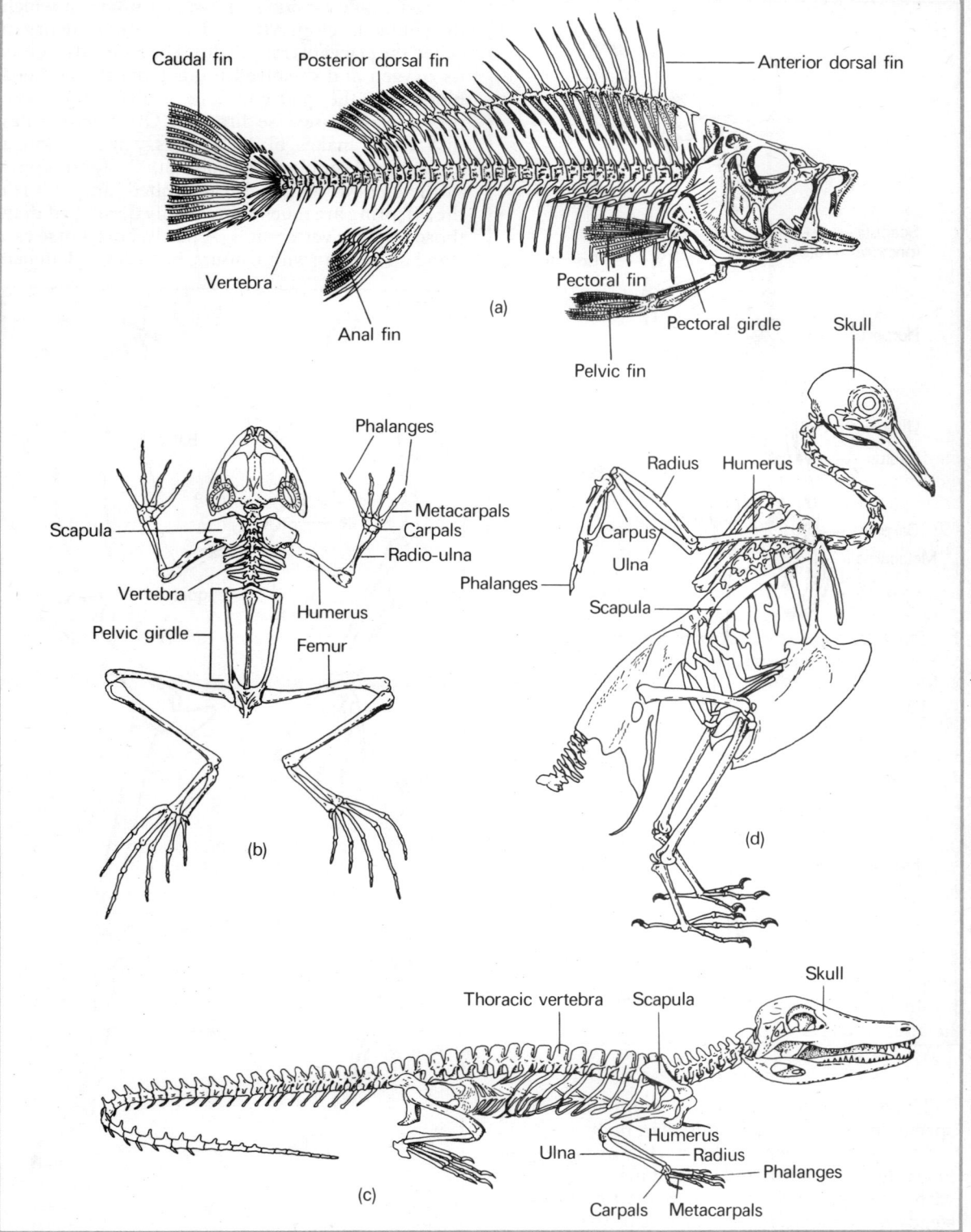

Caudal fin

Posterior dorsal fin

Anterior dorsal fin

Vertebra

(a)

Anal fin

Pectoral fin

Pectoral girdle

Pelvic fin

Skull

Phalanges

Metacarpals

Carpals

Radio-ulna

Scapula

Vertebra

Humerus

Pelvic girdle

Femur

(b)

Radius

Humerus

Carpus

Ulna

Phalanges

Scapula

(d)

Thoracic vertebra

Scapula

Skull

Ulna

Humerus

Radius

Phalanges

Carpals

Metacarpals

(c)

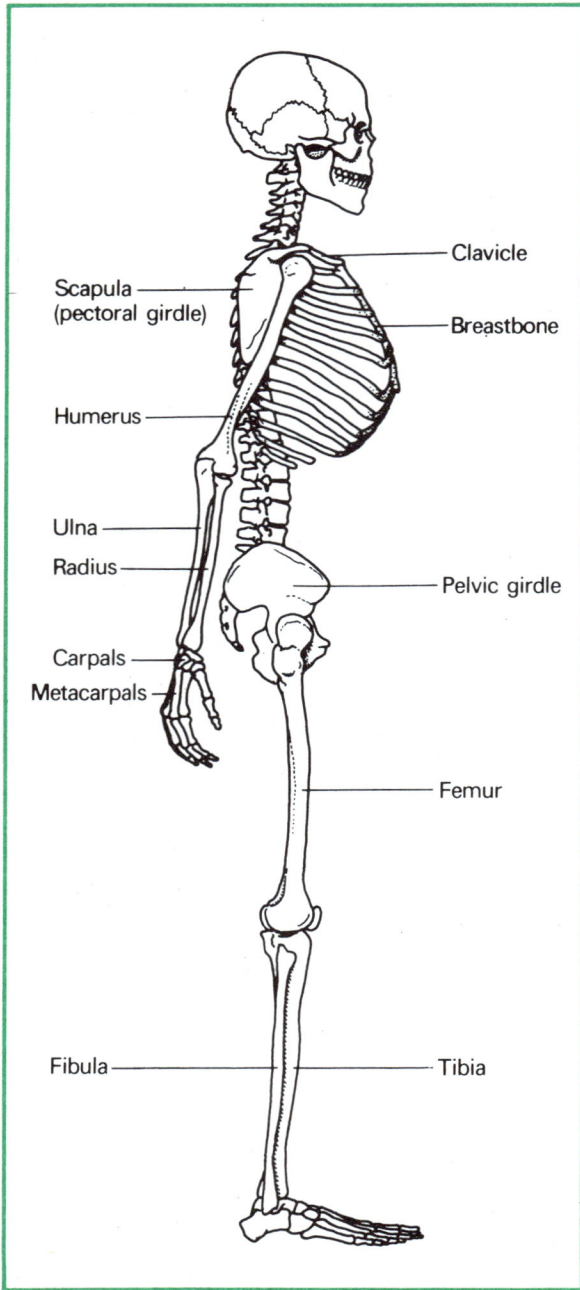

2.9(e) Human skeleton (mammal).

These characteristics follow from some of the special features of the mammalian body.

(i) Mammals possess hair. Hair is a very efficient heat insulator, and in most mammals, acting with other mechanisms, hair helps to maintain a steady high body temperature (see p. 168).

(ii) Female mammals possess a **womb** in which the young develop. Attached to its wall is an organ called the **placenta** (Fig. 12.20). The embryo receives its oxygen and dissolved foods from its mother's blood via the placenta and excretes its waste produces in the reverse direction. Other vertebrates (fish, frogs, snakes, birds) lay eggs, which contain a limited food supply in the yolk. But the food supply of the embryo mammal is unlimited, and at birth the offspring are much more highly developed than those of other vertebrates. A newly-born horse can stand up and run within hours, but a newly-hatched

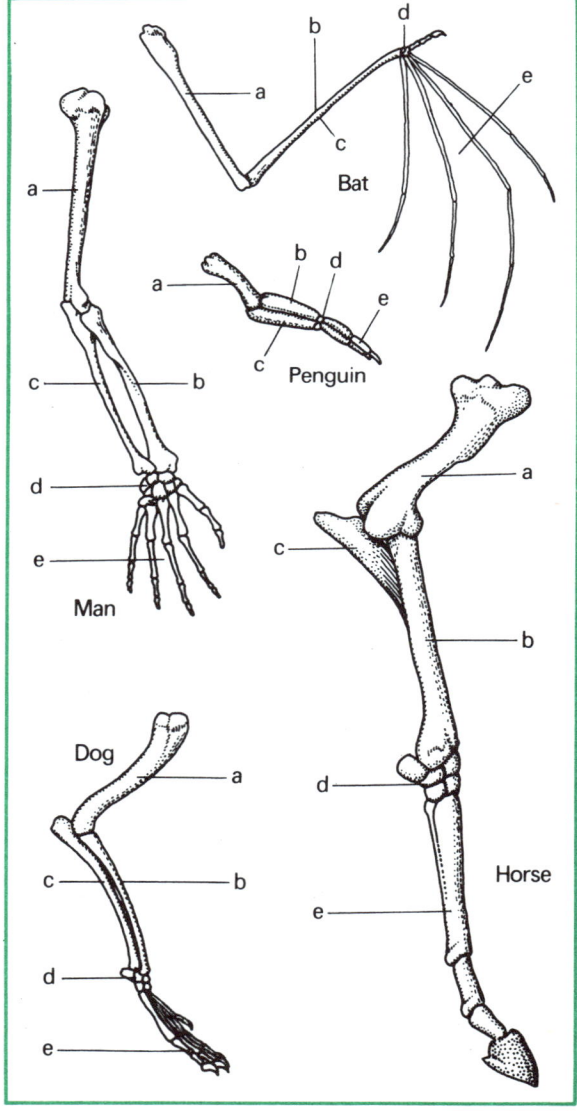

Fig. 2.10 Modifications of the basic pentadactyl limb pattern in different mammals.

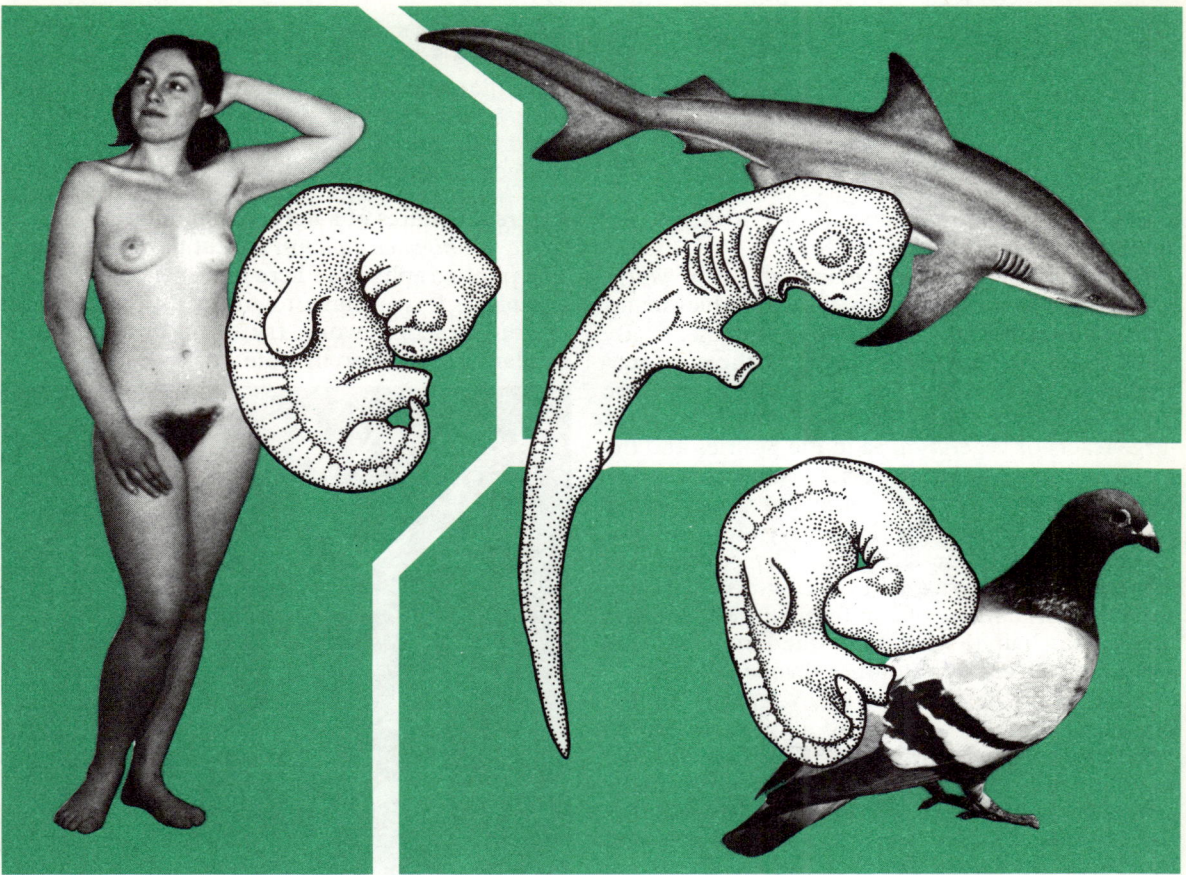

Fig. 2.11 Young embryos of fish, pigeon, and man. These show more clearly than do the adults the basic vertebrate body plan.

tadpole is a helpless, motionless food morsel for any predator. Eggs which are laid are likewise defenceless, but the embryo mammal is protected from enemies in its mother's womb.

(iii) Female mammals have **mammary glands**. These produce milk, which contains all the food substances necessary for life. Young mammals are first suckled at the mother's breast, then weaned on to solid food provided by the parents.

(iv) Mammals have a false roof, or **palate**, in their mouths. This separates the air passage from the food passage (Fig. 6.11), and enables the animal to chew and breathe at the same time. By contrast, a snake may take half an hour to swallow a victim, during which time it may not breathe at all, and at the end of which it is exhausted.

What features of the body of man show that he is a mammal?

3 The Flowering Plant Body

When considering mammals, two measures of success were put forward (p. 18):

(i) the large numbers of mammals in the world

(ii) the wide range of form which enables mammals to live in many different environments. By these standards, flowering plants are the most successful group of plants. They are found everywhere except in the sea and at the Poles. Often they form large tracts of vegetation as in the prairies and the tropical rain forests. By these standards other plant groups are not so successful. These other groups are: the algae, which nearly all grow submerged in water; the mosses and liverworts, very small plants; the ferns; and the conifers. Although there are large tracts of coniferous forest, conifers do not match up to the second standard. Nearly all conifers are trees, very few are shrubs or herbs.

Most of the land surface today, apart from the coniferous forests, is covered by a very varied vegetation of flowering plants. They fall into two main groups. First the **dicots**. There are so many different kinds of dicot that no short list would do them justice: the rose family, the vetch family, and the buttercup family are some. Many dicots are trees or woody shrubs. Others are herbs and have no woody parts that last all year above ground. Secondly, the **monocots**, which include the grasses, irises, crocuses, and lilies. Very few monocots are woody plants.

External features of a horse chestnut tree (*Aesculus hippocastanum*), a dicot flowering plant

The photographs used here are of *Aesculus hippocastanum*. If you have to use a different species in your practical work, try to find out how it is similar to and different from this species.

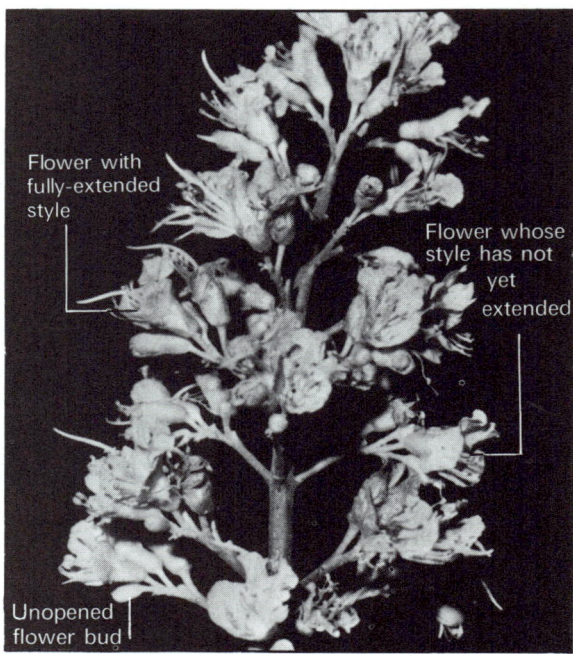

Fig. 3.1 Horsechestnut in full bloom.

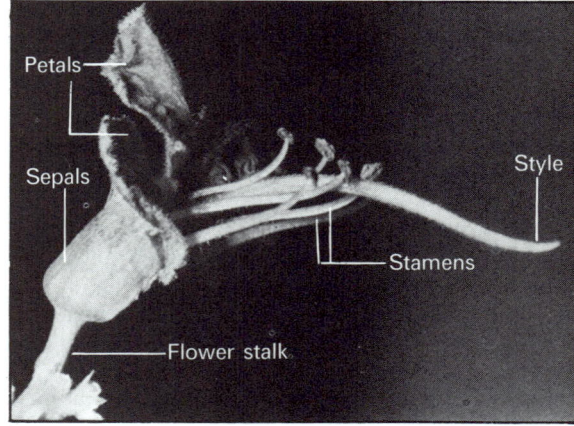

Fig. 3.2(a) Horsechestnut inflorescence.
(b) Horsechestnut single flower.

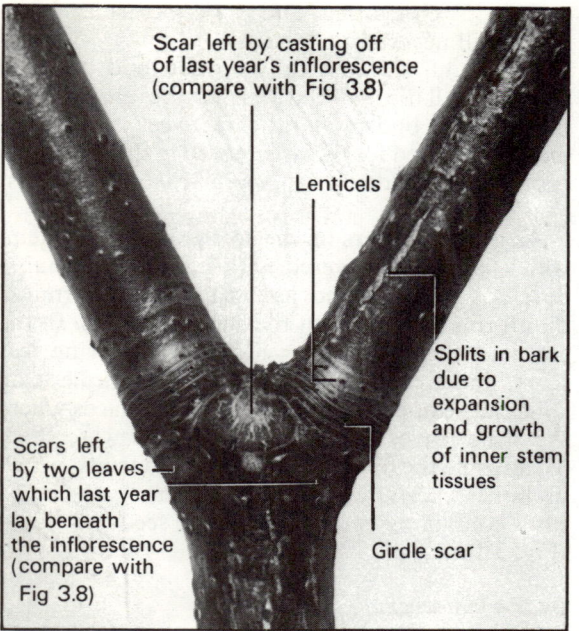

Fig. 3.3 Part of a horsechestnut twig.

Labels on Fig. 3.3:
- Scar left by casting off of last year's inflorescence (compare with Fig 3.8)
- Lenticels
- Splits in bark due to expansion and growth of inner stem tissues
- Scars left by two leaves which last year lay beneath the inflorescence (compare with Fig 3.8)
- Girdle scar

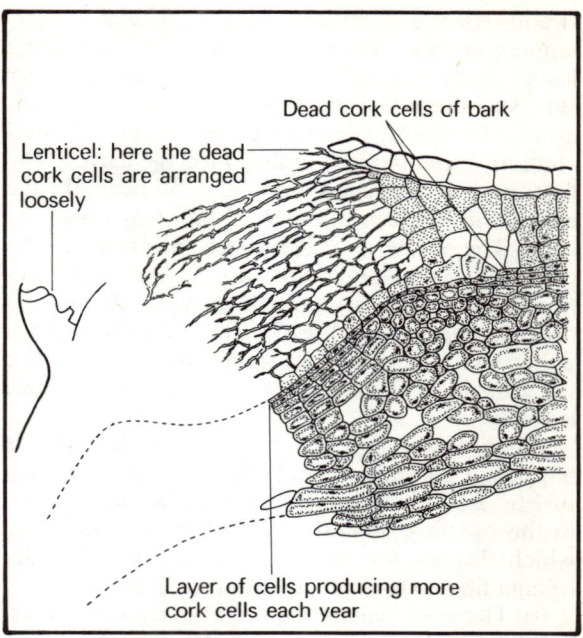

Fig. 3.4 Drawing of a transverse section of a lenticel.

Labels on Fig. 3.4:
- Lenticel: here the dead cork cells are arranged loosely
- Dead cork cells of bark
- Layer of cells producing more cork cells each year

Figure 3.1 shows a horse chestnut in full bloom in an English spring. The trunk and branches bear a large leaf canopy and the surface area of the plant is enormously greater than that of a mammal of comparable size (whale). There are also large numbers of **inflorescences**, each a collection of 20–30 flowers (Fig. 3.2). Figure 3.3 is a photograph of a twig in winter, after leaf fall. The outer surface of the stem is the bark and it bears a series of horseshoe shaped scars at the points where the leaf stalks were shed in the previous autumn. These give the tree its name.

One pair of leaf scars has been labelled. The pairs above it are at 90° to this pair. Examine a twig from a tree of another species and find out how its leaf scars are arranged. The place where the leaf stalk is attached to the stem is called a **node**, the length of stem between two nodes is an **internode** (see also Fig. 3.8).

The bark is hard and is mostly made of dead cork cells. It protects the living cells inside the stem from cold and prevents the entry of disease germs. All over the bark are small rough patches (**lenticels**) where the dead cork cells are arranged loosely and allow oxygen to pass and reach the living cells within the stem. Figure 3.4 is transverse section of a stem showing the crumbly arrangement of cork cells in a lenticel.

At the top of the stem is the **terminal bud**. Like other buds, this is a compact, undeveloped shoot.

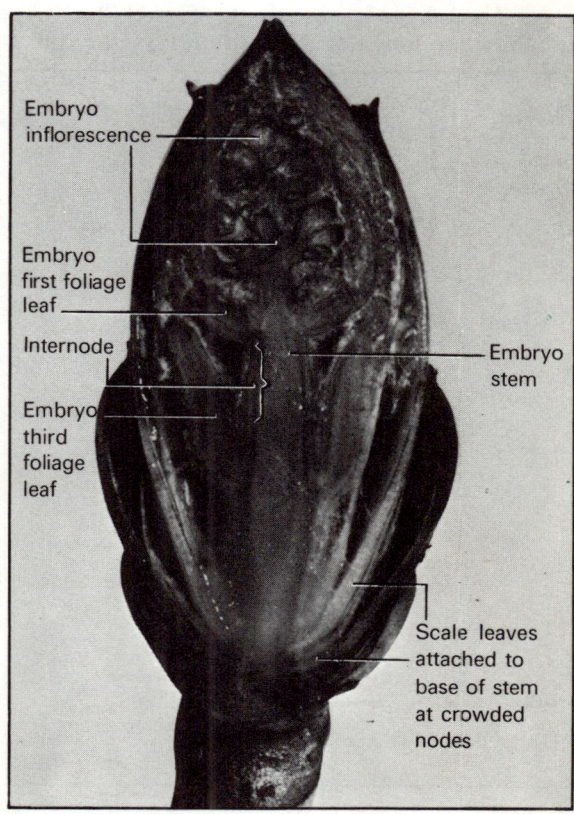

Fig. 3.5 Horsechestnut terminal bud sliced lengthways.

Labels on Fig. 3.5:
- Embryo inflorescence
- Embryo first foliage leaf
- Internode
- Embryo third foliage leaf
- Embryo stem
- Scale leaves attached to base of stem at crowded nodes

It consists of a short stem bearing crowded, overlapping embryo leaves (Fig. 3.5). The outer **scale leaves** are brown and sticky. They protect the inner embryo **foliage leaves** and in Fig. 3.5 can be seen attached to the stem at the base of the bud. The bud contains three pairs of embryo foliage leaves. The first pair is attached immediately beneath the embryo inflorescence, the third half way down the stem. The second pair is set at right angles to the other two pairs and so does not appear in the section, but Figs. 3.7(e) and 3.8 show all three pairs of foliage leaves. In Fig. 3.6 the leaves have been pulled off in turn and arranged in order. Notice the transition from scale leaves to green foliage leaves.

When the bud starts to open several things happen at the same time. Some of them are shown in Fig. 3.7. Identify the leaves shown in Fig. 3.6 in the opening buds shown in Fig. 3.7, and note which leaves are fully scale-like, which fully foliage-like, and which are intermediate.

(i) The scale leaves enlarge and separate (Figs. 3.7(a), (b)). Each scale leaf is attached separately to the stem at its node.

(ii) The foliage leaves emerge (Figs. 3.7(c), (d)), and the stem elongates. After some days, the young stem stops elongating, and these particular internodes will never grow longer.

Figure 3.8 shows a fully-opened bud of horse chestnut. All the material seen here developed from the terminal bud in about three weeks. The food material needed for this was stored in the stems and roots, being made by last year's leaves during last year's summer.

As the year wears on the soft green young stem, which is at first covered with light downy hairs, develops its bark and lenticels. A new terminal bud forms at the end of the stem (Fig. 3.9). In the autumn, the foliage leaves drop off leaving **leaf scars**, and we can now see where the **girdle scars** (Fig. 3.3) came from: they mark the places where the scale leaves of the previous buds were shed. Meanwhile the flowers have produced fruits, which in horse chestnut are spiny, green, ball-shaped cases containing two or three large seeds (conkers) (Fig. 3.10).

Buds and branching
Buds which develop at the end of the shoot are called terminal buds, but similar buds form in the angle between the leaf and the stem and are called **axillary buds**. Figure 3.8 shows developing axillary buds. Some buds contain an embryo inflorescence which develops when the bud opens (Figs. 3.5, 3.6, 3.8), but others contain only embryo leaves. In the case of horse chestnut the old inflorescence is cast off in the autumn leaving a permanent terminal scar (Figs. 3.3, 3.7). Where this happens there is no terminal bud, and next year's growth is from an axillary bud situated above one of the two terminal leaf scars. This is the development seen in Fig. 3.7. It means that at this point the stem cannot continue to grow straight on but branches. Using this knowledge, explain the development of the branched shoot shown in Fig. 3.3.

Roots
The function of roots is to anchor the plant in the soil and to absorb water and mineral salts by means of special cells called **root hairs**. Figure 3.11 is a photograph of root hairs on a broad bean root. They are thin and straight and grow vertically to the surface of the root. Each is a single elongated cell. This seedling was grown on blotting paper, but when a plant is growing in the soil, the root hair cells bend and twist round small soil particles.

Roots do not bear leaves or buds. They branch a great deal, but unlike the stem, this is irregular, forming no definite pattern. Old roots have an outer layer of cork, but there are no lenticels.

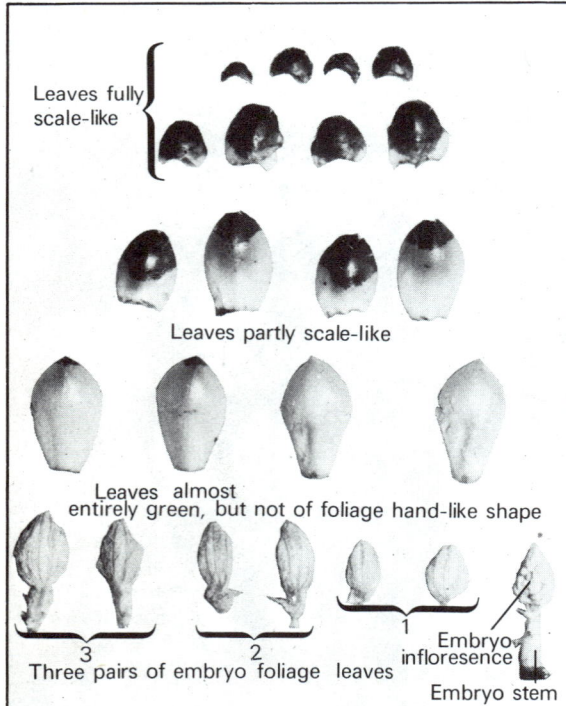

Leaves fully scale-like

Leaves partly scale-like

Leaves almost entirely green, but not of foliage hand-like shape

3 2 1
Three pairs of embryo foliage leaves

Embryo inflorescence

Embryo stem

Fig. 3.6 Leaves removed in succession from a horsechestnut terminal bud.

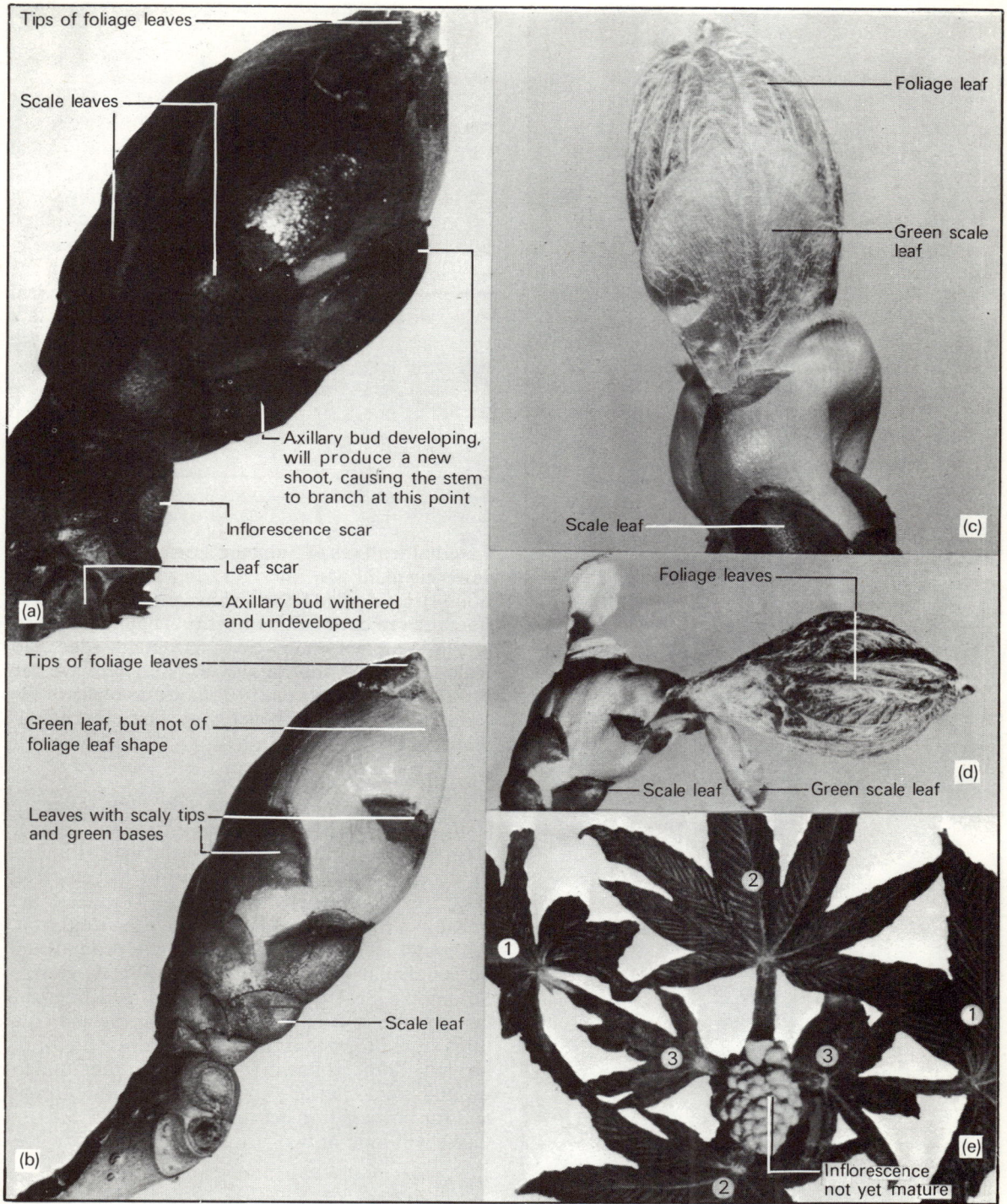

Tips of foliage leaves

Scale leaves

Axillary bud developing, will produce a new shoot, causing the stem to branch at this point

Inflorescence scar

Leaf scar

Axillary bud withered and undeveloped

(a)

Foliage leaf

Green scale leaf

Scale leaf

(c)

Tips of foliage leaves

Green leaf, but not of foliage leaf shape

Leaves with scaly tips and green bases

Scale leaf

(b)

Foliage leaves

Scale leaf

Green scale leaf

(d)

Inflorescence not yet mature

(e)

Fig. 3.7(a)–(d) Horsechestnut bud opening. **(e)** Young horsechestnut shoot seen from above. The three pairs of foliage leaves, 1,1; 2,2; 3,3 are borne mutually at right angles (compare with bud section, Fig. 3.5).

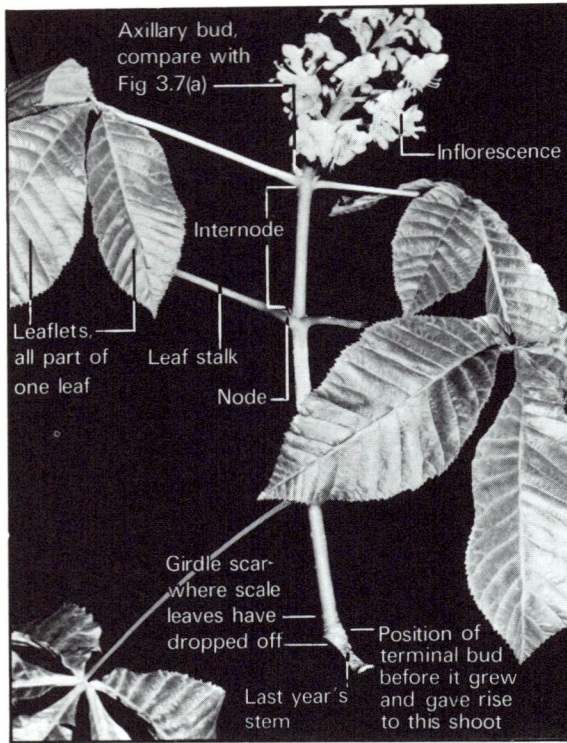

Fig. 3.8 Young shoot of horsechestnut developed from a terminal bud.

Fig. 3.9 New terminal bud formed soon after horsechestnut shoot has ceased to lengthen.

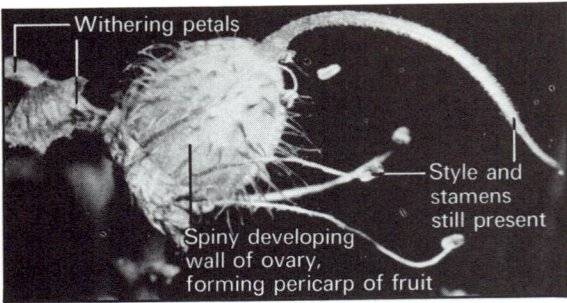

Fig. 3.10 Young developing fruit of horsechestnut.

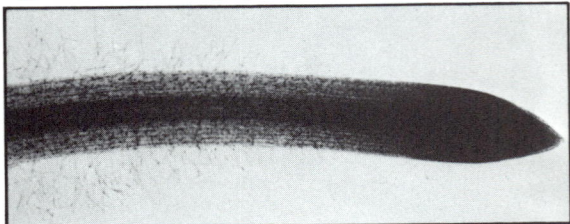

Fig. 3.11 Seedling of broad bean germinated on blotting paper, showing root hairs. (*Courtesy Carolina Biological Supply Co.*).

External features of Solomon's Seal, a monocot flowering plant

Long strap-shaped leaves which encircle the whole stem are typical of all monocots. They can be seen in Fig. 3.12, Solomon's Seal, and are divided into scale leaves and foliage leaves. Note here too the *horizontal* stem (**rhizome**) which grows beneath the soil and bears roots all over its surface. There is no main root.

Variations on the ground plan

The flowering plant body has only a few organs—stem, root, leaf, bud, flower, and fruit, but there is great variation in form amongst the many families of flowering plants. Cacti have no leaves at all (Fig. 3.13). Dandelion has no stem, all the leaves are borne as a rosette immediately above the stout root. Although stems usually grow above ground and perpendicular to the soil, we have already seen in Solomon's Seal a horizontal stem which grows beneath the soil. This is a thick structure and contains stored food that the plant can draw on for its growth in the following spring. Another underground stem specially modified for food storage and for propagation is that of potato (Fig. 3.14).

We will look at two more species.

Tulip

The **bulb** is a large swollen bud. The bulk of it consists of thick, swollen leaves whose function is to store food made in the previous year in readiness for the growth of the stem and foliage leaves in the

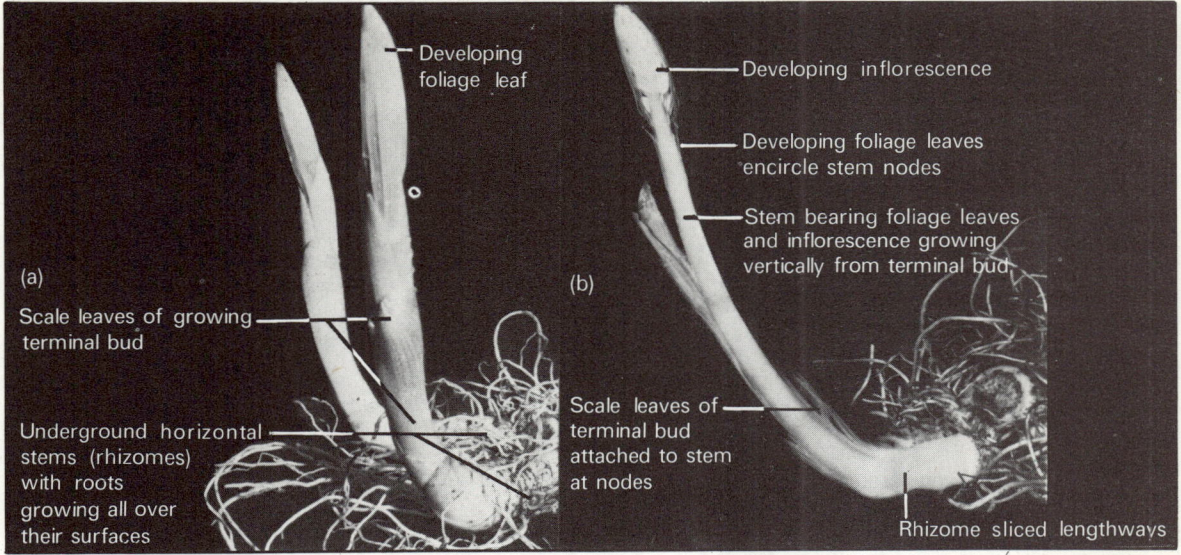

Fig. 3.12(a) Rhizome of Solomon's Seal with two developing terminal buds. (b) Part of the rhizome in Fig. 3.12(a), sliced lengthways.

spring. Each leaf encircles the stem, and is attached to the stem at its node. A large axillary bud can be seen between two of these leaves in Figs. 3.16(b) and (c).

In the centre of the bulb, at the top of the stem, is the terminal bud. This consists of a stem bearing embryo foliage leaves and ending in an embryo flower. In Fig. 3.16(a) these foliage leaves have grown out of the bulb and emerged above ground.

Crocus

Figure 3.15 shows two crocus corms and some of the outer papery structures have been removed from each. Try to answer the following questions:
1 What are the structures labelled *A*?
2 What are the regions labelled *B* and *C*?
3 What plant structure is the solid white mass of the corm? How do your answers to (1) and (2) support your view?
4 What is the structure *D* in Fig. 3.15(a)?
5 What is the structure *E* in Fig. 3.15(a)? What evidence in the photograph supports your answer?
6 Is this plant a dicot or a monocot? State at least two reasons in support of your answer.

Major differences between the body structures of vertebrates and of flowering plants

1 The open growth form The external features and shape of a man are much less variable than those of a flowering plant. Although men vary in height, weight, colour, and many other ways, they nearly all have two arms, two legs, ten fingers, ten toes, two eyes, and the same internal arrangement of organs. By contrast, while a horse chestnut tree always has stems, roots and leaves, these are not arranged with the same sort of definiteness. Branches and roots can grow in more or less any direction. Another way of putting this is to say that vertebrates have a **closed** growth form and that flowering plants have an **open** growth form.

2 Organ differentiation Flowering plants have no sensory organs, organs for moving the body, organs for coordinating the body's activities, or organs for digestion and excretion.

Fig. 3.13 Prickly pear.

27

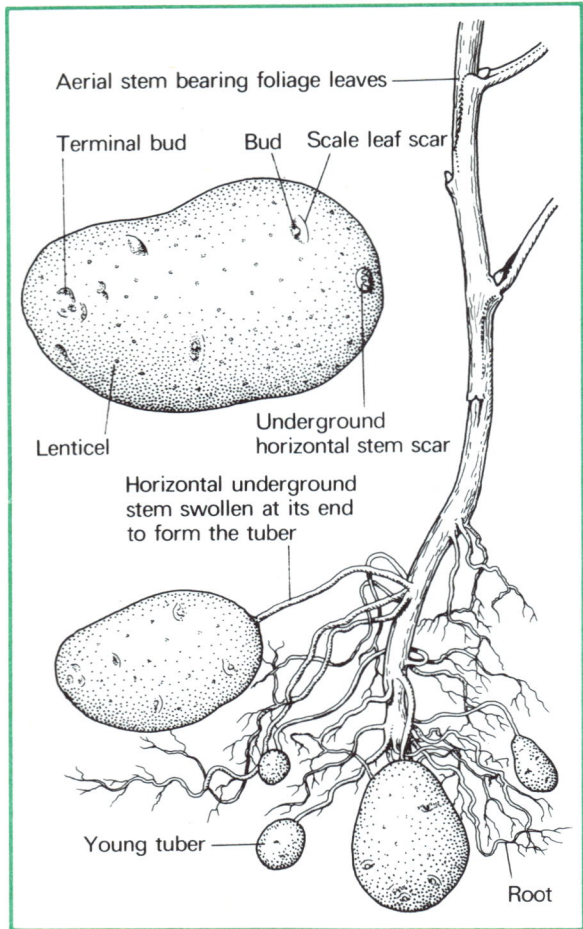

Aerial stem bearing foliage leaves

Terminal bud   Bud   Scale leaf scar

Lenticel

Underground horizontal stem scar

Horizontal underground stem swollen at its end to form the tuber

Young tuber

Root

Fig. 3.14 Base of a potato plant with tubers.
(inset) Potato tuber.

3 Surface area The external surface area of a vertebrate is much less than that of a flowering plant of the same overall size, and this is largely because plants have large numbers of flat leaves, each with two surfaces.

4 The fundamental difference: feeding The previous three differences are directly related to the fundamental difference between animals and plants, namely the ways in which they feed. Animals eat either plants or other animals: they obtain their cell chemicals such as proteins, fats, and carbohydrates ready made. This is called **holozoic nutrition**. Animals therefore have to be able to detect the presence of food; to be able to pursue and catch it; to eat and digest it. These activities require sense organs, muscles, bones for muscles to pull against, nerves to coordinate senses and movement, and a gut to digest and absorb food. Being much more

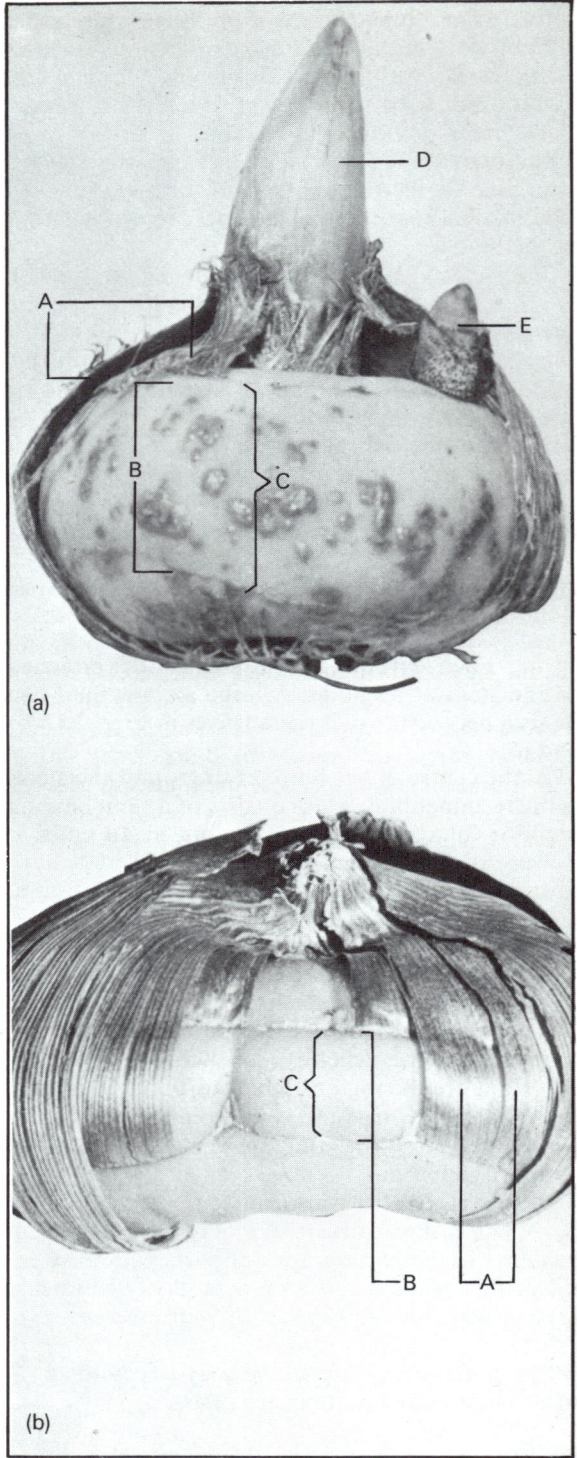

(a)

(b)

Fig. 3.15(a), (b) Crocus corms.

active than plants because of movement, most animal cells require more energy than do plant cells. The source of energy is glucose, obtained by digestion of food, and oxygen taken from the air. Vertebrates possess gills or lungs to absorb large quantities of oxygen. This passes into their blood stream and is distributed to all the cells of the body. The heart is the pump which causes rapid circulation of the blood.

Plants make their own life chemicals. Carbon dioxide from the air and water from the soil are converted into carbohydrates. Light energy is used to produce this change. Using ions, such as nitrate ions from the soil, all other cell chemicals, including proteins, are made by plant cells. This is called **holophytic nutrition**. Plants do not need organs for detecting prey or for moving. Being stationary they require less energy, and have no organs corresponding to the heart or the circulatory system. Their slow activities do not require rapid coordination, and they have no organs corresponding to a brain or nerves. However, they do have leaves which expose a large surface area to the air from which carbon dioxide is removed; and they possess extensive roots which enable them to absorb water and ions from the soil.

These broad differences of body form and of nutrition are true of animals and plants in general, though we have examined as particular examples a mammal and a variety of flowering plants.

Fig. 3.16(a) Young tulip plant.
(b) Tulip bulb sliced lengthways.
(c) Detail of Fig. 13.16(b).

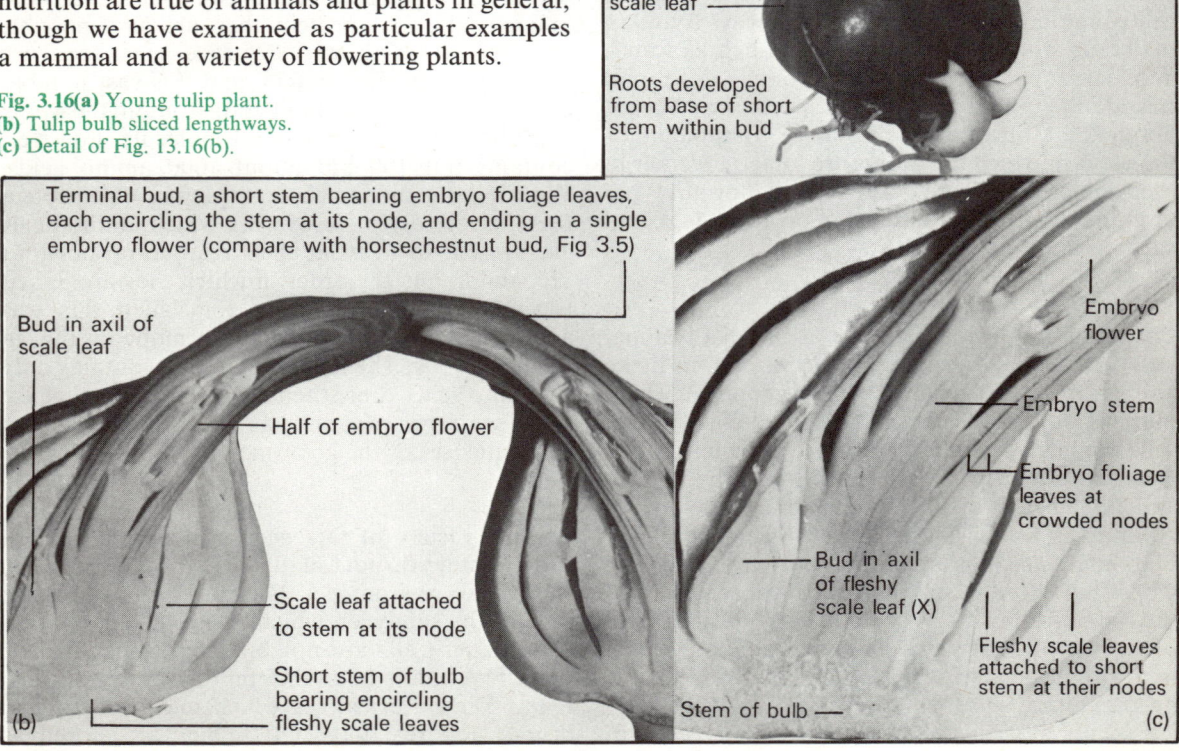

(a)

Foliage leaves developed from terminal bud

Foliage leaves developed from axillary buds such as X in Fig 3.15 (c)

Single outer dry scale leaf

Roots developed from base of short stem within bud

Terminal bud, a short stem bearing embryo foliage leaves, each encircling the stem at its node, and ending in a single embryo flower (compare with horsechestnut bud, Fig 3.5)

Bud in axil of scale leaf

Half of embryo flower

Scale leaf attached to stem at its node

Short stem of bulb bearing encircling fleshy scale leaves

(b)

Embryo flower

Embryo stem

Embryo foliage leaves at crowded nodes

Bud in axil of fleshy scale leaf (X)

Fleshy scale leaves attached to short stem at their nodes

Stem of bulb

(c)

4 Cell Chemicals and Chemical Reactions

There are many hundreds of different chemical compounds in cells. Nearly all of them do not exist in the world outside cells (except in decaying matter), and cells must therefore have made them. The chemical elements these compounds contain are the same as in materials which are non-living, such as rocks, water, and air. Most cell compounds contain the element carbon whose atoms, more than those of any other element, have the ability to join together forming chains or rings of atoms. Less than 1 % of the earth's substance is carbon. It is remarkably concentrated in living material because of the very large number of different molecules that can be built using chains or rings of its atoms as 'skeletons'. Nitrogen, sulphur and phosphorus are three other scarce elements that are found as parts of molecules in all living cells. So are oxygen and hydrogen, but oxygen accounts for 50 % of the earth's crust, and hydrogen 1 %.

The ways in which atoms of these and other elements are linked together to form molecules of cell compounds are the same as the ways found in materials from non-living sources, such as water. While cells contain special compounds not found outside them, there is nothing special or different about the atoms composing these compounds or the ways in which the atoms are joined together in their molecules. Three of the most important types of compound found in cells are proteins, fats, and carbohydrates.

Proteins

Proteins form the main structural part of all animal and plant cells. Muscles, whose contractions make animals move, are made of protein; bone, on which muscles pull, is largely made of protein. Animal cells are held together by fibres of a protein called **collagen** (this forms the white fibres seen in Fig. 2.5), and alone makes up 30 % of the human body mass. Antibodies, which give protection against microbial infections, are made of protein. Enzymes, which regulate the speed of chemical reactions in the cell, are all made of protein. The liquid portion of the blood has proteins dissolved in it, and oxygen is transported in vertebrate animals by means of haemoglobin, a protein.

Protein molecules are very big. They contain thousands of atoms, but they have a fairly simple underlying pattern of structure. They all consist of small molecular units called **amino acids**, linked together in long chains like a string of poppet beads. Large molecules formed from smaller molecules linked together are called *polymers*.

There are about 20 different kinds of amino acids, and the number of possible arrangements of them in a protein is very large. Consider part of a protein with a small molecule consisting of only 100 amino acids in the chain:

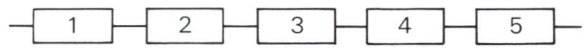

$\boxed{1}$ could be any one of them. Similarly, there are 20 possibilities for $\boxed{2}$, making 20×20 or 400 possible sequences of the first two amino acids. $\boxed{3}$ could also be any one of 20 amino acids, so for the first three positions there are $20 \times 20 \times 20 = 800$ or 20^3 possible sequences. Taking all 100 amino acids in this protein, there are 20^{100} possible varieties of amino acid sequence, each sequence forming a different protein, and this vast number exceeds the supposed number of atoms in the whole universe. An average protein molecule, however, contains not 100 but about 1000 amino acids. Working out the molecular structure of a protein therefore involves discovering which amino acids are present in it, what their relative proportions are, and finally the order in which they are linked together. The first protein for which this was achieved was the mammalian hormone insulin, in the mid 1950s. The insulin molecule contains only 51 amino acids. Since then the amino acid sequence of a few larger proteins has been discovered, but the sequences of the great majority is unknown.

Fats

A wide variety of fats occurs in nature. Unlike proteins, they do not usually form *structures* in cells, though we saw on p. 5 that fats form the middle part of the sandwich which makes up the plasma membrane. Cells are not built from them. Their main function is quite different from that of proteins. They are sources of chemical energy for driving cell reactions. Often they are stored as reserves, for example in seeds. In mammals they are stored as fat droplets in special cells and these form an important heat-insulating layer in the skin.

A good example of fat storage is in suet. Next time you are in a butcher's shop look at the masses of suet which embed and probably hide the kidneys of a sheep or cow carcass.

Some natural fats are liquids, e.g., olive oil; some are soft solids, e.g., butter; some are hard solids like suet. Although all fats differ slightly from all other fats, they all have a basically similar structure. The molecule has a 'backbone' with three 'arms' sticking out from it:

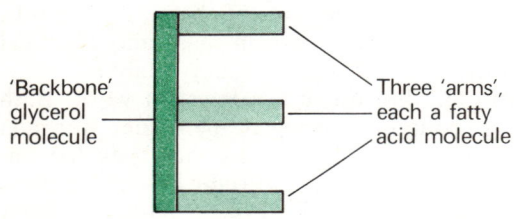

The 'backbone' is a molecule of a chemical substance called **glycerol**. The 'arms' are molecules of substances called **fatty acids**. Glycerol and fatty acids contain only carbon, hydrogen, and oxygen in their molecules. There are many different kinds of fatty acids and so there is a very large number of different kinds of fat.

Carbohydrates

The word 'carbohydrate' describes the structure of the molecule. It means 'hydrated carbon', that is, a compound whose basic structure is CH_2O. Like a fat molecule, it contains only carbon, hydrogen, and oxygen, but the hydrogen and oxygen are present in the same proportions as in water (H_2O).

Sugars and starches are carbohydrates and, like fats, are vital sources of stored chemical energy for cell reactions. Sugars are made by green plants from carbon dioxide and water, using light energy from sunlight:

$$6CO_2 + 6H_2O \xrightarrow{\text{light energy}} \underset{\text{sugar}}{C_6H_{12}O_6} + 6CO_2.$$

In this way, light energy is changed into chemical energy and becomes available to cells, both the plant cells which made the sugar and the cells of animals which eat plants.

The bulk of sugar made by plants is used to make another carbohydrate, **cellulose**, which is a building material. It forms the cell wall boundary of all plant cells. Figure 5.6 is an electron micrograph of a section through some leaf cells; the cellulose cell walls can be clearly seen. Figure 1.16 is an electron micrograph showing a surface view of a cellulose cell wall. The cellulose consists of long thin threads lying parallel to each other and arranged in layers.

In mammals, carbohydrates are stored in the liver and the muscles as **glycogen** or **animal starch**. Animal cells are held together by connective tissue (p. 14), one of the main components of which is ground substance, which contains a carbohydrate. **Chitin**, which forms the exoskeleton of insects and crustaceans (the largest group of animals) is a modified carbohydrate. Like proteins and fats, carbohydrates are found everywhere in living things.

Glucose
The simplest way of writing the formula of one molecule of this important sugar is $C_6H_{12}O_6$. One molecule therefore contains six atoms of carbon, twelve atoms of hydrogen, and six atoms of oxygen: this is a ratio of CH_2O.

Starch, glycogen, and cellulose
These three important carbohydrates are all polymers, consisting of chains of glucose molecules joined together. Sometimes the resulting molecule is long and straight, sometimes branched, as in Fig. 4.1, where each glucose molecule is represented as a circle. Cellulose is built in a similar way.

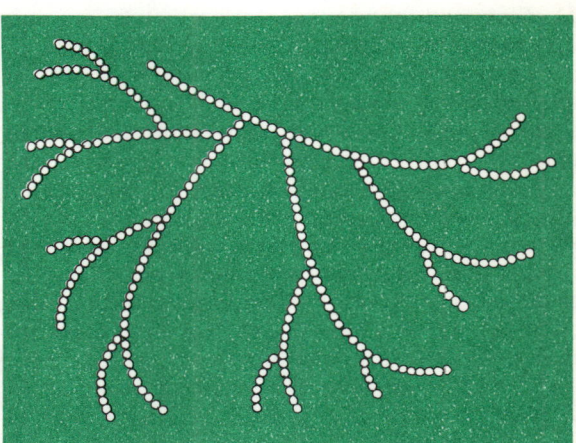

Fig. 4.1 Diagrammatic structure of part of a starch molecule. Each circle represents a glucose molecule. Note the branching of the chains.

Chemical reactions in the living cell

A multitude of chemical reactions is constantly taking place in all living cells. For example, in the cells of green leaves, glucose ($C_6H_{12}O_6$) is made from carbon dioxide (CO_2) and water (H_2O). Using glucose and ions such as nitrate (NO_3^-), fats and proteins are made by these cells. Red blood cells are made in an adult man at the rate of some 200 000 million per 24 hours, and each possesses a membrane made of fat and protein; dissolved in the liquid part of the cell is the protein haemoglobin. A great deal of chemical activity has to take place in the bone marrow to form so many cells.

Collision theory of chemical reaction

Atoms, ions, and molecules are all in a constant state of motion, and if this is free motion they collide with each other. In gases and liquids the movement is free; in solids these particles are only able to vibrate. When two cars collide and rebound the kinetic energy (energy of motion) that they possess is used to perform work in damaging the cars. When two atoms, molecules, or ions collide part of their kinetic energy may be used to form a new chemical compound in which the particles are united. The chemical and physical properties of the new compound are quite different from those of the particles which formed it. For example, water (H_2O) is quite different from the two elements hydrogen and oxygen, which are gases at room temperature:

$$2H_2 + O_2 \longrightarrow 2H_2O$$

Factors affecting the rates of chemical reactions

1 Temperature

The *rate* of a chemical reaction is the amount of chemical change taking place in a unit time. A rise in temperature always increases the rate, and often a rise of 10°C doubles it. For example, if in the reaction

$$A + B \longrightarrow C + D$$

x grams of A and B react at 5°C, 2x grams would react at 15°C. Heat energy causes ions and molecules to move faster. At a higher temperature the particles collide more frequently and, because they are moving faster, with more impact.

2 Concentration

The concentration of a solution of substance X is the mass of it that is dissolved in a unit volume of water, and it is usually expressed in g/cm^3. Doubling the concentration of a substance doubles the number of molecules per unit volume and therefore doubles the number of molecular collisions per second. Thus increase in concentration increases rate of reaction.

3 Catalysts

It is often found that the rates of chemical reactions are increased by the presence of certain other chemicals which are not themselves reactants. These substances are called **catalysts**, and they are particularly important in the chemical reactions taking place in living cells. All but a very few of these are speeded up by catalysts called **enzymes**.

Properties of catalysts

Catalysts increase the rates of chemical reactions. They are *unchanged*, both in mass and chemical composition, by the reaction they catalyse.

Figure 4.2 illustrates a catalyst at work. Both beakers contain hydrogen peroxide (formula H_2O_2). The beaker on the left has been left untreated; the beaker on the right has had a piece of liver added to it. Hydrogen peroxide slowly decomposes at room temperature to form water and oxygen, which is liberated:

$$2H_2O_2 \longrightarrow 2H_2O + O_2$$

Liver cells contain an enzyme called **catalase** which catalyses this reaction, as shown by the masses of oxygen bubbles in the right hand beaker.

The rates of chemical reactions are increased by rise in temperature and by rise in the concentration of the reacting chemicals. Both these methods of speeding up chemical reactions are of limited use to

(a) (b)

Fig. 4.2 Release of oxygen bubbles from hydrogen peroxide when a piece of liver, whose cells contain the enzyme catalase, is added to it.

Fig. 4.3 Drawing of a molecule of the protein myoglobin. The helix of amino acids is folded into a complicated and precise shape. The disc on the right of centre is a non-protein part of this particular protein molecule.

living organisms, for at temperatures above 45°C the proteins of which living cells are made become drastically changed. We see this every time we boil an egg. The slimy white of a fresh egg is a solution of the protein albumin. When it is heated it coagulates and its physical properties change. It is no longer a solution but solid and insoluble. In the same way temperatures above 45°C change the proteins of living cells (many of which are enzymes), killing the cell.

Importance of enzyme catalysts in living cells
Speeding up of reaction rate by increasing the concentration of the reactants is also of little value to living cells, for they contain many hundreds of chemicals and there would not be room for high concentrations of all of them. Furthermore high concentrations of dissolved chemicals would raise the osmotic pressure of the cell (see p. 84), and this would have far-reaching, harmful consequences. It is known, indeed, that some of the most important chemicals in the cell are present in concentrations of only a few hundred molecules per cell.

However, most of the hundreds of chemical reactions that take place in the cell proceed slowly at room temperature (like the decomposition of hydrogen peroxide in Fig. 4.2), and the rates of nearly all of them are speeded up by enzymes. This means that there are hundreds of enzymes regulating reaction rates, and a large part of the cell protein is enzyme material.

Some 600 enzymes are known; many have been isolated from cells, purified, and even crystallized. There are undoubtedly many more to be discovered, and the study of enzymes—enzymology—is a whole branch of science in itself. All the enzymes that have been extracted from cells have proved to be made of proteins. Protein molecules are often very large. They consist of amino acid units linked together in chains in the manner described on p. 30. There may be hundreds or even thousands of amino acid units in a single molecule of protein, and in enzyme proteins the chain is coiled into a helix, like a spiral staircase. The helix may become folded to produce a very large molecule with a strange but precise shape. Figure 4.3 shows the shape of a molecule of myoglobin, a protein which resembles haemoglobin, the red pigment of red blood cells. Myoglobin is not an enzyme, but the shapes of one or two enzyme molecules have also been worked out.

Effect of temperature on enzyme activity
The precise molecular shape is thought to play an important part in the way that enzymes work, and it explains the effect of rise in temperature on enzyme activity. When proteins are heated their molecular shapes change, and this results in a change in their physical and chemical properties. Dissolved egg albumin becomes insoluble when heated; enzymes loose their catalytic activity when heated.

Figure 4.4 shows how the activity of enzymes varies with the change in temperature of the medium in which they are working. From 0°C to about 40°C, rise in temperature produces a rise in activity. This is because collisions between the enzyme molecules and molecules of the chemical substance whose change they are catalysing (e.g. hydrogen peroxide in the case of catalase) are more frequent at higher temperatures. But at about 40°C there is a complete reversal; activity falls rapidly, and at about 60°C has almost ceased. The higher temperatures put more and more enzyme molecules out of action by denaturing the proteins of which they are made. For this reason very few living organisms can live at temperatures above 40–45°C. The body temperatures of many birds and mammals is between 37° and 40°C. These are very favourable temperatures for high rates of enzyme activity.

Where enzymes are situated in the cell
In chapter 1 we saw electron micrographs that revealed much fine detail in the cell. Is it possible to see the enzymes? Before attempting to answer this question we must consider the relative sizes of cells, cell organelles, cell particles, and molecules.

Very roughly speaking, cells are 20 μm in diameter (though there is much variation, and some cells are very long). 1 μm = 0.001 mm. Cell organelles, such as mitochondria and lysosomes, are about 4 μm–1 μm in diameter; and the cell particles, such as ribosomes are about 0.1 μm in diameter. Ribosomes can be seen in Fig. 1.11. Another kind of cell particle is seen covering the inside of a mitochondrion (Fig. 1.10(b)). Figure 4.5 is an

electron micrograph of them, showing that each is attached to the inner membrane of the mitochondrion by a short stalk. The sizes of these particles approaches that of very large molecules, such as those of enzyme proteins. Very roughly, most molecules are much smaller than this, with an average diameter of 1 nm (1 nm = 0.001 μm).

Enzyme sites

(a) Protein-synthesizing enzymes
The manufacture of proteins, including enzymes, takes place on the ribosomes. These particles are not single enzymes, but they contain enzymes.

(b) Digestive enzymes
These are made on the ribosomes, packaged into lysosomes by the Golgi apparatus, and then used to digest material in the cell (Fig. 1.15 route 1), or are exported from the cell (Fig. 1.15 route 2). Salivary amylase is an example of a digestive enzyme that is made and exported by salivary gland cells. Thus while it is not possible to see individual molecules of digestive enzymes with the electron microscope, it is possible to follow the paths they traverse in the cell.

(c) Energy-liberating enzymes
Chemical energy is liberated in the cell by the oxidation of glucose. This process is called **tissue respiration** (p. 67) and is summarized by the equation:

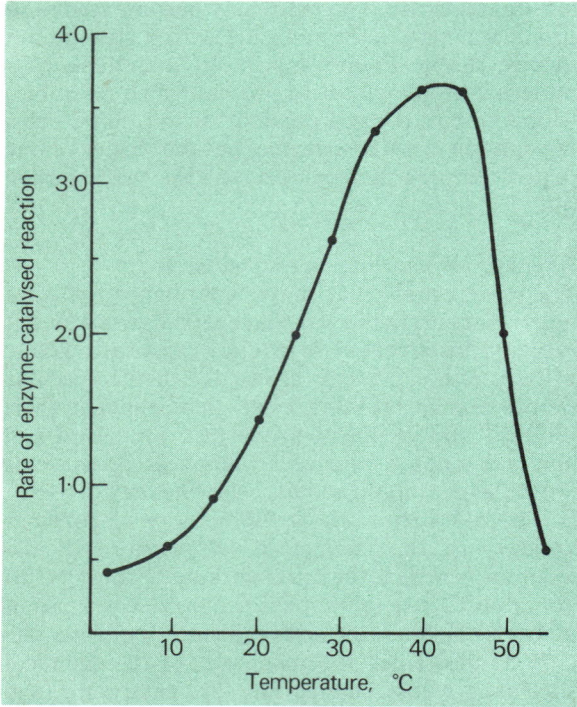

Fig. 4.4 Effect of temperature on the rate of an enzyme-catalysed reaction.

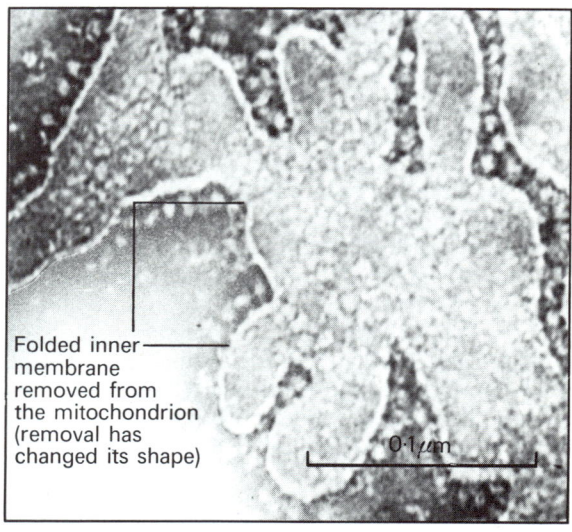

Folded inner membrane removed from the mitochondrion (removal has changed its shape)

0·1 μm

Fig. 4.5 Part of the inner membrane of a mitochondrion, showing the enzyme-containing cell particles attached to the membrane by stalks.

$$C_6H_{12}O_6 + 6O_2 \longrightarrow 6CO_2 + 6H_2O + \text{ENERGY}$$
glucose

There are many individual chemical steps in this change, but there is a 'halfway-house' stage where the glucose molecule is split into two molecules each containing three carbon atoms:

glucose (6C) $\longrightarrow$ (2 × 3) C molecules + energy

The enzymes that bring about this change are dissolved in the fluid ground substance that lies between the cell organelles (Fig. 1.7). Their molecules cannot be seen with the electron microscope, though their location is known.

The 3 C molecules then pass into the mitochondria where they undergo a series of chemical changes liberating much more chemical energy for cell use and producing carbon dioxide and water as final waste materials:

2 × 3C molecules + 6O$_2$ $\longrightarrow$ 6CO$_2$ + 6H$_2$O + ENERGY

The mitochondrial enzymes that catalyse these changes are partly dissolved in the fluid-filled interior of the mitochondria (Fig. 1.10(b)). Many of them are also attached to the inner mitochondrial membrane, forming the cell particles seen in Fig. 4.5. Thus again the sites of the enzymes are precisely known though their molecules are not revealed by the electron microscope.

5 Plant Nutrition

Importance of plant nutrition

The fundamental difference between animals and plants lies in the ways in which they make or obtain their food (p. 28). Animals have to obtain their life chemicals such as proteins, fats, and carbohydrates, ready made, by eating plants or other animals (holozoic nutrition). Plants are able to make life chemicals from carbon dioxide, water and ions, using light energy (holophytic nutrition). All life thus depends on the life chemicals made by plants. To give an idea of the immensity of this activity, it has been estimated that during the history of the earth some 10^{26}g of carbon from carbon dioxide gas has been turned into life chemicals by plants. This mass equals one fifth that of the planet itself.

We must note too that carbon dioxide is a rare gas, forming only 0.03% of the atmosphere. Although most of the land surface is covered with vegetation which is consuming carbon dioxide (as is the floating plankton in the surface layer of the oceans), the gas cannot vanish or plant life would come to a stop. Some other process must be putting it back into the atmosphere as fast as plants are taking it out. This process is the oxidation of carbohydrates and fats by cells to produce carbon dioxide, water and chemical energy. The process is tissue respiration:

glucose $\longrightarrow$ (2×3)C molecules + chemical energy

(2×3)C molecules $+ 6O_2 \longrightarrow 6CO_2 + 6H_2O$
$+$ CHEMICAL ENERGY.

Since these chemical changes liberate energy, the reverse process of making carbohydrates such as glucose from carbon dioxide and water must involve the *input* of energy. Green plants will not grow in the dark, so it is a reasonable guess that light is the form of energy involved. A possible equation for the building-up process (**synthesis**) might be:

CO_2 + light energy $\longrightarrow$ carbohydrates
equation (1)

This is called **photosynthesis**, which means 'building up in light'.

Production of oxygen

In 1779 an Austrian, Jan Ingenhousz, made an important discovery about photosynthesis. He plac-ed green leaves in a sealed glass vessel containing air whose oxygen had been replaced by carbon dioxide. After exposing the leaves to sunlight he found that the vessel contained oxygen, which must have been produced by the leaves. We must therefore modify equation (1) as follows:

CO_2 + light energy $\longrightarrow$ carbohydrates $+ O_2$
equation (2)

The earth's original atmosphere is thought to have contained no oxygen. Today it contains about 20% oxygen, all of which has been produced by photosynthesis. Photosynthesis is not, therefore, only important because it turns unlimited supplies of light energy into chemical energy which can be used by cells. It also led to the establishment of an oxygen-containing atmosphere, and this had a profound effect on the evolution of life (Chapter 20). The 'invention' of photosynthesis must be reckoned as one of the most important events in the history of life on earth.

Fig. 5.1 A starch print on a geranium leaf. The plant was first kept in the dark for 10 hours so that any starch present in the leaf was removed and not replaced by photosynthesis. Then a photographic negative was clipped to the leaf, which was exposed to the light. The amount of starch produced was proportional to the amount of light penetrating the various parts of the negative. The leaf was then killed and treated with iodine solution which stained the starch black.

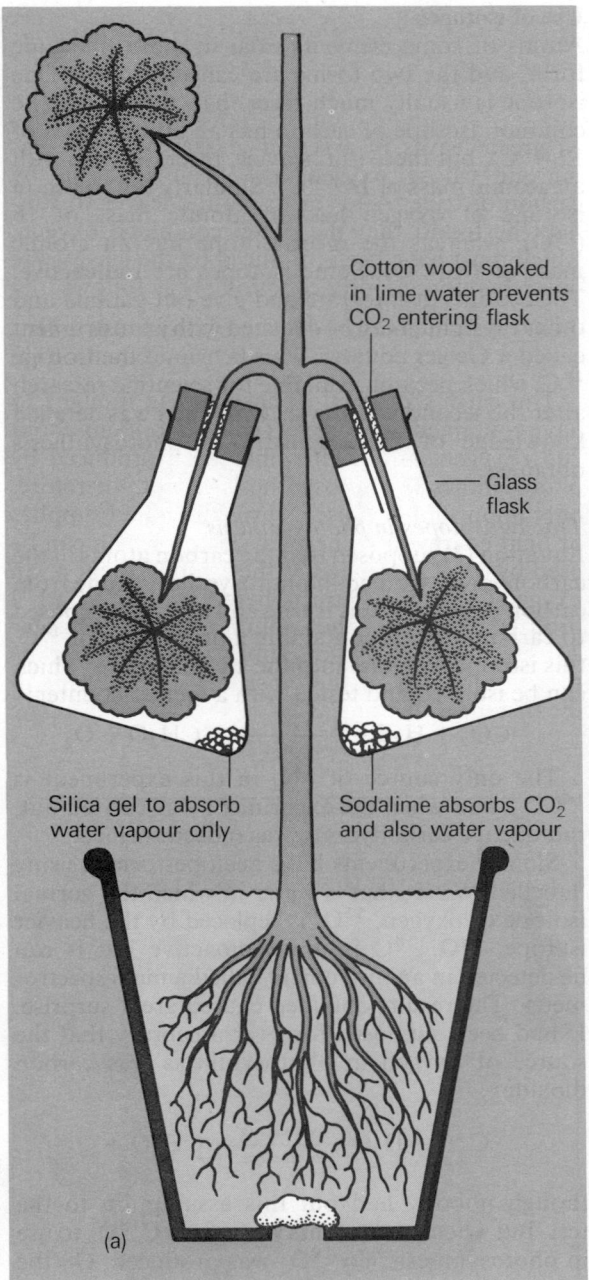

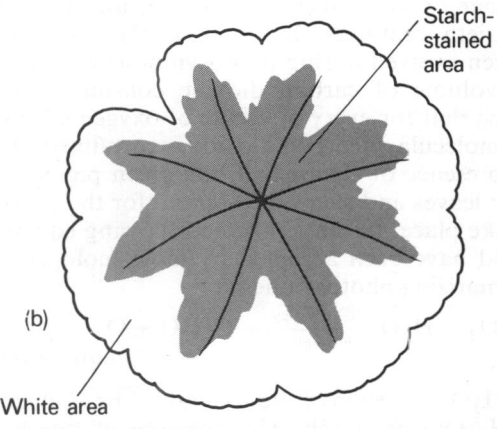

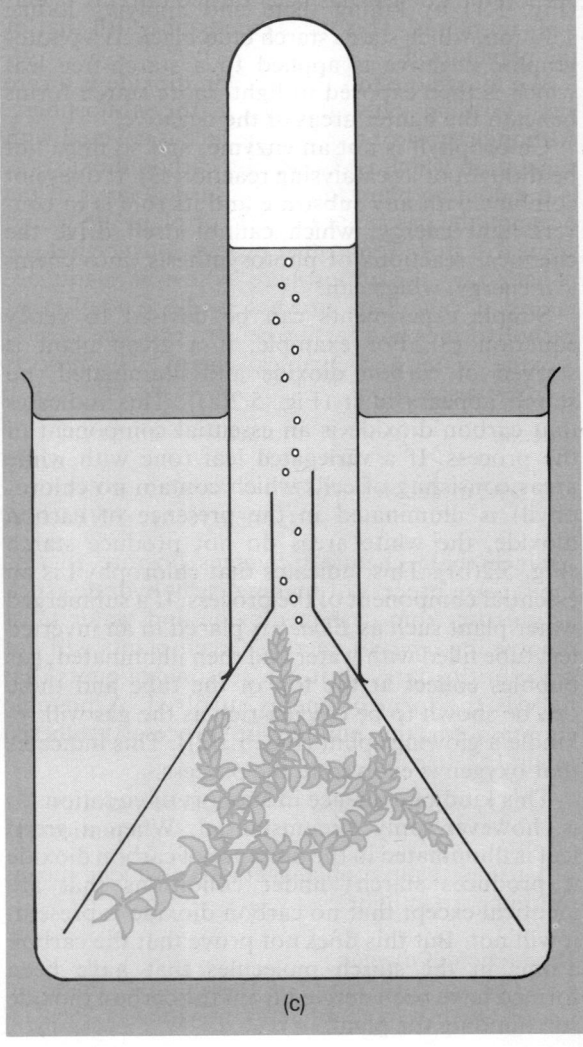

Fig. 5.2(a) Demonstration that in a CO_2-free atmosphere an illuminated green leaf produces no starch. Soda lime in the right hand flask absorbs CO_2. The left hand flask is a *control*. Why is this a necessary part of the experiment?
(b) When a variegated leaf is exposed to light and then tested for starch, starch is found only in the areas which were green. None appears in the formerly white areas.
(c) Demonstration that *Elodea* gives off a gas when illuminated. Since this gas will re-kindle a glowing splint, it must be oxygen-rich.

Chemistry of photosynthesis

Further experiments in the nineteenth and early twentieth centuries showed that the volume of oxygen evolved during photosynthesis is equal to the volume of carbon dioxide consumed. This means that for every molecule of oxygen evolved, one molecule of carbon dioxide is consumed. Also the presence of **chlorophyll**, the green pigment in plant leaves and stems, is essential for the process to take place. In the 1940s the following equation would have been accepted by most biologists as summarizing photosynthesis:

$$CO_2 + H_2O \xrightarrow[\text{chlorophyll}]{\text{light energy}} \underset{\text{carbohydrate}}{CH_2O} + O_2 \qquad \text{equation (3)}$$

CH_2O, carbohydrate, could be glucose ($C_6H_{12}O_6$) or starch. The presence of starch in leaves that have been illuminated is easily shown (Fig. 5.1) by killing them and applying iodine solution, which stains starch blue-black. If a photographic negative is applied to a starch-free leaf which is then exposed to light, more starch forms beneath the lighter areas of the negative.

Chlorophyll is not an enzyme, and so must not be thought of as catalysing reaction (3). It does not combine with any substrate and its role is to convert light energy, which cannot itself drive the chemical reactions of photosynthesis, into chemical energy, which can.

Simple experiments can be devised to verify equation (3). For example, if a green plant is starved of carbon dioxide and illuminated, no starch appears in it (Fig. 5.2(a)). This indicates that carbon dioxide is an essential component of the process. If a **variegated** leaf (one with white areas consisting of cells which contain no chlorophyll) is illuminated in the presence of carbon dioxide, the white areas do not produce starch (Fig. 5.2(b)). This indicates that chlorophyll is an essential component of the process. If a submerged water plant such as *Elodea* is placed in an inverted test tube filled with water and then illuminated, gas bubbles collect at the top of the tube and these can be shown to be oxygen-rich as the gas will re-kindle a glowing splint (Fig. 5.2(c)). This indicates that oxygen is evolved in the process.

This kind of evidence in support of equation (3) is, however, only circumstantial. When a green leaf is illuminated in the presence of carbon dioxide it produces starch; under conditions that are identical except that no carbon dioxide is present, it will not. But this does not prove that the carbon atoms in the starch molecules that have been formed have been derived from the carbon dioxide surrounding the plant.

Use of isotopes

Atoms of some elements exist in more than one form, and the two forms are called **isotopes**. One isotope is usually much rarer than the other. The common isotope of carbon has an atomic mass of 12 (^{12}C), but there is a heavier, rarer, isotope with an atomic mass of 14 (^{14}C). Similarly the common isotope of oxygen has an atomic mass of 16 (^{16}O), whereas the rarer isotope has an atomic mass of 18 (^{18}O). Some isotopes are **radioactive**. Their atoms disintegrate and give out gamma and other rays which can be detected with an instrument called a Geiger counter. This is true of the isotope ^{14}C, which became available for scientific research after the second world war. Only then was detailed knowledge of the chemistry of photosynthesis obtained.

Tracing isotopes in photosynthesis

Equation (3) supposed that the carbon atoms in the carbohydrate formed in photosynthesis came from carbon dioxide. If this is true and a plant is exposed to carbon dioxide containing a proportion of ^{14}C, this isotope will pass into the carbohydrate, which can be isolated and tested with a Geiger counter:

$$^{14}CO_2 + H_2O \xrightarrow[\text{chlorophyll}]{\text{light energy}} {}^{14}CH_2O + O_2$$

The only source of ^{14}C in this experiment is $^{14}CO_2$, and when the experiment was carried out, radioactive carbohydrate was indeed formed.

Similar experiments have been performed using 'labelled' water, that is water in which the normal isotope of oxygen, ^{16}O, is replaced by the heavier isotope, ^{18}O. ^{18}O is not radioactive but it can be detected in an instrument called a mass spectrometer. The results obtained caused great surprise. It had been supposed for over a century that the source of oxygen in photosynthesis was carbon dioxide:

$$C O_2 + H_2O \xrightarrow[\text{chlorophyll}]{\text{light energy}} CH_2O + O_2$$

though nobody had put this assumption to the test. But when green plants were given $C^{18}O_2$ to use in photosynthesis, no $^{18}O_2$ was produced. On the other hand, when supplied with $H_2^{18}O$ during photosynthesis, the oxygen produced did contain ^{18}O, and this shows that the source of oxygen in photosynthesis is not carbon dioxide but water. To fit this data, equation (3) must be modified as follows:

$$CO_2 + 2H_2{}^{18}O \xrightarrow[\text{chlorophyll}]{\text{light energy}} CH_2O + {}^{18}O_2 + H_2O \qquad \text{equation (4)}$$

Examine equation (4) carefully. What is the

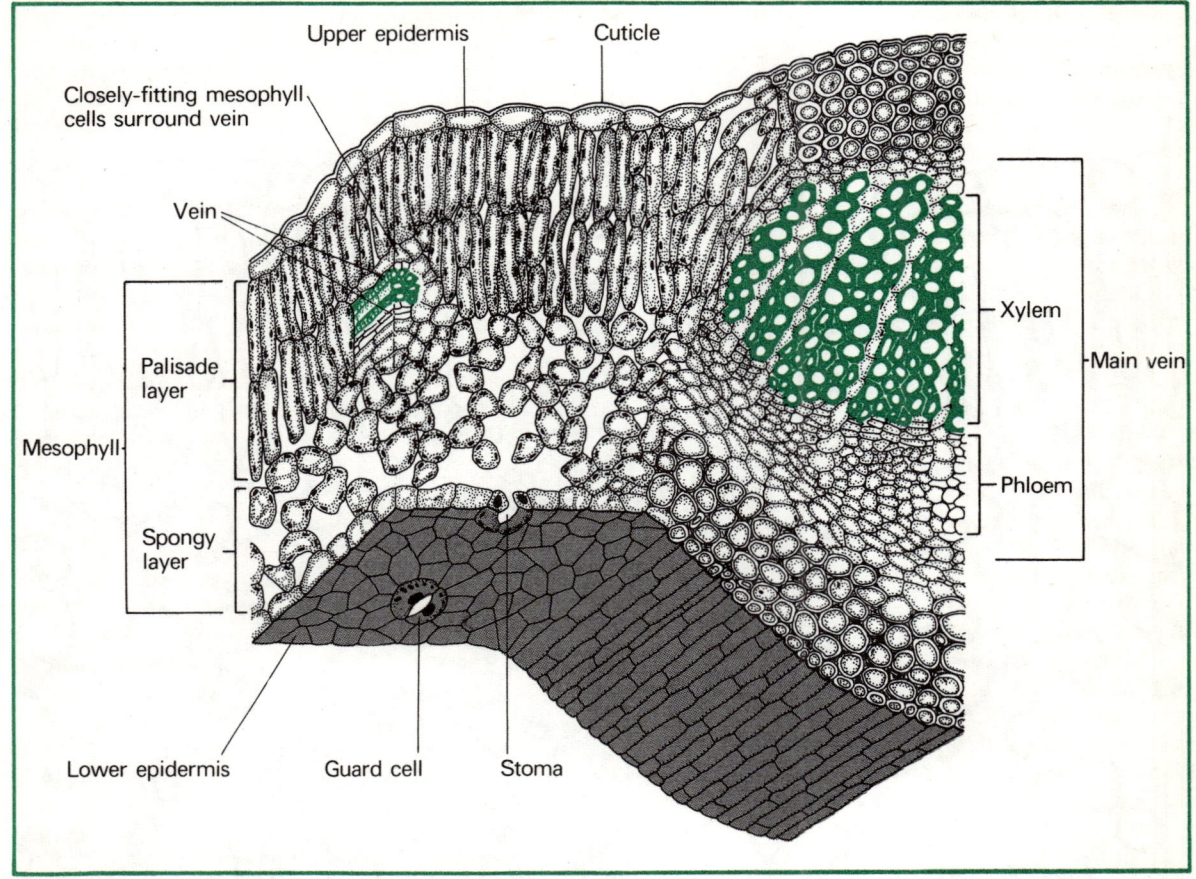

Fig. 5.3 Stereogram showing part of the internal structure of a dicot leaf and illustrating the loose arrangement of cells.

source of the hydrogen in the carbohydrate? How could it be demonstrated using 'heavy' hydrogen (the isotope ^{2}H, which is twice as heavy as the common isotope ^{1}H) that the hydrogen atoms in the carbohydrate had come from the hydrogen atoms in the water?

'Splitting' of water molecules
Equation (4) shows that during photosynthesis water molecules are 'split' giving oxygen atoms which appear as oxygen gas, and hydrogen atoms which are used to convert carbon dioxide into carbohydrate (CH_2O). The role of chlorophyll is to make light energy available in a form that can be used for this molecule-splitting process.

The biochemistry of carbohydrate production was worked out by Calvin in the 1950s and this feat is one of the triumphs of biochemical research. Between carbon dioxide and carbohydrate lie at least ten chemical steps, and it was the use of ^{14}C as a tracer that enabled the synthetic pathway of

conversion of one substance to the next to be followed.

The structure of the leaf and how it is adapted to photosynthesis

The large surface areas of plants
The plant body is nearly always very expansive and has a large external surface area, mainly accounted for by large numbers of thin, flat, green leaves. This adaptation exposes the plant to a large catchment area of carbon dioxide and light, two of the essential ingredients of photosynthesis.

Most plants hold their leaves roughly parallel to the soil, which means that light strikes their upper surfaces. Related to this is the fact that most leaves are darker green on their upper than on their lower surface, indicating that chlorophyll is concentrated in the position where the highest intensity of light falls.

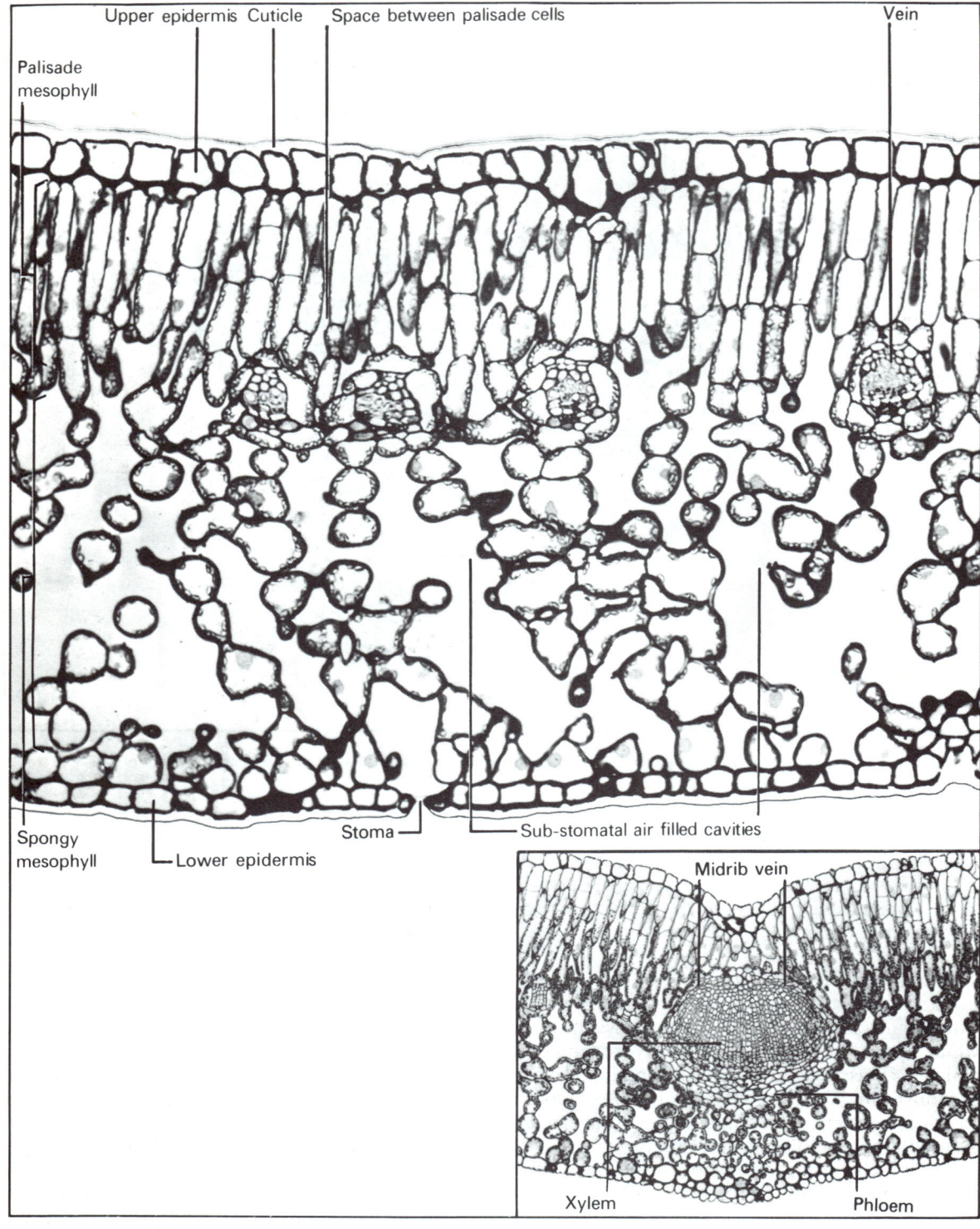

Palisade
mesophyll

Upper epidermis Cuticle   Space between palisade cells   Vein

Spongy
mesophyll   Lower epidermis   Stoma   Sub-stomatal air filled cavities

Midrib vein

Xylem   Phloem

Fig. 5.4(a), (b) Vertical sections through a privet leaf. (Courtesy of Carolina Biological Supply Company.)

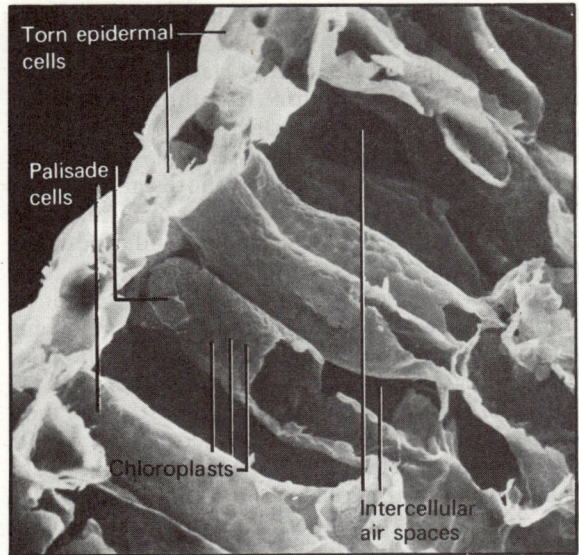

Fig. 5.5 Scanning electron micrograph of palisade cells in leaf of spinnach.

Inside the leaf the cells are arranged very loosely with air spaces between them (Fig. 5.3). This means that a large part of the surface of each cell is exposed to the air inside the leaf. Counting this as well, the true surface area of the plant is very great indeed, and this is undoubtedly an adaptation to provide the largest possible catchment area for carbon dioxide, which forms only 0.03% of the atmosphere. It has the drawback that it also provides a very large surface area for the evaporation of water from the leaf cells. Water is obtained from the soil by the roots, and a plant which cannot obtain enough water to cope with this loss wilts and dies. Plants loose at least one hundred times more water vapour by evaporation than they use in photosynthesis.

Inside the leaf
Figure 5.4 shows two vertical sections through parts of a privet leaf, and the arrangement of leaf cells here is typical of most flowering plants. There is an **upper** and a **lower epidermis** between which lies the **mesophyll** (Greek, middle of the leaf). This is divided into two regions. The upper is the **palisade mesophyll**, and most of its cells are elongated in a plane perpendicular to the epidermis. Although there are air spaces between the cells, these are more closely packed than the cells in the lower, **spongy mesophyll**. Here the arrangement is extremely loose and large areas of the cell surfaces are exposed to the air in the leaf. There is an especially large space behind a **stoma**, the **substomatal cavity**, which is continuous with all the other leaf air

spaces. The only places in the mesophyll where the cells are tightly packed with no intercellular spaces are around the veins and vein endings (see also Fig. 5.3). Veins consist of xylem vessels which bring water and mineral salts up to the leaf from the roots, and phloem sieve tubes which remove the synthesized sugars, amino acids, and other products from the leaf to the growing points and storage organs.

The shape of the palisade cells and the spaces between them are shown in surface view (a scanning electron micrograph) in Fig. 5.5. The section in Fig. 5.6 was cut *parallel* to the surface of the leaf. It shows that the bulk of each cell is occupied by the central, permanent vacuole, filled with non-living cell sap. The cytoplasm is a thin layer between the vacuole and the non-living cellulose cell wall.

The section has passed through the nucleus of one of the cells; but perhaps the most striking feature of the cytoplasm is the large number of chloroplasts. These are circular in surface view but biconvex when, as here, they are cut across. Although they are embedded in cytoplasm, this forms a very thin layer on either side of them (Fig. 5.7), and they are very close to the intercellular air

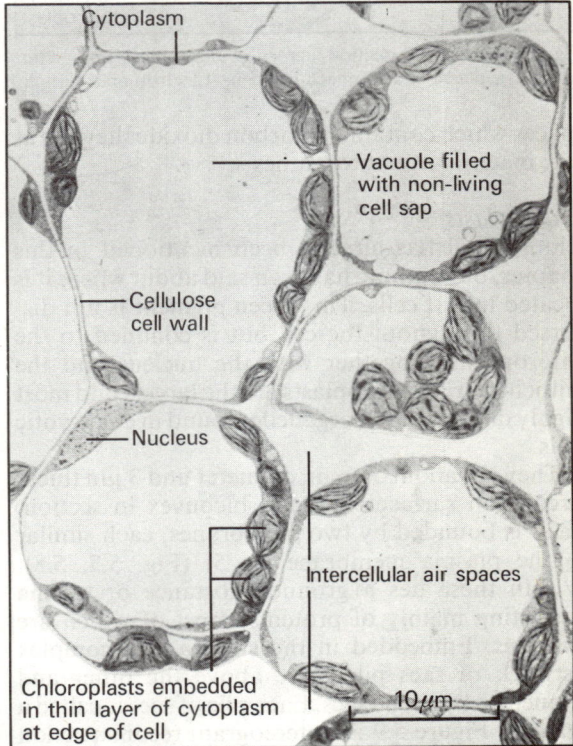

Fig. 5.6 Horizontal section through several cells of the palisade mesophyll of spinach. Note the intercellular spaces. (Electron micrograph.)

41

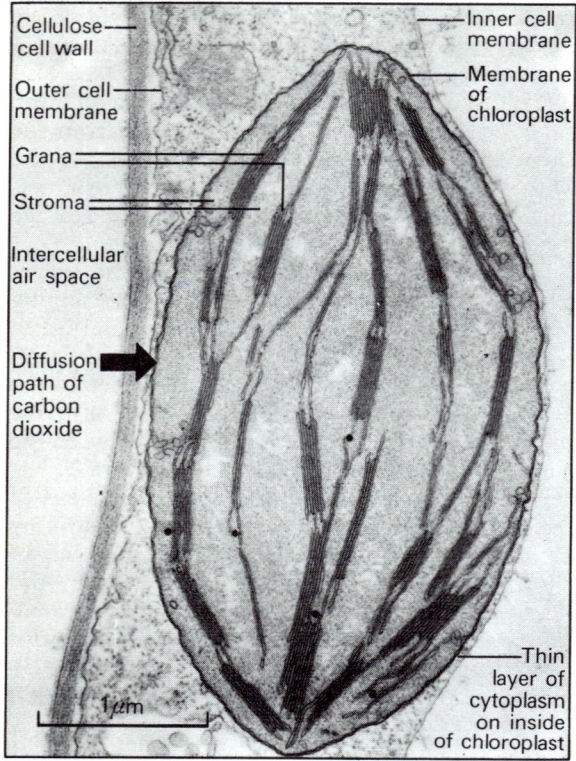

Cellulose cell wall
Outer cell membrane
Grana
Stroma
Intercellular air space
Diffusion path of carbon dioxide
Inner cell membrane
Membrane of chloroplast
Thin layer of cytoplasm on inside of chloroplast
1 μm

Fig. 5.7 Horizontal section across a chloroplast in position in the cytoplasm of a spinnach leaf cell. (Electron micrograph.)

spaces which contain the carbon dioxide they use as raw material in photosynthesis.

Chloroplasts

Chlorophyll has already been mentioned in this chapter, but nothing has been said about where it is located in leaf cells. This green pigment is not dispersed throughout the cell but is confined to the chloroplasts. Together with the nucleus and the mitochondria, chloroplasts are the largest and most highly differentiated organelles found in eukaryotic cells.

They are about 5 μm in diameter and 3 μm thick, circular in surface view and biconvex in section. Each is bounded by two membranes, each similar to the plasma membrane (p. 5) (Fig. 5.7, 5.8). Within these lies a ground substance or **stroma** consisting mainly of proteins, most of which are enzymes. Embedded in the stroma is a complex network of sacs piled one above the other and connected by single sacs. Each pile of sacs is called a **granum**. Figure 5.9 is a stereogram reconstruction of three grana. In Figs. 5.7 and 5.8, the grana have been sectioned down the length of the pile, and so appear as a series of pairs of membranes. Embedded

in these membranes are the chlorophyll molecules.

It seems likely that the water-splitting reaction, which consumes light energy, takes place in the membranes of the granum. It is thought that the reactions producing carbohydrate from carbon dioxide occur in the stroma. Certainly the stroma contains the final products of photosynthesis, for Fig. 5.8 shows a large starch grain in it, and also oil droplets which are an alternative end product of photosynthesis.

The arrangement of mesophyll cells and of chloroplasts within them ensures the most efficient exposure of chlorophyll to light and carbon dioxide. It is on the upper side of the leaf that direct sunlight falls. Here the cells are close together (though they still have air spaces between them), and within the palisade cells the chloroplasts are more densely packed than in the cells of the spongy mesophyll. The chloroplasts have their flat faces parallel to the cell wall, displaying their maximum surface to the light. Within the chloroplasts the chlorophyll molecules are arranged on the flat membranes of the grana and these are also parallel to the surface of the cell.

How carbon dioxide reaches the chloroplasts: diffusion

Carbon dioxide moves towards the chloroplasts from the air outside the leaf by **diffusion**. Diffusion

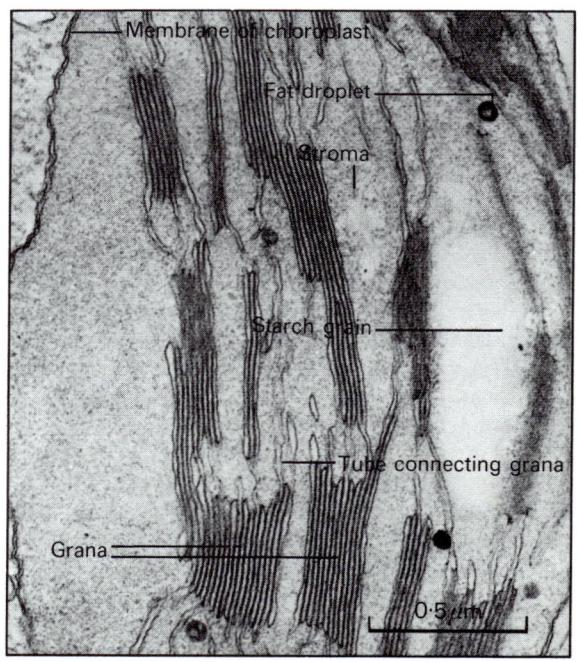

Membrane of chloroplast
Fat droplet
Stroma
Starch grain
Tube connecting grana
Grana
0.5 μm

Fig. 5.8 Detail of grana within a chloroplast (horizontal section). (Electron micrograph.)

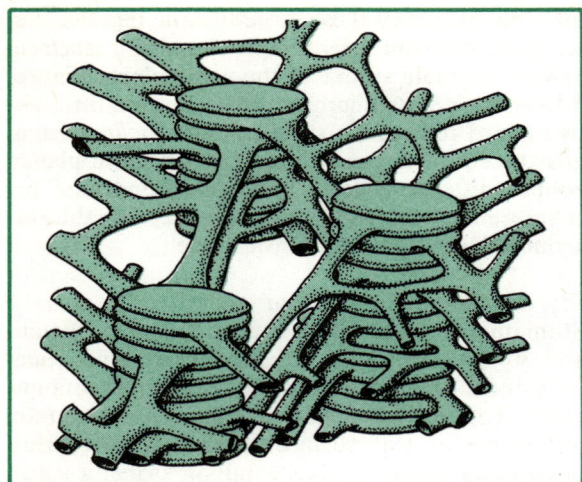

Fig. 5.9 Stereogram reconstruction of three grana from a chloroplast.

cannot be observed directly in leaves because carbon dioxide is a colourless and invisible gas. All gases and all substances dissolved in liquids are capable of diffusion. It is a phenomenon of very great importance in living cells. It is best observed with bromine vapour, a richly-coloured brown vapour which is easy to see.

Figure 5.10 shows an experiment in which the lower gas jar was filled with bromine vapour. A second, empty gas jar was placed above it and the lid separating the jars was removed. The photograph shows the appearance of the jars three or four minutes later. Bromine is moving upwards into the top jar. It is a dense gas, and sinks rapidly in air, so this movement is against the force of gravity.

The explanation is that all atoms, ions, and molecules are in a constant state of motion (p. 32). In gases and liquids the movement is free. Before the lid was removed from between the jars there were no bromine molecules in the top jar. When the lid was removed, bromine molecules moved into the top jar and molecules of oxygen and nitrogen in the air in the top jar moved into the bottom jar. A state would eventually have been reached where the concentration of bromine in the top jar equalled that in the bottom jar. There would then still have been constant movement from one jar to the other, but the concentration of bromine would have remained the same throughout.

Diffusion defined

Figure 5.10 shows that *if a situation exists where there is a region of high concentration of a gas or vapour adjacent to a region of lower concentration of the same substance, molecules of the substance will pass from the region of high concentration into the region of low concentration until its concentration is uniform throughout.* This is called diffusion. It applies equally to dissolved substances, for example sugar dissolved in water.

How does diffusion apply to the leaf?

Air contains about 0.03 % of carbon dioxide. In illuminated chloroplasts carbon dioxide is converted into carbohydrate and so removed: the chloroplast is a 'sink' in which the carbon dioxide concentration is nil. The gas therefore diffuses from the air outside the leaf to the chloroplasts along a path which is called its **diffusion gradient**. Some is used by the chloroplasts in the spongy mesophyll, but the very loose arrangement of cells here offers little resistance to its passage to the palisade cells. The walls of all the mesophyll cells are saturated with water in which the carbon dioxide dissolves. It then diffuses across the cytoplasm into the chloroplast. We have already noted in Fig. 5.7 that the layer of cytoplasm separating the chloroplast from the cell wall is very thin, and as the rate of diffusion of carbon dioxide through water is 10 000 times slower than through still air, this is another adaptation ensuring rapid and efficient access of carbon dioxide to the chloroplast.

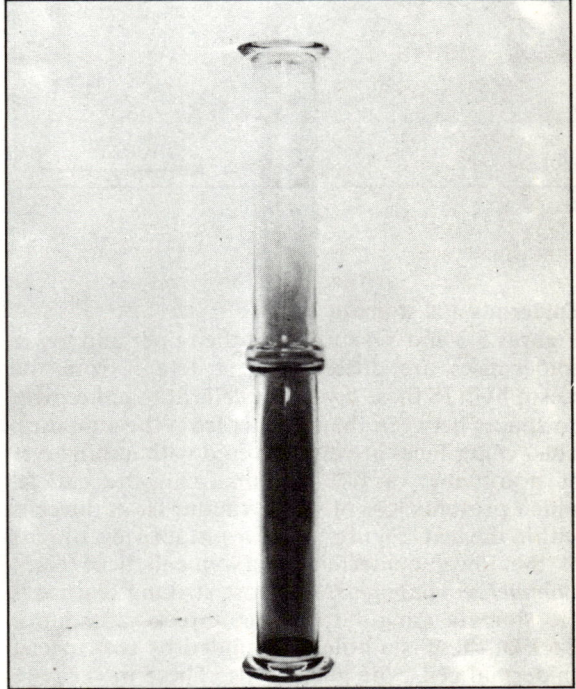

Fig. 5.10 Diffusion of bromine from the lower into the upper jar.

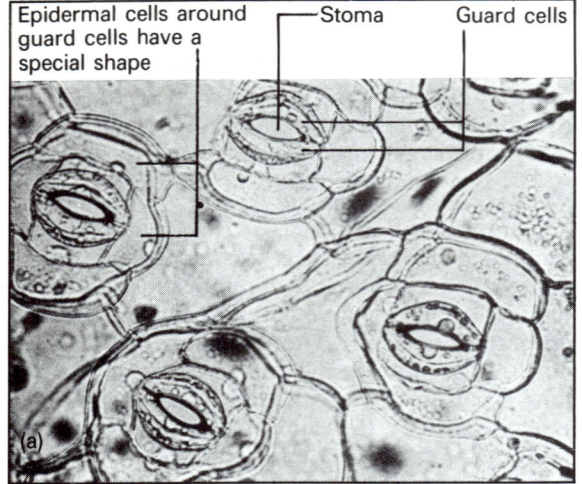

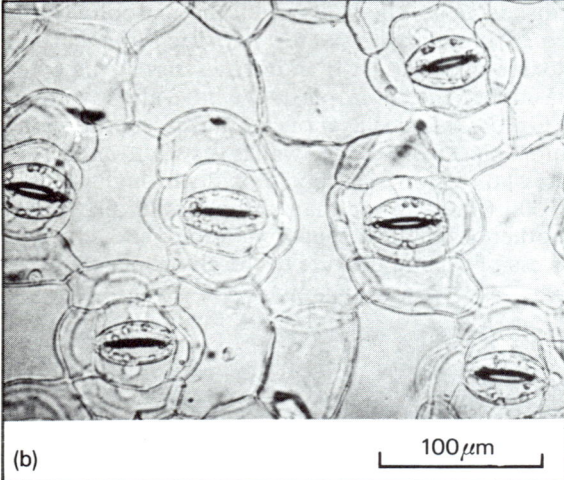

Fig. 5.11(a), (b) Surface views of the lower epidermis of *Commelina communis:* (a) stomata open; (b) stomata shut. (Light microscope.)

Epidermis and stomata

Figures 5.3 and 5.4 show that the upper and lower epidermises are different in structure from the mesophyll. In these layers the cells fit together with no spaces between them (apart from the stomata). Their outer faces are also covered with a thin layer of non-living varnish-like material, the **cuticle**, which prevents loss of water vapour from the cells within the leaf. Figure 5.11 is a surface view of part of the lower epidermis of a typical dicot plant *Commelina communis*. Its most striking feature is the **stomata** (singular, stoma, Greek, a mouth). Each of these is a hole surrounded by two special epidermal cells, the **guard cells**. These in turn are surrounded by small epidermal cells, then come the irregularly-shaped but closely fitting main epider-

mal cells. Identify these structures in Fig. 5.12, a scanning electron micrograph showing a surface view of a single stoma of the same plant. Figure 5.13 is an electron micrograph of a section through two guard cells and a stoma of *Commelina*. Each guard cell possesses a nucleus and cytoplasm containing chloroplasts and mitochondria. The cellulose wall is not evenly thickened, being thicker around the margin of the stoma itself.

Size and numbers of stomata

Stomata are small in all plants, between 10 μm and 30 μm long and 5 μm and 10 μm wide when fully open. They occur in very large numbers, commonly between fifty and two hundred per square millimetre of leaf surface. (One sunflower leaf possesses approximately 13 million stomata.) But for all their numbers the total area of the open pores is only between 0.5 and 1.5% of the total leaf area.

Their function is to admit carbon dioxide to the sub-stomatal cavities (Fig. 5.4) and so to the chloroplasts of the mesophyll cells. Measurements show that they are very efficient at this, and although leaves have a cuticle that is impermeable to gas and perforations which add up to only about 1% of the leaf area, the rate at which carbon dioxide penetrates into leaves is as much as one half what it would be if there were no epidermis at all.

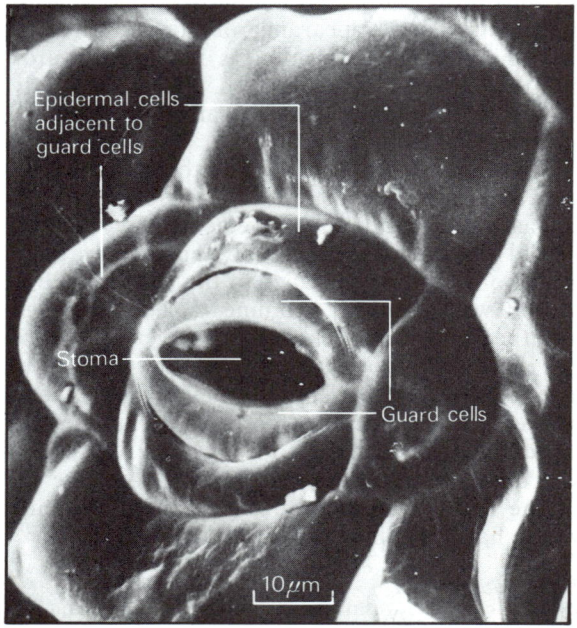

Fig. 5.12 Stoma of *Commelina communis* (scanning electron micrograph).

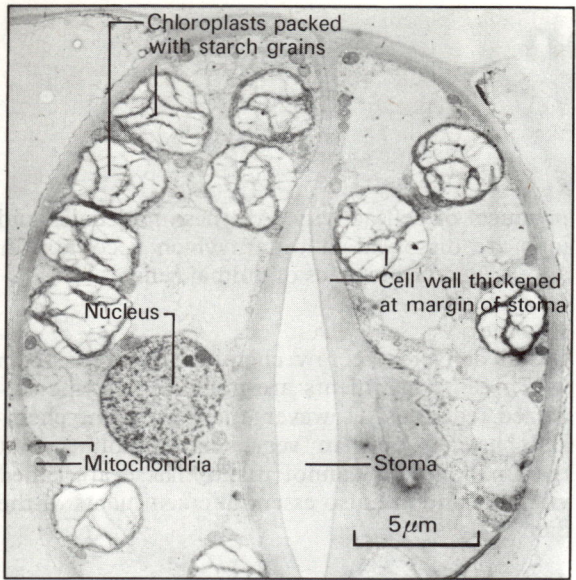

Fig. 5.13 Section through guard cells and stoma of *Commelina communis* (electron micrograph).

other life chemicals needed by cells, notably proteins and fats. Figure 5.8 shows fat droplets inside a chloroplast, whose stroma is made of protein.

Fats and carbohydrates contain the elements carbon, hydrogen, and oxygen only. Amino acids, and therefore proteins, contain nitrogen, which plants absorb from soil water through their root hairs in the form of nitrate ions, NO_3^-. Many other ions are needed in small quantities as constituents of life chemicals. For example, some proteins contain sulphur and phosphorus; chlorophyll contains magnesium; and the all-important ATP (see p. 67) contains phosphorus. These elements and others are also absorbed as ions from soil water by root hairs.

Although one thinks of leaves primarily as organs of photosynthesis, proteins and other life chemicals are made in them as well. These are then transported to the growing points and storage organs by the phloem.

Opening and closing of stomata

Stomata also provide an efficient means by which water vapour can leave the leaf interior (see p. 103). This loss is almost entirely detrimental to the plant. Since photosynthesis cannot proceed in darkness, it is not surprising to find that guard cells are capable of changing their shapes so that, in darkness, the stoma is closed, thereby cutting down loss of water vapour (Fig. 5.11(b)).

The mechanism of closure is associated with loss of water from the guard cells to the surrounding epidermal cells. Conversely, opening is caused by an uptake of water into the guard cells from the surrounding epidermal cells. Because of the uneven thickening of the cellulose walls of the guard cells, the stoma gapes when the guard cells swell. Most stomata open in the light (to admit carbon dioxide to the leaf interior), and as they are the only epidermal cells to contain chloroplasts (Fig. 5.13), opening is probably associated with photosynthesis and the production of sugar in the guard cells. But many factors are known to affect the opening and closing of stomata, and there is no simple explanation of their behaviour.

Other life chemicals made by the plant

Photosynthesis produces carbohydrates. Plants use these to synthesise cellulose for their cell walls and as a source of chemical energy (via tissue respiration) for cell chemical reactions. Using chemicals produced from carbohydrates, they can make all the

6 Animal Nutrition

Animals eat plants or other animals, thus obtaining their proteins, fats, and carbohydrates ready made. These they digest, absorb, and make into their own substance. All the examples of these processes in this chapter are mammalian, but the principles involved apply to animals in general.

Chemicals needed for the life of mammalian cells

Proteins

Proteins are needed by mammalian cells to build cell organelles and enzymes. Some are therefore solid, structural materials, others are dissolved in the ground substance of the cell (Fig. 1.7).

Fats and carbohydrates

Fats are constituents of all cell membranes. They also occur as droplets in cells, forming a reserve of chemical energy. Muscle and liver cells store carbohydrate in the form of insoluble starch grains (called animal starch, or glycogen). Some are shown in Fig. 6.1.

Water

The ground substance of the cell is a solution of proteins, carbohydrates, and hundreds of other chemicals. Some are passing into the cell as raw materials for growth; others are passing out as waste

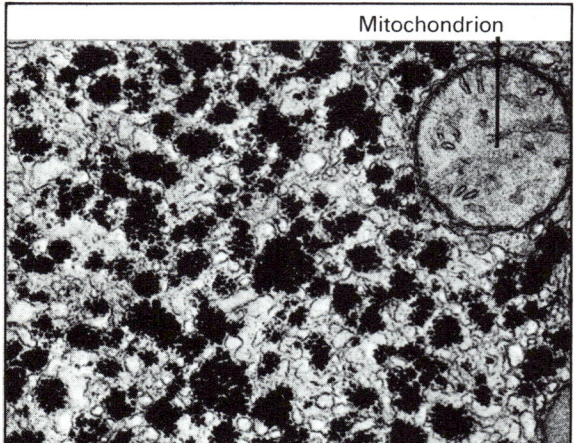

Mitochondrion

Fig. 6.1 Electron micrograph of glycogen granules in cytoplasm of a hamster liver cell. The single granules cluster into aggregates.

products of cell activity. All these molecules and ions are dissolved in water, which accounts for about 70 % of the mass of animal cells.

Vitamins

Blood delivers the raw chemical materials from which cell constituents are made by enzyme-catalyzed reactions. However, there are some chemicals needed only in very small amounts that mammalian cells cannot make. These are called vitamins and are also essential constituents of the diet.

Mineral ions

These are required to make some of the essential cell chemicals. For example, iron is needed to make haemoglobin, the red pigment of red blood cells; calcium and phosphorus are needed to make bone; phosphorus is needed to make ATP.

Essential amino acids

There are 22 different kinds of amino acids in proteins. Animal cells can make some of them but others either cannot be made at all or else cannot be made fast enough for the animal cell's needs. These are said to be *essential* amino acids since they are an essential ingredient of the animal's food. About 10 of the 22 different kinds of amino acids fall into this category for man, and the value of a particular protein food depends on its essential amino acid content. Most animal proteins, e.g., meat, eggs, contain all the essential amino acids, and they are called **first class proteins**. Plant proteins such as those contained in cereals, are deficient in various essential amino acids and are called **second class proteins**.

Proteins for growth and replacement

An animal starts its life as a single cell, but the mature animal contains billions of cells. Proteins are obviously needed for cell multiplication and growth. It is less obvious that at all stages of its life, the animal's cells are wearing out and being replaced in enormous numbers. In an adult human, the red blood cells are replaced at the rate of about 200 000 million in twenty four hours. About 2 m^2 of skin cells are shed during a night's sleep. The cells

Foodstuff	Waste	Calorific value (kcal per g)	Protein	Fat	Carbohydrates	Calcium	Iron	Vitamin A	Vitamin B$_1$ (Thiamine)	Vitamin B$_2$ (Riboflavin)	Nicotinic acid	Vitamin C	Vitamin D
	per cent	kcal	g	g	g	mg	mg	i.u.	mg	mg	mg	mg	i.u.
Cereals													
Sweet biscuits	0	44.7	0.45	2.46	5.35	6.80	0.08	0	0.008	0.003	0.08	0	0
Bread, white	0	19.5	0.62	0.11	4.22	7.36	0.14	0	0.014	0.003	0.14	0	0
brown	0	19.5	0.70	0.17	4.02	7.64	0.20	0	0.016	0.006	0.20	0	0
Rice	0	28.9	0.51	0.085	6.97	0.28	0.3	0	0.006	0.003	0.11	0	0
Cornflakes	0	29.5	0.54	0.06	7.14	0.57	0.23	0		0.003	0.008	0.14	0	0
Dairy products													
Butter	0	64.0	0.03	68.0	0	1.13	0	280.5	0	0	0	0	3.12
Cheese, cheddar	0	34.0	2.04	2.78	0	65.2	0.06	113.3	0.003	0.004	0	0	1.13
Cream, single	0	17.6	0.20	1.70	0.25	6.23	0.03	56.7	0.003	0.008	0	0	0.57
Eggs, fresh	3.40	13.0	0.99	0.99	0	4.53	0.2	79.3	0.008	0.03	0	0	13.6
Milk, fresh	0	5.38	0.25	0.31	0.40	9.63	0	11.3	0.003	0.01	0	0.17	0.11
Fats													
Cooking fat, lard	0	74.2	0	7.96	0	0	0	0	0	0	0	0	0
Margarine	0	64.0	0	6.90	0	0.28	0.03	240.8	0	0	0	0	25.5
Fish													
Fresh fish, white	0	16.4	1.59	0.82	0.59	6.51	0.03	0	0.006	0.008	0.23	0	0
Herring	10.5	19.0	1.33	1.44	0	8.2	0.11	11.3	0.003	0.03	0.28	0	73.7
Meat													
Beef (stewed steak)	76.7	4.81	4.53	0	2.83	1.13	0	0	0	0	0	0	0
Mutton	87.3	4.0	7.93	0	2.83	0.57	0	0	0	0	0	0	0

Fig. 6.2 Composition of food. All values per, 100 g edible portion, raw unless otherwise stated.
1 international unit (i.u.) Vitamin A = 0.0006 mg
1 international unit (i.u.) Vitamin D = 0.000025 mg

lining the gut function for about 48 hours and are then replaced, and this amounts to about 250 g of intestinal cells per day. Hairs are constantly growing, falling out, and being replaced. In a man, sperms and semen are produced all the time. Proteins in the form of enzymes are also secreted into the gut from various glands, such as the salivary glands, the stomach glands and the pancreas. All these replacement processes require a constant supply of proteins in the food.

Energy (calorific) values of foods

Figure 6.2 shows the protein, fat, carbohydrate, mineral and vitamin contents of some human foods. Examine it and compare their nutritional values.

The second column is headed 'calorific value'. This provides a rough measure of the amount of chemical energy possessed by one gramme of the food. A calorie is a unit of heat, and the calorific value of a food is the number of kilocalories given out when 1 g of it is completely burned. The standard method of expressing energy is in joules (1 calorie = 4.2 joules), but nutritionists are not agreed on dropping the calorie terminology as it is so widely and popularly known.

What the chemical energy derived from foods is used for

Chemical energy is needed for growth and replacement and for movement. But if an adult is lying in bed in a mentally and physically relaxed state he still needs energy for basic body processes such as breathing movements, the heartbeat and the circulation of the blood, the maintenance of body temperature, and the maintenance of muscle tone. These processes constitute the **basal metabolism** or basic chemical turnover and it consumes about 1.3 kcal/min. Further activity requires the provision of more energy. Dressing consumes a further 2.5 to 4.0 kcal/min; walking on the level at 3 km/hr requires a further 2.7 kcal/min. Women's domestic work takes an extra 2 to 8 kcal/min and hard exercise may require up to an extra 20 kcal/min. But mental work, such as reading and understanding this page requires practically none!

Daily calorific needs

Figure 6.3 shows the daily energy needs of teenagers (15–19) and adults (over 23). The teenagers need more kilocalories per day than the adults and this reaches a peak of 3500 kcal per day for boys and

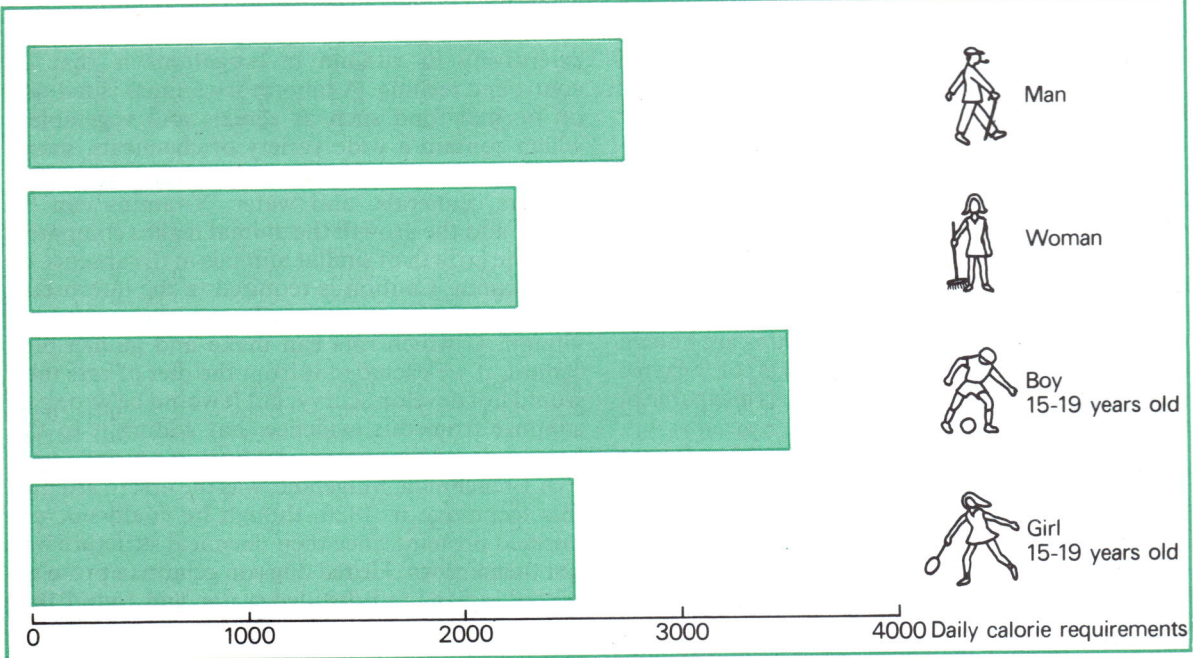

Fig. 6.3 Daily calorie requirements.

2500 for girls. An adult man doing light work needs 2750 kcal per day, and a woman 2250. At present over half the human population of the world receive less than this amount.

Proteins as fuels

We have encountered the idea of fats and carbohydrates acting as fuels for the cell, providing it with chemical energy (p. 31). Examination of the protein foods in Fig. 6.2, such as eggs, fish, and meat, shows that they too contain chemical energy, though not so much as fats and carbohydrates.

Amino acids that are not needed for constructing proteins are converted by the liver into the same 3-carbon molecules that are produced from fats and carbohydrates and which serve as fuel to produce chemical energy inside mitochondria. This is shown in Fig. 6.4.

Similarly the 3-carbon molecules derived from fats and carbohydrates can be used by animal cells to make the non-essential amino acids. Thus although the three major classes of life chemicals (proteins, fats, and carbohydrates) are very dissimilar in basic chemical structure, they are all inter-convertible in the cell, and all can be used as sources of chemical energy.

Double role of proteins

Proteins are used by cells both for growth and replacement and as an energy source. They are the expensive foods (e.g. meat, fish, eggs), and it might seem that the full energy requirements of the body could be met from the cheaper fats and carbohydrates (e.g. bread, potatoes). In practice this is not so, however, and the Committee on Nutrition of the British Medical Association recommended in

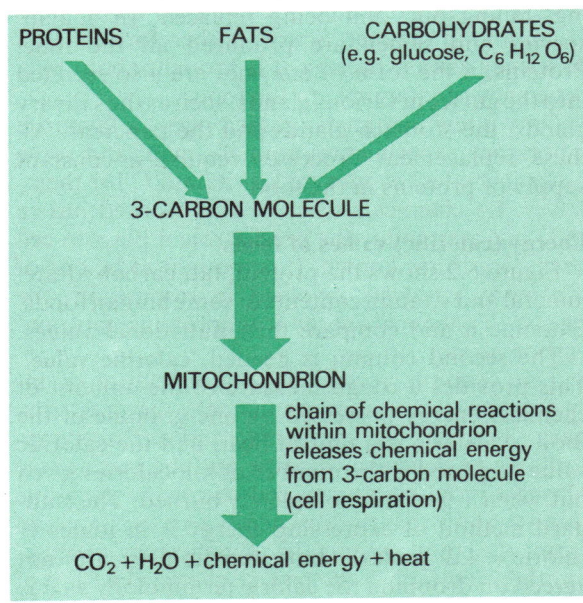

Fig. 6.4 Common pathway for releasing chemical energy from proteins, fats, and carbohydrates.

1950 that adults engaged in hard physical work should obtain not less than 11% of their energy intake from proteins. For pregnant women and young people under the age of 19, the figure should be 14%. An unbalanced diet in which the cheaper foods are taken in excess leads to obesity.

Vitamins

Animal cells are not able to make all the chemicals that they require. Some of these chemicals, the essential amino acids, are needed in large quantities. Others are needed only in very small quantities and are called vitamins. Many vitamins are known to form essential non-protein parts of enzyme molecules without which the enzyme is incapable of working.

The first observations about a particular vitamin have often been that eating certain foods *prevents* the occurrence of certain illnesses. Thus the need to include citrus fruits (especially lemons) in the human diet was noted early in the seventeenth century in connection with the long sea voyages that were then common. Sailors often developed scurvy, a disease in which the patient becomes weak, tired, short of breath, and suffers from extensive internal bleeding, especially beneath the skin and into the joints. The gums become swollen and tender, making eating painful, and they bleed. When lemons were kept on the ship and given with rations, this disease did not occur.

We now know that this is because nearly all fruits and vegetables contain vitamin C, and the disease can be prevented and cured by taking pure vitamin C in the diet. Scurvy is therefore called a **deficiency disease** because it is brought about by a deficiency of vitamin C in the food.

Only a few mammals (guinea pigs—a favourite laboratory experimental animal—man, and some other primates) cannot make vitamin C for themselves. Its chemical structure was worked out in 1933 but even today the precise role it plays in cell chemistry is not known. It is thought to be one of the many catalysts involved in respiration, the release of chemical energy from proteins, fats, and carbohydrates.

This example reveals a general feature of investigations into vitamins, namely the great disparity between the widespread and diffuse symptoms of the deficiency disease, and the precise role played by the vitamin in cell chemistry. It is one thing to show that the absence of a certain material from the food results in a deficiency disease, another to establish the precise chemical nature of the substance, and yet another to discover the role this substance plays in the functioning of cells.

Controlled feeding experiments

An experimental technique that has been used extensively in vitamin investigations is that of controlled feeding. A laboratory animal is fed not on normal food such as cereals and vegetables, which contain a wide variety of chemicals, many perhaps unknown, but on pure protein, fat, carbohydrate, minerals, and water. Vitamins can be added and the growth the animal makes compared with the growth of similar animals in the absence of the vitamin. Caution is required in the interpretation of the results, however. To take the example of vitamin C which rats can make and guinea pigs cannot, if we excluded it from the diet of rats they would not develop scurvy, but it would be wrong to suppose from this evidence that adding it to the human diet does not prevent human scurvy.

A Frenchman, Magendie, was the first to attempt this technique in 1820, though he could not use purified proteins since their chemical structure was not then known. He fed dogs on gelatin (a protein), butter, sugar, and distilled water and found that they died within a month. This showed that these substances were not enough to sustain the animals. However, they probably died not from lack of vitamins but because gelatin is a protein entirely lacking in two amino acids, one of which is an essential amino acid for dogs.

Sir F. G. Hopkins

When the chemical nature of proteins, fats, and carbohydrates became known at the end of the nineteenth century it was possible to improve this technique by feeding known pure substances. In

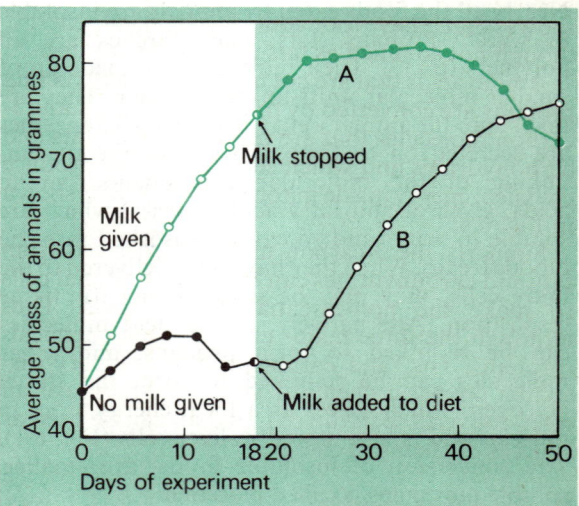

Fig. 6.5 F.G. Hopkins' experiments on feeding milk to rats kept on a purified diet.

1906 Sir F. G. Hopkins conducted the first precise experiments along these lines. He fed rats on purified casein (cheese protein), starch, sucrose, lard, inorganic salts, and water. He experimented with two *groups* of eight young rats, that is rats whose growth had not ceased. Groups are necessary because the results obtained on a single rat might be exceptional—for instance, one particular rat might not feed well. By taking a group and averaging the results from them, individual variation is less likely to be a source of experimental error.

One group received 3 cm³ of cow's milk per day in addition to the diet mentioned. Their average mass increased rapidly (Fig. 6.5 line A). After 18 days the milk was dropped from their diet and although they continued to put on weight for a few more days, this soon ceased.

The second group were given no milk for the first 18 days of the experiment and their average mass hardly increased at all. On adding milk to their diet on the 18th day, however, there was a rapid change (Fig. 6.5 line B).

Hopkins concluded that protein, fat, carbohydrate, mineral salts, and water were not enough to permit the growth of rats; milk appears to contain a substance or substances which are essential for growth, though this experiment does not reveal what it/they are. It has been found since that milk contains vitamins A and D and at least eight others, and the chemical structures of all these substances was worked out in the first fifty years of this century.

Digestion and absorption of food

The insoluble nature of food

Nearly all the foods eaten by animals are insoluble in water. Proteins, fats, and starches are all polymers (p. 30). Their molecules are made up of smaller molecular units linked together. They are taken into the animal's gut where digestive enzymes are secreted on to them. These catalyse the unlinking of the molecular components (amino acids, glycerol and fatty acids, sugars) which are soluble in water and so can be absorbed into the blood stream. When they have been delivered to the body cells, they may be used as fuel for tissue respiration. Alternatively, amino acid molecules can be re-linked to form proteins, and sugar molecules can be re-linked to form the starch glycogen, which serves as a local energy store in muscles and is also stored in liver cells (Fig. 6.1). The conversion of insoluble foods into smaller, soluble molecules is called **digestion**.

Thus, for example, the protein which forms the

Enzyme molecules prevent each other from reaching the limited surface of the food particle

Food particle broken up: far more enzyme molecules can now make contact with the increased surface area of the food

○ Food

● Enzyme

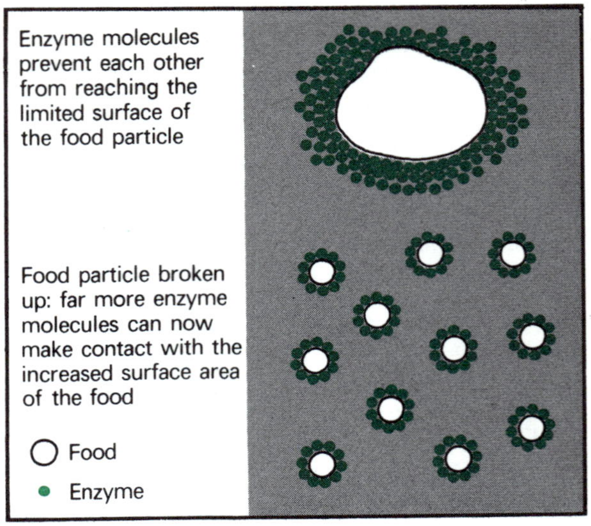

Fig. 6.6 Effect of surface area of food on the activity of digestive enzymes.

muscle of a cow is unlinked into its constituent amino acids and reassembled to form the muscles of a human being. Although cow and human muscles are nearly identical in structure and function, they differ slightly in the sequence of amino acids that compose them.

An important factor in enzyme action: surface to volume ratio

Food is taken into the mouth in large pieces. As it passes along the gut it becomes converted into particles of much smaller size. The process starts in the mouth where food is bitten and chewed, and continues in the stomach, where a great deal of churning and pulping takes place. This increases the surface area of the food considerably.

For enzymes to act their molecules must make physical contact with their substrates, and the greater the surface area of the food, the more opportunity there is for doing so (Fig. 6.6).

The following table illustrates this. It shows that the smaller an object (such as a cube), the greater

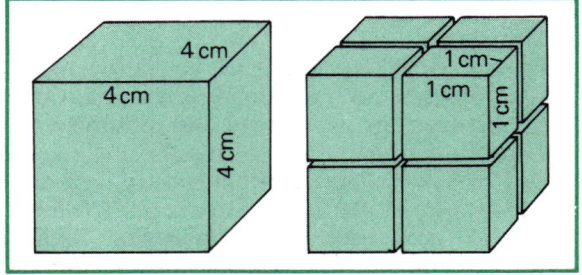

Fig. 6.7 A cube cut into eight parts has a larger surface area.

its *relative* surface area (column 4, ratio of surface area to volume).

side (cm)	volume (cm³)	surface area (cm²)	ratio of surface area to volume
1	1	6	6:1
2	8	24	3:1
3	27	54	2:1
4	64	96	1.5:1

A cube of side 4 cm has a volume of 64 cm³ and a surface area of 96 cm². Suppose it is cut into eight equal parts (Fig. 6.7). What is its surface area now? How many times has its surface area increased?

The plan of the gut

Figure 6.8 shows the gut of a rat removed from the animal and unravelled. Compare it with Fig. 2.6 where the organs are shown in place. Figure 6.9 shows the human gut in position, and in Fig. 6.10 the parts are shown separated from each other.

Compare the rat and human gut figures. Do both animals possess the same organs, and if so are they arranged in the same order?

The gut is a long continuous tube and it has to be folded considerably to fit it into the abdomen. The mouth is followed by the **oesophagus**, a muscular tube which conveys chewed food to the **stomach**, where it is churned and partly digested. It then passes to the **duodenum** and **small intestine**, where digestion is completed and soluble foods are absorbed. Then follows the **colon (large intestine)**, where water is re-absorbed from the indigestible remains. These consist in man mainly of cellulose and constitute the faeces which are passed into the **rectum** immediately before defaecation at the **anus**.

Activity at the front and the rear ends of the gut can be controlled voluntarily—chewing and swallowing by the cheek, tongue, and throat muscles, and defaecation by the circular anal sphincter muscle. Along the length of the gut, however, movement and activity is involuntary, being controlled partly by nerves beyond the domain of the thinking area of the brain, and partly by chemical messengers (hormones).

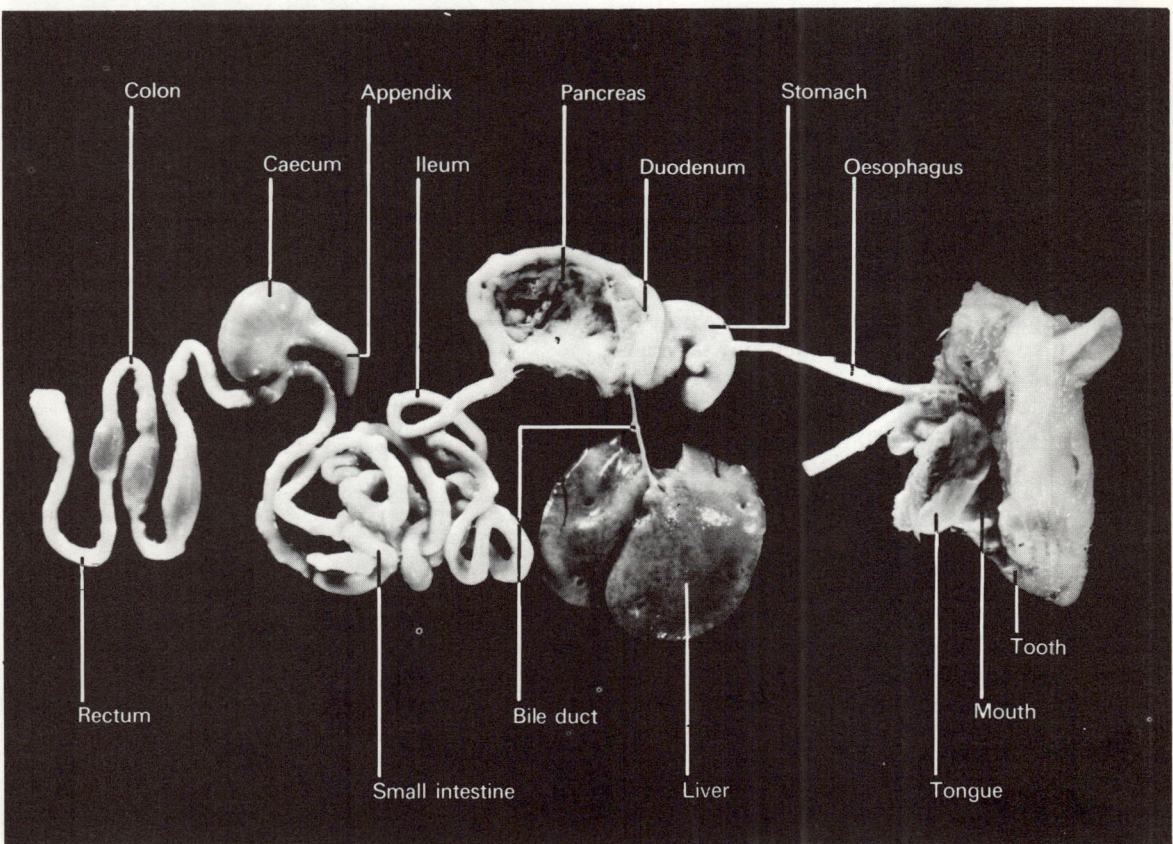

Fig. 6.8 Gut of a rat unravelled.

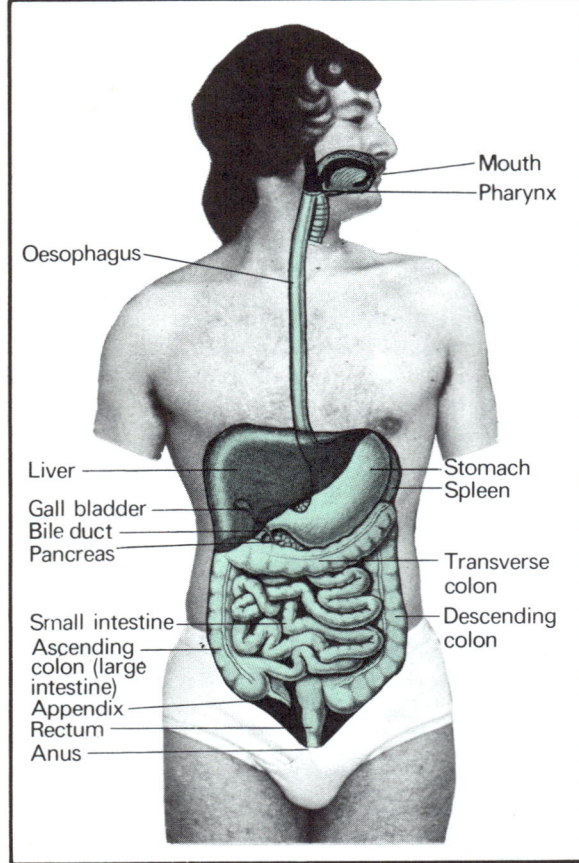

Fig. 6.9 The human gut in position in the body.

Two large glands are associated with the gut: the **liver** and the **pancreas**. Their secretions are poured into the gut cavity via the **bile duct** and the **pancreatic duct** respectively (Fig. 6.10). The gut, liver and pancreas are also connected to each other by the blood which is circulating through them.

Detail of the gut

Mammalian guts vary considerably in detail, but the basic structure and functions are the same for all. For medical reasons most detail is known about the human gut.

The mouth, palate, tongue, and teeth

Figure 6.11 is a section through the human head and shows the two cavities separated by the palate, which forms the roof of the mouth. Above the palate lies the **nasal cavity** which contains three shell-shaped bones called **conchae**. The lower of these is covered by a thin skin containing many blood capillaries; the blood which circulates here warms the air entering the nasal cavity and the

lungs. The upper concha possesses an area of skin about 10 cm^2 which contains cells sensitive to chemical vapours; these are responsible for the sense of smell. The remainder of the conchae are covered with skin containing mucus-secreting cells; the mucus traps dust and is then wafted down the oesophagus. Thus the nasal cavity has two functions: first it warms and cleans the air entering the lungs; secondly the smell-sensitive cells help to detect food and dangerous chemicals. The smell of food initiates the secretion of digestive juices in the stomach.

Advantages of the palate

Dividing the mouth cavity into two has several biological advantages. As we have just seen, the

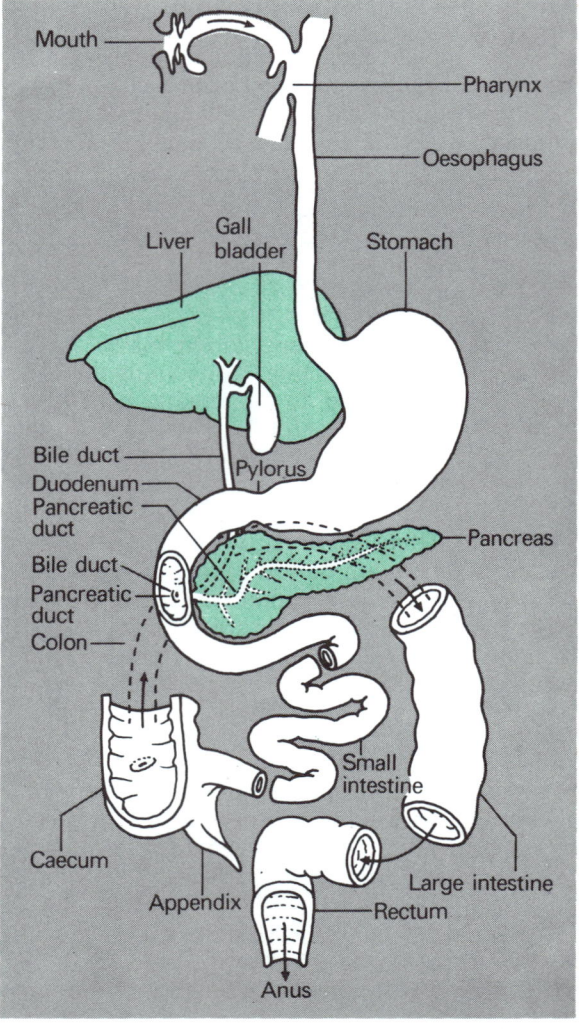

Fig. 6.10 The human gut separated to show the parts in order.

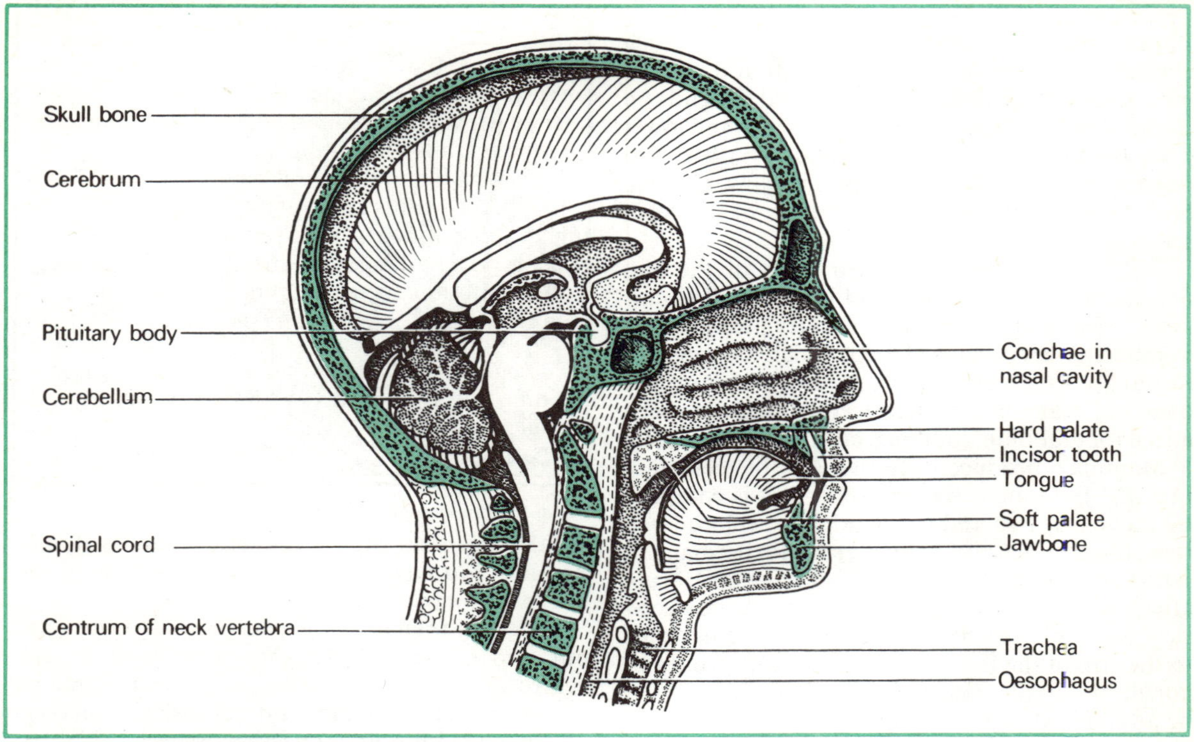

Skull bone

Cerebrum

Pituitary body

Cerebellum

Spinal cord

Centrum of neck vertebra

Conchae in nasal cavity

Hard palate
Incisor tooth
Tongue

Soft palate
Jawbone

Trachea
Oesophagus

Fig. 6.11 Vertical section through the human head.

upper part can become specialized for breathing and smelling. The warming of incoming air is of particular importance to mammals, which are warm blooded. Because food is shut off from the air passage it can be chewed without interfering with breathing. Food grinding and digestion start in the mouth. The functional length of the gut is thus greater than in other vertebrates like fish which use the mouth only for grabbing and swallowing. There is no doubt that the evolution of the palate has been a major feature in the success of mammals, and they are the only vertebrates apart from one reptilian order (Crocodilia) to possess it.

The tongue

The muscular tongue pushes food under the teeth. A constant stream of messages is fed from the tongue to the brain which uses the information to adjust the position of the tongue and prevent it from being bitten.

The tongue contains nerve fibres that help with food selection. Some are sensitive to touch, pain, and temperature differences. There are also groups of special cells sensitive to chemicals in solution; these are the **taste buds** which enable mammals to

distinguish good food from poisonous substances. However, in man at least, much of the sense of taste comes from smells. When the nose is blocked, as when a person has a heavy cold, food is relatively tasteless.

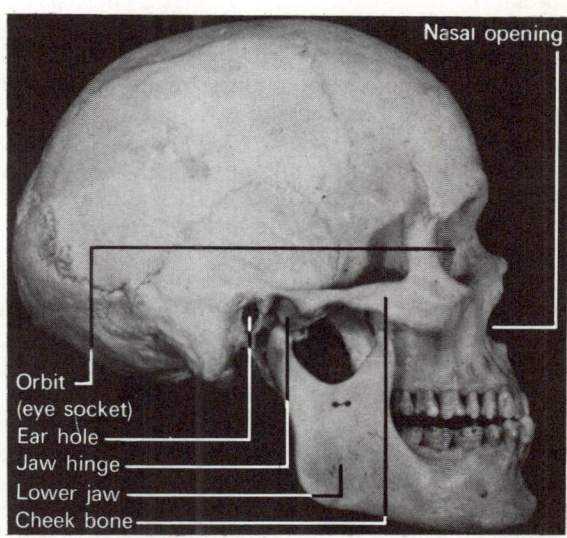

Nasal opening

Orbit
(eye socket)
Ear hole
Jaw hinge
Lower jaw
Cheek bone

Fig. 6.12 Human skull from the side.

53

Teeth

The teeth are rooted in the jaws, the upper of which is fixed. The lower jaw, and the teeth in it, can be moved, and the freedom of movement depends on the kind of bony joint between the upper and lower jaws. In humans (Fig. 6.12) this allows both up and down and sideways movements of the lower jaw, so both biting and grinding movements of the teeth are possible.

The teeth are the only hard parts of the gut, providing its only cutting and hard grinding surfaces. Fish, amphibians, and reptiles possess large numbers of unspecialized teeth which are replaced throughout the animal's life as they wear out. By contrast, mammals possess teeth which are specialized for grabbing, stabbing, cutting, and grinding.

Mammals produce only two sets of teeth in a lifetime. In humans the first (milk) set is complete by the age of two and it is gradually replaced between the ages of 6 and 12 by the second (permanent) set. The hindmost molars (wisdom teeth) erupt after the age of 17. Two sets is the minimum necessary to enable the size of the teeth to be matched to the size of the jaws. Just as milk teeth would be hopelessly small in an adult mouth, so there would be no room for a permanent set of teeth in a child's jaws.

Types of mammalian teeth

Four kinds of mammalian teeth can be distinguished on the basis of shape and function. These are: (i) **incisors** (biting and grabbing teeth); (ii) **canines** (fighting and stabbing teeth, especially well developed in carnivores); (iii) **premolars**; and (iv) **molars**. The last two are grinding teeth with broad crowns which are especially elaborate in herbivores. In most carnivores the pre-molars and molars are not broad and are used like scissor blades for slicing. In the milk dentition there are no molars.

Identify these four kinds of teeth in the human skull (Fig. 6.12), sheep skull (Fig. 6.25(a)), dog skull (Fig. 6.28), and in your own mouth.

Tooth structure

Mammalian teeth are embedded in sockets in the jaw bone (Fig. 6.13). Molars have three roots, premolars two, and incisors and canines, one. The root is attached to the jawbone by a special layer of bone-like material called **cement**.

The exposed crown is covered by a layer of **enamel**, which resembles bone in structure (p. 18), though it contains 97% inorganic calcium phosphate compared with only 41% in bone. It is the hardest substance in the body (it gives sparks with steel) and so does not chip easily and is difficult to

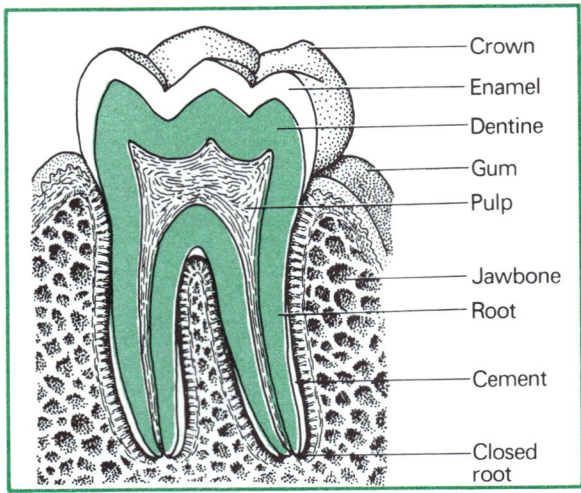

Fig. 6.13 Human molar tooth *in situ*.

break. It can cut substances softer than itself, and perhaps you have watched a lion consuming the entire limb of an animal, bone and all.

Beneath the enamel is the **dentine**, also bone-like, but also harder and tougher than bone. In the centre of the tooth lies the **pulp** which contains blood capillaries, lymphatic vessels, and nerves, all embedded in soft connective tissue.

Once a human tooth has erupted it ceases to grow. This fact is related to the very poor food supply that human teeth receive through the 'closed' (i.e. very narrow) roots (Fig. 6.13). In herbivorous mammals, such as sheep, the teeth are subject to constant wear through grinding. They are adapted to this by having very long crowns which erupt slowly. Herbivorous rodents (rats, coypu) and rabbits have incisors with unlimited,

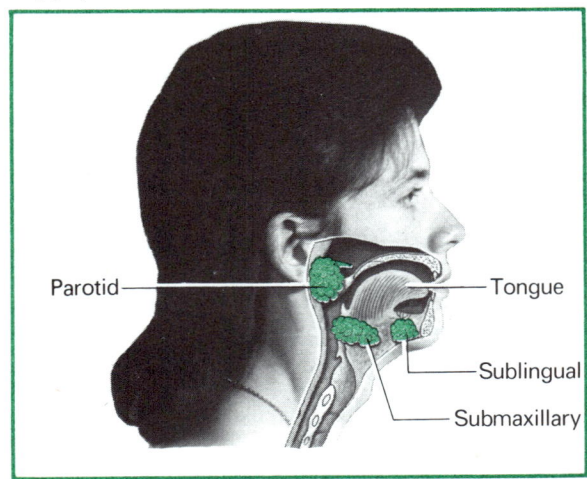

Fig. 6.14 Human salivary glands on one side.

continuous growth. Here the roots are 'open' and the teeth are well supplied with food and oxygen from the blood vessels passing into the roots.

Mammalian dentitions are so highly adapted to the nutritional needs of the species that a great deal can be said about its life style and habits from an inspection of its skull and teeth. This fact is much used in reconstructing dead mammals from fossil material.

Salivary glands
The mouth contains three pairs of large salivary glands and many smaller ones. The largest lie below the ear (Fig. 6.14). The other two pairs are situated below the tongue.

The sight, smell, thought, and especially the taste of food stimulates secretion of saliva, which makes the food moist and slippery. It contains the enzyme **salivary amylase** which catalyses the conversion of starch to glucose (p. 31).

Swallowing
Whilst in the mouth, food is chewed and moved about voluntarily. The decision to swallow is also voluntary, but once swallowing has started it goes on involuntarily and cannot be stopped by an effort of will.

Two tubes lead down the throat (Fig. 6.11). At the front is the **trachea** (windpipe), topped by the **larynx** (voice box or Adam's apple) and passing to the lungs. Behind lies the oesophagus (gullet), a muscular tube whose sole function is to convey food to the stomach. Thus the mouth is a common entrance to both windpipe and gullet. This means that the deep breathing necessary during exercise, which cannot take place via the nose alone, is possible through the mouth. Speech also requires an open mouth.

It is important that when food is being swallowed it should not enter and block the trachea. To prevent this the tongue rises to force the bolus of food up against the palate (Fig. 6.15). This causes a reflex holding of the breath, and the soft palate rises, shutting off the rear entrance to the nose and preventing food from entering it. At the same time the larynx rises and bulges into the lower part of the back of the mouth. The epiglottis, a flap of gristle, flicks forward closing the opening to the windpipe.

Swallowed food passes down on either side of the larynx, not in the midline. Occasionally the timing of all these events may get out of phase and a food particle may enter the windpipe. If it does there is a further reflex action, a vigorous coughing which expels the misplaced particle.

Structure of the gut

How food moves along the gut: peristalsis
Although the oesophagus, stomach, duodenum, small intestine, and colon are different from each other in appearance and function, they are all constructed on the same ground plan (Fig. 6.16(a)).

The inner surface, which makes contact with food in the gut cavity, is very much folded and lined with glands of various kinds. The wall contains two layers of muscle. In the inner layer the muscle cells, which are long and thin (Fig. 6.16(b)),

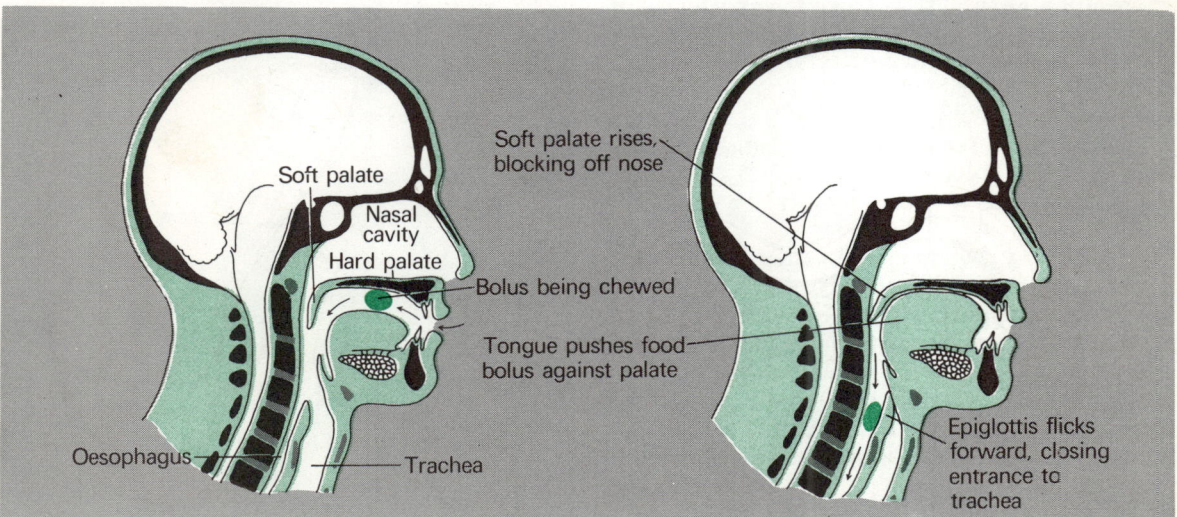

Fig. 6.15 Reflex movements of the soft palate and epiglottis during swallowing.

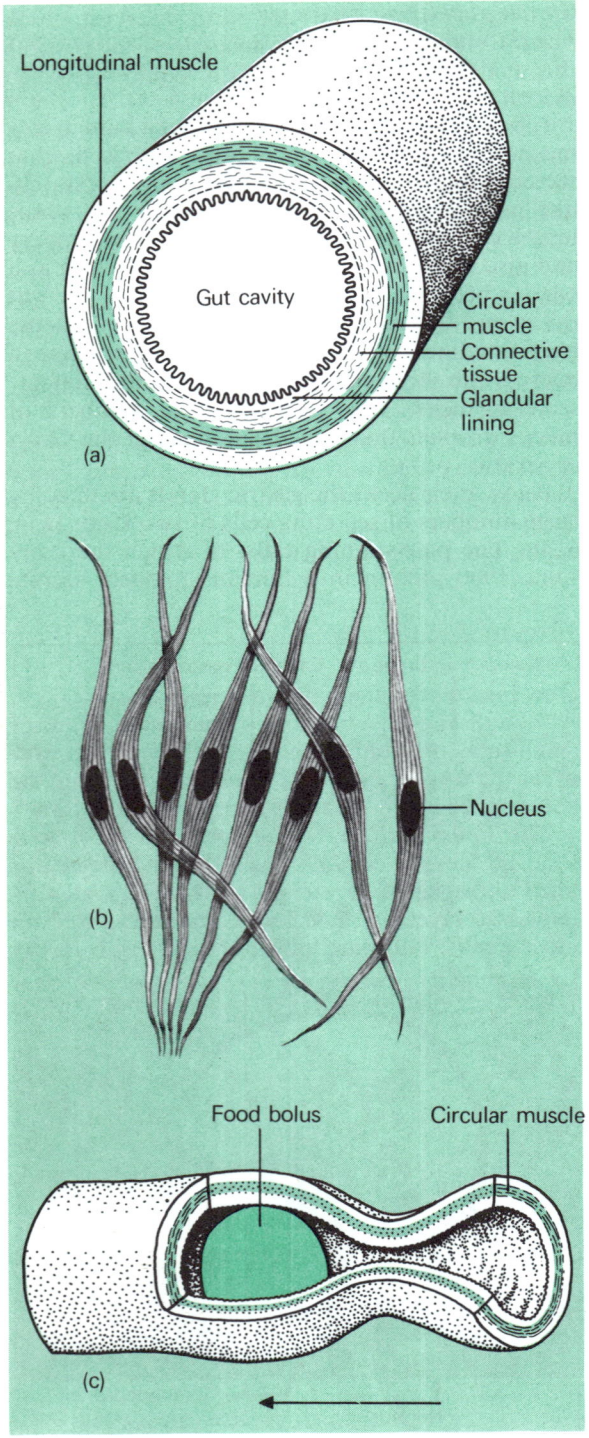

(a)

(b)

(c)

Fig. 6.16(a) The ground-plan of the mammalian gut.
(b) Muscle cells from the gut.
(c) Peristalsis.

lie in a circular fashion parallel to the circumference of the gut tube, while in the outer layer the cells are arranged parallel to the length of the gut tube.

To propel the food bolus, a band of the circular muscle cells contracts, narrowing the tube and pressing the bolus forwards. At the same time the circular muscle cells immediately in front of the bolus relax, allowing the food to move on (Fig. 6.16 (c)). This coordinated muscular activity is called **peristalsis** and is used elsewhere in the body to move liquids or semi-liquids along tubes.

The stomach
The human stomach is a J-shaped sac with the oesophagus leading in to it at the top and the duodenum leading out of it at the bottom (Fig. 6.17). Find its position in Fig. 6.9. Its lining produces secretions which are mixed with the food by churning peristaltic movements of the stomach wall, resulting in a porridge-like sludge called **chyme**; but it is essentially a reservoir. Most digestion and nearly all absorption of food takes place in the duodenum and the small intestine, and humans who have had their stomachs removed for medical reasons can live comfortably as long as they take small and frequent meals.

The stomach is separated from the duodenum by the **pylorus** (Greek, a gatekeeper), which is a local thickening of the circular muscle in the stomach wall. When this contracts the local gut cavity is obliterated and the stomach cavity and duodenal cavity are separated. Most of the time, however, the pylorus is relaxed. Its function is not to prevent food moving out of the stomach (into a duodenum

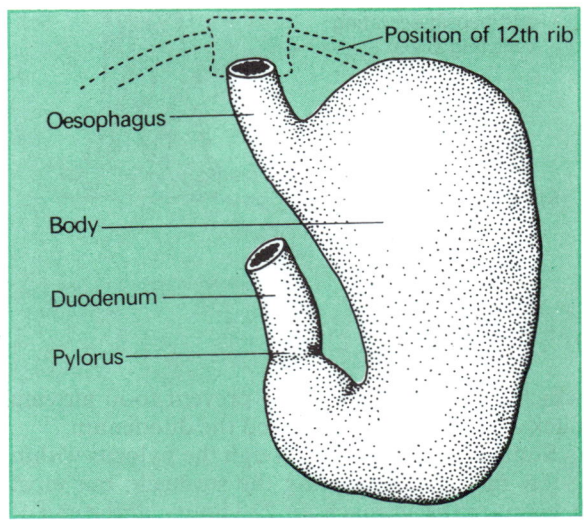

Fig. 6.17 The human stomach.

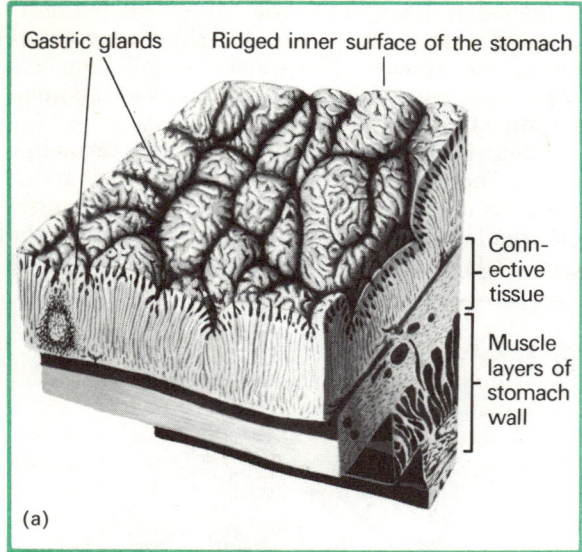

Gastric glands   Ridged inner surface of the stomach

Connective tissue

Muscle layers of stomach wall

(a)

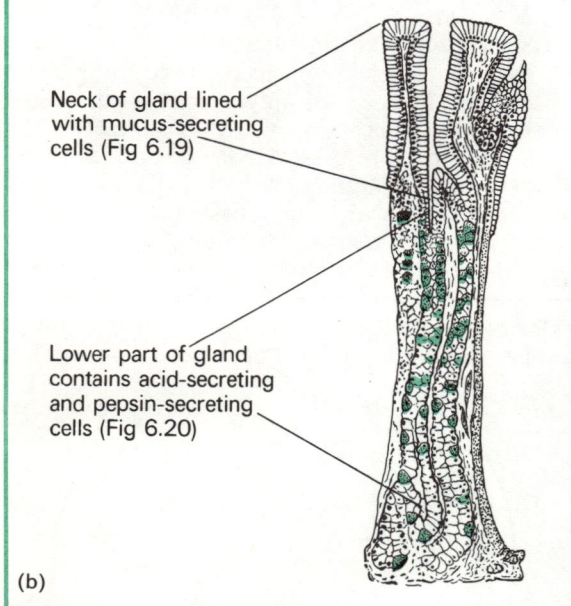

Neck of gland lined with mucus-secreting cells (Fig 6.19)

Lower part of gland contains acid-secreting and pepsin-secreting cells (Fig 6.20)

(b)

Fig. 6.18(a) Internal surface of the human stomach, showing the gastric glands.

(b) One of the gastric glands of the stomach seen in longitudinal section.

Gastric glands and gastric juice

The glandular lining of the stomach is deeply ridged (Fig. 6.18(a)) and its surface is pierced by test-tube-like pits called **gastric glands** (Fig. 6.18(b)). There are about 35 million gastric glands in a human stomach. The neck of each is lined with cells that secrete **mucus**. This is a slimy solution of a starch-like material and it is secreted along the whole length of the gut. It makes the food mass slippery and also has the important function of acting as a barrier between the digestive enzymes of the gut cavity and the gut cells themselves: it prevents the gut from digesting itself. Figure 6.19 is an electron micrograph showing a section through four mucus-secreting cells. They contain large numbers of mucus droplets that have been made at the Golgi apparatus (p. 7).

Below their necks the gastric glands are lined by large numbers of secreting cells of two kinds (Fig. 6.20). The pale-staining cells, of which there are some 1000 million in a human stomach, secrete

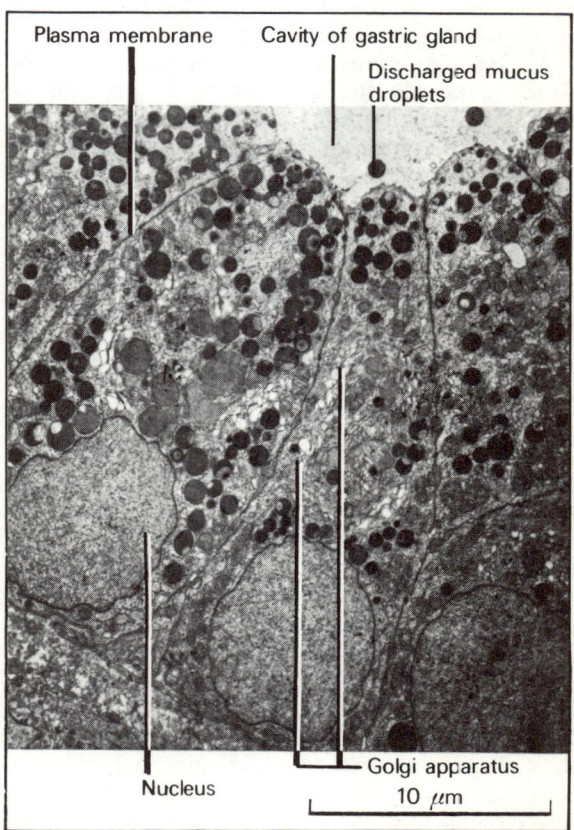

Plasma membrane   Cavity of gastric gland

Discharged mucus droplets

Nucleus   Golgi apparatus

10 μm

Fig. 6.19 Section through four mucus-secreting cells from a gastric gland (electron micrograph).

which is already full) but to prevent food moving back again once it has entered the duodenum.

Some food may pass through the pylorus within a few minutes of entering the stomach, but most remains in the stomach for 2–3 hours, at the end of which time there is very little left.

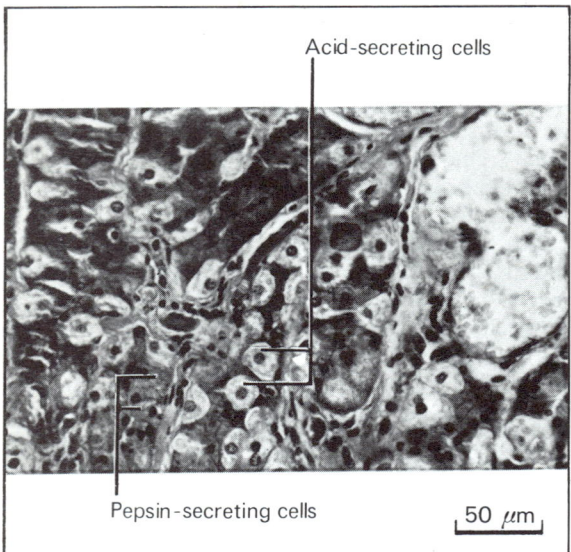

Fig. 6.20 Section through gastric glands showing acid- and enzyme-secreting cells (light microscope).

hydrochloric acid (HCl). This kills most of the germs which are taken in with food.

The dark-staining cells secrete the enzyme **pepsin**, which catalyses the conversion of proteins to shorter chains of amino acids called **peptides**. The combined secretion of mucus, acid, and pepsin is called **gastric juice**.

Acidity has a very strong effect on the activity of enzymes. Pepsin works best in strongly acid conditions, but the enzymes of the duodenum and small intestine work best in slightly alkaline conditions.

The duodenum and the small intestine

The large internal surface area
The human small intestine is 3–5 m long and its external surface area is about 5000 cm². Digested foods are absorbed through the inner, glandular lining. The larger this surface is, therefore, the more efficient absorption will be. In fact it is about two million cm², and Fig. 6.21 shows how this large increase in the internal, compared with the external, surface area is achieved.

Figure 6.21(a) shows the intestine of a young pig. Notice that in various places it is pinched in where the circular muscles in the wall have contracted, as they do during peristalsis. Figure 6.21(b) is a drawing showing that the inner surface is thrown up into **circular folds**, and these are covered with projections called **villi**. Villi have a finger-like or leaf-like appearance; they are about 1 mm long and 1 μm

in diameter, there being about 20–40 villi per mm² of small intestine. Figure 6.21(c) shows a group of villi in surface view. The presence of villi increases the surface area of the glandular lining of the small intestine to about 10 m².

Figure 6.21(d) is a longitudinal section through a villus. There is a central core of connective tissue which contains blood capillaries, a special vessel called a lacteal, and muscle cells. Surrounding the core is a layer of cells of two types, mucus-secreting

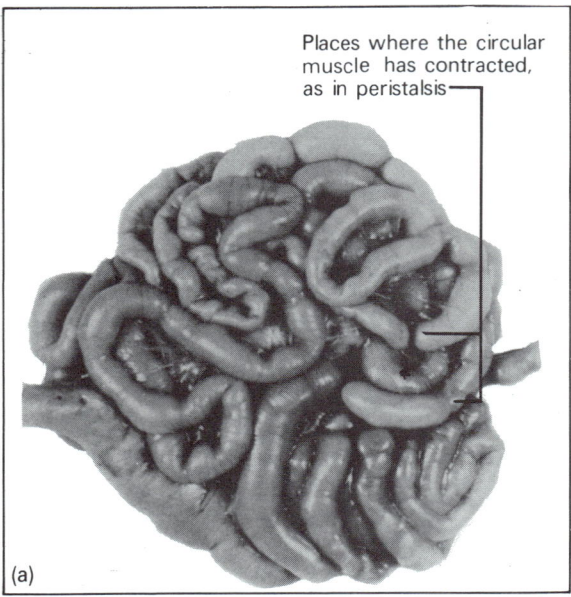

(a)

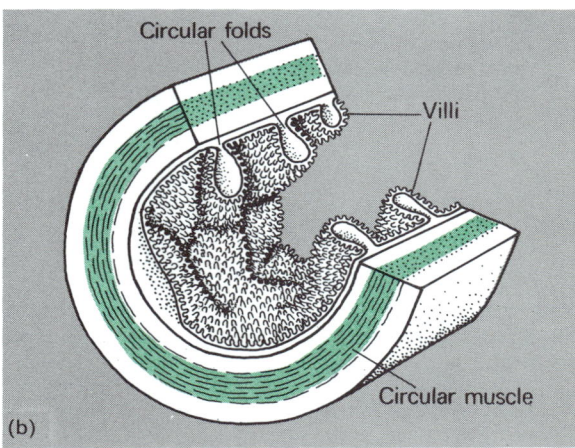

(b)

Fig. 6.21(a) Small intestine of a four-week old pig.
(b) Part of the intestine showing the circular folds and the villi.
(c) Glandular lining of the small intestine seen with the scanning electron microscope.
(d) Longitudinal section through one villus (light microscope).
(e) Electron micrograph of a group of microvilli on the surface of a villus cell. The material adhering to them is mucus.

Convoluted villus   Leaf-like villus   Finger-like villus

(c)   100 μm

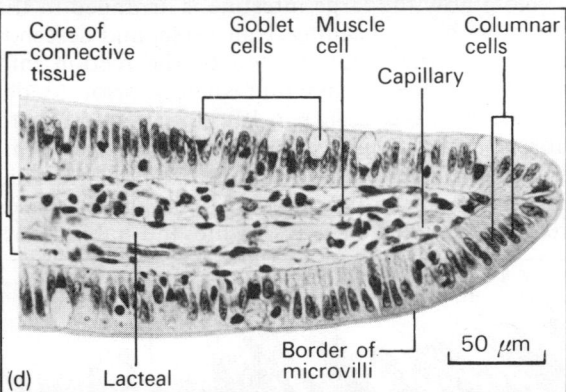

Core of connective tissue   Goblet cells   Muscle cell   Columnar cells

Capillary

(d)   Lacteal   Border of microvilli   50 μm

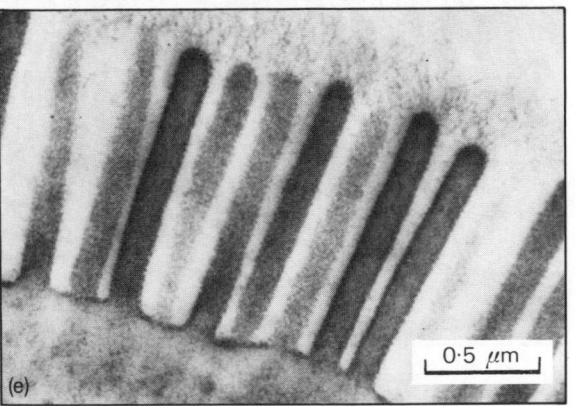

(e)   0.5 μm

goblet cells, and columnar cells. Both types bear on their outer surfaces 2–3000 microvilli, each one about 1 μm long and 0.1 μm in diameter. The presence of microvilli increases the surface area of the glandular lining by a further thirty times. Figure 6.21(e) is an electron micrograph of a few microvilli on the surface of one of these cells.

Bile and pancreatic juice
About 9 cm from the pylorus two ducts join and enter the duodenum; the **bile duct** from the liver, and the **pancreatic duct** from the pancreas (Fig. 6.10).

Bile
Some 0.5–1.0 litres of bile are secreted by the liver each day, to be concentrated and stored in the **gall bladder**. From here it passes down the bile duct into the small intestine. Bile is slightly alkaline and neutralizes some of the hydrochloric acid from the stomach, helping to create a non-acid environment in which the enzymes of the small intestine can work.

Bile also contains a detergent whose function is to disperse fats. These aggregate into large blobs which have relatively small surface areas (Figs. 6.6, 6.7). When bile detergent is added the fat disperses into large numbers of small droplets, and the surface area available for the activity of the fat-splitting enzyme **lipase** is thus greatly increased.

Pancreatic juice
The pancreas lies in the loop of the duodenum (Fig. 6.10) and secretes about 1.2 litres of pancreatic juice per day. Like bile, this is slightly alkaline and it, too, helps to neutralize the hydrochloric acid from the stomach. If differs from bile in containing digestive enzymes: **trypsin**, which converts proteins to shorter chains of amino acids (peptides); lipase, which converts fats to glycerol and fatty acids; and amylase, which converts starch to the sugar, maltose.

Control of the flow of bile and pancreatic juice
These juices provide a large volume of water in which digested foods can dissolve, digestive enzymes, and an adjustment of the acidity of stomach contents so that the intestinal enzymes can work.

They are discharged into the duodenum just as food moves into it from the stomach. This coordinated timing is not brought about by nerves but by chemical messengers (hormones, see p. 181). The villi of the duodenum contain scattered cells which, when they make contact with acidic stomach contents, secrete hormones into the blood capillaries of

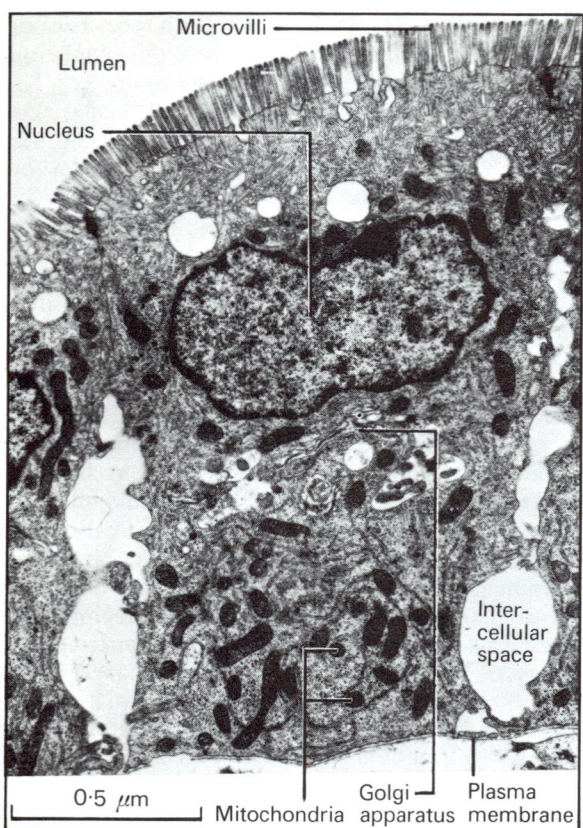

Microvilli

Lumen

Nucleus

Inter-cellular space

0·5 μm

Golgi apparatus
Mitochondria
Plasma membrane

Fig. 6.22 Longitudinal section of columnar villus cells (electron micrograph).

the villus. These hormones pass into the general blood circulation, and so are distributed all over the body. On reaching the gall bladder, one of them causes it to contract and expel bile through the bile duct into the duodenum. A second (**secretin**), reaching the pancreas, stimulates secretion of pancreatic juice. Thus bile and pancreatic juice are delivered to the duodenum at the right time.

Digestion and absorption: the columnar cells of the villi

Trypsin, lipase, and amylase convert insoluble proteins, fats, and starches into soluble peptides, glycerol, fatty acids, and maltose. Blood leaving the intestine, however, contains amino acids, not peptides, and glucose, not maltose. (A maltose molecule consists of two glucose molecules joined together.) Glycerol and fatty acids enter the lacteal vessel in the centre of the villus and do not pass directly into the blood stream.

This means that further digestion must take place, digestion of peptides to single amino acids, and of maltose to glucose. Until recently it was thought that this occurred in solution in the cavity of the intestine. It is now known that this digestion takes place within the columnar cells of the glandular lining, perhaps on the surface membranes of the microvilli, which contain the appropriate enzymes as constituent parts.

Figure 6.22 is an electron micrograph of a section through three columnar villus cells. Note the very large nucleus and the large numbers of mitochondria. These produce the chemical energy needed for the uptake (**absorption**) of dissolved foods, though to some extent this takes place by diffusion.

Amino acids, and glucose pass into the capillaries with which the villi are richly supplied (Fig. 6.23) and are removed to the liver.

The large intestine (colon)

After spending about 4.5 hours in the small intestine, digestion and absorption of foods is complete. The remaining gut contents consist mainly of water and cellulose (from the cell walls of plant material in the food). This provides roughage, solid material against which the muscular wall of the intestine can contract.

Note how the large intestine is arranged in the body, with its ascending, transverse, and descending limbs, the latter leading to the rectum into which the faeces pass immediately prior to defaecation (Figs. 2.4(a), 6.10). Its function is to absorb and so conserve water, so forming the semi-solid faeces. Obtaining water is an activity of vital importance to all land-living animals. It places restrictions on their ways of life and habitations,

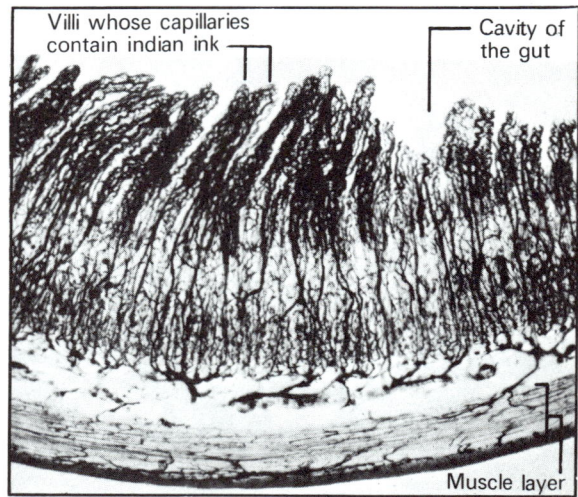

Villi whose capillaries contain indian ink

Cavity of the gut

Muscle layer

Fig. 6.23 Transverse section of part of the intestine of a cat showing capillaries injected with indian ink.

and all land-living animals possess devices of one sort or another for conserving water.

The liver

The liver is the largest organ in the body, weighing about 1.5 kg, some 3–4% of the total body mass. Blood laden with digested foods leaves the gut and is delivered directly to the liver. This part of the circulation is shown in Fig. 6.24.

Arterial blood, which is depleted of dissolved foods, arrives in the small intestine from the aorta (Fig. 6.24). Here the artery bringing it divides to form the many capillaries of the villi seen in Fig. 6.23. These re-join to form the hepatic portal vein; this passes to the liver where it too divides forming many capillaries. So the cells of the liver come into intimate contact with blood freshly delivered from the intestine. What is the purpose of delivering newly-absorbed foods directly to the liver?

Regulation of blood sugar level

Since meals are taken intermittently (with a particularly long gap between supper and breakfast) it might appear that digestion and absorption produce a sudden flush of amino acids and glucose in the circulation followed by an increasing dearth until the next meal is taken. But the cells of the body would not work well with such a fluctuating supply of nutrients. In particular the proper functioning of nerve cells requires that the level of glucose in the blood should remain at 0.1%, neither more nor less.

After a meal, blood entering the liver contains a higher level of glucose than this, and the liver cells remove excess and store it as the insoluble starch glycogen. The liver of an adult man contains about 100 g of glycogen. This can be mobilized (converted back to glucose) when the level of sugar in the blood starts to fall. Thus the liver provides both a reservoir and a valve for controlling blood sugar level. The balance between blood glucose and liver glycogen is controlled by several hormones.

Deamination of excess amino acids

Some of the amino acids derived from the proteins in the food are used for making new proteins both for growth and for replacement. If too much protein is eaten for this purpose, the excess amino acids cannot be stored in a way similar to the storage of excess glucose as glycogen. Instead they are changed by liver cells into the 3-carbon molecules (noted in Fig. 6.4) and urea, a non-poisonous waste material which is carried in solution in blood plasma to the kidney where it is excreted in the urine.

Three of the atoms in an amino acid are grouped together forming the amino group ($-NH_2$), which gives the acids their name. In the liver cells the amino group is removed and converted into urea, so the reaction is called **deamination**.

The liver has many other functions, one of which is the production of bile. Others are: the synthesis of vitamin A; rendering poisons, such as alcohol, harmless; the removal of bacteria, broken down red blood cells, and other debris from the blood; the storage and distribution of iron; and the synthesis of the plasma proteins globulin and fibrinogen.

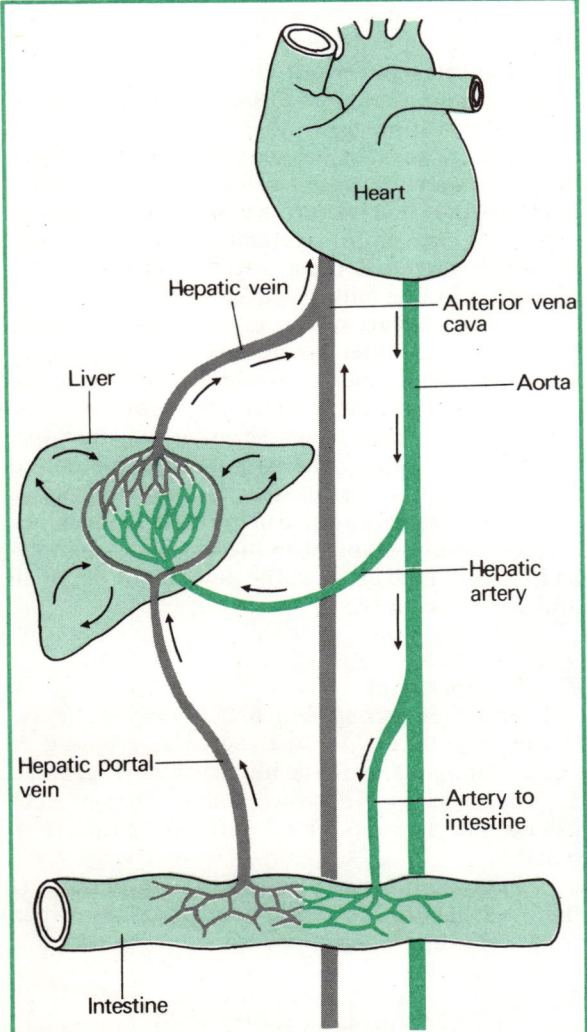

Fig. 6.24 Part of the circulation showing how blood laden with absorbed food is led directly to the liver.

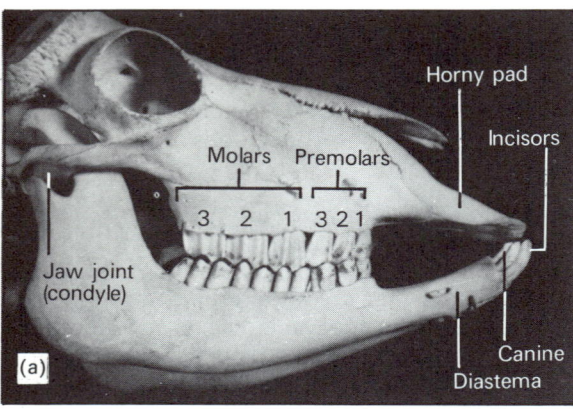

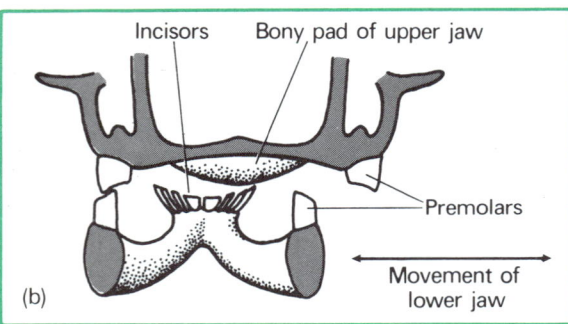

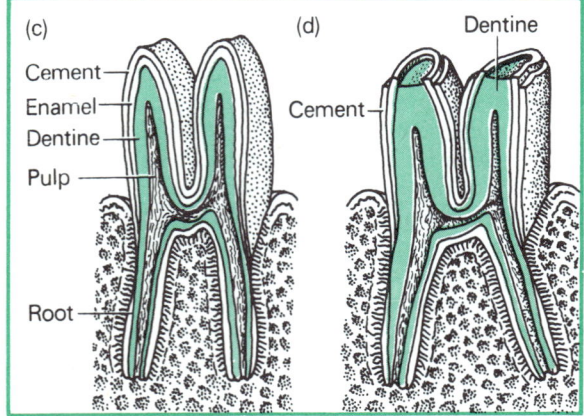

Fig. 6.25(a) Adult sheep skull from the side.
(b) Horizontal movement of sheep lower jaw.
(c) Newly-erupted molar tooth of sheep.
(d) Worn sheep molar.

Feeding adaptations in some other mammals
Herbivores

The first mammals were small creatures which fed on insects, as do the present day shrews and hedgehogs. Very early in their evolutionary history (about 60 million years ago), many plant-eating forms appeared. Most of these fed on grass, which is a tough, hard material containing little protein and a high proportion of cellulose. Cellulose is a polymer of glucose (p. 30), differing from starch only in the way in which the constituent glucose molecules are linked together. It is therefore a potential source of chemical energy but this is not immediately available to mammals because none of them produces the digestive enzyme **cellulase** which unlinks the glucose molecules of cellulose.

Bacteria and also **protozoa** (single-celled animals) do produce cellulase, and herbivorous mammals have evolved ways which enable them to employ these microorganisms to digest cellulose for them. First, the grass has to be thoroughly ground. This breaks open leaf cells, makes available their protoplasmic contents, and tears the cellulose walls apart so that a large surface area of cellulose is exposed to the enzyme cellulase. All herbivorous mammals therefore possess teeth which make an efficient grinding mill.

Secondly, a part of the gut is used as a large fermenting chamber where the microorganisms secrete cellulase which then works on the cellulose.

In one group of mammalian herbivores, the Artiodactyls (cattle, bison, sheep, goats, antelope) this chamber is an enormous and highly specialized stomach. In the Perissodactyls (horses, zebras, rhinoceros, tapir), and also in the rabbits and hares, the fermenting chamber is the **caecum**, a blindly-ending sac which arises at the junction of the small and large intestines.

The herbivore skull

The skull of a sheep (Fig. 6.25(a)) shows the following adaptations for the thorough grinding of grass. Compare it with the human skull (Fig. 6.12).
1 The lower jaw is narrower than the upper. Both sides of the jaw cannot meet at the same time (Fig. 6.25(b)).
2 The jaw joint (condyl) is flat and oval, permitting the lower jaw to move in a circular fashion when food is being ground and forwards when food is being cropped.
3 There are no incisors or canines in the upper jaw. The lower incisors and canines look alike; they are blunt and press against a bony pad in the upper jaw. When cropping, the animal grips grass

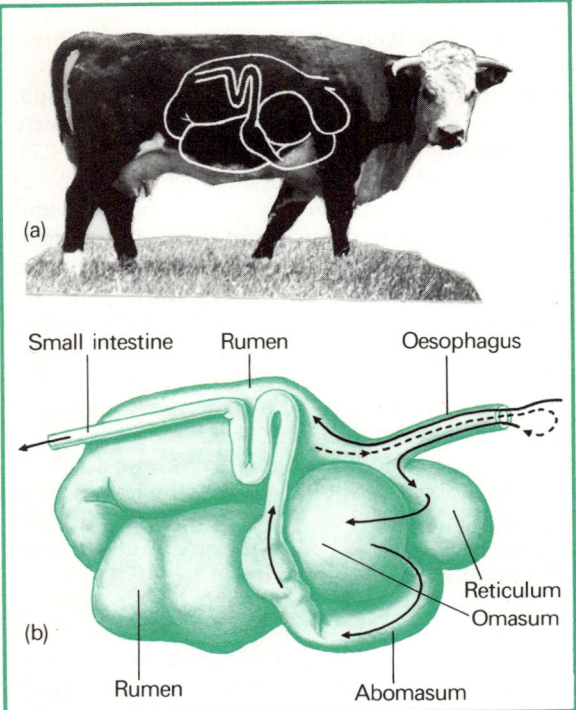

(a)

Small intestine   Rumen   Oesophagus

Reticulum
Omasum

(b)

Rumen   Abomasum

Fig. 6.26(a) Cow stomach *in situ*, showing its relatively enormous size.
(b) Cow stomach removed.

between the front teeth and the pad and tears it off with a tug.

4 There is a gap (**diastema**) between the cropping teeth and the grinding battery of premolars and molars. The premolars and molars all look alike.

5 The grinding teeth are very different from human molars. Where the crowns meet, the lower molars are worn into an M shape and the upper into a W shape, and the teeth fit neatly together. When the jaw is moved sideways the lower grinding teeth slide across the upper ones. Confirm this by moving the lower jaw of a sheep skull against its upper jaw.

6 The exposed surface of each molar tooth is not smooth but ridged, consisting of layers of enamel, dentine, and cement. In the teeth of herbivorous mammals, cement covers the whole tooth before it erupts (Fig. 6.25(c)). Enamel is the hardest of these substances, dentine the next hardest, and cement the softest. They wear away at different rates, producing the rough ridged pattern seen in Fig. 6.25(d). The resulting surface of the grinding battery is hard, sharp, and abrasive; well suited for cutting and grinding grass.

The cow stomach

The stomach of a cow, which is illustrated in Fig. 6.26, is enormous relative to the size of the animal. It holds 180 litres and is divided into four parts. Grass is cropped and swallowed unchewed into the **reticulum** and the **rumen**, the largest part, which forms the fermenting chamber. When feeding is over material is regurgitated from the rumen into the mouth, the liquid in it is re-swallowed, and the solid material is systematically ground on the premolars and molars. This is called chewing the cud and takes place for about eight hours per day.

The food is now finely divided and is re-swallowed into the rumen, the contents of which contain tens of millions of bacteria and protozoa per cm^3—one of the most concentrated populations of microorganisms found anywhere in nature. These secrete the enzyme cellulase on to the food, unlinking the polymer molecules of cellulose and liberating glucose. This is broken down further to two 3-carbon molecules which can be used either as a source of chemical energy or as a molecular 'backbone' from which to construct essential amino acids and so proteins (Fig. 6.4).

The food now passes from the rumen first into the **omasum**, where water is removed from it, then into the **abomasum**, where hydrochloric acid and protein-digesting enzymes are secreted. Here the bacteria and protozoa are killed. The protein they contain, and that of the grass cells, is digested and absorbed in the abomasum and in the small intestine.

Fermentation in the rumen also produces some two litres of waste gases (methane (CH_4), and carbon dioxide) per minute, and the cow gets rid of this by belching. Occasionally, if the animal feeds on fresh legumes, such as clover and beans, the gas may not be able to escape fast enough, causing 'bloat'. The pressure of the gas in the stomach rises to several atmospheres and it swells, pressing against the arteries and blocking them. This quickly kills the animal. Bloat can be relieved by piercing the stomach wall, and this operation forms a memorable episode in Thomas Hardy's novel *Far From the Madding Crowd*.

It is suggested that the Artiodactyls are peaceful, non-aggresive animals because they spend so much of their lives gathering and chewing food. The ability to gather now and digest later makes them keen competitors with the Perissodactyls, who feed on the same material but cannot do this. Today the Artiodactyls are a diverse and flourishing group in nature, but the Perissodactyls are dwindling and may soon (speaking in geological terms) become extinct.

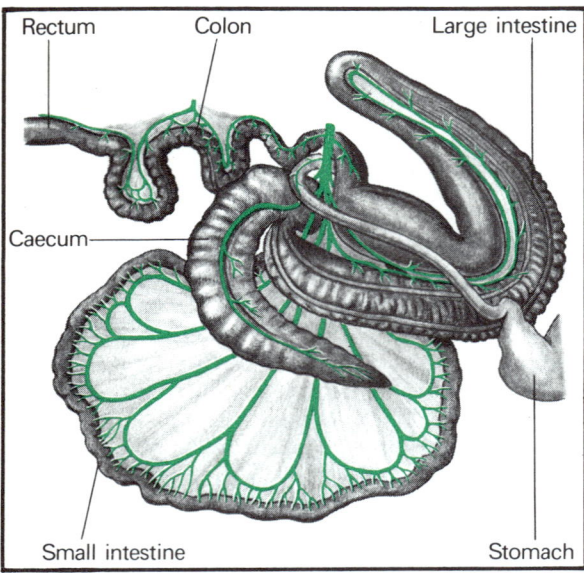

Fig. 6.27 Gut of a horse unravelled.

Perissodactyls: the caecum as a fermenting chamber

In the horse group there is no such elaboration of the stomach, which is quite small, and the animal does not chew the cud. Food is ground on molar teeth similar to those of the sheep and cow, though this is not as effective as chewing the cud because it is much briefer. On swallowing, food enters the

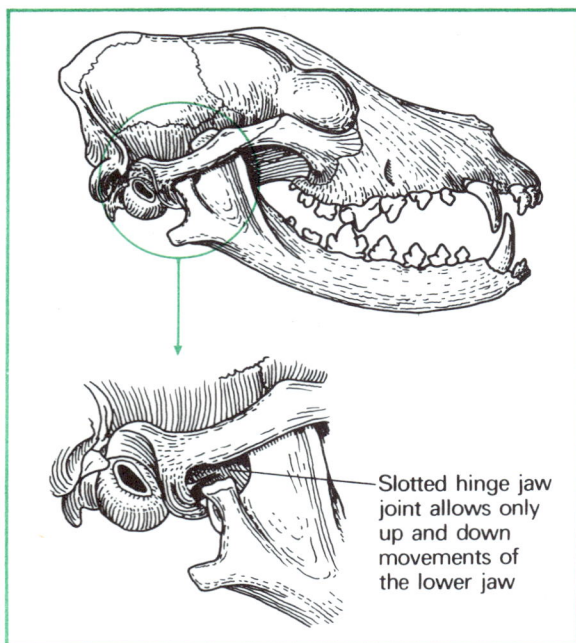

Fig. 6.28 Dog skull from the side.

Slotted hinge jaw joint allows only up and down movements of the lower jaw

stomach where digestion of liberated grass cell contents starts, to be continued in the small intestine.

Instead of passing on into the large intestine, food now enters the caecum, which is relatively large, having a capacity of 22 litres (Fig. 6.27). Compare its relative size with that in man (Fig. 6.10) and rat (Fig. 6.8). In the caecum are bacteria and protozoa similar to those found in the cow rumen, and fermentation takes place here. After this the only route for the food is into the large intestine and rectum, so absorption of the digested cellulose must take place either in the caecum or the large intestine, but surprisingly very little research has been done on this and there is no certain knowledge of the absorption of digested food in horses.

In rabbits and hares (Lagomorpha) the arrangement of the gut is similar to that in horses, but these animals eat their own droppings and the digested food is absorbed in the small intestine on its second time round the gut.

Symbiosis

These herbivorous mammals provide examples of two organisms living together in close association for mutual benefit. The cow (for example) benefits from the presence of the bacteria and protozoa to such an extent that it could not exist without them. The microorganisms benefit because they are provided with a safe warm environment and a constant supply of food. The state of *two distinct*

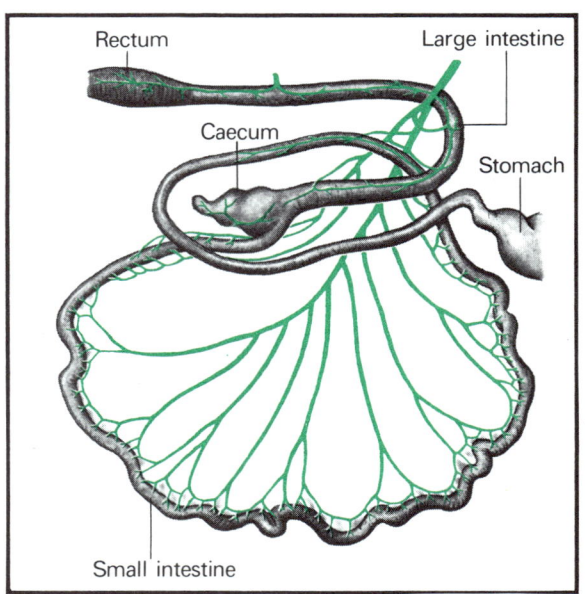

Fig. 6.29 Gut of a dog unravelled.

organisms living together for mutual benefit is called **symbiosis**, and the partners are called **symbionts**.

There are many examples of mutual living in nature apart from symbiosis. In some, one of the organisms harms the other (they are *parasites* e.g. tapeworm in the gut of mammals). In others the two organisms live in close association but do not depend on each other. This association is described as *commensalism*, for example sea anemone living on the shell of a hermit crab.

Although the symbiotic relationship we have examined is clearly of benefit to both partners, there are many degrees of partnership in nature and it is not possible to draw sharp distinctions between symbiosis, parasitism, and commensalism (see chapter 19).

Carnivores

Flesh-eating was only a short step from the insectivorous diet of the ancestral mammals. The habit led to a large order of mammals, the Carnivora, which includes the dogs (wolf, fox, jackal), cats (lion, tiger, leopard), the bears, and the pandas.

There are many adaptations to meat eating in carnivore skulls and teeth (Fig. 6.28). The canines are enlarged and cone-shaped, adapted for slashing prey when fighting, a specialization which reached its peak in the extinct sabre-toothed tigers. The premolars and molars are not flat-crowned as in herbivores but pointed and blade-like for slicing like a pair of scissors. Related to this is the slotted hinge-type jaw joint which allows movement only in one plane, up and down. Sideways movements would reduce the efficiency of the jaw: the slicing apparatus would then be like a pair of scissors with a loose rivet. In dogs (Fig. 6.28) the fourth upper premolar and the first lower molar are especially well developed as slicing teeth, and are called **carnassial** teeth.

Fig. 6.29 shows the gut of a dog unravelled. In what ways does it differ from the guts of a horse and a cow, and how does it appear to be adapted to a diet of meat?

7 Energy for the Cell: Respiration

Meanings of the word respiration

In nearly all cells, animal and plant, chemical energy is liberated from glucose by means of a series of chemical reactions whose ultimate waste products are carbon dioxide and water. This takes place partly in the ground substance of the cell and partly in the mitochondria (p. 35). It is called *tissue respiration* and can be summarized by the following equation:

$$CH_2O + O_2 \longrightarrow CO_2 + H_2O + ENERGY$$
carbo-
hydrate

It is of central importance to the cell since it provides energy for all its chemical reactions. These include synthesis of new cell material, the contraction of muscle cells, the passage of electrical messages along nerve cells, and the absorption of materials across plasma membranes (p. 95).

As the equation shows, oxygen is consumed during tissue respiration and carbon dioxide is liberated. Most animals have evolved special mechanisms to enable them to obtain oxygen rapidly enough, and to eliminate the carbon dioxide produced. These are called respiratory organs (e.g. lungs, gills, tracheae in insects), and their operation is also spoken of as respiration, which is a word with more than one meaning.

Burning (combustion)

Some substances, such as wood, coal, and paper, burn when they are heated in air. This means that a chemical reaction takes place between the material and oxygen in the air. The atoms which compose the material, and oxygen atoms, contain chemical energy. The molecules formed as a result of burning contain less chemical energy than the molecules of the reacting substances, and the surplus energy appears as heat and light. Thus, when glucose is burned in air, oxygen is consumed and the products, carbon dioxide and water, contain less chemical energy in their molecules than did the reacting glucose and oxygen. The surplus chemical energy appears as heat and light:

$$C_6H_{12}O_6 + 6O_2 \longrightarrow 6CO_2 + 6H_2O + Heat$$
glucose
$$+ Light$$
equation (1)

Experiments have been performed to discover how much energy (measured in kilojoules) is liberated in this reaction, with the following results:

180g glucose + 192g oxygen $\longrightarrow$ 264g carbon dioxide + 108g water + 2892 kJ
equation (2)

This means that the chemical energy contained in the molecules of 180 g glucose and 192 g oxygen is greater by 2892 kJ than the chemical energy contained in the molecules of 264 g carbon dioxide and 108 g water. This energy is liberated when the glucose is completely burned. It is sufficient to raise the temperature of seven litres of water from 0°c to 100°c.

The energy needed by cells is chemical energy

The burning of glucose in air is a sudden and violent reaction liberating a large amount of energy in a short time. Living cells do not obtain energy from glucose in this way, for it would be largely valueless to them. Light energy cannot be used to drive cell reactions (except the splitting of water molecules in photosynthesis, p. 39). Heat energy speeds up the rates of chemical reactions by making molecules move faster and so increasing the numbers of molecular collisions in unit time (p. 32), but the extent to which cells can take advantage of this is limited by the fact that they are largely composed of proteins which are denatured at about 45°C (p. 34).

Over the course of 2–3000 million years, cells have evolved a chemical machinery into which glucose is fed to be 'burned' not in one step, as in equations (1) and (2) but in a consecutive series of about 20 enzyme-controlled steps. The chemical energy in the glucose molecules is liberated gradually along the line.

Some (about 30–40 %) appears as heat. This helps to maintain a warm environment in which enzymes work best, but part of this heat is to be regarded as wasted energy. Man-made machines, such as steam turbines, also waste energy as heat, and a machine that wastes only 40 % of the energy fed into it is regarded as highly efficient.

ATP

It is the remaining 60–70 % of the chemical energy which is of most importance to the cell. To understand this we must consider two chemicals which enter into almost all cell chemical reactions, **adenosine diphosphate, ADP**, and **adenosine triphosphate, ATP**.

Adenosine is a complicated substance whose chemical structure need not concern us. In ATP it is linked to three phosphate groups, and a simple way of depicting it is:

$$A - P \sim P \sim P$$

where **A** represents adenosine and **P** represents a phosphate group.

Notice that the bond linking adenosine to the first phosphate group is written as a dash, but that joining the second phosphate group to the first and the third to the second is written as a squiggle, $\sim$. This squiggle represents a special property of ATP. It is able to enter into a very large number of different kinds of cell reactions in which it passes on to other chemicals its third phosphate group *plus* a large amount of chemical energy. The $\sim$ is called an **energy-rich** bond, and the compound which receives it becomes enriched in chemical energy and is thereby made more chemically reactive.

Let us consider an example of energy donation by ATP. Suppose the cell makes chemical substance C by linking together substances A and B (for example, the linking together of two amino acids to form part of a protein). In nearly every case this reaction will require an input of energy to make it take place; that is, A and B will not react together unless energy is supplied:

$$A + B + energy \longrightarrow C$$

ATP may enter into this reaction by reacting with substance B, passing on to it its third phosphate group and becoming ADP in the process:

$$ATP + B \longrightarrow ADP + B \sim P$$

The energy-rich form of B, $B \sim P$, now contains enough chemical energy to enable it to react with A, producing C:

$$A + B \sim P \longrightarrow C + P$$

The overall reaction for ATP is that it is converted into ADP and a phosphate group P, and has lost an energy-rich bond.

For more C to be made, more ATP is necessary, and unless the ATP supply of the cell is to be quickly exhausted, it must be replenished by the re-joining of ADP and P. However, since ATP gave out energy (as an energy-rich bond) when it formed ADP and P, the reverse process, synthesis of ATP, cannot take place unless energy is supplied:

$$A - P \sim P + P + energy \longrightarrow A - P \sim P \sim P$$

Synthesis of ATP via tissue respiration

The remaining 60–70 % of energy deriving from the oxidation of glucose in the cell is used to make ATP from ADP and P. Along the course of the 20 enzyme-controlled steps between glucose and the waste products carbon dioxide and water, about 40 ATP molecules are generated for each glucose molecule consumed.

We can now re-write equation (1) in a more meaningful way:

$$C_6H_{12}O_6 + 6O_2 \qquad 6CO_2 + 6H_2O$$
$$40\ ADP + 40\ P \qquad 40\ ATP + heat$$
$$\text{equation (3)}$$

The double bent arrows in this equation indicate that the reactions in the top line are linked to those in the bottom line.

Where ATP is produced in the cell

The first stages of glucose breakdown produce two 3-carbon molecules (p. 48, Fig. 6.4). This takes place in the ground substance of the cell and generates two ATP molecules per molecule of glucose consumed.

The remaining stages of breakdown occur within the mitochondria and generate about 38 molecules of ATP per molecule of glucose consumed. ATP provides the chemical energy for cell reactions in a useable form, and for this reason the mitochondria have been called 'the power-house of the cell'.

ATP is an energy currency

The cell can use any of the three major classes of foodstuffs, proteins, fats, and carbohydrates, as sources of chemical energy for all its diverse activities. This is possible because ATP forms a common intermediate between the sources of energy and the energy-consuming processes. It is generated by tissue respiration and consumed in cell reactions. In a similar way, money forms a currency for men who are paid for their labours not in goods but in cash with which they can purchase goods or services.

ATP as a cyclic intermediate

ATP is a key substance produced by one cell process and consumed by another. Its molecules are passing through a constant cycle of synthesis and break-down:

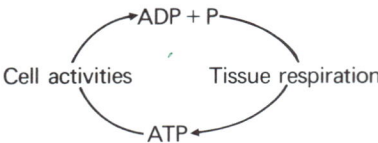

It is like a cogwheel in the chemical machine. This arrangement provides a considerable economy of space and material, for there is no need for large quantities of ATP to be present in the cell. To go back to the previous example of ATP making possible the synthesis of substance C from substances A and B, there may be hundreds, thousands, or even millions of molecules of C made during the life of the cell, but it is only necessary for one molecule of ATP to participate in the synthesis of all of them:

Respiration without oxygen

All the arguments in this chapter so far have been based on the idea of respiration as a *controlled oxidation of glucose*, summarized in equations (1), (2), and (3). This is very generally true, but there are some organisms that can respire *without* oxygen. Glucose is still the fuel that they use, though the end products are not carbon dioxide and water as they are when oxygen is present, but carbon dioxide and a compound such as ethanol (alcohol), C_2H_5OH, or lactic acid. Respiration in the presence of oxygen is called **aerobic**, and in its absence, **anaerobic**.

Giving separate names to these processes is a convenience when talking about them, but do not make the mistake of thinking that there are two 'kinds' of respiration, aerobic and anaerobic. They are not alternative processes but consecutive ones.

All cells possess the chemical machinery for converting glucose into two 3-carbon molecules.

This generates two molecules of ATP and does not involve the participation of oxygen (Fig. 7.1). In anaerobically-respiring cells, the 3-carbon molecules are converted into waste end-products such as ethanol or lactic acid. In aerobically-respiring cells, the 3-carbon molecules are fed into a further chemical machinery in the mitochondria which oxidises them to carbon dioxide and water, generates 19 molecules of ATP for each 3-carbon molecule consumed, and *does* involve the participation of oxygen.

Relative efficiencies of aerobic and anaerobic respiration

Aerobic respiration liberates 2892 kJ of energy from 180 g glucose and generates 40 molecules of ATP. Anaerobic respiration is much less efficient than this. It liberates only 210 kJ from 180 g glucose and generates only 2 molecules of ATP per molecule of glucose consumed:

$$C_6H_{12}O_6 \longrightarrow 2CO_2 + 2\,C_2H_5OH$$
$$2\,ADP + 2P \longrightarrow 2\,ATP \qquad \text{ethanol}$$

equation (4)

Aerobic respiration yields about twenty times as much useful energy (in the form of ATP) as does anaerobic. It is not surprising, therefore, that most organisms in the world today are aerobic. An aerobically-respiring organism has to find twenty times less food to meet its energy needs than does an anaerobically respiring one.

Being able to respire aerobically depends on the possession of mitochondria. These are complicated organelles (p. 6, Fig. 1.10), but once they had evolved the cells containing them would have had a great advantage over those that did not contain them. The development of mitochondria was thus an important event in the evolution of life.

Evolution of aerobic respiration

The earth's original atmosphere is thought to have been devoid of oxygen, and the present-day level of 20% has been generated by photosynthesis. The age of the earth is about 4500 million years. Fossils of unicellular organisms and their products exist from about 2000 million years ago, but there was probably no oxygen in the atmosphere until 1500 million years ago, and no appreciable quantity until 600 million years ago.

Thus the first organisms must have respired anaerobically and aerobic respiration must have evolved later. Evolution did not, however, take the course of scrapping the anaerobic chemical

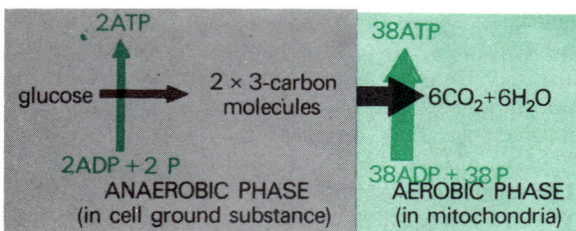

Fig. 7.1 Overall chemical steps in anaerobic and in aerobic respirations.

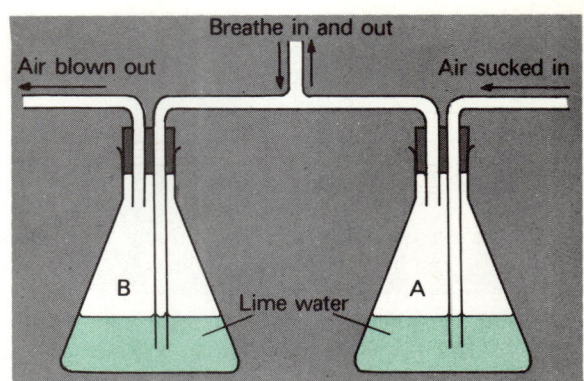

Fig. 7.3 Apparatus to demonstrate that carbon dioxide is present in expired breath.

machinery. Instead it retained it and developed a second set of chemical machinery which takes in the end product of anaerobic respiration as a fuel. In a sense, all aerobically respiring cells contain the ancestral anaerobic machinery as well as the more recently developed aerobic machinery. Those organisms alive today that can respire *only* anaerobically are certain bacteria—organisms of the type which are the first to appear in the fossil record.

Anaerobic respiration in mammalian muscle cells
When mammalian muscle cells are working hard (for example, if the animal is running), the oxygen-carrying power of the blood is not adequate to meet the demand for oxygen in the contracting muscle cells. Under these conditions the cells are able to respire anaerobically. They generate two molecules of ATP per glucose molecule consumed and produce a 3-carbon molecule end product, lactic acid:

$$glucose \longrightarrow 2 \text{ molecules of lactic acid}$$
$$2ADP + 2P \diagdown 2ATP$$

equation (5)

In doing this the muscle incurs an **oxygen debt** (see p. 78).

Some class demonstrations
The only parts of equation (3) that it is possible to investigate in a school laboratory are the evolution of carbon dioxide, the production of heat, and the consumption of oxygen. To show that glucose is

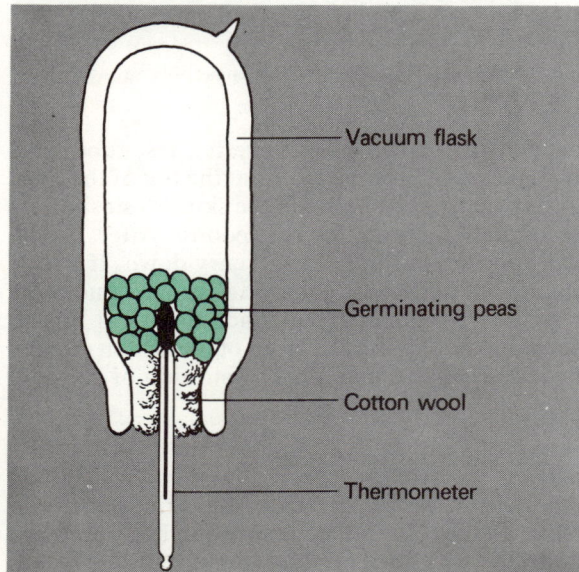

Vacuum flask

Germinating peas

Cotton wool

Thermometer

Fig. 7.2 Apparatus to demonstrate the evolution of heat by germinating seeds.

indeed the fuel used by living organisms requires the use of isotopically labelled glucose.

Production of heat by respiring plant tissues
The apparatus used is shown in Fig. 7.2. About 100 germinating pea seeds are used. The flask must be inverted as shown so that the dense gas carbon dioxide produced by respiration sinks out of the flask. If it were allowed to remain in the flask it would first slow down and then stop the respiration of the seeds.

A second, control flask must be set up using the same number of germinating seeds that have been killed by boiling them. The only difference between the two treatments is that in one flask the seeds are alive and in the other, dead. Any differences between the two flasks will be due only to the activity of the living seeds. The dead seeds must be sterilized by washing them in 2 % formalin solution, otherwise bacteria and fungi will grow on them and, by their respiration, raise the temperature in the control flask.

Thermos flasks are used to conserve the small amount of heat that is given off. If plain glass flasks were used the heat would be dissipated and it would not be possible to record a rise in temperature.

The temperatures in the two flasks should be recorded at two-hourly intervals over a period of two days and the results plotted on a graph.
(N.B. It is interesting to note that the heat released by a tonne of ripening apples stored at 0°C during shipment melts six tonnes of ice per day.)

Carbon dioxide is given off by living tissue
The apparatus shown in Fig. 7.3 can be used to show that breath contains carbon dioxide. Holding the nose, some thirty breaths are taken into and out of the breathing tube without removing it from

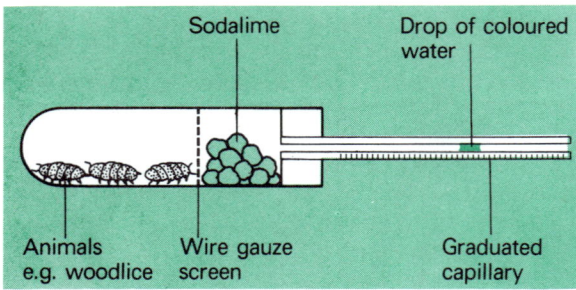

Fig. 7.4 Apparatus for investigating the rate of oxygen consumption by woodlice.

the lips. Air is drawn through the limewater in flask A and this becomes slightly milky because of the small proportion (about 0.03%) of carbon dioxide in the air. Breathing out blows the same air out of the lungs into flask B, where the carbon dioxide it now contains causes the limewater to become very milky.

Oxygen is consumed by living tissue
The apparatus shown in Fig. 7.4 can be used to investigate how oxygen consumption varies with change in temperature.

The animals (earthworms or woodlice) respire, taking in oxygen and giving out carbon dioxide which is absorbed by the soda lime. The volume of gas in the boiling tube thus decreases and the drop of indicator solution moves inwards towards the cork. Results should be plotted as rate of respiration (time taken for the drop to move 3 cm) on the Y axis against temperature on the X axis.

Anaerobic respiration of yeast
The microscopic fungus yeast (Fig. 18.9(a)) can

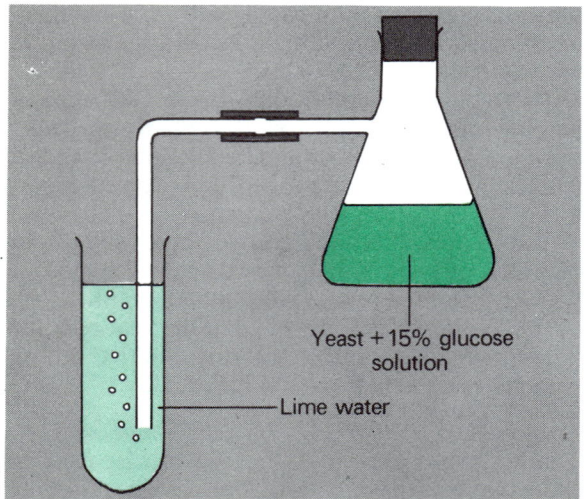

Fig. 7.5 Apparatus to demonstrate anaerobic respiration (fermentation) by yeast.

respire both aerobically and anaerobically, when it produces carbon dioxide and ethanol (equation (4)). This is the basis of the production of beers and wines as well as the raising of dough in bread-making. This anaerobic respiration is called **fermentation**.

The apparatus shown in Fig. 7.5 shows that carbon dioxide is given off during fermentation. The conical flask has no inlet tube. It contains a small amount of oxygen in the air above the liquid, but the yeast quickly uses this up in aerobic respiration. If the apparatus is kept at about 25°C, bubbles are evolved for some 24 hours, and as these turn the limewater milky, they are carbon dioxide.

Anaerobic respiration in plant tissue
The apparatus is shown in Fig. 7.6. At the start of the experiment, skinned germinating peas are

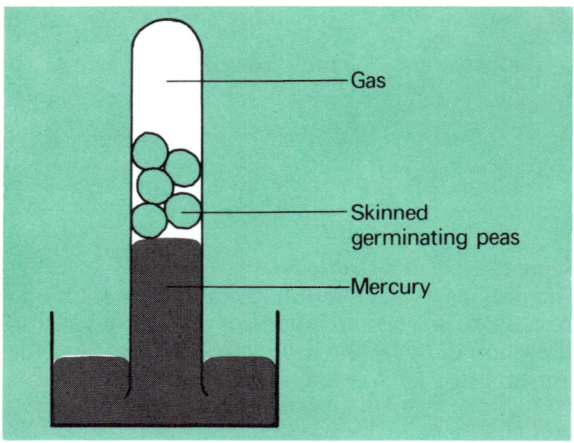

Fig. 7.6 Apparatus to demonstrate anaerobic respiration by germinating seeds.

slipped to the top of an inverted test tube filled with mercury. There is no air at the top of the tube, and none trapped between the skins (testas) of the seeds and their fleshy cotyledons. After a day, gas appears, pushing the mercury down. If a little potassium hydroxide solution is slipped under the test tube, the mercury at once rises and fills it, showing that all the gas given off is carbon dioxide produced by the anaerobic respiration of pea seed cells.

Respiration in mammals

Apart from the dead cells of the skin, hairs, and nails, all the cells in the mammalian body are alive and respire. Oxygen for this is supplied by the blood, which also removes the carbon dioxide they produce.

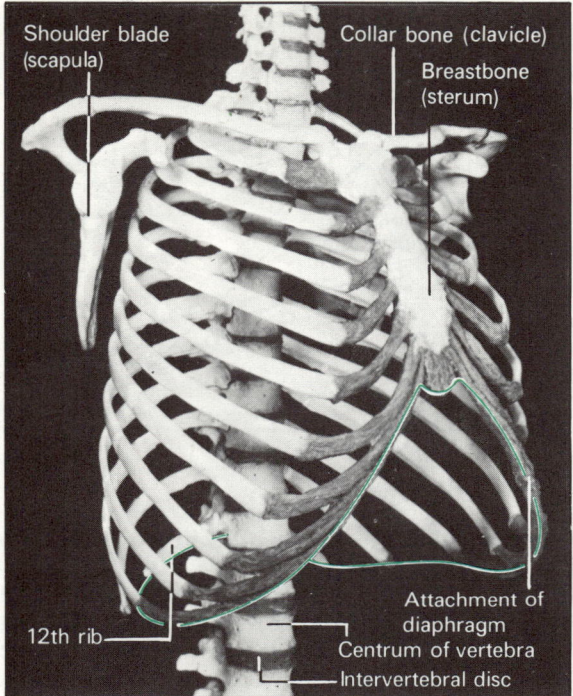

Fig. 7.7 Human chest skeleton.

Shoulder blade (scapula)
Collar bone (clavicle)
Breastbone (sterum)
Attachment of diaphragm
12th rib
Centrum of vertebra
Intervertebral disc

Air enters and leaves the lungs because of the breathing movements of the chest. Oxygen is taken up from this air by the red blood cells and is carried by them to the respiring body cells.

The chest (thorax)

Figure 7.7 is a photograph of the chest skeleton of a man. Look back to Fig. 2.4 and note how these bones lie in relation to the organs underlying them.

Each rib is firmly attached to the thoracic vertebrae of the backbone, from where it bends outwards, forwards, and downwards. Most of the rib is made of bone, but a small part of each of the first ten (counting from the top down) is made of **cartilage** (gristle). In Fig. 7.7 this part appears dark because it has been painted with varnish.

In the living animal each rib is joined to the next by two sheets of muscle (Fig. 7.8). The outer of these is the **external intercostal** muscle and it slopes diagonally forwards. This muscle causes the chest to change shape during breathing, so helping to ventilate the lungs.

The top of the chest is formed by the muscles, organs, and tissues of the neck. Lying across the floor of the chest and separating it from the abdomen is the **diaphragm**, a dome-shaped sheet (Fig. 7.9). In its centre is a crescent-shaped region made of tough unstretchable material, the **central tendon**.

From this radiate muscle fibres which are anchored at their lower ends to the margins of the ribs. When these contract the central tendon is pulled downwards and the slope of the sides of the dome changes as shown in Fig. 7.10. This causes an increase in the volume of the chest and accounts for about 60% of the air taken into it during each breath.

Passing through the diaphragm are the oesophagus, the **aorta** (main artery leaving the heart), and the **posterior vena cava** (main vein returning blood to the heart from all regions of the body below the heart).

Movements of the chest during breathing

Breathing in (inspiration)

The movement of the chest during inspiration is caused by the contraction of the external intercostal muscles. This makes the whole of the chest skeleton move upwards and outwards. To make this clear, use the Meccano model shown in Fig. 7.11. The 'backbone' is held firm, and the elastic band representing the external intercostal muscles is allowed to contract. The 'ribs' and 'sternum' at once rise.

Figure 7.12 shows how the chest and abdomen change their shapes during inspiration. Notice how the width of the chest from backbone to sternum increases, particularly during deep inspir-

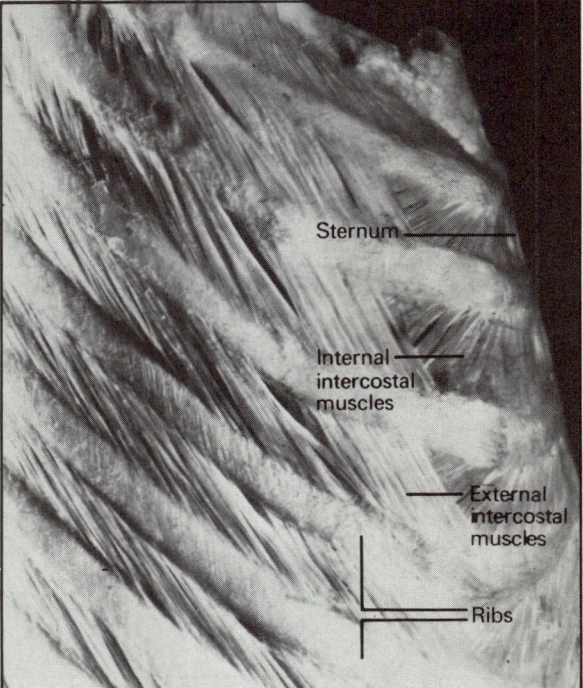

Sternum
Internal intercostal muscles
External intercostal muscles
Ribs

Fig. 7.8 Human chest flayed and showing the intercostal muscles.

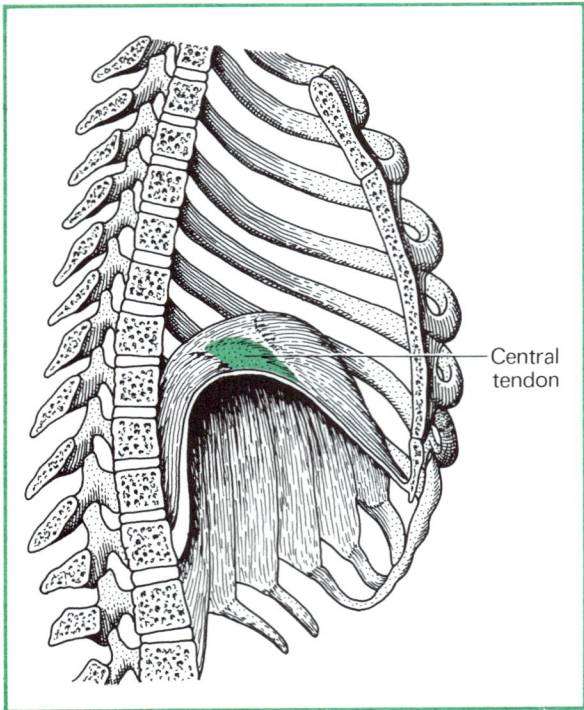

Fig. 7.9 The human diaphragm, showing the shape, central tendon, and attachment of the ribs.

ation. The movements of the chest skeleton during inspiration are shown in Fig. 7.13. Note that the rib cage is firmly attached to the backbone, but the ribs and sternum are free to move upwards, forwards, and outwards.

Increase in the volume of the chest draws air into the lungs. When the chest increases in volume the pressure of the air in it decreases. Atmospheric air is then at a higher pressure than the air in the

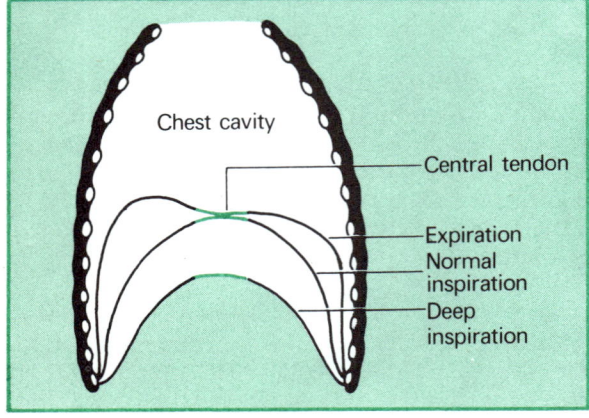

Fig. 7.10 The positions of the diaphragm in expiration, normal inspiration, and deep inspiration.

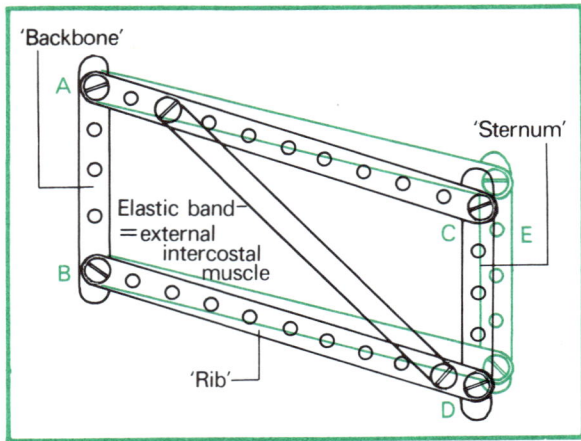

Fig. 7.11 Meccano model to illustrate the operation of the external intercostal muscles.

lungs, and the difference in pressures forces air into the lungs.

At the end of inspiration the lungs are again filled with air at atmospheric pressure. Expiration causes the volume of the lungs to decrease, and the pressure of the air in them rises above atmospheric pressure. Some of the air in the lungs is therefore forced out.

During a quiet inspiration a man takes in about 0.5 litres of air at each breath. This is diluted by the 1.5 litres of air which are already present in the lungs, and lung air therefore always contains a lower proportion of oxygen than does atmospheric air. The rate of breathing of an adult man at rest is about 16 breaths per minute, so about 8 litres of air are breathed in and out per minute. During exercise this volume may rise to as much as 70 litres per minute.

Breathing out (expiration)
This is largely a passive process. The rib cage is pulled downwards by gravity; and the lungs, which are very elastic, pull the chest inwards.

During inspiration the stomach, intestines, and liver are pushed down by the contraction of the diaphragm, and this causes the belly (abdominal) muscles to stretch outwards, as shown in Fig. 7.12. During expiration, the abdominal muscles contract, pressing the stomach etc. against the diaphragm, which is stretched back to its uncontracted position ready for the next inspiration.

The air passages and the lungs
Figure 6.11 shows the nasal (air) passage lying above the hard and soft palates and the conchae that warm the incoming air. In the throat the windpipe (trachea), which leads to the lungs, lies in front of the oesophagus.

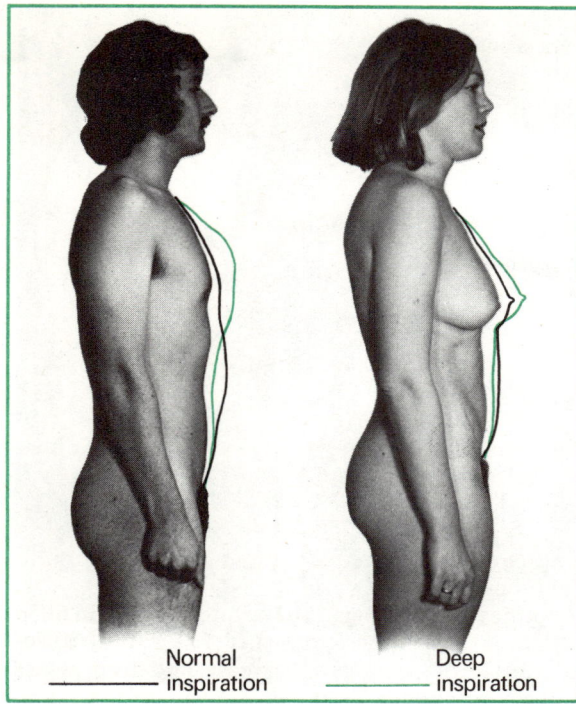

Fig. 7.12 Movement of the chest and abdomen in normal and deep respiration.

Figure 7.14 shows the chest with the skin removed (left side) and with the ribs removed (right side). Notice how the lungs fill the chest and are lined by a thin skin, the pleura. Figure 7.15 shows

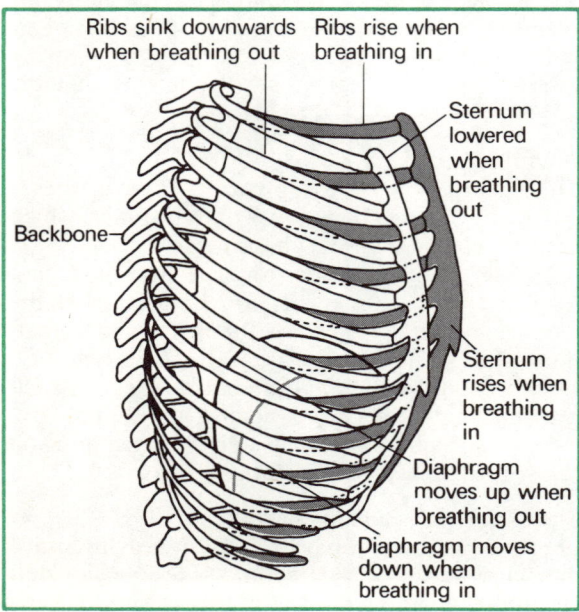

Fig. 7.13 Movements of the rib cage and diaphragm in inspiration.

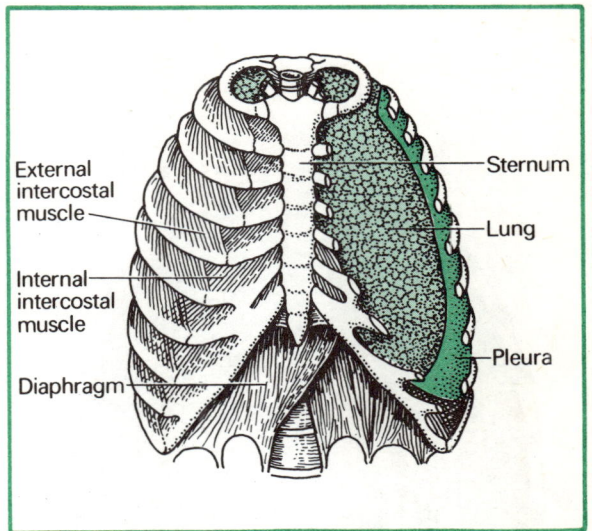

Fig. 7.14 The human chest and diaphragm. On the right the ribs and intercostal muscles are shown removed, revealing the lungs.

more detail of the contents of the chest. The heart is in the centre, resting on the diaphragm and surrounded by the lungs into which lead the branches of the trachea, the **bronchi** and the **bronchioles**. These are all made of cartilage (gristle).

The bronchial tree
It appears from Fig. 7.15 that the bronchioles disappear into the lung tissue, but this is not really so, as can be seen from Figs. 7.16(a) and (b). These are corrosion preparations of human lungs. Liquid plastic was poured into the lungs of two corpses so that it penetrated to every part. The lung tissue was then dissolved away, leaving the casts shown in these photographs. In Fig. 7.16(b) the endmost portions of the branches have been cut off, making the branching of the bronchi and bronchioles, the so-called bronchial tree, easier to see.

Alveolar ducts and alveoli
Careful inspection of Fig. 7.16 shows that the end branches of the bronchi, the **terminal bronchioles**, whose diameter is about 1 mm, lead to blindly-ending sacs. Details of these are seen in Fig. 7.17. They are called **alveolar ducts** and their walls bulge out into the multitudes of cup-shaped **alveoli**.

Alveoli   These are the ultimate ends of the lungs. They are thin walled cavities with one side open to the air entering the alveolar ducts from the bronchioles (Fig. 7.18). In an adult man there are about 300 million alveoli and their total surface area is about 80 m^2, approximately the same as that of a

73

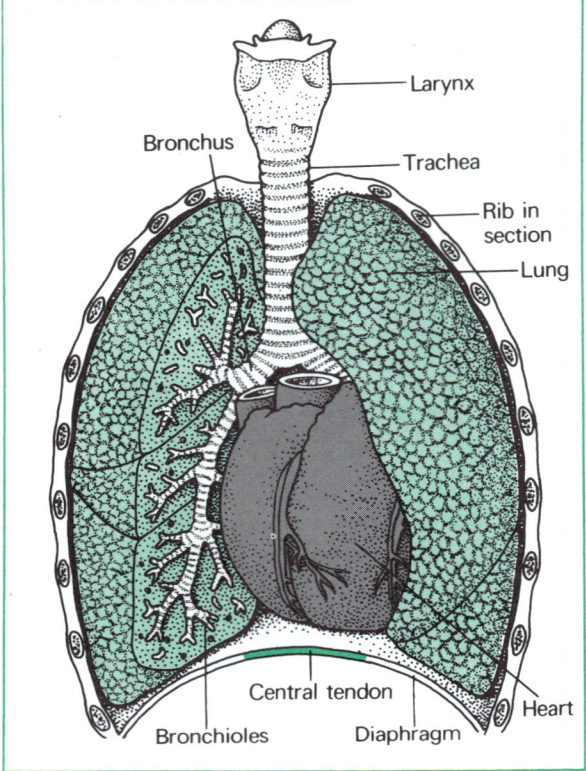

Fig. 7.15 Some of the contents of the human thorax.

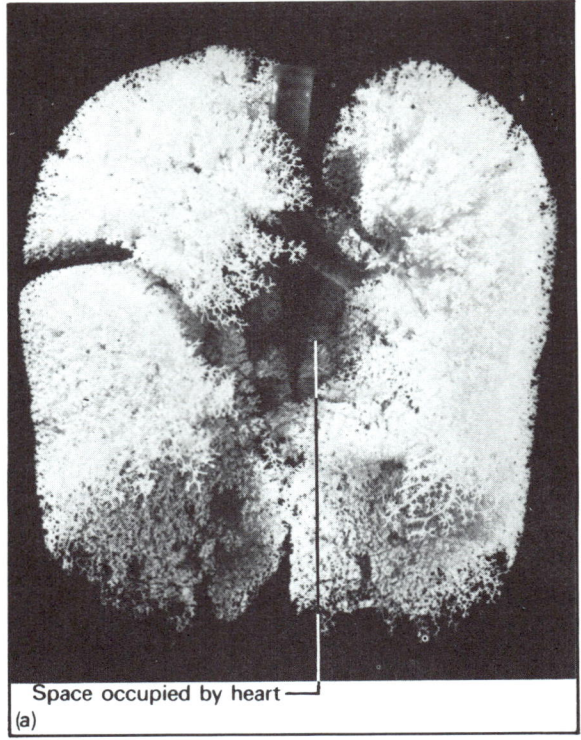

Space occupied by heart
(a)

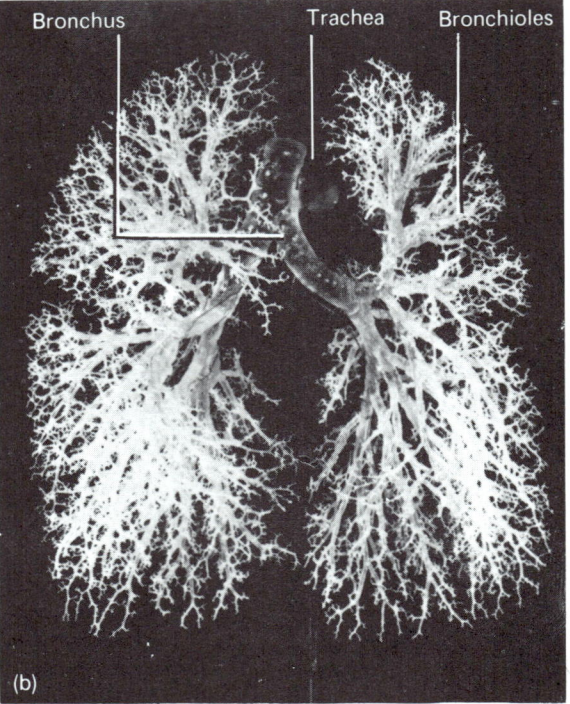

Bronchus   Trachea   Bronchioles

(b)

Fig. 7.16(a) Corrosion preparation of a human lung.
(b) Corrosion preparation of a human bronchial tree.

tennis court. The surface area of the skin of an adult man is about 1.8 m², and comparison of these figures shows how well the lungs are adapted to provide a large surface area across which sufficient oxygen can diffuse to meet his respiratory needs.

Note in Fig. 7.18 the many blood capillaries containing red blood cells.

Compare this drawing with Fig. 7.19 which is an electron micrograph of mammalian lung tissue. Both figures show how the **alveolar wall cells** are intimately connected with the **alveolar capillaries**. These contain oxygen-deficient blood that has just arrived from the heart via the **pulmonary artery**. They convey blood to the branches of the **pulmonary vein** which takes it back to the left auricle of the heart (see p. 86).

Large parts of the surfaces of these capillaries are exposed to the alveolar air. They are separated from it by a part of the alveolar wall cell which is so thin that it is only in recent years that it has been definitely established that it exists. It can be seen clearly in Fig. 7.20. The total area of capillaries exposed to the alveolar air in the human lungs is about 100 m².

Uptake of oxygen in the alveoli

Each alveolus is about 40 μm in diameter and at any one time there are some 2300 red blood cells in the capillaries in its wall. The red blood pigment in these has a strong affinity for oxygen, and a strong diffusion gradient (see p. 43) therefore exists between the alveoli and the red blood cells. Oxygen is present in high concentration in the alveolar air space; in the red blood cells its concentration is nil. It therefore diffuses across the very thin alveolar wall cell, across the very thin capillary wall, into the blood plasma in the capillary and so into the red blood cells. Trace this path on Fig. 7.20. The total thickness of the barrier between the alveolus and the red blood cell is only 0.2 μm, and it therefore offers little obstruction to diffusion. (How does this compare with the diffusion path shown in Fig. 5.7?).

Inside the red blood cells the oxygen combines with haemoglobin to form **oxyhaemoglobin**. The red blood cells are instantly whisked off on their way to the heart, to be replaced by more red blood cells which in turn contain no oxygen. Thus the red blood cells form a 'sink' for oxygen, and a constant diffusion gradient is maintained causing oxygen to pass continuously from the alveolar air spaces into the red blood cells.

Pulmonary artery and pulmonary vein

Blood is delivered to the lungs from the heart in the pulmonary artery. After its journey through the lungs it returns to the heart in the pulmonary vein (see p. 86). Fig. 7.21 is a corrosion preparation of the human pulmonary artery and its branches (shown in white) and the pulmonary vein and its branches (shown in grey). The ways in which these vessels branch corresponds so closely to the branching of the bronchial tree that the photograph might be mistaken for the bronchial tree (Fig. 7.16(b)). It shows none of the lung capillaries, but it is clear from it that all the gas-exchanging surfaces of the lung are served by branches of the pulmonary artery and the pulmonary vein.

The elastic nature of the lung

Figure 7.18 shows what appear to be tufts of hairs arising from the walls of some of the alveoli. These are very fine **elastic protein fibres** which form a basket-work round the alveolar walls. They connect with the **pleura**, the thin skin surrounding the lung shown in Figs. 7.14 and 7.17.

Lining the inside wall of the chest is a second pleura, and the two pleurae stick to each other because there is a thin film of liquid between them. During inspiration the chest wall expands and because the lungs stick to it via the pleurae and are stretchable, the lungs expand as well. During expiration the elastic pull which the lungs exert on the chest wall is the main force causing the chest to sink down to its expiratory position.

If air is introduced between the pleurae (for example, when a person is shot in the chest) the adhesion between the pleurae is lost, as it is when air is introduced under a rubber sucker stuck on to a glass plate. The lung now shrinks away from the chest wall (it collapses) and is functionless.

Transport of oxygen

Oxygen is transported in the form of oxyhaemoglobin by **red blood cells**. These are shaped like biconcave discs (Fig. 7.22). They are very small, on average 7.2 μm in diameter and 1.5 μm thick.

Unlike all the other cells of the body, red blood cells possess no nucleus or any other organelles. They are thus unable to divide or to repair themselves since they contain no coded instructions concerning the conduct of cell reactions. Their

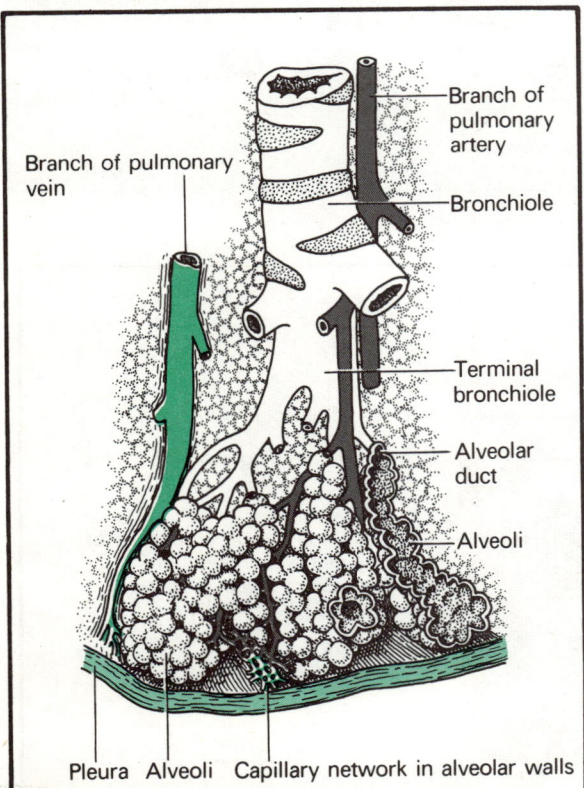

Fig. 7.17 The ultimate branches of the lung and their blood supply.

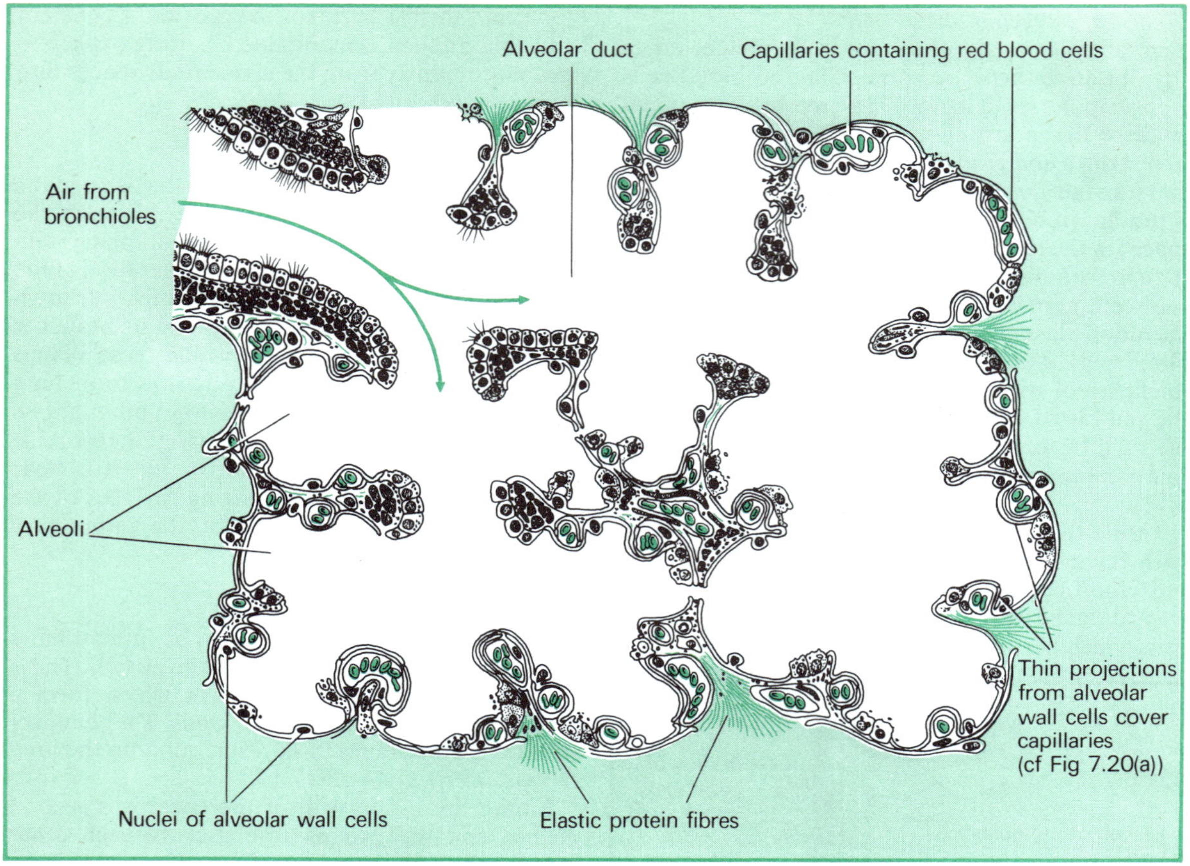

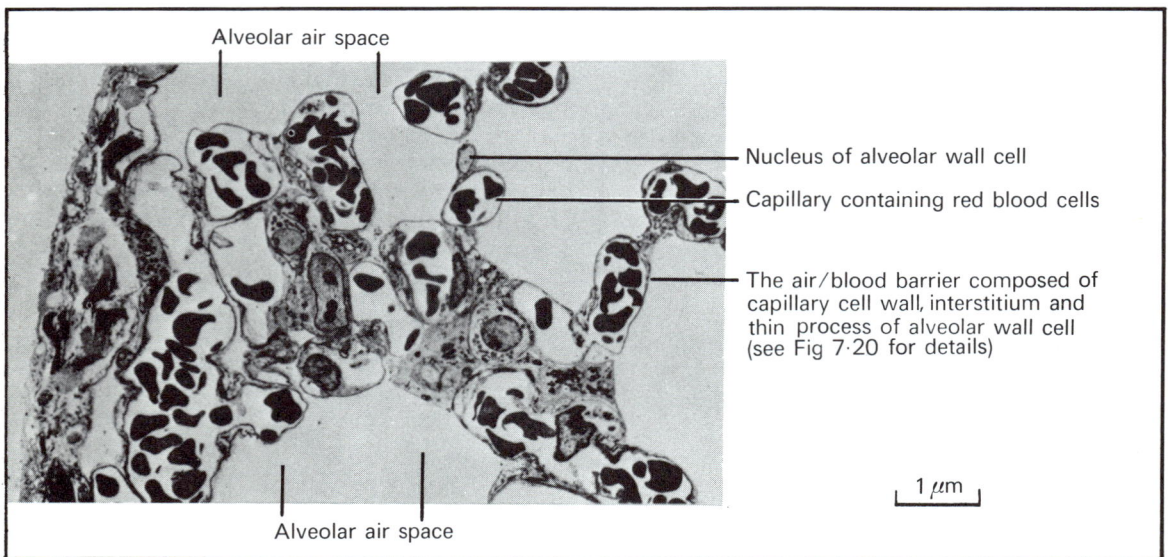

Alveolar duct

Capillaries containing red blood cells

Air from bronchioles

Alveoli

Thin projections from alveolar wall cells cover capillaries (cf Fig 7.20(a))

Nuclei of alveolar wall cells

Elastic protein fibres

Fig. 7.18 Alveolar ducts and alveoli.

Alveolar air space

Nucleus of alveolar wall cell

Capillary containing red blood cells

The air/blood barrier composed of capillary cell wall, interstitium and thin process of alveolar wall cell (see Fig 7·20 for details)

1 μm

Alveolar air space

Fig. 7.19 Electron micrograph of a section through mammalian lung tissue.

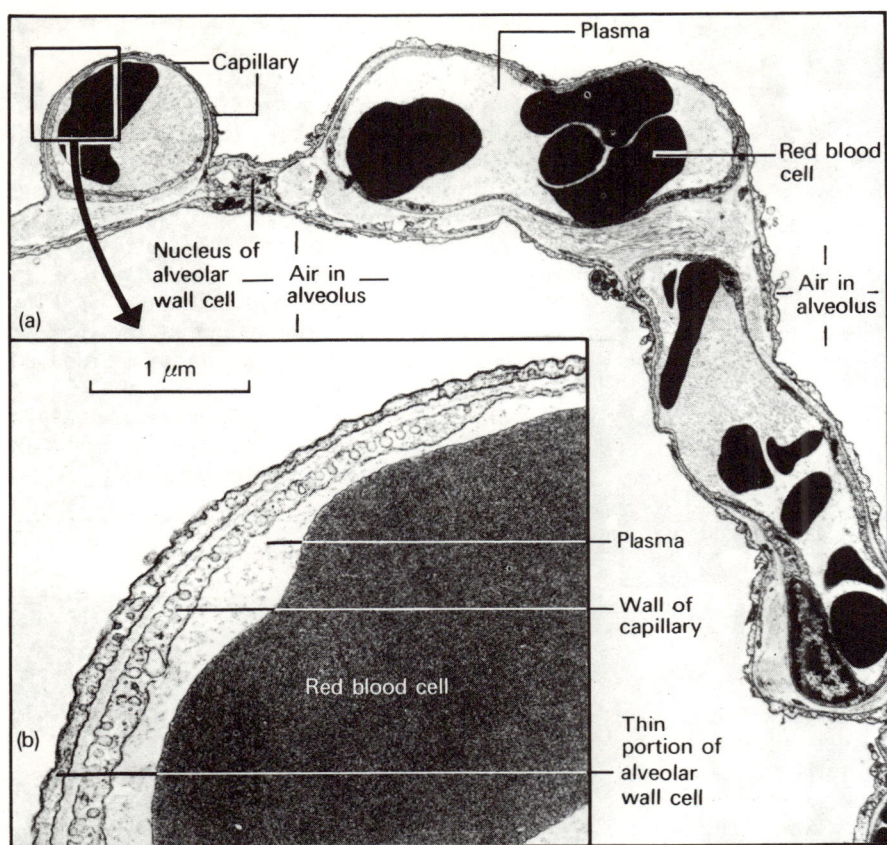

(a)

1 μm

(b)

Capillary

Plasma

Red blood cell

Nucleus of alveolar wall cell

Air in alveolus

Air in alveolus

Plasma

Wall of capillary

Red blood cell

Thin portion of alveolar wall cell

Fig. 7.20 Sections of the lung of a dog showing capillaries and alveolar wall cells. Note in (b) the barrier between oxygen in the alveolar cell and the red blood cell consists of (i) the thin layer of alveolar wall cell; (ii) the wall of the capillary; (iii) blood plasma. Electron micrographs.

cytoplasm looks structureless in electron micrographs. Note the contrast in Fig. 7.23 between them and the capillary wall cells which surround them. These are rich in organelles.

The shape of the red blood cell is well adapted to its function of absorbing oxygen. Its surface area is 20–30% larger than that of a sphere of the same volume, and the total surface area of all the red blood cells in the human body is about 3500 m². The cell is thin, which permits rapid diffusion of oxygen into its centre.

Haemoglobin
95% of the dry mass of red blood cells is due to the pigment haemoglobin. Haemoglobin has the remarkable property of combining readily but loosely with oxygen to form oxyhaemoglobin:

$$\text{haemoglobin} + \text{oxygen} \underset{\text{lack of oxygen in surroundings}}{\overset{\text{plenty of oxygen in surroundings}}{\rightleftharpoons}}$$

$$\text{oxyhaemoglobin}$$

As the equation shows, the association of oxygen with haemoglobin is freely reversible. In the presence of oxygen, oxyhaemoglobin is formed; in the absence of oxygen, oxyhaemoglobin decomposes, liberating oxygen and forming haemoglobin once more.

The left hand side of this equation shows the situation that exists in the alveolar capillary. There is a strong diffusion gradient for oxygen between the alveolar air and haemoglobin in the red blood cells (p. 75).

The right hand side of the equation shows the situation that exists in the respiring tissues. Here red blood cells are passing through capillaries which are surrounded by regions very deficient in oxygen (Fig. 7.23). This causes oxyhaemoglobin to decompose; the oxygen liberated dissolves in the blood plasma and passes along its diffusion gradient into the respiring tissue cells.

Because of the presence of haemoglobin, one volume of blood absorbs sixty times as much oxygen as does one volume of water. The 5–6 litres of blood in an adult man can hold between 1 and 1.2 litres of oxygen, enough to last for five minutes when the body is at rest or for a fraction of a minute during vigorous exercise.

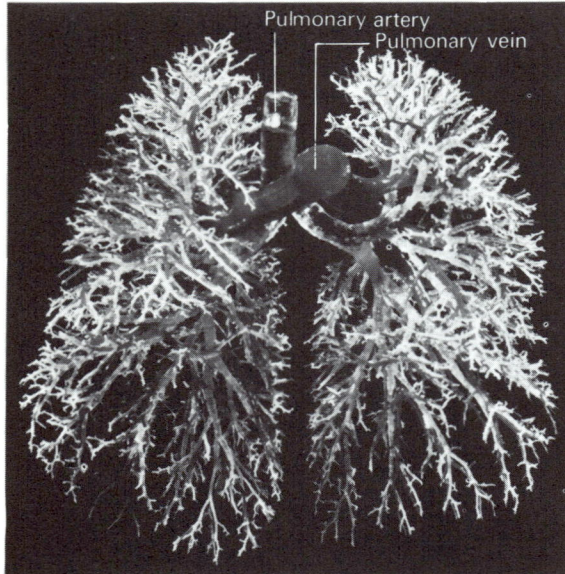

Fig. 7.21 Corrosion preparation of the human pulmonary artery (white) and the pulmonary vein (grey).

Oxygen debt

As body activity increases, for example during exercise, more oxygen is needed, particularly by muscle cells which need more chemical energy. The body responds by breathing deeper and faster, and by speeding up the circulation of the blood. Often, however, this is inadequate to meet the oxygen demands; the body cells then respire anaerobically (equation (5), p. 69), and this gives rise to lactic acid.

This is shown by an experiment summarized in Fig. 7.24. The concentration of lactic acid was measured in the blood of a resting subject. It was about 10 mg/100 cm³. He then pedalled vigorously on a stationary bicycle for two minutes (the period labelled in Fig. 7.24). During this time there was little change in the lactic acid concentration in his blood, but in the first five minutes after exer-

cise ended the lactic acid produced in his previously exercising muscles diffused into his blood stream causing a rapid rise in concentration to 95 mg/100 cm³. Over the next hour, this concentration gradually fell until it almost regained its original level of 10 mg/100 cm³.

Thus during exercise glucose was converted to lactic acid in the muscles by anaerobic respiration. This subsequently disappeared because much of it was oxidised *after the exercise had finished*. In other words the muscles had built up a need for oxygen which could not be met at the time of exercise—an **oxygen debt** which had to be repaid after the exercise finished.

A trained athlete can run a 100 metre race whilst holding his breath. When the oxygen supply in his blood is exhausted, all the energy for his move-

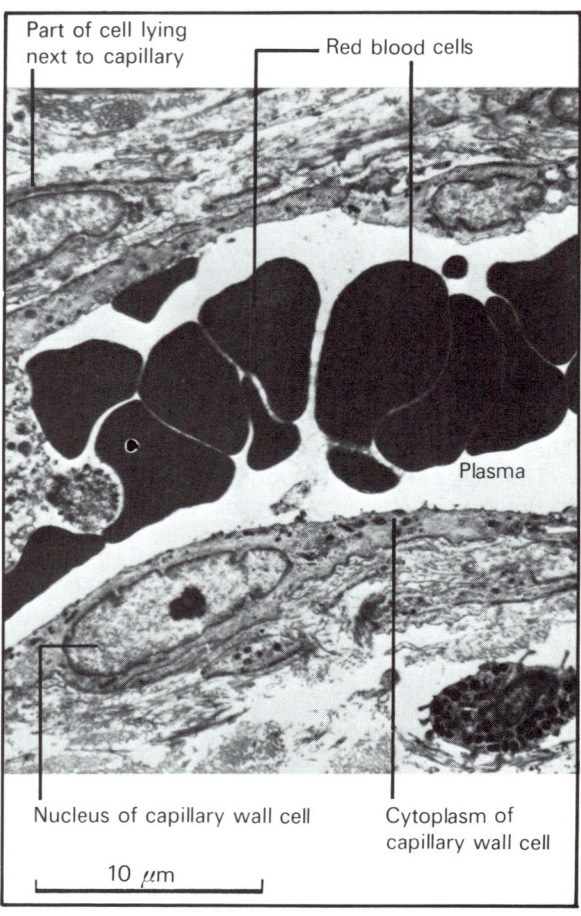

Fig. 7.23 Longitudinal section of a capillary in respiring tissue. The red blood cells (regions of high O_2 concentration) lie very close to the tissue cells (regions of low O_2 concentration). (Electron micrograph).

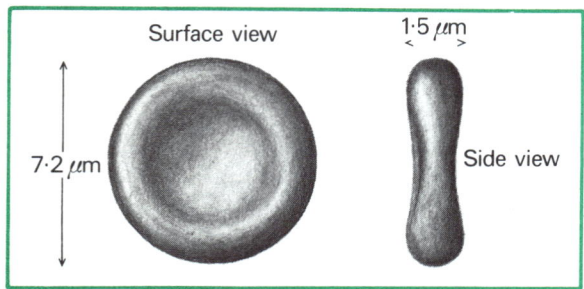

Fig. 7.22 Human red blood cells.

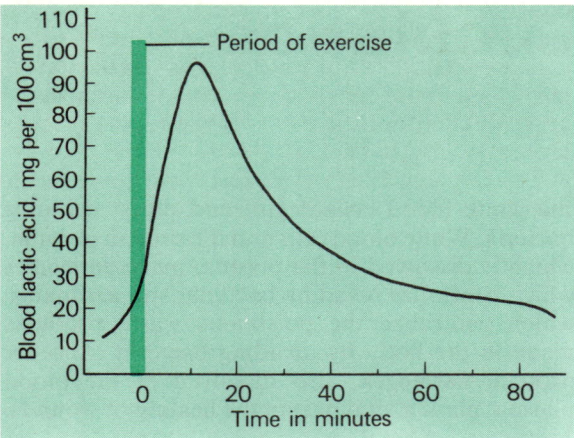

Fig. 7.24 Appearance of lactic acid in the blood after exercise.

ment comes from the anaerobic respiration of glucose to lactic acid.

Respiration in plants

The chemical processes in cell respiration summarized in Fig. 6.4 apply equally to animal and to plant cells. Plant cells (except for the blue-green algae) are eukaryotic and contain mitochondria. However, because plants are much less active organisms than animals, a plant of a given size respires much less rapidly than an animal of comparable size. This difference is related to the basic difference between plants and animals— difference in nutrition, for animals have to move in order to obtain food.

The majority of cells in a tree (the xylem) are dead, and therefore do not respire. Plants possess nothing corresponding to blood or a circulatory system. Their cells are arranged loosely (Fig. 5.4) with spaces between them; oxygen from the air passes through stomata (in leaves and soft green stems) or lenticels (in woody shoots, Figs. 3.3 and 3.4). Since the living cells of leaf and stem are consuming oxygen in respiration, there is a diffusion gradient for oxygen from the air into the stem. Reaching the wet surfaces of the living plant cells, the oxygen dissolves and passes into solution in their cytoplasm.

Stem and root cells are not as loosely packed as leaf cells, but there are always fine intercellular spaces between them (Fig. 1.3). Roots have neither stomata nor lenticels. Young roots obtain oxygen from the soil air (present between the soil particles). This dissolves on the surface cells of the root and diffuses inwards in solution. Many plants die in waterlogged soils, where there is no soil air, because their roots cannot obtain enough oxygen.

A complication: green plants photosynthesize

All living plant cells respire and produce carbon dioxide. Some contain chloroplasts, and these cells also photosynthesize and consume carbon dioxide. If the amount of carbon dioxide consumed is used as a measure of the rate of photosynthesis, therefore, a false value, the *apparent* rate of photosynthesis, will be obtained. This is because some of the photosynthesis will use the carbon dioxide that the leaf cells produce in their respiration:

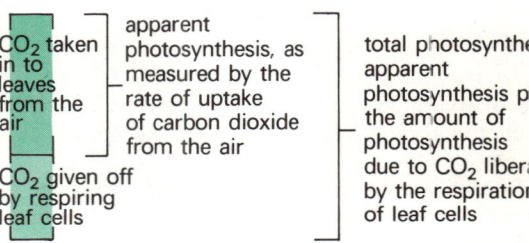

CO_2 taken in to leaves from the air — apparent photosynthesis, as measured by the rate of uptake of carbon dioxide from the air

CO_2 given off by respiring leaf cells

total photosynthesis: apparent photosynthesis plus the amount of photosynthesis due to CO_2 liberated by the respiration of leaf cells

Compensation point

There must be a particular light intensity at which the amount of carbon dioxide given off by the respiring leaf cells of a certain plant exactly equals the amount being used up by them in photosynthesis. This light intensity is called the **compensation point**.

Compensation point varies from species to species, and is an important factor in determining the distribution of plants. Shade-loving plants, such as those that grow on the woodland floor, have lower compensation points than sun-loving plants, such as the trees which form the woodland crown. This means that shade-loving plants need a lower light intensity in order to reach the point where photosynthesis exeeds respiration, the point at which there is a net gain in carbohydrate. Such plants are clearly more effective photosynthesizers, and this could be because they have a more efficient internal leaf structure or a more efficient chemical machinery for converting carbon dioxide into carbohydrate, or both.

8 Blood and the Circulation

Functions of the blood

It is convenient to draw up a table of the more important functions of the blood, but we must remember that it carries them out simultaneously and not separately. Its many functions integrate the life of the whole body.

Blood serves:

1 To transport water, which forms 70 % of the bulk of living cells and is the solvent in which all the chemical reactions of the body take place.

2 In nutrition by carrying dissolved foods in the plasma and oxygen in the red blood cells. The main foods are amino acids, from which proteins are built, and glucose (blood sugar), the source of chemical energy for cells.

3 In respiration by carrying oxygen from the lungs to the tissues and carbon dioxide from the tissues to the lungs.

4 In excretion by carrying waste substances to the kidneys, lungs, and skin, where they are expelled from the body. The main waste substances are **urea** and carbon dioxide. Urea is made in the liver by the deamination of excess amino acids eaten as protein in the food (p. 61).

5 In temperature regulation Blood serves the body like the water-cooling system of an internal combustion engine. It is warmed in regions where the respiration rate is high and is cooled in the skin, where it comes close to the air surrounding the animal. For example, the temperature of blood leaving the liver is 1°C higher than that entering it. Blood also supplies the sweat glands with water, and so contributes to an important cooling mechanism (see p. 169).

6 In transporting hormones These are the chemical coordinators of the body. They are carried from the glands which make them to the cells on which they produce a particular effect (see chapter 15).

7 In protecting the body from infection Most of the white blood cells engulf and digest invading bacteria. White blood cells and the protein **globulin**, which is dissolved in the plasma, make **antibodies** which neutralize invading bacteria, and **antitoxins**, which neutralize the poisonous waste products made in the body by invading bacteria. Another protein, **fibrinogen**, also dissolved in the blood plasma, plays a vital part in the healing of wounds.

Constituents of the blood and their functions

If mammalian blood is allowed to stand in a glass vessel and clotting is prevented, it separates out under the influence of gravity into three layers. These are:
1   **Plasma**. This is the liquid in which the blood cells are suspended and accounts for 55 % of the volume of the blood.
2   **Red blood cells**. These occupy 45 % of the volume of the blood.
3   **White blood cells**. These are scanty and make up only a fraction of one percent of the volume of the blood.

Plasma

91 % of the volume of plasma is water. The dissolved foods and waste substances that it is transporting from one part of the body to another occupy 2 % of its volume. The remaining 7 % is taken up by the **plasma proteins**. These are **albumin**, **globulin**, and **fibrinogen**. Each has its particular function and is a constituent part of the blood: the plasma proteins are not dissolved foods in transit.

Plasma gives rise to tissue fluid (see p. 83).

Red blood cells

Some of the features of red blood cells were considered on p. 78.

There are about six million red blood cells in one cubic millimetre of blood, so an adult man with 5.5 litres of blood in him possesses about 33 million million red cells; they are one of the major constituents of the body.

In an adult they are made in the **red bone marrow** of the ribs, vertebrae, sternum and skull bones; before birth they are also made in the bone marrow of the limb bones. The rate of production, about

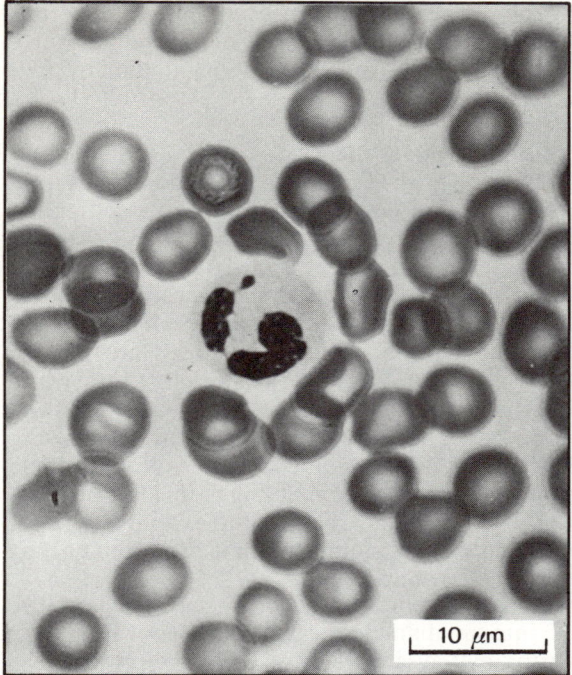

Fig. 8.1 Human blood smear showing red blood cells (no nuclei) and a white cell (polymorph).

a bridge across which the rest of the polymorph flows (Fig. 8.2). Their cytoplasm contains many lysosomes, and their function to engulf and digest bacteria that have invaded the body or other particles such as blood clots. The process is illustrated in Fig. 1.15, route 1. If a foreign particle such as a splinter enters the body, large numbers of poly-

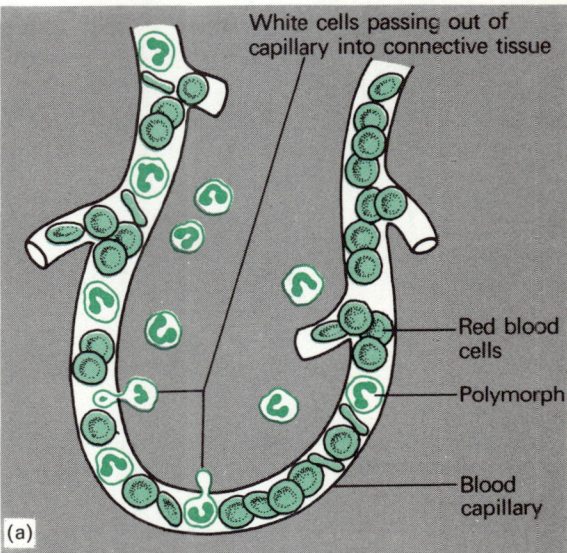

(a)

White cells passing out of capillary into connective tissue

Red blood cells

Polymorph

Blood capillary

one million per second, equals the rate at which they wear out. Their life span is about three months. The frictional wear of passing through capillaries narrower than themselves eventually tears the plasma membrane. This membrane is unusual in having the special property of elasticity; although the red blood cells can be squeezed and have their shapes distorted, when pressure is removed from them they always resume the characteristic biconcave disc shape seen in Fig. 7.22.

White blood cells

There is only one white blood cell to about 600 red blood cells, which makes about 8000 in one cubic millimetre of blood. Their life is short, mostly only a few days, and they too are made in the red bone marrow.

There are at least five different kinds of white blood cell, but Fig. 8.1 shows the commonest, the **polymorph**. The name means 'many shapes' and refers to the extraordinary appearance of the nuclei of these cells.

Whereas red blood cells only leave the circulation if a blood vessel is broken, polymorphs leave it easily and frequently. A small part of the cell 'creeps' between the abutting edges of the cells which make the walls of the smallest veins (venules). This forms

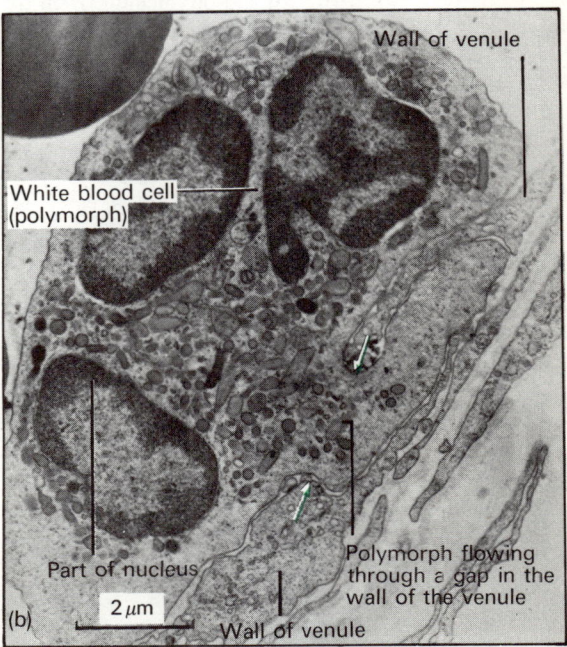

Wall of venule

White blood cell (polymorph)

Part of nucleus

2 μm

(b)

Polymorph flowing through a gap in the wall of the venule

Wall of venule

Fig. 8.2(a) Polymorphs leaving a capillary.
(b) Electron micrograph showing a polymorph poking its way between the edges of a venule (arrows) prior to leaving the circulation.

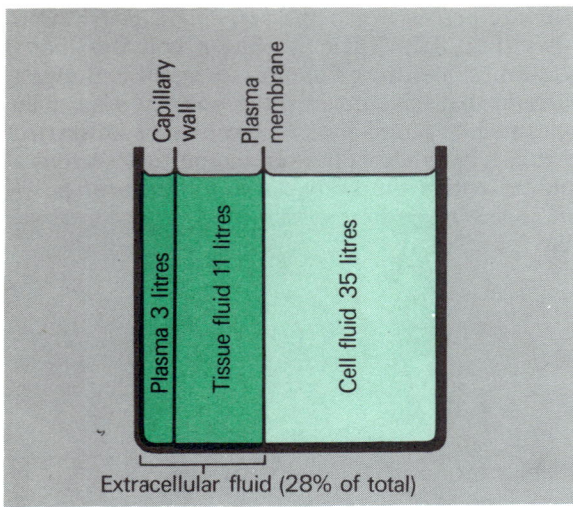

Fig. 8.3 The body fluids of an adult man.

Fig. 8.4 Vertical section of the outer skin of a pig, showing how areolar connective tissue forms a continuum between blood capillaries and skin cells.

morphs congregate at once at the site. The pus of boils consists almost entirely of dead polymorphs, which erode the skin from underneath until the boil bursts.

Blood platelets: clotting

Another defence mechanism provided by the blood is **clotting**. A wound forms an entry point for disease-causing germs, and, since blood is under pressure, a point at which blood can be lost from the circulation. When a wound is received, millions of tiny insoluble fibres of the protein **fibrin** derived from the plasma protein fibrinogen are thrown down over it forming a mesh in which red and white blood cells become trapped, thus staunching the blood loss. The fibrin fibres then contract, drawing the margins of the wound together. Wound tissue grows on top of this, completing the healing process. Blood clotting must be a complex and delicately balanced system, so that fibrin is produced only at the right time and place.

As well as cells, blood contains minute particles, 2–3 μm in diameter, called **platelets**. There are approximately a quarter of a million present in a cubic millimetre of blood. When a wound is

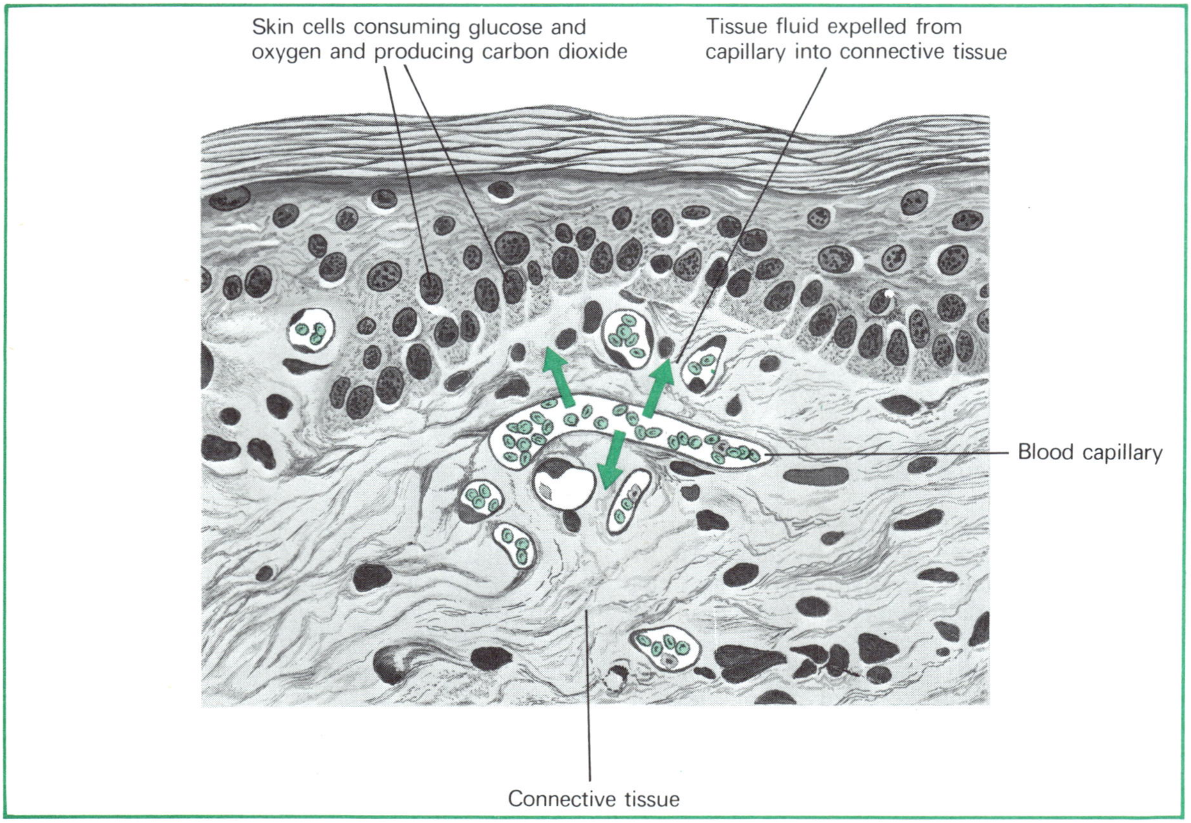

Skin cells consuming glucose and oxygen and producing carbon dioxide

Tissue fluid expelled from capillary into connective tissue

Blood capillary

Connective tissue

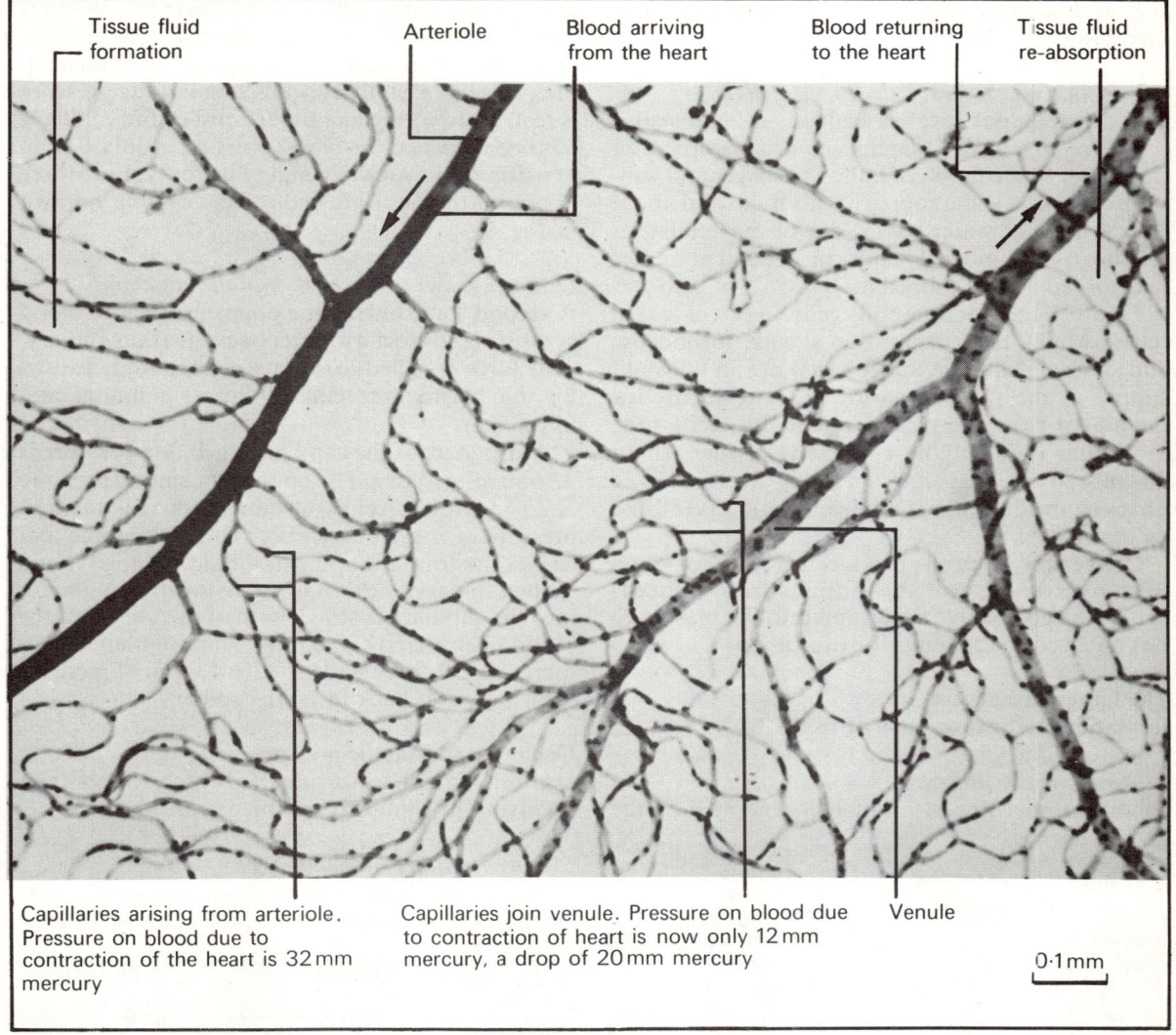

Tissue fluid formation

Arteriole

Blood arriving from the heart

Blood returning to the heart

Tissue fluid re-absorption

Capillaries arising from arteriole. Pressure on blood due to contraction of the heart is 32 mm mercury

Capillaries join venule. Pressure on blood due to contraction of heart is now only 12 mm mercury, a drop of 20 mm mercury

Venule

0·1 mm

Fig. 8.5 Capillaries of the human retina, showing how tissue fluid is formed and re-absorbed.

received, the platelets trigger the clotting mechanism which results in the dissolved plasma protein fibrinogen being converted into insoluble fibres of fibrin at the site of the wound only.

Tissue fluid

Water is able to pass from the plasma through the walls of the blood capillaries, forming **tissue fluid**. In some parts of the circulation it is forced out of the capillaries by the pressure exerted on the blood by the heart. In other parts it is reabsorbed. Dissolved foods and wastes are able to enter and leave

the capillaries in solution, but the plasma proteins and red blood cells are retained. The capillary wall acts as a sieve, allowing small but not large molecules to pass.

Figure 8.3 shows that of the 49 litres of water present in the body of an adult man, 3 litres is part of the plasma, 11 litres is tissue fluid, and 35 litres is cell fluid—the water constituent of the ground substance of cells (Fig. 1.7). Thus although blood circulates in a system of tubes it is not to be thought of as separate from the cells which it serves. The liquids in the plasma, tissue fluid, and cell fluid are continuous, although there are various barriers such as the capillary wall and the plasma membrane, along the route.

Tissue fluid and food supply for the cells
Blood capillaries are embedded in connective tissue (p. 16, Fig. 2.5), and a major constituent of this is the tissue fluid contained in its ground substance. Plasma is the source of this fluid, and about 80 000 litres of water passes out of the capillaries into the tissue fluid in 24 hours in an adult man. It is reabsorbed almost immediately.

Figure 8.4 shows how body cells, in this case skin cells, form by their respiration a 'sink' for glucose and oxygen. Glucose and oxygen are in constant supply in the plasma, so diffusion gradients for them exist from the plasma to the cell, across the 'no man's land' of the connective tissue. About 20 000 g of glucose passes out along this route in 24 hours in an adult man. Of this only 400 g is utilized by the cells.

Conversely, the cells produce carbon dioxide in their respiration, and this diffuses from the cells across the connective tissue and into the plasma to be conveyed to the lungs for expiration.

The forces causing the formation and reabsorption of tissue fluid

Filtration under pressure
Blood is under pressure which is generated by the contraction of the ventricles of the heart. When it arrives at the start of a capillary (Fig. 8.5) its pressure is 32 mm of mercury. In forcing liquid through the permeable capillary walls some of the pressure is lost, and by the time the far end of the capillary has been reached the blood pressure is only 12 mm of mercury, a drop of 20 mm of mercury. Find these figures at the opposite ends of the capillary drawn in Fig. 8.6.

Osmosis draws water back into the capillary
A second force having the opposite effect to blood pressure tends to draw water back into the capillary. This force is called **osmotic pressure** and is caused by the plasma proteins albumin, globulin, and fibrinogen.

At the start of the capillary the blood pressure is 32 mm of mercury. The opposing osmotic pressure is 25 mm of mercury, producing a net outward pressure of 7 mm of mercury. This is the force that causes the formation of tissue fluid. At the far end of the capillary, the osmotic pressure drawing water into the capillary is still 25 mm of mercury, but the blood pressure has fallen to 12 mm of mercury, so there is a net force of $25 - 12 = 13$ mm of mercury drawing water back into the capillary.

Osmosis and osmotic pressure
Osmosis and the pressures to which it gives rise involves diffusion through membranes.

The permeability of membranes
Visking tubing can be used to investigate the permeability of a membrane to molecules of various

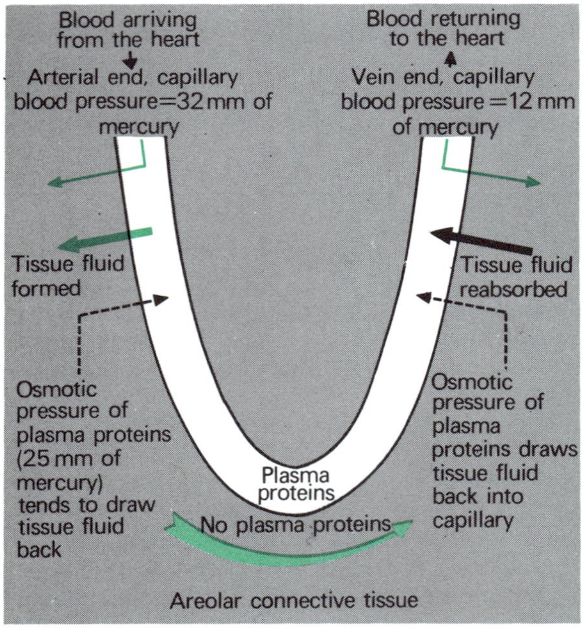

Fig. 8.6 Formation and re-absorption of tissue fluid.

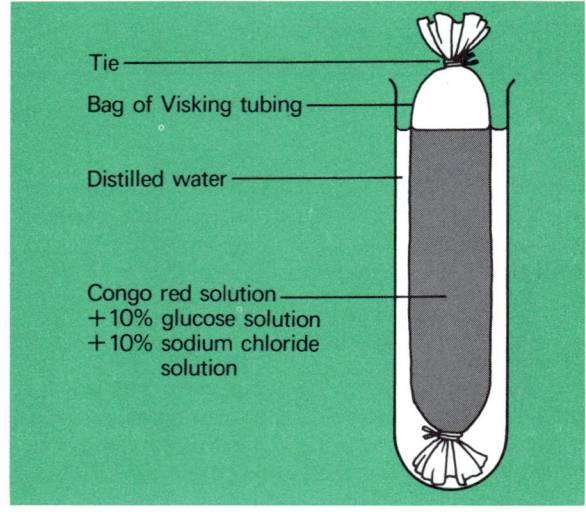

Fig. 8.7 Investigating the permeability of a Visking membrane.

sizes. It is made of cellulose and its molecules are arranged in clusters with minute gaps about 10 nm in diameter between them. The diameter of a water molecule is about 0.3 nm, so it will pass freely through these holes.

If the apparatus shown in Fig. 8.7 is set up with a solution of the dye Congo red, glucose, and sodium chloride inside the Visking tube and pure water outside, it is possible, by periodically testing the water, to see whether any of the enclosed substances is penetrating the membrane. Congo red will stain the water red; glucose can be tested with Benedict's solution; and chloride ions can be detected with silver nitrate solution.

Diffusion gradients for all the substances exist across the membrane and into the water. However, Congo red does not pass out of the membrane at all. Glucose molecules penetrate slowly, but chloride ions penetrate quickly. Since Congo red molecules are the largest of these particles, glucose molecules next largest, and chloride ions the smallest, this indicates that Visking tubing is acting as a sieve, allowing small but not large particles to pass across it. Such a membrane is said to be **partially permeable**. The plasma membrane has a very different composition from Visking tubing (Fig. 1.8), but it too is partially permeable.

Water molecules pass both ways across the Visking tubing and across the plasma membrane, but to show this experimentally it would be neces-sary to use water containing isotopic hydrogen, ^2H_2O, or isotopic oxygen $H_2^{18}O$ (see p. 38).

Demonstration of osmosis
A slight modification of the apparatus shown in Fig. 8.7 demonstrates osmosis and indicates that osmosis is connected with diffusion across a partially permeable membrane.

When the apparatus shown in Fig. 8.8 is assembled and observed over a period, the level of liquid in the capillary tube rises. The only possible cause of this is that water molecules are passing across the Visking tubing into the sucrose solution.

The passage of water molecules across a partially permeable membrane into a solution is called **osmosis**. Osmosis also takes place when a strong solution is separated from a weaker one by a partially permeable membrane. Water molecules then pass from the weaker into the stronger solution.

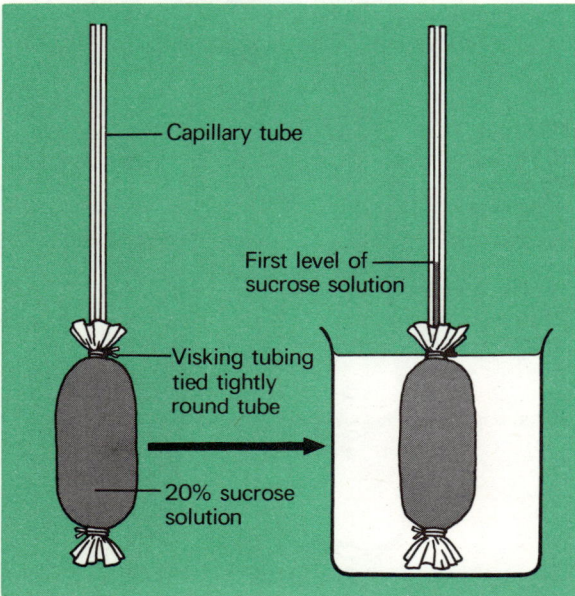

Fig. 8.8 Demonstration of osmosis.

Capillary tube

First level of sucrose solution

Visking tubing tied tightly round tube

20% sucrose solution

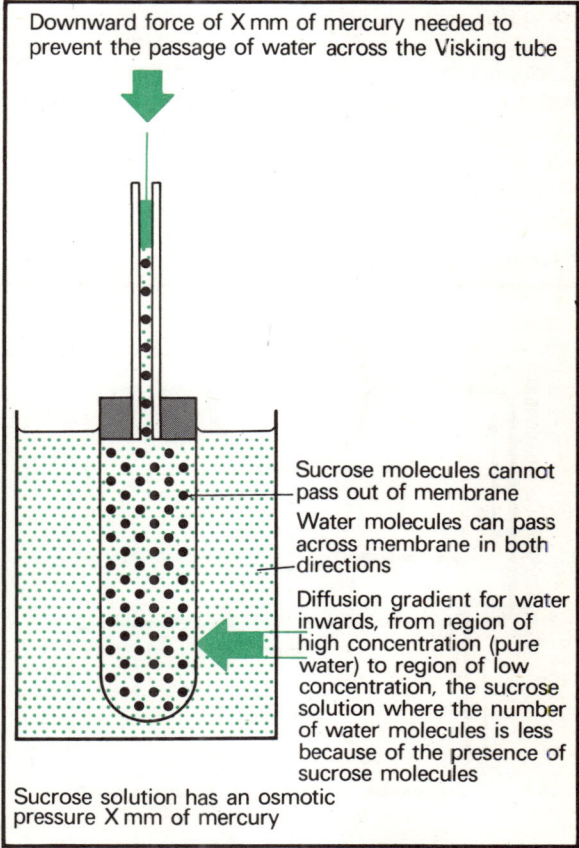

Fig. 8.9 Explanation of osmosis.

Downward force of X mm of mercury needed to prevent the passage of water across the Visking tube

Sucrose molecules cannot pass out of membrane

Water molecules can pass across membrane in both directions

Diffusion gradient for water inwards, from region of high concentration (pure water) to region of low concentration, the sucrose solution where the number of water molecules is less because of the presence of sucrose molecules

Sucrose solution has an osmotic pressure X mm of mercury

Osmotic pressure

Since liquid is being forced up the capillary tube, osmosis is exerting a pressure. Imagine a plunger inserted into the bore of the capillary tube of this apparatus. It would then be possible to prevent the rise of liquid by pushing down on the plunger with a force equal and opposite to that with which water molecules are passing through the Visking membrane. The solution in the Visking tube is said to have an **osmotic pressure** equal to this, the pressure needed to prevent osmosis taking place.

The cause of osmosis

Figure 8.9 shows what is happening in the osmosis experiment. The solution of sucrose inside the Visking tubing contains less water molecules per unit volume than does the pure water outside it, for some of the space in the sucrose solution is taken up by sucrose molecules. If the membrane were permeable to sucrose molecules, these would diffuse outwards along their diffusion gradient until the concentration of sucrose molecules was uniform throughout the apparatus.

But sucrose molecules are too large to pass through the pores in the Visking tubing. Instead, water molecules pass inwards along their diffusion gradient from the pure water, where they are most concentrated, into the sucrose solution, where they are diluted by sucrose molecules.

The circulation, the heart, and the blood vessels

The double circulation in mammals

Blood is circulated by the heart, which consists of two pumps. The left side of the heart is a pump which receives oxygenated blood from the lungs and expels it to all other parts of the body. This is called the **systemic** circulation. The right side of the heart is a pump which receives blood that has finished its journey round the systemic circulation. It sends it to the lungs for re-oxygenation, and this is called the **pulmonary** circulation (Fig. 8.10).

Details of the circulation

Figure 8.11 is a more detailed plan of the mammalian circulation. It does not attempt to show the

Fig. 8.11 Plan of the mammalian double circulation.

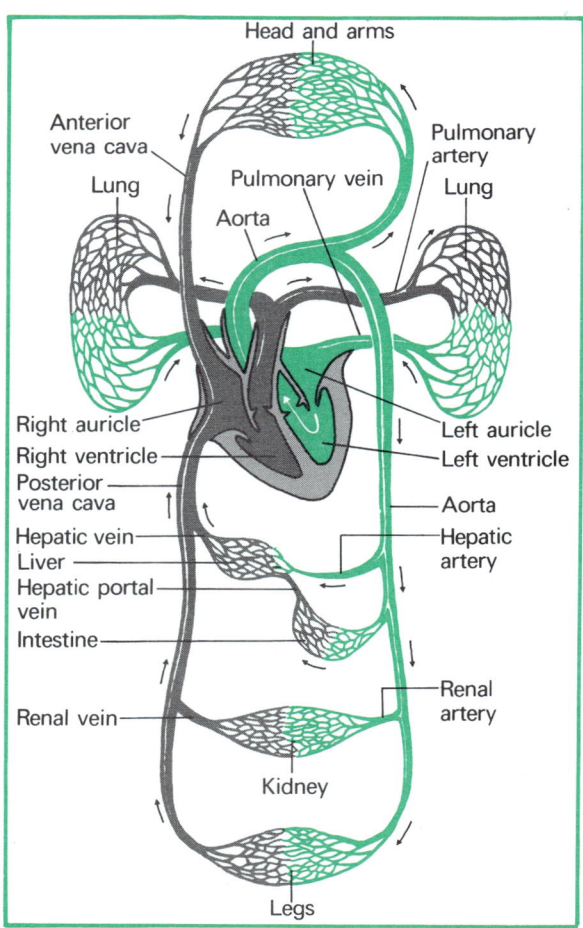

Fig. 8.10 The double circulation in mammals.

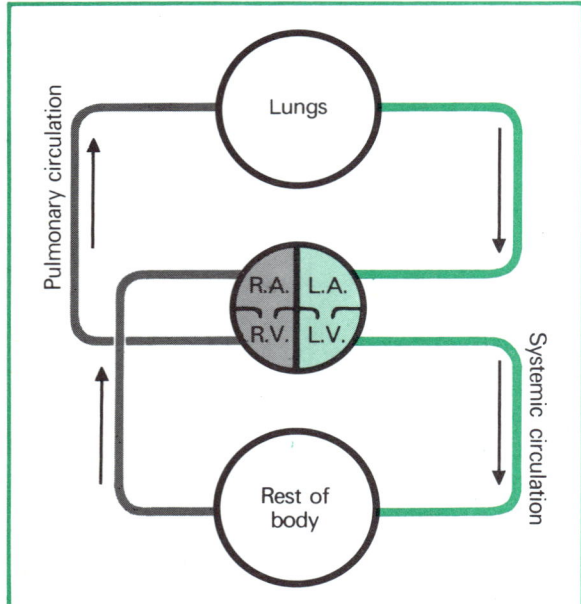

actual positions of the parts of the body but only the correct order in which blood flows through them. (In a similar way, a map of the London Underground shows which station follows which but does not relate the track to the streets or any other topographical features.)

Showing the aorta on the right side and the posterior vena cava on the left makes the diagram easier to follow, though in fact they both lie in the midline, as Fig. 8.12 shows. Figure 8.13 is a corrosion preparation of the arteries of a baby. No veins or capillaries appear in it. Note especially the aorta, the rich blood supply to the heart, kidneys, and brain, and the main vessels to the arms and legs.

Is this specimen shown from the front or from behind?

You should know the names and relative positions of the following blood vessels. Find them in Figs. 8.11 and 8.12 and trace out the systemic and pulmonary circulations.

Fig. 8.13 Corrosion preparation of the arteries of a baby.

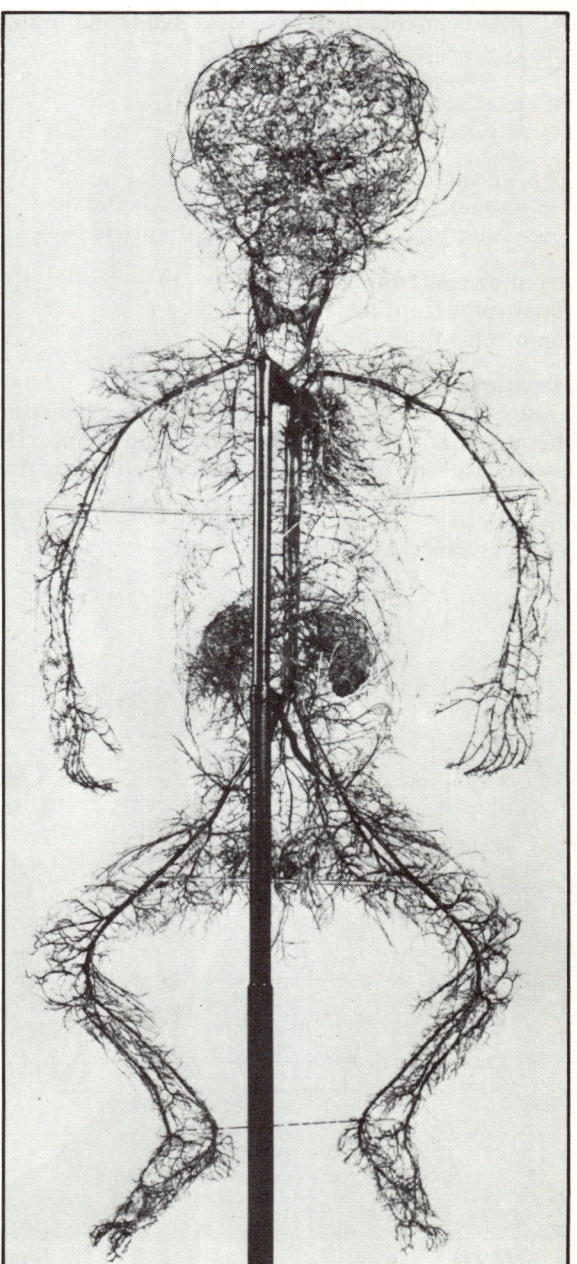

Fig. 8.12 The main arteries and veins of the human body.

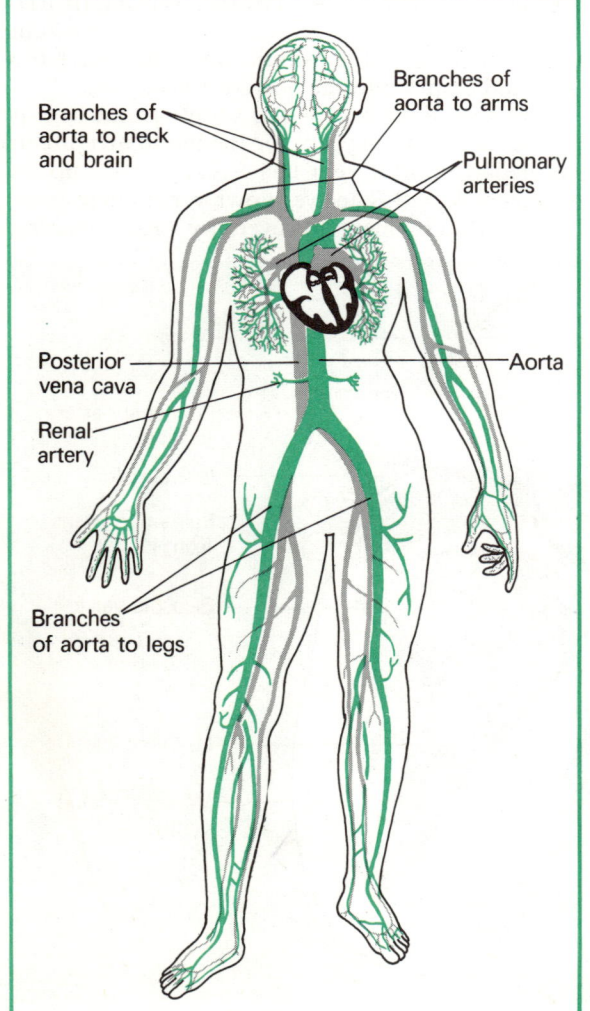

Branches of aorta to neck and brain

Branches of aorta to arms

Pulmonary arteries

Posterior vena cava

Renal artery

Aorta

Branches of aorta to legs

Pulmonary circulation:

⟶ right auricle ⟶ right ventricle ⟶ pulmonary artery ⟶ capillaries of the lung ⟶ pulmonary vein ⟶

Systemic circulation
(a) *Leading from the heart*
⟶ left auricle ⟶ left ventricle ⟶ aorta

⟶ main arteries to neck, head, and brain
⟶ hepatic artery to liver
⟶ arteries to the gut
⟶ renal arteries to the kidneys
⟶ main arteries to the legs

(b) *Leading back to the heart*
⟶ main veins from the brain, head, and neck
⟶ anterior vena cava ⟶ right auricle ⟶

hepatic veins from the liver ⟶
renal veins from the kidneys ⟶
main veins from the legs ⟶

Hepatic portal system
Figure 8.11 shows beds of capillaries lying between the arteries and their corresponding veins. In every case but one a main artery leads into the capillary bed and a corresponding vein leads out of it into the vena cava. The exception is the blood supply to the small intestine and the liver and this has already been mentioned on page 61 and in Fig. 6.24.

On arriving at the small intestine, arterial blood from the heart passes into the extensive capillary bed in the villi (Fig. 6.23). Here it delivers oxygen and receives the digested foods (amino acids, fatty acids, glycerol, and simple sugars) that have been absorbed by the columnar cells. Leaving the villi, the capillaries join to form small veins which in turn join and form the **hepatic portal vein**.

This does not lead to the posterior vena cava but to the liver. Here it divides to form a very extensive capillary bed and during its passage through the liver the concentrations of amino acids and glucose are monitored and adjusted by the liver cells (p. 61). The capillaries re-join to form small veins which in turn join to form the **hepatic veins** that leave the liver and empty into the posterior vena cava.

If this were the only blood supply to the liver it would receive no oxygen, for the blood arriving in the hepatic portal vein has already given up its oxygen in the gut. The liver's own very high respiratory needs are met, however, by blood passing

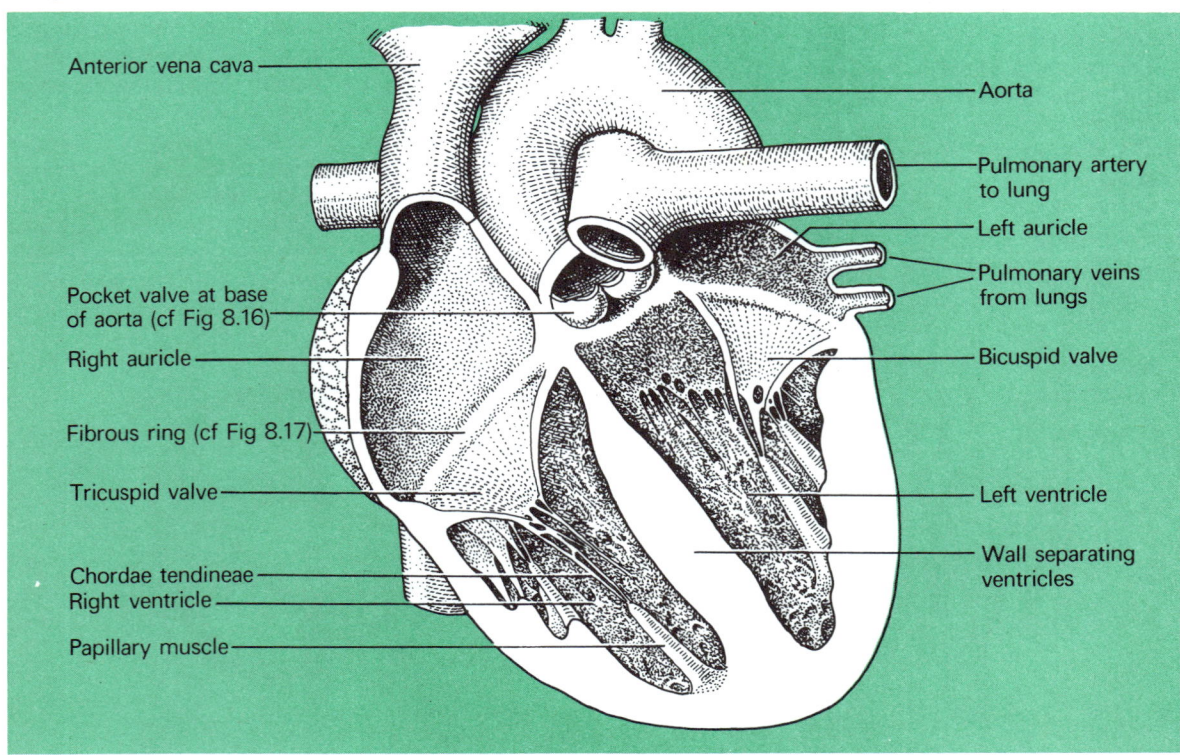

Fig. 8.14 A dissected human heart.

directly to it from the aorta in the **hepatic artery**.

The mammalian heart

The position of the heart in the thorax is shown in Fig. 7.15. Considering the work it does, it is a surprisingly small organ, being at all stages of life about the size of the owner's clenched fist. The mature human heart weighs about 300 g, approximately 0.5 % of the total body mass.

It beats non-stop from the time that it is formed (three weeks after conception) to the moment of death. When an adult man is sitting at rest the rate of beat is about 75 per minute, some 4500 beats per hour. Each beat expels 120 cm^3 of blood (60 cm^3 from each ventricle), an hourly volume of 540 litres. This is an equivalent amount of work to raising a mass of 100 kg to a height of 3.5 metres.

Figure 8.14 shows that it is divided into four chambers. The upper right and left **auricles** are thin walled, and their job is to squeeze blood into the ventricles. The lower right and left **ventricles** have much thicker walls, that of the left being three times thicker than that of the right. This is related to the fact that the left ventricle has to pump blood round the entire body excepting the lungs, whereas the right ventricle has the smaller task of pumping blood through the lungs, which immediately surround the heart.

The course followed by blood as it is pumped through the left (systemic) and the right (pulmonary) sides is shown in Fig. 8.15. Check the great blood vessels entering and leaving the heart in Fig. 8.14.

Heart valves

Blood must flow only one way through each side of

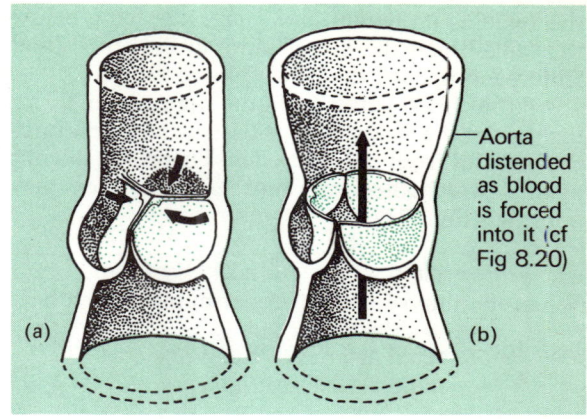

Fig. 8.16 Aortic valve (a) closed (b) during ejection of blood.

the heart, and this is ensured by the presence of valves. No valves guard the entrances to the auricles. The left auricle and ventricle are separated by the **bicuspid (two-flap) valve**, and the right auricle and ventricle by the **tricuspid (three-flap) valve**. These are thin sheets of tough skin which are flattened against the ventricle walls when blood enters from the auricles. When the ventricles contract the blood pushes the valve flaps across and separates the auricles from the ventricles. They balloon up into the auricles, but as blood cannot reflux past them it has to leave the heart through the pulmonary artery and the aorta. The bi- and tricuspid valves are held in place by tendons (the **heart strings** or **chordae tendineae**, Fig. 8.14) which are pulled downwards by the cone-shaped **papillary muscles**.

The aorta and pulmonary artery are supplied with '**pocket**' **valves**. These are simple flaps of skin which are pressed against the artery wall when blood is forced past (Fig. 8.16). When the blood in

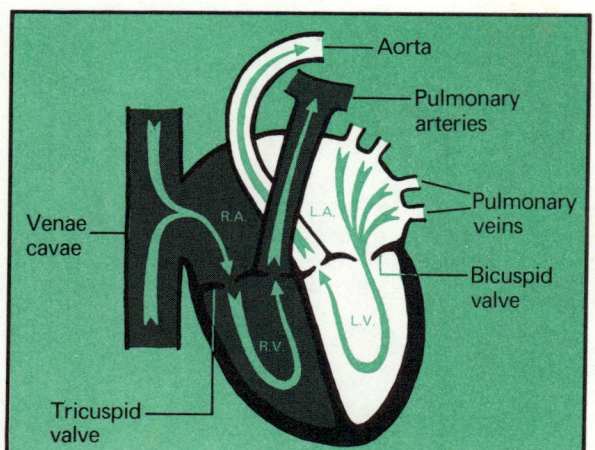

Fig. 8.15 Flow of blood through the mammalian heart.

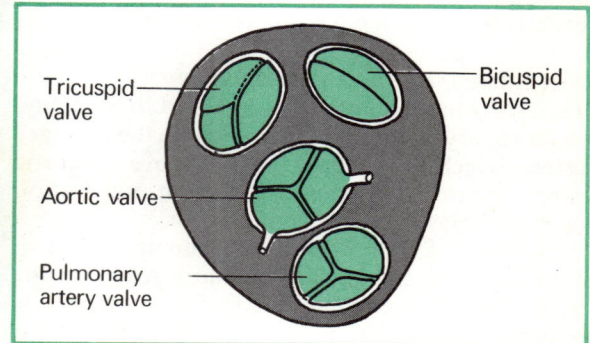

Fig. 8.17 The heart valves seen from above, lying in the sheet of fibrous tissue that separates the auricles from the ventricles. (compare with Fig. 8.14).

the vessel is no longer under pressure from below they fill, block the vessel, and so prevent blood from passing back the way it has come.

All four heart valves lie in the same plane, forming part of the tough fibrous tissue which separates the auricles from the ventricles. This is the anchor against which the heart muscles pull when they contract (Fig. 8.17).

The heart cycle
When the rate of beat is 75 per minute, each beat lasts for $\frac{60}{75}$ sec or 0.8 sec (Fig. 8.18).

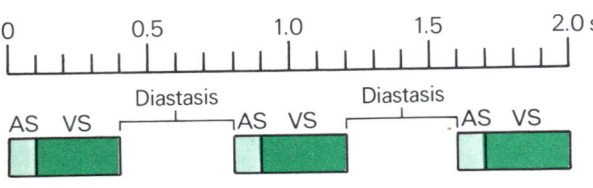

AS - auricular systole
VS - ventricular systole

Fig. 8.18 Events in the heart cycle.

First the auricles contract together (**auricular systole**), and this takes 0.1 sec. The auricles then rest for the remaining 0.7 sec of the heartbeat. Next the ventricles contract together (**ventricular systole**), which takes 0.3 sec, after which they rest for 0.5 sec.

Fig. 8.18 shows that after ventricular systole all parts of the heart rest for 0.4 sec (half the length of the heartbeat), when the next auricular systole follows. This rest is called the **period of diastasis**.

When the heart beats faster (for example during exercise), the contraction times remain about the same but the period of diastasis is shortened, to as little as 0.1 sec when the rate is 180 beats/min.

Nourishing the heart during diastasis
The heart muscle has its own extensive blood supply, the coronary circulation. It arises from the **coronary artery**, which leaves the aorta just above the aortic valve. Figure 8.19 shows a corrosion preparation of the arteries in the heart of a horse.

Half the mass of the heart is made up of its own arteries, veins, and capillaries. During exercise one litre of blood, one sixth of the total body supply, is pumped through the human coronary circulation every minute. But when the heart muscle has contracted, blood is squeezed out of all its capillaries and the period of diastasis is essential for the heart

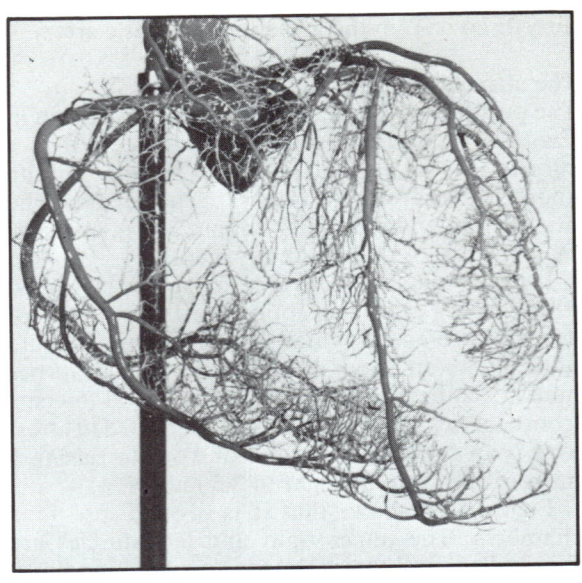

Fig. 8.19 Corrosion preparation of the arteries in the heart of a horse.

muscles to be supplied with oxygen, glucose, and amino acids, and for their waste products (mainly carbon dioxide) to be removed.

Appearance of the heart during its cycle
Figure 8.20 shows the human heart (a) during auricular systole (b) during ventricular systole. Notice how much the aorta is stretched during ventricular systole. What other changes in appearance can you correlate with the events in the heart cycle?

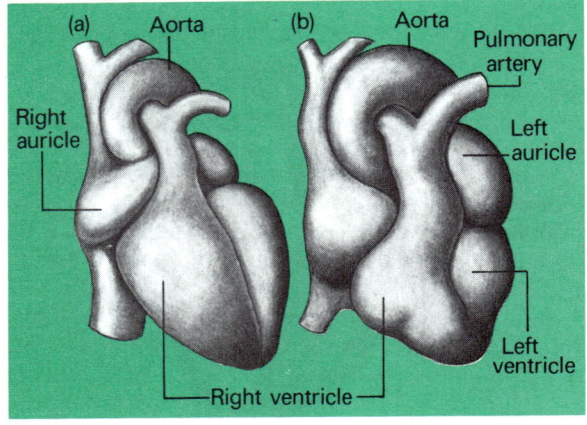

Fig. 8.20 Appearance of the human heart (a) before and (b) after contraction of the ventricles (note the change in the aorta).

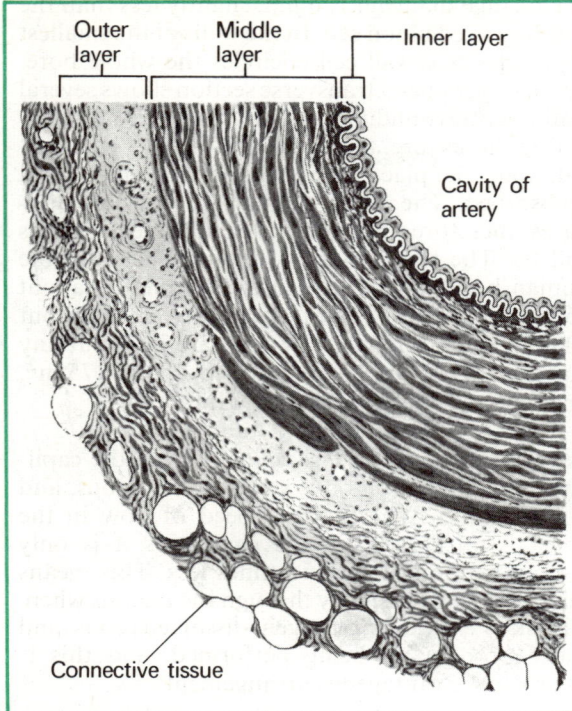

Fig. 8.21 Transverse section of part of a human artery.

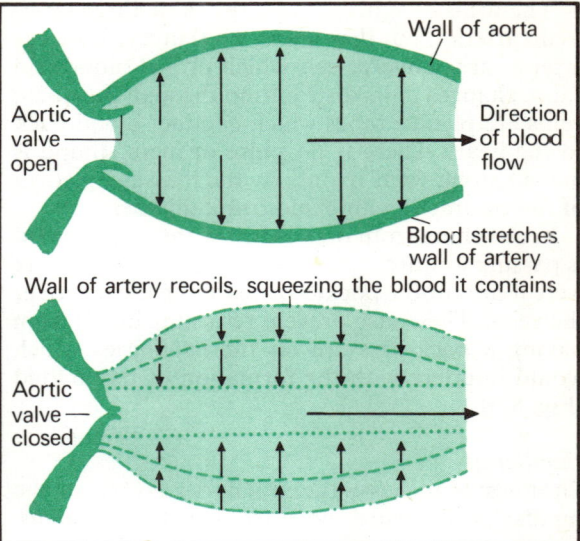

Fig. 8.22 Diagram to show how blood forced into an artery by contraction of the heart stretches the artery wall, which then recoils.

The Blood vessels

Arteries

Vessels in which blood is moving away from the heart are called **arteries**. Most arteries arise from the aorta and contain **oxygenated blood** (that is, blood whose red cells contain oxyhaemoglobin). There are two exceptions: the pulmonary artery, leading from the right ventricle, contains deoxygenated blood; so does the umbilical artery leading to the placenta from the developing baby (p. 125).

Structure of the artery wall: blood pressure

The walls of arteries are made of three layers, the inner, middle, and outer layers. The inner layer is composed of elastic protein fibres. When there is no blood in the vessel, as in Fig. 8.21, it crinkles. The middle layer contains elastic protein fibres but also muscle cells arranged round the circumference of the arterial tube.

Blood is forced into the aorta, which is already full, by the left ventricle. To accommodate the extra blood, the walls of the aorta stretch (Fig. 8.22). Because of the stretched elastic protein fibres in its inner and middle layers, the aorta recoils, squeezing the blood and forcing it onwards. After each ventricular systole a wave of stretch and recoil passes down the whole arterial system. This is called the **pulse**, and can be felt in arteries that lie near the surface of the body, for example at the wrist. The pumping action of the heart is thus supplemented by the squeezing action of the arterial walls.

The muscle cells in the walls of the arteries are served by nerves which can cause them to contract. This narrows the bore of the artery, slowing the flow of blood through it, or perhaps stopping it altogether. By this means, wholesale changes can be made in the distribution of the blood, according to the body's current needs. After a meal, when digestion is proceeding, the arteries leading to the gut are relaxed, their bores are wide, and much blood flows through them to collect the dissolved foods from the villi. On the other hand, if a person is running a race the blood supply to the gut is all but cut off and a higher proportion flows to the heart, to the muscles of the limbs, and to the trunk. The only organ to receive a steady, unaltering quantity of blood no matter what the body's activity, is the brain.

Veins

Vessels in which blood is moving back to the heart are called **veins**. In most veins the blood is deoxygenated (for it has just left the respiring tissues), but in the pulmonary veins (leading from the lungs to the heart) and the umbilical veins (Fig. 12.22) it is oxygenated.

The walls of veins contain the same three layers as do arteries, but they are much thinner. Whereas arteries are active vessels which propel blood and cause changes in its distribution through the body, veins are passive tubes which conduct blood back to the heart. There is no pulse in them. Blood is moved along veins mainly by the massaging effect of the contracting limb and trunk muscles.

In returning from organs below the heart, blood is passing upwards and gravity would make it sink were it not for the pocket valves along the length of the veins. These also prevent returning blood from having a back-pressure on the capillaries which would interfere with the formation of tissue fluid (Fig. 8.6).

Capillaries

The finest branches of the circulatory system are the **capillaries**. The word comes from the Latin *capillus*, meaning a hair, but they are much finer than hairs.

An average diameter is 6 μm, slightly less than the width of a red blood cell. In the walls of the smallest capillaries one wall cell encircles the whole bore, but in larger ones a transverse section shows several wall cells surrounding the bore (Fig. 8.23).

Capillaries are the 'business end' of the circulation—the place where tissue fluid is made and reabsorbed. The larger the surface area of capillaries, therefore, the more efficient these processes will be. The total length of the capillaries in the human body, if laid end to end, would be about 96 000 kilometres. Their total surface area is about 700 m², and although only 25% are in use at any one time, these still have a surface area of 175 m², one hundred times that of the adult body.

Blood flows slowly in the capillaries In the capillaries the flow of blood is steady, continuous, and non-pulsatile. Whereas the speed of flow in the aorta is 40 cm/sec, in the capillaries it is only 0.4 mm/sec, one thousand times less. This means that blood moves slowly through the regions where its task of exchanging oxygen, dissolved foods, and waste products is being performed, and this is clearly an advantageous arrangement.

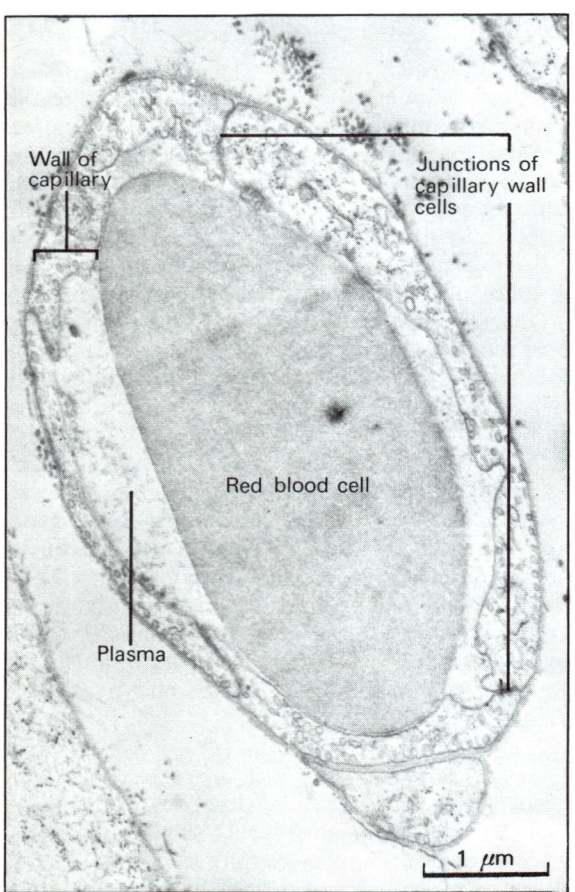

Fig. 8.23 Transverse section of a human capillary containing a red blood cell. (Electron micrograph.)

9 Soil

Soil is formed from the weathering of rocks. Three of the commonest rocks are **sandstone**, **limestone**, and **granite**. Granite is solidified material from the molten interior of the earth.

Weathering processes

The earth and the moon are both approximately 4500 million years old. Although the rocks on the surface of the moon are more or less in the condition in which they were formed, nowhere on the earth's surface is there any trace of the original rock which composed our newly-formed planet. The oldest rock sample so far found (buried deep below other rocks) is about 3500 million years old. The cause of the difference between earth and moon is that the earth has an atmosphere containing both water and oxygen.

Water

Rain weathers limestone because it contains small quantities of dissolved **carbonic acid** (H_2CO_3), formed by the combination of carbon dioxide in the air with water:

$$H_2O + CO_2 \longrightarrow H_2CO_3$$

Limestone is calcium carbonate, $CaCO_3$, and it dissolves in carbonic acid, forming soluble calcium hydrogen carbonate, which is washed (**leached**) out of the soil. This is how limestone caves are formed:

$$H_2CO_3 + CaCO_3 \longrightarrow Ca(HCO_3)_2$$
calcium hydrogen carbonate

Other soluble materials from decomposing rocks are leached in rainwater, perhaps being taken away in rivers to the sea, or being redeposited at lower levels in the soil.

Ice

Most liquids, such as molten metals, contract when they solidify. That is why coins have to be stamped out and not cast. One of the remarkable properties of water is that when it solidifies it *expands*, and 10 cm^3 of water become 11 cm^3 of ice. If water is not allowed to expand when it freezes, it exerts an enormous force on the walls of its container. Steel water pipes can burst for this reason in frosty weather, and water trapped in pockets in

porous rocks splits and fragments the toughest rock material when it freezes.

Oxygen

Large quantities of atmospheric oxygen are consumed in the oxidation of rocks, whose chemical composition is thus changed. The atmosphere is also continuously disturbed, making storms, gales, and hurricanes. These blow away dusty soil and, by hurling the sea at cliffs, cause coastal weathering.

Constituents of the soil

In these and other ways rocks are weathered, producing two of the main constituents of the soil, **sand particles** and **clay particles**.

Together with **humus**, a dark-coloured material formed mainly from the decomposition of cellulose and lignin of plant remains, they form most of the solid matter of the soil. Between the particles there are **air pockets**, and surrounding them films of **soil solution**. The soil is also teeming with a great variety of life, invertebrate, algal, fungal, and bacterial. If a dilute suspension of soil is used to inoculate a culture medium, several million bacterial colonies and tens of thousands of fungal colonies develop from each gramme of soil. Each of these colonies is probably derived from one bacterium or one fungal spore.

Sand

Sand is silicon dioxide, SiO_2, a chemically very unreactive and insoluble substance. It is not porous and water can only be associated with it by forming a film round the surface of sand particles. Being hard and chemically inert, individual grains retain jagged outlines and do not fit closely together. Their presence therefore helps to make the soil more porous to air and to water.

Clay

Clay is a complex chemical substance containing aluminium, magnesium, and silicon. It is formed by the chemical weathering of granite. Individual particles are small, flat, and platelike, fitting closely together and making the soil impervious to air and to water. This can be a disadvantage if the soil contains a large proportion of clay; but clay is an essential component of all fertile soils because of its water-holding and cation-holding powers.

Water-holding power   Soil water is held in the form of films on the surfaces of clay particles. As these are very small they have a large surface area relative to their volume (see p. 50, Fig. 6.7). One sand particle 2 mm in diamter has the same volume as 1000 clay particles each 0.002 mm in diameter, but the clay particles have 1000 times the surface area of the sand particle. This means that much more water is held on the surface of a unit mass of clay than on the surface of the same unit mass of sand.

Water molecules are also able to penetrate into the crystals of which clay particles are composed. They are said to be *imbibed*. Such water is held very strongly and cannot be absorbed by plant roots.

Cation-holding power   Cations are positively-charged ions such as H^+, Ca^{2+}, K^+ and NH_4^+. Many of them are essential plant nutrients.

Clay particles are negatively charged, and they attract and bind cations in the soil solution, preventing them from being leached out of the soil by drainage. Root hairs are able to absorb cations by excreting hydrogen ions 'in exchange' (Fig. 9.1). Thus clay particles form an important reservoir of mineral cations needed for plant growth.

Continuous leaching of soil over long periods may lead to the replacement of calcium, potassium, ammonium, and other cations by hydrogen ions. Such a soil is acid, which prevents the growth of plants in it; it lacks essential cations for plant growth. It is common practice to add quicklime (calcium oxide, CaO) or chalk (calcium carbonate,

$CaCO_3$) to such soils. This neutralizes the acid and exchanges calcium ions for the hydrogen ions on the clay particles.

Humus

Humus is composed of biological materials in all stages of chemical degradation. It is the food of the vast number of micro-organisms in the soil and most of the carbon it contains is eventually returned to the atmosphere as carbon dioxide produced by their respiration. Other materials in it ultimately become ions which can be absorbed by plant roots and used for new plant growth.

Like clay, humus holds water within its particles as well as on their surfaces. Its dark colour means that it absorbs both light and far-red radiation, making the soil warm and favourable for plant growth.

Humus plays an important part in binding together sand and clay particles. This produces a structure which combines the advantages conferred by the large (sand) particles—drainage and aeration—with those provided by the small (clay) particles—water-holding capacity and cation exchange. In a good soil, the particles are not compacted into a hard mass, impenetrable to roots, but are arranged in an open, crumb structure. Such well-balanced soils are called **loams**.

Sand and clay particles are also held together by the growth of large numbers of fungi and by gummy organic materials produced by the bacteria growing in the soil. Roots, especially root hairs, also have a binding effect on soil particles.

Soil air

Most of the spaces between soil crumbs are occupied by soil air. Its extent depends on rainfall and drainage. Nearly all the cells of young roots are alive and use oxygen from the soil air for their respiration.

The total length of roots of quite small herbaceous plants is to be measured in kilometres. For example, one rye plant growing in a container 30 cm square and 550 cm deep developed 6192 km of roots in four months, an average growth of 496 km per day. It was estimated that in addition to this, 816 km of new root hairs were developed each day, being spread out over some 13 million roots. This vast amount of growth requires a good oxygen supply.

Figure 9.2 shows the effect of aeration on the growth of barley roots in nutrient solution. All experimental conditions were identical for the two plants except that air was bubbled through the solution in which the plant on the left grew.

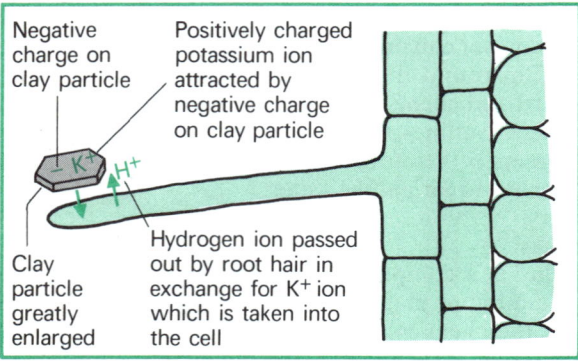

Fig. 9.1 Cation exchange between root hair and clay particle in the soil. H^+ ions are passed out from the root hair cell and K^+ ions (for example) are 'exchanged' for them.

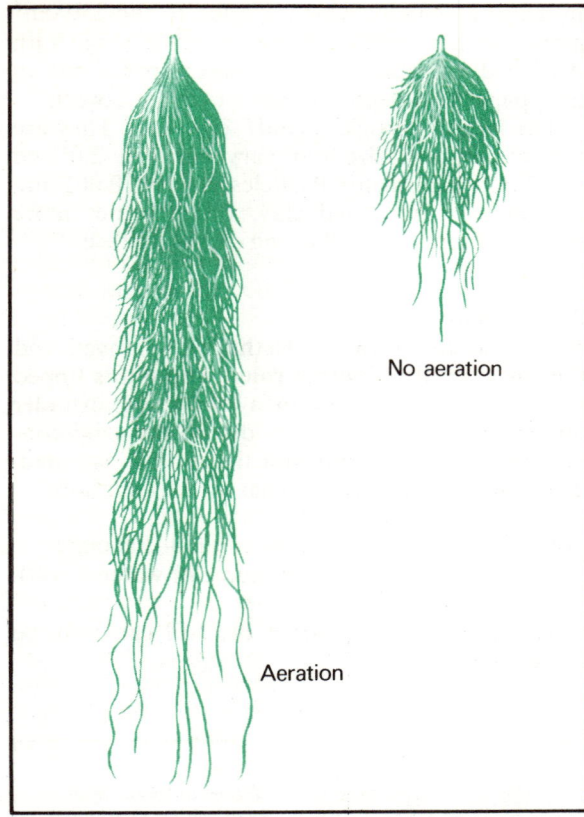

Fig. 9.2 Effect of aeration on the growth of barley roots in nutrient solution.

Oxygen in the soil air is essential for the uptake of ions by plants Water and ions are absorbed into the plant through its root hairs, but soil solution is not taken in as such. Root hairs are able to absorb certain ions which the plant needs for growth in preference to others, and this process of **selective uptake** requires the expenditure of energy derived from cell respiration. It is therefore dependent on oxygen in the soil air.

Figure 9.3 illustrates selective uptake by roots. Barley plants were grown in a made-up solution, and the extent to which various ions were removed from the solution was measured. If the root hairs had been freely permeable to ions there would have been a uniform fall in the concentrations of all the ions present. But in fact some ions (e.g. K^+, NO_3^-) were absorbed strongly. Others were absorbed less strongly, and some (e.g. Mg^{2+}) were actually excreted from the root hairs into the medium, so their final concentration in the solution was higher than their initial concentration.

Selective uptake is brought about by the plasma membranes of the root hair cells. As a result of it some ions are piled up in the root hair cell sap and so in the plant, while others are excluded. This should be contrasted with the behaviour of a non-living membrane, such as Visking tubing (Fig. 8.7). Ions pass across this by diffusion until the ionic concentrations on both sides are equal—there is no question of build up or exclusion.

Interference with respiration of the root hairs slows down selective uptake, which must therefore depend on the supply of chemical energy in the root hair cell. Figure 9.4 summarizes an experiment illustrating this. It shows that when barley roots were starved of oxygen (anaerobic conditions), the rate at which radioactive sulphate ions were absorbed from culture solution dropped considerably. Root hair cells contain large numbers of mitochondria, the main source of ATP (p. 67), exactly what would be expected of cells which consume large amounts of chemical energy.

Poorly aerated soils In well-aerated soils the soil air has a composition similar to that of atmospheric air. This includes 20% oxygen and 0.03% carbon dioxide. In a poorly-aerated soil, however, the oxygen concentration may fall to zero and the carbon dioxide concentration may be as high as 10%. Such conditions are impossible for the majority of plants, which die if the soil becomes permanently waterlogged.

Soil water
Almost three quarters of the surface of the earth is covered by water. It is constantly evaporating from oceans, lakes, and rivers. Huge amounts are lost from the soil by the transpiration of plants (see p. 103).

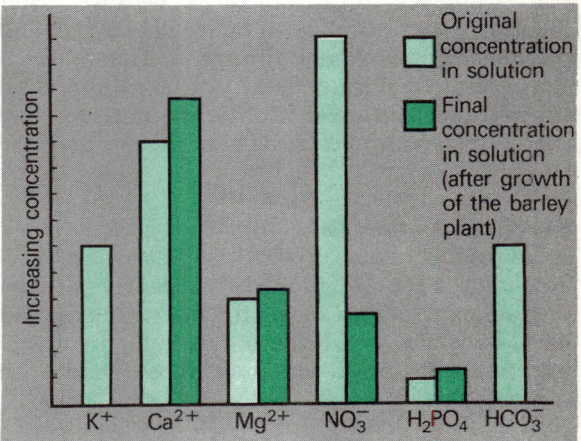

Fig. 9.3 Changes in concentration of various ions in nutrient solution due to the uptake of ions by a barley plant.

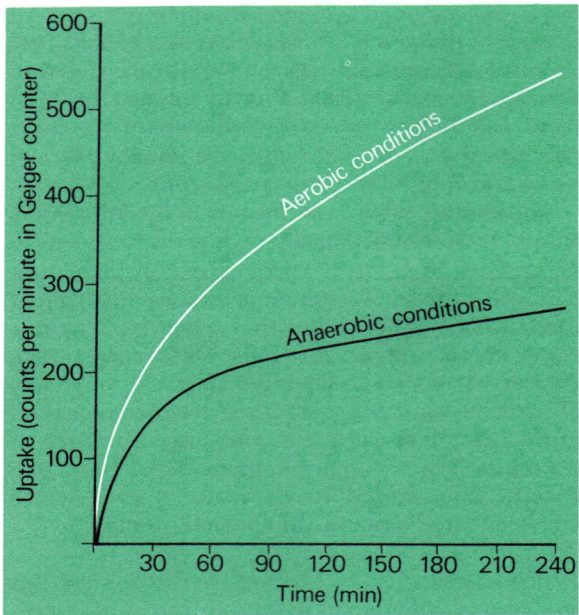

Fig. 9.4 Influence of oxygen on the uptake of radioactive sulphate ions by barley plants.

Water vapour condenses into clouds and falls again on the land and sea as rain and snow. It percolates the soil at a rate which depends on soil porosity and on the water-retaining properties of the soil.

Gravity pulls it downwards; but a far greater force causing it to move is that of **capillary attraction**. The finer spaces between soil particles (**soil capillaries**) attract and become filled with water which is retained. It does not drain away, and is the main supply of water on which roots draw (Fig. 9.5). Capillary attraction exists because water molecules stick strongly to molecules of clay and of humus, and also stick strongly to each other.

The amount of water held by capillary attraction varies from as little as 5 % of the dry mass for very sandy soils to as much as 45 % of the dry mass for clays.

Experiments with soils
A **sandy soil** is one in which sand particles predominate; a **clay soil** is one in which clay particles predominate; and in a **loam** these constituents and humus form a well-balanced mixture. In experimenting with soils, different types should be compared.

1 Constituents
20 cm³ of air-dried soil is placed in a 250 cm³

measuring cylinder which is filled to the 250 cm³ mark, shaken, and allowed to settle (Fig. 9.6). Air-dried soil is soil which has been spread out on newspaper to a depth of about 1 cm for a week.

The largest particles (sand) settle first. They are not of uniform size but vary between 2.0 and 0.002 mm in diameter. Particles less than 0.002 mm in diameter are called clay, which settles more slowly. Humus is left floating on the surface.

2 Air content
A tobacco tin is pushed into the soil, removed, and the soil leveled off with a ruler. The soil is tipped into 200 cm³ of water in a measuring cylinder and the increase in volume, due to the solid constituents of the soil, but not the soil air, is noted. The volume of soil air is given by the equation:

$$\text{volume of soil air} = (\text{volume of tin} + 200 \text{ cm}^3) - (\text{volume of water} + \text{soil}).$$

The percentage of air in the soil can thus be calculated.

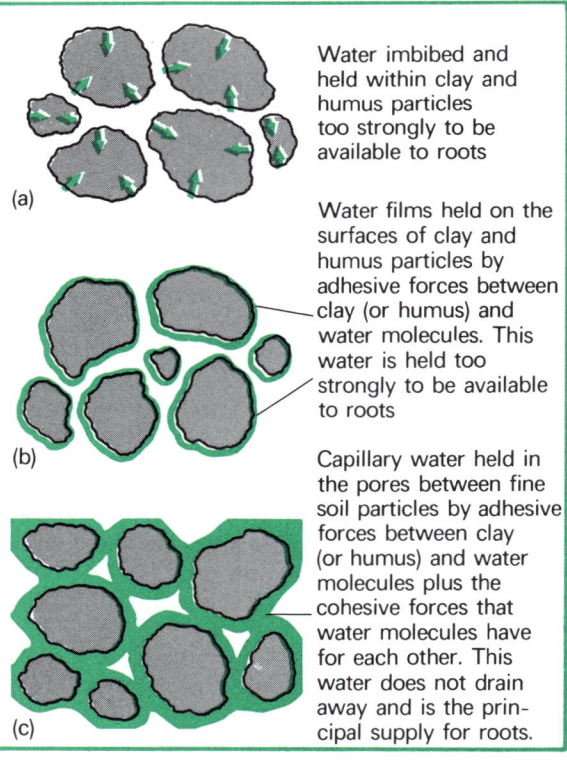

Water imbibed and held within clay and humus particles too strongly to be available to roots

Water films held on the surfaces of clay and humus particles by adhesive forces between clay (or humus) and water molecules. This water is held too strongly to be available to roots

Capillary water held in the pores between fine soil particles by adhesive forces between clay (or humus) and water molecules plus the cohesive forces that water molecules have for each other. This water does not drain away and is the principal supply for roots.

Fig. 9.5 Forces retaining water in the soil.

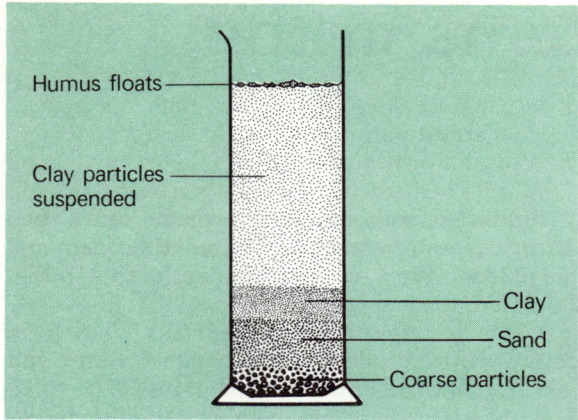

Fig. 9.6 Separation of soil constituents by flotation.

3 Water content

This can be determined by weighing soil in a crucible and heating it to constant mass at 100°C in an oven. If a temperature above 100°C is used, humus will burn off, giving an inaccurate result. After heating and before weighing, the crucible and the soil it contains must be cooled in a desiccator so that the soil does not absorb moisture from the air in cooling.

4 Humus content

Using dried soil from (3), the same procedure is followed except that the soil is now heated to redness until a constant mass is obtained.

5 Permeability of soil to air

The apparatus used is shown in Fig. 9.7(a). The glass wool plug is to prevent soil from falling down the funnel. 25 cm³ of air-dried soil should be used.

The tap is opened and the time taken for the water to fall to the 50 cm³ mark noted. This time is a measure of the permeability of the soil to the air which is being drawn through it. An average of three readings should be taken.

6 Permeability of soil to water

The apparatus is shown in Fig. 9.7(b). 50 cm³ of water is poured onto the soil, and the time taken for ·10 cm³ of water to collect in the burette noted.

7 Capillarity of soil

Air-dried soil is used to fill a glass tube 30 cm long and 1.5 cm in diameter. One end is blocked with glass wool. This end is stood in water and the rise due to capillarity (as shown by darkening of the soil) is noted each day at the same time for a week.

8 Presence of micro-organisms in the soil

Two test tubes are 1/3 filled with milk and sterilized in an autoclave. A few crumbs of soil are added to one after cooling, using forceps so that bacteria from the fingers do not contaminate it. After incubation at 37°C for a day, bacteria in the soil will have caused the milk in the inoculated tube to curdle and smell. The milk in the control tube should remain unchanged.

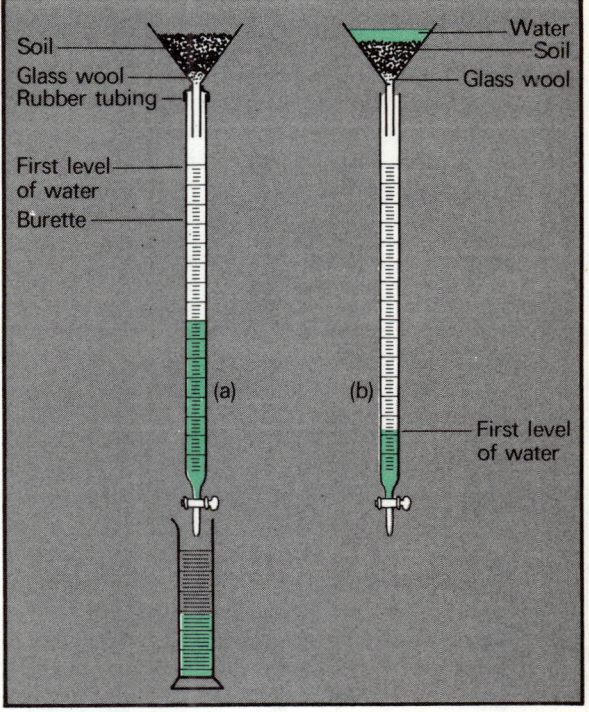

Fig. 9.7 Apparatus for determining the permeability of soil to (a) air and (b) water.

10 Transport Systems in the Flowering Plant

Major differences between the transport systems of mammals and flowering plants

Since plants do not move to pursue food, their energy requirements are modest compared with those of animals. The plant body contains nothing corresponding to the lungs, for oxygen reaches the living cells by simple diffusion (p. 43). Neither is there any pump corresponding to the heart, and the conducting tubes in flowering plants do not have contractile or elastic walls.

Unlike animals, flowering plants have two distinct types of conducting tissue, the xylem and the phloem. We have seen them in the root in Fig. 1.3 and in the leaf in Fig. 5.3.

Xylem vessels consist of dead cells. They form empty wooden tubes which conduct water and ions from the soil upwards to the leaves (Fig. 10.1). The phloem is living and conducts food materials made in the leaves to the growing regions, namely the elongating shoots and root tips. It also conducts

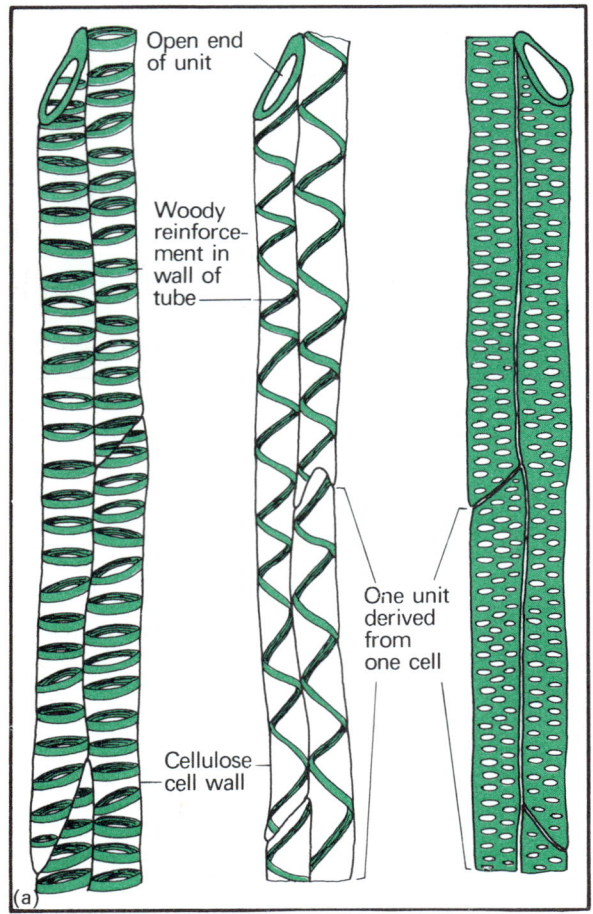

Open end of unit

Woody reinforcement in wall of tube

One unit derived from one cell

Cellulose cell wall

(a)

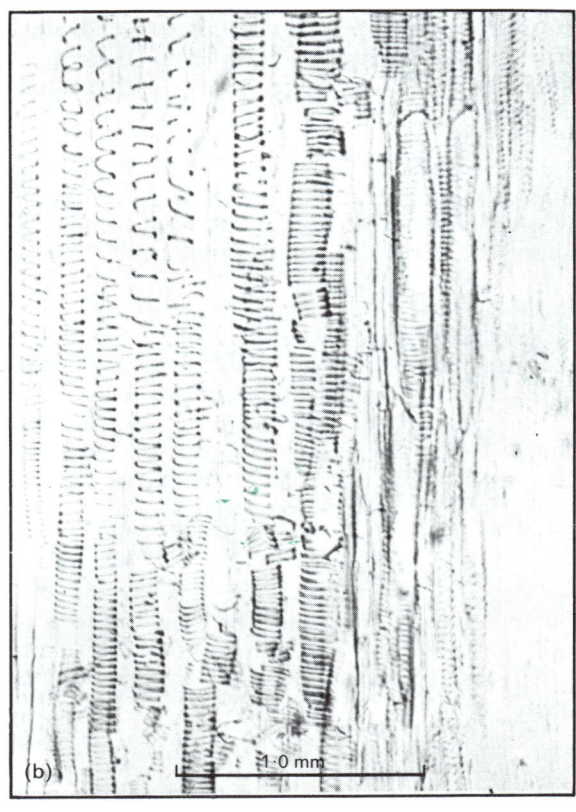

(b)

1·0 mm

Fig. 10.1(a) Xylem vessels. Units of the tubes are derived from single cells. When mature, as here, these are dead and their walls are reinforced with various patterns of woody thickening. **(b)** Xylem vessels in longitudinal section. (Light microscope.)

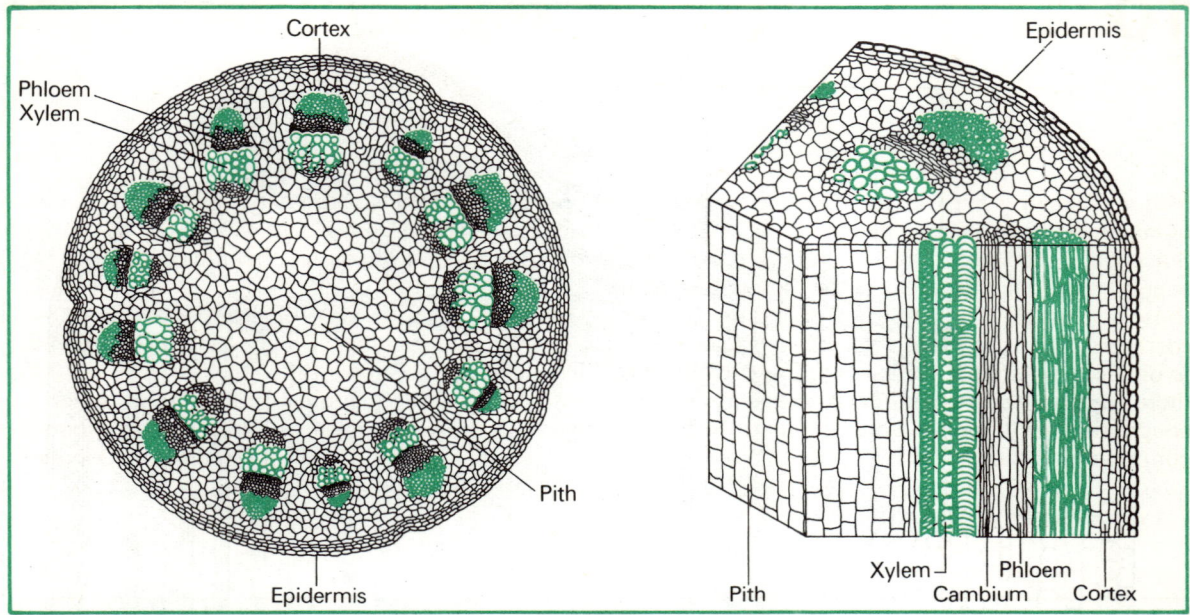

Fig. 10.2 Transverse section and stereogram of the tissues in a sunflower stem.

food to storage regions such as rhizomes, tubers, bulbs, corms, and the woody parts of stems and roots.

The arrangement of xylem and phloem

In the young root (Fig. 1.3), the xylem is centrally placed in a star-like arrangement, with small patches of phloem lying in the bays between the arms of the star. Xylem and phloem lie within the **endodemis**; outside this is the **cortex**, composed of living packing cells (**parenchyma**).

In a herbaceous stem, such as that of sunflower (Fig. 10.2) the xylem and phloem are associated to form **veins**, but they are still distinct strands separated by a layer of cells, the cambium. This can divide and produce more xylem and more phloem. However, the conducting tissue is not central but disposed towards the edge of the stem. This arrangement means that the hard, woody xylem is placed in the best position for resisting the stresses that fall on the stem when it is blown by the wind from the side.

Outside the veins, between them, and in the centre of the stem, is living packing tissue. The outermost layer is the **epidermis**, and as in the leaf (p. 44) these cells fit closely with no spaces between them. They too are covered by a varnish-like cuticle which cuts down loss of water vapour from the stem cells. The epidermis of the root, by contrast, has no

cuticle, for it is one of the functions of the roots to absorb soil water.

The stems of dicot shrubs and trees

When the seed of a tree, such as lime or horse chestnut, germinates, it produces a young thin stem whose internal structure is similar to that of a sunflower. Unlike sunflower, it may go on growing for two or three hundred years, producing a large woody tree, and increasing in diameter from a few millimetres to as much as two metres.

It is beyond the scope of this book to consider in detail how this happens. Each year the cambium makes more cells, more xylem towards the centre of the stem, more phloem towards the outside. In the outer parts of the cortex, layers of dead cork cells are formed, making the protective bark (Fig. 3.3).

Figure 10.3 is a stereogram of a two year old stem of lime (*Tilia vulgaris*). Try to answer the following questions about it:

1 What kind of cell composes the cortex and the pith?
2 Considering the general plan of cortex, phloem, cambium, xylem, and pith, in what ways is this stem (a) like (b) unlike the sunflower stem?
3 Find the position of the cambium in Fig. 10.3. Each year the cambium produces more xylem towards the centre of the stem. What, therefore,

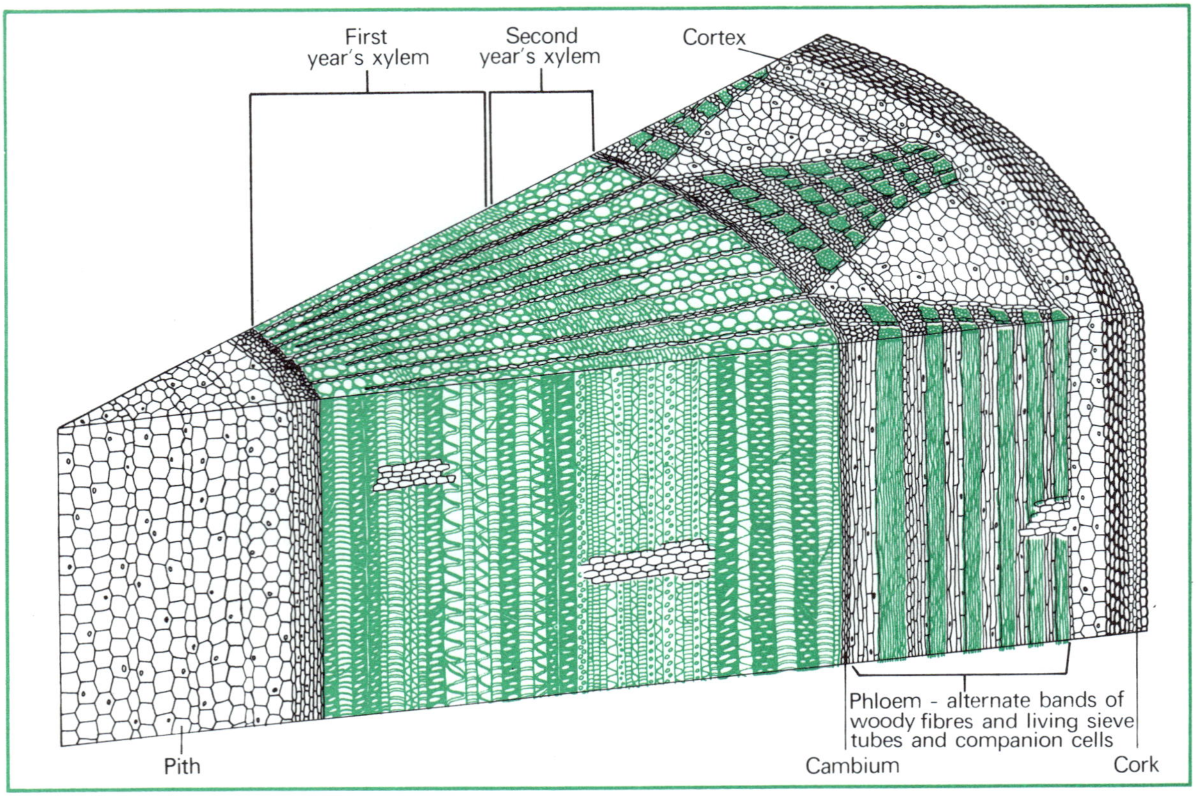

First
year's xylem

Second
year's xylem

Cortex

Pith

Phloem - alternate bands of
woody fibres and living sieve
tubes and companion cells

Cambium

Cork

Fig. 10.3 Stereogram showing the arrangement of tissues in a two year old stem of lime *(Tilia vulgaris)*.

would have been the location of the cambium in this stem at the end of its first year of growth?

Fig. 10.4 Longitudinal section of a young root, from the tip to the root hair zone.

Structure of the root tip and root hairs

Root hairs are confined to a definite region, starting a few millimetres behind the **root tip** (Fig. 3.11).

The root tip, which grows forwards between rough soil particles, is protected by the **root cap** of degenerating packing cells (Fig. 10.4). Just behind this lies the **region of cell division**. Immediately

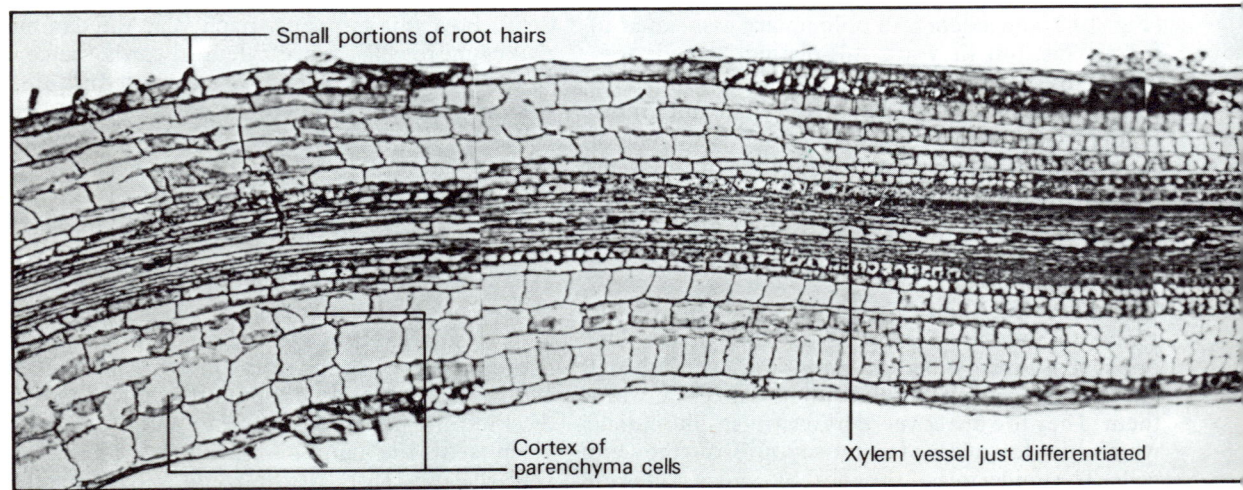

Small portions of root hairs

Cortex of
parenchyma cells

Xylem vessel just differentiated

behind this the newly-formed cells start to differentiate, so that a very few millimetres from the
root tip the plan of the young root (Fig. 1.3) can be
seen. Cells in the centre elongate and at the level of
the root hair zone, the first xylem and phloem cells
can be seen.

On roots grown in water culture, root hairs are
straight and arise at 90° to the surface of the root.
But in the soil they are very contorted, forming
intimate associations with individual soil particles,
which adhere strongly to them and help to anchor
the plant.

Each hair is a tubular outgrowth of an epidermal
cell and its structure is that of a living plant cell
as described on p. 9. It is about 1 cm long when
fully elongated, and 10 μm in diameter. There are
as many as one hundred root hairs per cm² of
root surface. Their life span is difficult to investigate
but is estimated to be a few weeks. Each hair has a
cell wall made of cellulose and is lined by cytoplasm.
There is a central permanent vacuole, and the
cytoplasm is bounded on both its faces by a plasma
membrane (Fig. 10.5).

Absorption, transport, and loss of water by plants
Almost all the water which enters the plant does
so from the soil through the root hairs. Since the
xylem, which conveys water to the leaves, lies in
the centre of the root, water has to pass from the
root hairs across the cortex to reach it. Water
follows two distinct pathways across the cortex.

First pathway
Parenchyma cells of the cortex have the same
structure as root hair cells, though their shape is
different, and water can be sucked along them by

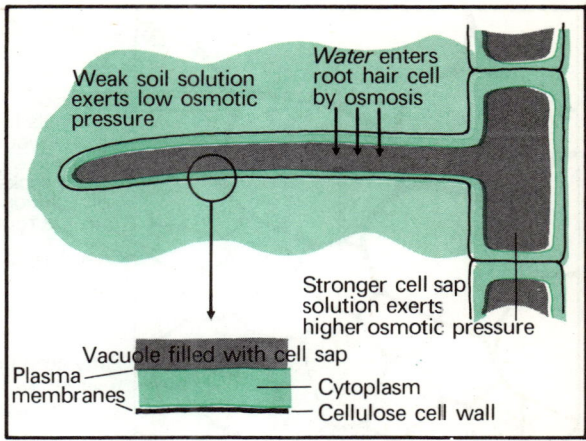

Fig. 10.5 Structure of a root hair cell.

capillary attraction, by-passing the living cell
contents.

This pathway is shown in Fig. 10.6(a), where the
cellulose cell walls are shown in black and the cell
contents in white. If this were the only pathway
followed, soil solution would be carried into the
xylem, and so into the whole plant, unaltered. But
we have seen on page 95 that this is not so: ions are
absorbed selectively, and cannot, therefore, be
merely swept into the plant in solution. How is
this prevented?

Careful inspection of Fig. 1.3 shows that the
cells in the innermost layer of the cortex (the endodermis) differ from other cortical cells in two ways.
First, they are brick-shaped, and fit closely together
with no spaces between them. Secondly, deposited
in the cellulose wall of each endodermal cell is a
fatty strip, a barrier which does not allow water to
pass.

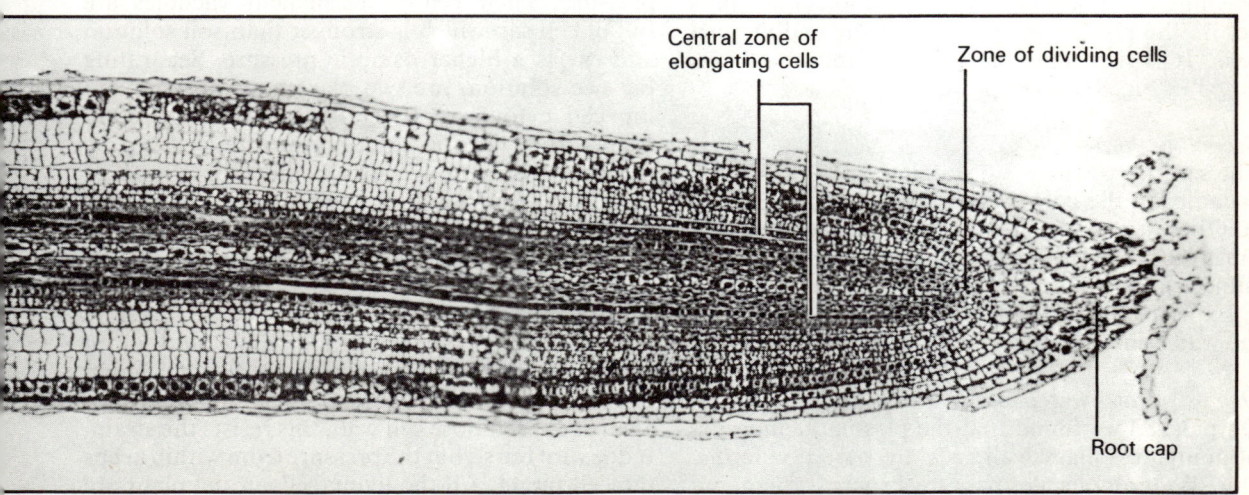

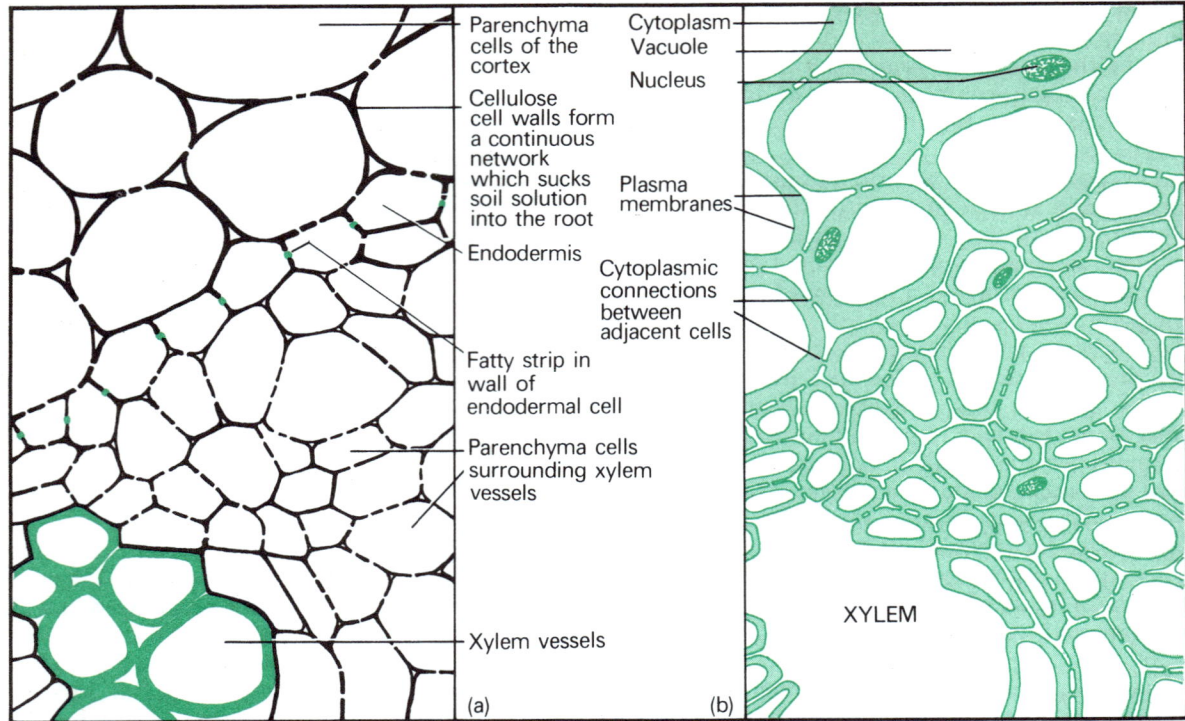

Fig. 10.6(a) The cell walls in a transverse section of a buttercup root. **(b)** The living cell contents seen in a transverse section of a buttercup root.

Figure 10.7 shows how these features of the endodermal cells prevent the passage of soil solution along the cell walls from cortex to xylem. Up as far as the endodermis soil solution moves unchanged through the cellulose cell walls; but on reaching this point both water and ions must pass across the plasma membranes of the endodermal cells. It is here that the selective uptake of ions noted in Fig. 9.3 takes place.

Second pathway
The second pathway is through the cytoplasm and vacuoles of the cortex cells. This is shown in Fig. 10.6(b). In passing into each cell, water and ions must cross the plasma membranes where ions are subject to selection.

Experiments show that about three quarters of the water entering the xylem follows this pathway.

Osmosis draws **water** *into root hair cells*
On page 83 we found that the plasma membranes of animal capillary wall cells are partially permeable. Water molecules pass freely across them, but

whether dissolved substances will pass depends on the sizes of their molecules. The same applies to the plasma membranes of root hair cells and cortical cells. Root hair cells are surrounded by weak soil solution which exerts a low osmotic pressure. Their central permanent vacuoles are full of cell sap which is stronger than soil solution and exerts a higher osmotic pressure. Separating the two solutions are two plasma membranes and the cell cytoplasm (Fig. 10.5). Water therefore passes from the soil solution into the cell sap by osmosis. It is important to remember that dissolved ions in the soil solution are not swept in on this osmotic tide, for they are subject to selection at the plasma membranes.

Turgor and plasmolysis
It might appear that root hair cells would go on taking in soil water by osmosis until they burst. But although the cell contents swell and press hard against the cellulose cell wall, this resists the strain. It does not burst, but the pressure from within keeps the cell **turgid**. All the living cells in the plant are

102

turgid. Turgor is the principal force holding plants erect. When plants wilt they lose water by evaporation and the cytoplasm of each cell no longer presses against its cellulose wall. Since every living cell loses its rigidity, so does the entire plant.

The correctness of regarding living plant cells as osmotic systems is shown by the phenomenon of **plasmolysis**. When such cells are placed in a sugar solution which is stronger than the solution in the cell sap, the osmotic situation is the reverse of that just considered for the uptake of water from soil solution. Water passes out of the cell sap, across the plasma membranes and into the surrounding sugar solution.

The living contents of the cell thus shrink and eventually draw away from the cellulose cell wall (Fig. 10.8). Water and substances dissolved in it can pass freely between the cellulose fibres which make up the wall, so the space between the shrinking cytoplasm and the cellulose wall becomes filled with the sugar solution in which the cells are immersed. Such cells are said to be **plasmolyzed**. They differ from **wilted** cells because in these the space between the shrunken cytoplasm and the cellulose wall is filled with air.

Xylem vessels are the water-conducting channels of the plant
The only plant tissues which possess elongated tube-like cells of a kind likely to conduct water are the xylem and the phloem. In all the other tissues the cells are more or less spherical or brick shaped.

That it is the xylem and not the phloem that conducts solution up the stem to the leaves can be shown by placing a cut shoot of Busy Lizzie (*Impatiens sultani*), or any other herbaceous shoot, in a dilute aqueous solution of a red dye. After a few hours, thin transverse sections cut at different distances up the stem and examined under the microscope show that only the xylem vessels are stained red.

Transpiration
Water taken from the soil by roots is lost from the leaves as vapour, and this process is called **transpiration**. Very large quantities of water are involved. A small beech tree weighing 22 kg was found to lose 105 kg of water per day, five times its own mass. A small beech wood would lose something like twenty million litres of water per day. Water transpired by plants affects the formation of clouds and the flow of rivers.

We noted on p. 41 that the large internal surface area of leaves, while an adaptation for efficient photosynthesis, makes it inevitable that the plant will lose a great deal of water vapour. A diffusion gradient for water vapour exists from the wet walls of the leaf cells to the air outside the leaf. Ten times

Fig. 10.7 The endodermis of brick-shaped close-fitting cells, with fatty strips in their walls, forms a barrier preventing soil solution from seeping into the centre of the root.

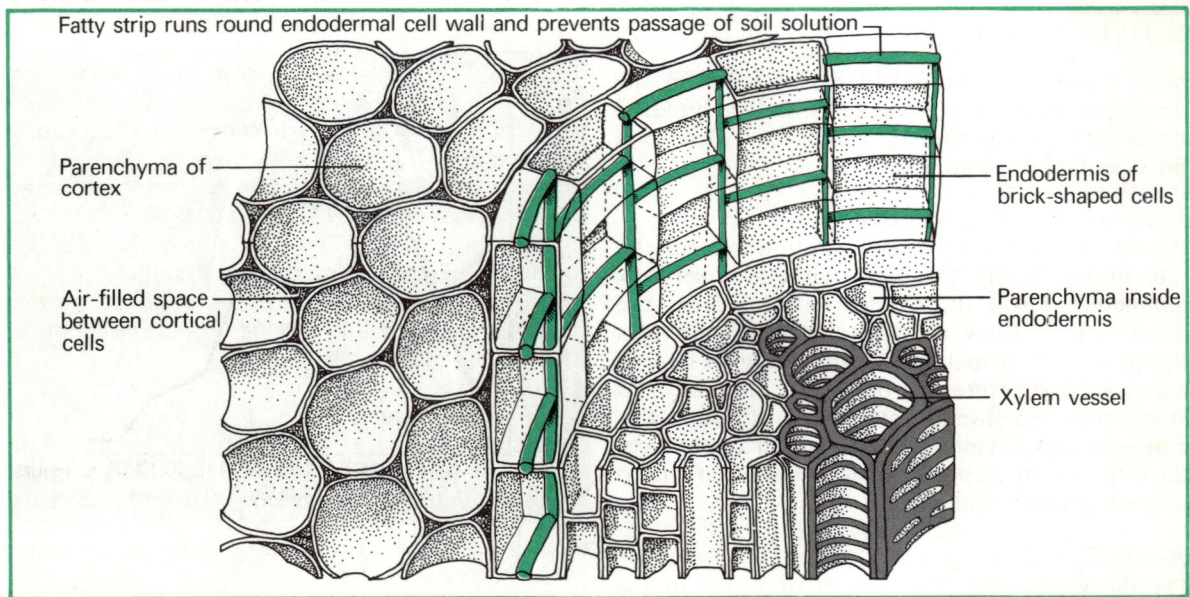

Fatty strip runs round endodermal cell wall and prevents passage of soil solution

Parenchyma of cortex

Air-filled space between cortical cells

Endodermis of brick-shaped cells

Parenchyma inside endodermis

Xylem vessel

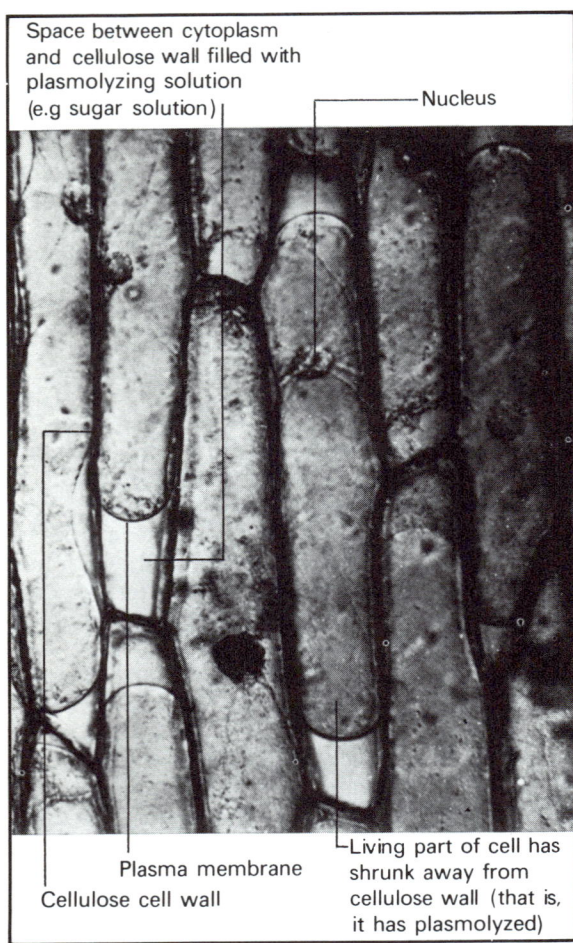

Space between cytoplasm and cellulose wall filled with plasmolyzing solution (e.g sugar solution)

Nucleus

Living part of cell has shrunk away from cellulose wall (that is, it has plasmolyzed)

Plasma membrane

Cellulose cell wall

Fig. 10.8 Plasmolyzed onion cells.

more water is transpired than is needed for the development of plant cell vacuoles, and one hundred times more than is needed to combine with carbon dioxide in photosynthesis.

Disadvantages of transpiration

If rate of water loss by transpiration exceeds the rate of uptake by the roots, growth is severely hampered. There is no water available for the development of new cell vacuoles, which provides the plant's chief method of increasing in size. If losses are too severe, the plant wilts and dies. To check this, the stomata close, and although this may reduce loss of more water vapour it also reduces the diffusion of carbon dioxide into the leaf and so restricts photosynthesis.

Benefits of transpiration

On the credit side, the stream of water in the xylem transports ions absorbed by the roots to the leaves where the manufacture of proteins and other vital chemicals proceeds. But the rate of transpiration is much in excess of that needed for this purpose.

Transpiration cools the leaves, whose cells, especially in hot climates, might otherwise be killed when exposed to the sun. Figure 10.9 shows how the internal temperature of a transpiring leaf stayed below that of an adjacent leaf which had been prevented from transpiring by smearing it with a thin layer of wax. However, other factors also cause loss of heat from leaves—radiation, and, especially in wind, convection.

Resistance to transpiration

The outer surfaces of leaf epidermal cells are covered with a cuticle (Figure 5.4). Though thin, this cuts down a lot of evaporation from the inside of the leaf. Plants growing in deserts have thick cuticles.

Open stomata offer surprisingly little resistance to transpiration, even though they occupy only about 1% of the total leaf surface area. Experiments have shown that the rate at which leaves with wide open stomata tanspire is up to 50% of the rate of evaporation from a wet filter paper or open dish of water of the same shape and size as the leaf.

Boundary layer

Even in the comparative stillness of a room the air is in turbulent motion, as can be seen from the

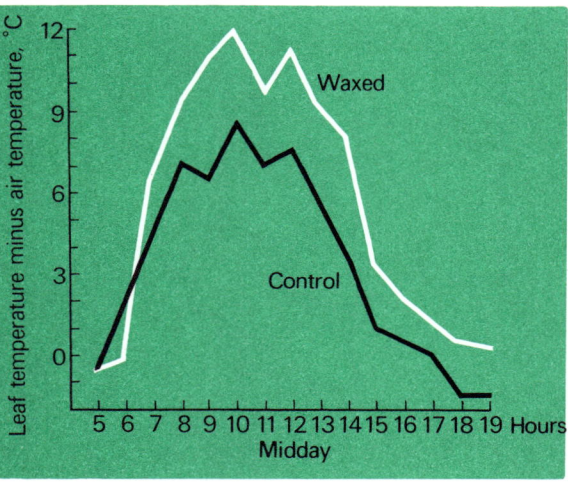

Fig. 10.9 Temperatures of two adjacent leaves on a plant of white goosefoot (*Chenopodium album*). One leaf has been coated with wax to prevent transpiration.

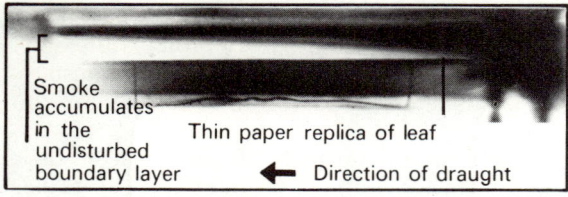

Fig. 10.10 Boundary layer on the upper side of a paper replica of a leaf.

rapid swirling and disappearance of tobacco smoke. Out of doors, where there are usually breezes, this effect is more pronounced. Water molecules that have left the leaf are rapidly incorporated into the general body of the air, thus maintaining a high diffusion gradient for water vapour between the inside of the leaf and the surrounding air.

But immediately next to the leaf surface is a **boundary layer**, about 5 mm deep, in which the air is still. A similar boundary layer is shown in Fig. 10.10 where a paper replica of a leaf is seen edge on in a smoke chamber. Air is being blown over the leaf from right to left, and a drop of smoke-forming liquid has been placed on the right hand upper surface of the 'leaf'. Most of the smoke has blown away, but a layer of undisturbed smoke has accumulated in the still boundary layer immediately next to the 'leaf' surface.

In real life this layer is a local region of high concentration of water vapour. It offers a resistance to the passage of more water molecules into it and

Fig. 10.11 The response of transpiration to wind and to stomatal aperture in *Zebrina pendula*.

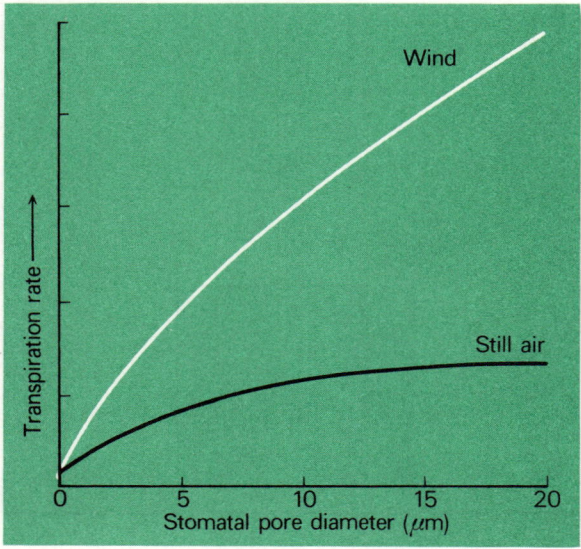

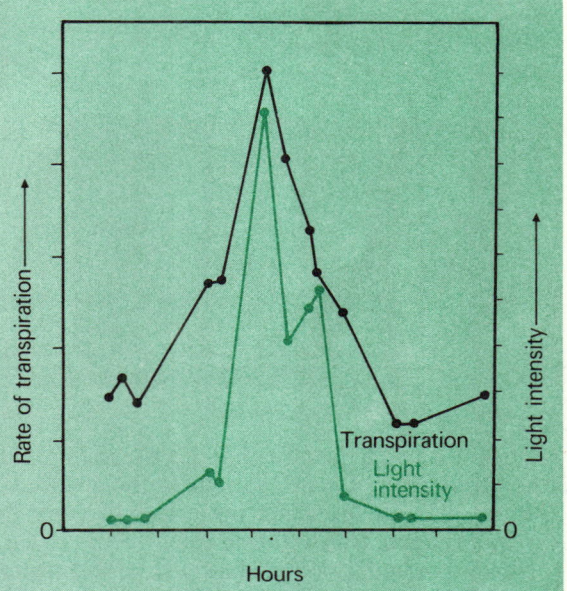

Fig. 10.12 Daily fluctuations in transpiration rate of hairy willowherb (*Chamaenerion hirsutum*) are closely related to changes in light intensity.

places as much restriction on transpiration rate as do the stomata.

Closing of the stomata

The effect of the boundary layer is well illustrated by Fig. 10.11. The transpiration rate of a plant called *Zebrina pendula* was measured in wind and in still air when the stomata were open to varying extents. The Y (vertical) axis shows transpiration rate and the X (horizontal) axis shows the width of the stomatal pores in μm.

In still air, closing the stomata had almost no effect in slowing down transpiration rate until they were almost shut (less than 5 μm in diameter). This was because most of the resistance to transpiration was imposed by the boundary layer, which was thick in still air. However, in wind this thickness diminished, and the resistance the layer offered was correspondingly reduced. Under these conditions transpiration rate was controlled by stomatal aperture and it fell steadily as the stomata closed.

The very low transpiration rate recorded when the stomata were shut shows that there was almost no transpiration through the cuticle.

Factors affecting the rate of transpiration

A rise in temperature causes more water inside the leaf to evaporate and so increases the rate of

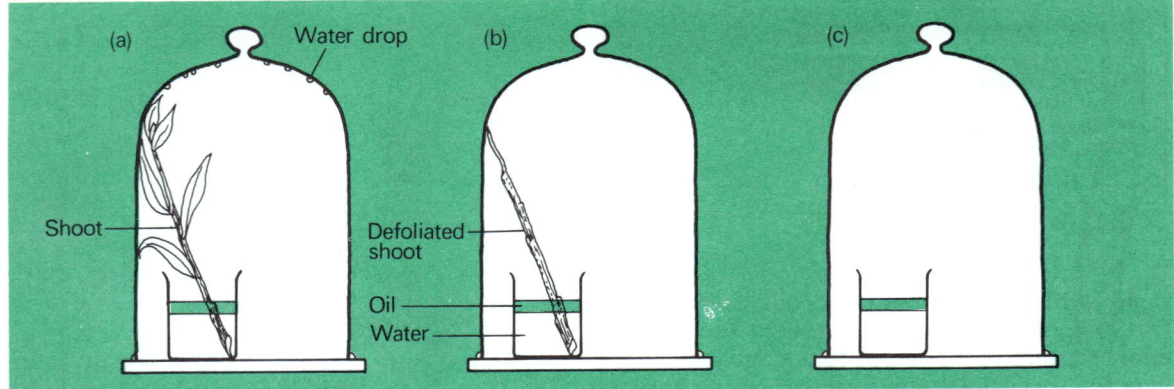

Fig. 10.13 Demonstration of transpiration.

transpiration. On the other hand, a rise in humidity (the proportion of water vapour in the air) makes it more difficult for water molecules to enter the air, and reduces transpiration rate.

Since stomata close in darkness, transpiration would be expected to decline in darkness, and Fig. 10.12 shows how transpiration rate mirrored the increase and decrease in light intensity over a 24 hour period.

The effects of temperature, humidity, and light are all complicated by the effect of wind, which exerts a powerful influences on the thickness of the boundary layer.

Class experiments on transpiration

The existence of transpiration can be demonstrated with the apparatus shown in Fig. 10.13. Two twigs from any tree will do. The layers of oil should be poured on after the shoots have been inserted into the water, otherwise it might enter and block the xylem vessels. The effectiveness of the oil in preventing evaporation of water is demonstrated by the control C, which has no shoot.

The **potometer** (Fig. 10.14) can be used to compare the *rate of uptake* of water by a *cut shoot* under varying environmental conditions. We must remember that water passing into the stem in a potometer has not come via the roots. A potometer could not therefore give much idea of transpiration by an intact plant.

The potometer measures rate of water uptake. Only if the shoot is turgid can this be assumed to equal the rate of water loss (i.e. transpiration). If the shoot is not turgid, some of the water it takes up may be retained in its cell vacuoles and less water will be given off than is absorbed.

It is essential to get a tight fit between the shoot and the rubber and between the glass tube and the

rubber. As soon as the shoot is cut from the plant it should be plunged under water to prevent air bubbles being drawn into the xylem vessels. These

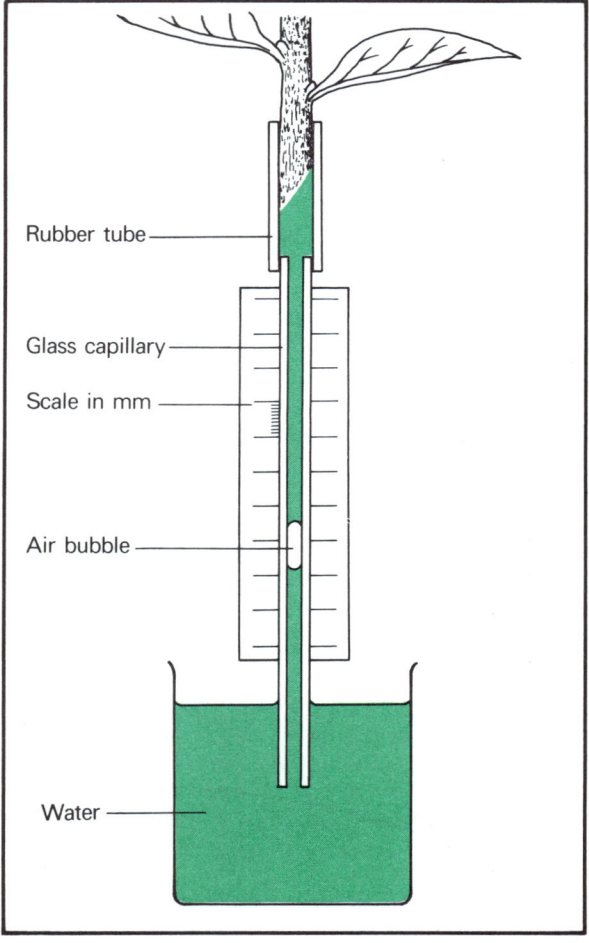

Fig. 10.14 A potometer.

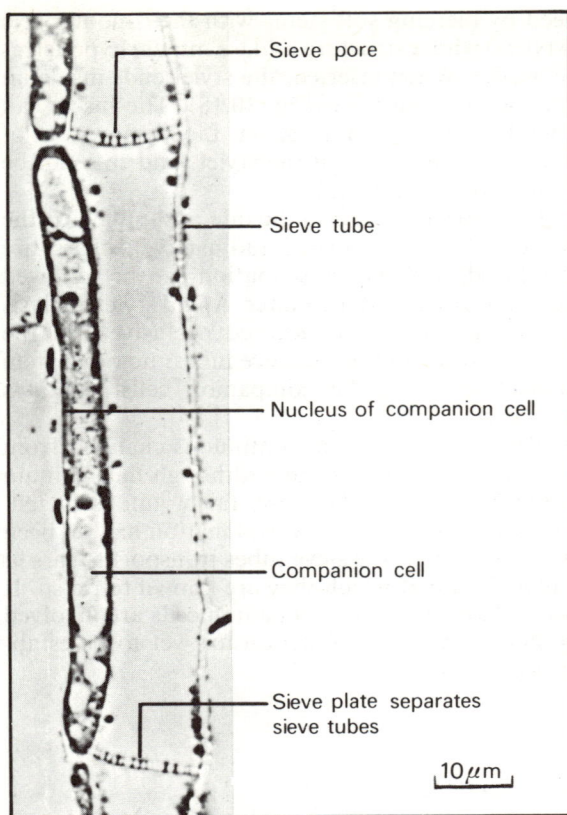

Fig. 10.15 Longitudinal section through an entire sieve tube and parts of two others, and through a companion cell (light micrograph).

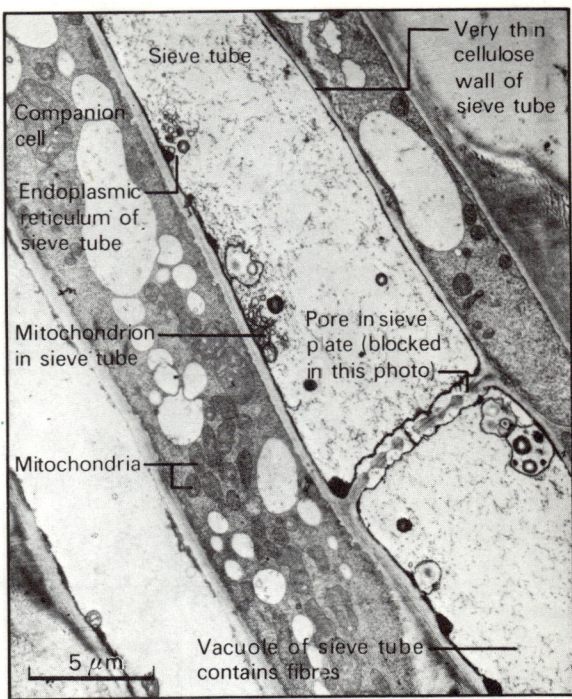

Fig. 10.16 Longitudinal section through parts of two sieve tubes and two companion cells (electron micrograph).

would block the xylem vessels and prevent the passage of water, so the shoot should be fitted to the potometer under water.

Assuming the shoot is turgid, an average of two readings of the time taken for the bubble to traverse a unit section of the capillary tube gives a measure of transpiration rate. This should be expressed as cm^3 water loss per cm^2 leaf area per min.

Sugars and amino acids are transported in the phloem

A great deal of experimental evidence of various kinds shows that sugars, amino acids, and other chemicals made in the leaves are transported in the phloem and not in the xylem; but despite much research the forces which cause movement and the mechanism of movement are largely unknown.

What phloem looks like

Phloem contains two types of elongated cell, each about 0.5 mm long. They are called **sieve tubes**

and **companion cells**. Figure 10.15 is a longitudinal section of phloem seen under the light microscope. The sieve tube is very empty looking. The **sieve plates**, which give these cells their name, can be seen at each end. They are thin cellulose walls pierced by large holes through which the cytoplasm

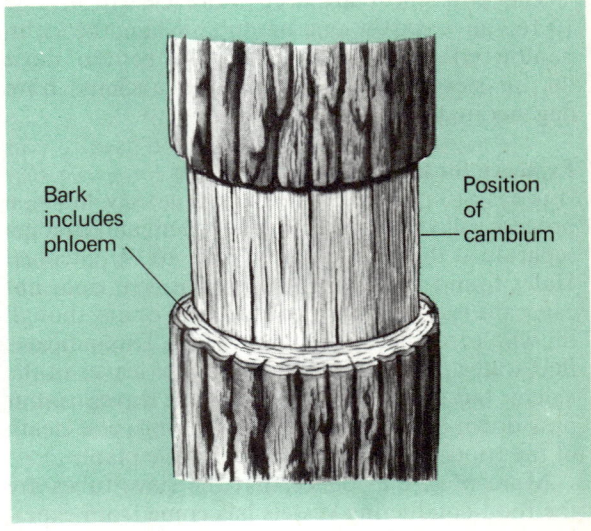

Fig. 10.17 Ringing a woody shoot.

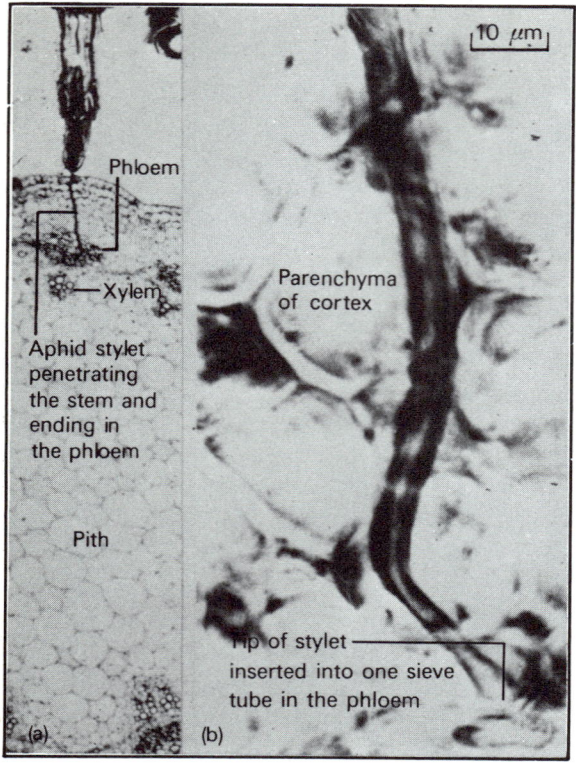

10 μm

Phloem

Xylem

Parenchyma
of cortex

Aphid stylet
penetrating
the stem and
ending in
the phloem

Pith

Tip of stylet
inserted into one sieve
tube in the phloem

(a)   (b)

Fig. 10.18(a) Aphid stylet inserted into a phloem sieve tube.

(b) Tip of stylet inserted into one sieve tube in the phloem.

of adjacent cells makes contact. The companion cell is very different and contains both a nucleus and granular cytoplasm.

The differences are seen more clearly in Fig. 10.16, an electron micrograph. While the companion cell contains many organelles, some of those in the sieve tube, including the nucleus, have degenerated.

Evidence for transport in the phloem
It is easy to peel the bark from a woody shoot, a process called 'ringing', for the xylem and phloem separate at the cambial layer (Fig. 10.17). Stephen Hales found in 1727 that this treatment does not cause the leaves above the ring to wilt or die, though the whole tree dies after some months. This indicates that water passes up the trunk to the leaves in the xylem; but food substances from the leaves cannot pass down to the roots across the ring; and death of the roots causes death of the whole plant.

More precise evidence that the sieve tubes are the food-conducting vessels has come from experiments using plant lice (aphids). These insects feed by piercing soft stems with their mouthparts (stylets) whose structure is like a minute hypodermic needle. When inserted, the stylet ends inside an individual sieve tube (Fig. 10.18). The insect can then be cut off from its stylet. Liquid flows out of the sieve tube through the stylet, and this can be collected and analyzed.

If radioactive carbon dioxide is supplied to the leaves of such a plant, radioactive sugars are produced, and their distribution can be followed by the use of a Geiger counter. Aphid stylet analysis shows that these are located exclusively in the liquid flowing out of the sieve tubes: no other stem tissue, not even the companion cells, contains any at all.

The sieve tubes form continuous channels from the leaves to the roots, and although they contain some organelles, they are fairly unobstructed. However, no satisfactory explanation has yet been given as to how the sieve tubes transport sugars at the high rate at which they are known to do so. It seems likely that the companion cells are involved somehow, but their function has yet to be established clearly.

11 Excretion

Excretion is *the elimination from the body of waste products of cell activity, which are often poisonous*. For example, aerobic respiration produces carbon dioxide, water, and heat as waste products in every living cell:

$$CH_2O \underset{\text{carbohydrate}}{} + O_2 \longrightarrow CO_2 + H_2O + \text{heat}$$

If carbon dioxide accumulated it would cause the acidity of the cell to rise, creating an environment in which most of the cell's enzymes could not work efficiently.

The need for excretion therefore applies equally to animals and plants, but the means of achieving it are very different in the two groups because animals are much more active than plants.

Major differences in excretion between animals and plants

Carbon dioxide

In mammals the speedy removal of carbon dioxide is one of the principal functions of the blood and lungs, but in plants carbon dioxide merely diffuses away from respiring cells into the spaces between the cells and so into the atmosphere via the stomata and lenticels. Some of the carbon dioxide produced by a plant is used up again by the same plant in photosynthesis (see page 79).

Water

Plants produce only a little water by respiration, and this is lost in transpiration. By contrast an adult man produces a daily average of 350 cm^3 of water by tissue respiration, one seventh of his daily requirement. Man and other land-living vertebrates have to excrete water, partly from the lungs in breath, partly from the skin in sweat, partly from the kidneys in urine.

Some animals, such as the kangaroo rat (*Dipodomys spectabilis*, a small mammal) and the flour beetle (*Tribolium spp.*) never drink water during the whole of their lives. They use water derived from tissue respiration as their sole supply.

Breakdown wastes

Other waste chemicals are produced from the breakdown of dead cells. This is a far more active process in animals than in plants, and most of the material ends up as the chemical substance **creatinine** which is disposed of in the urine. Figure 11.1 shows that creatinine is one hundred times more concentrated in human urine than in blood plasma.

Herbaceous (non-woody) plants retain such wastes until they die at the end of the season. Shrubs and trees either store them in the heartwood or pass them into the leaves just before leaf fall.

Ammonia and urea

The major difference in the excretory products of animals and plants comes from their fundamentally different modes of nutrition. The holozoic nutrition of animals gives rise to large quantities of **ammonia** in the body. This is a very poisonous substance and must be excreted. One part in 20 000 is toxic to cells.

Source of ammonia Ammonia (chemical formula NH_3) is derived from excess amino acids taken in as protein in the food (see page 61). Carnivorous mammals such as dogs eat far more protein than is

	Plasma (g per 100 cm^3)	Urine (g per 100 cm^3)	Ratio Urinary conc. / Plasma conc.
Water	90 to 93	95	—
Glucose	0.1	0.0	—
Proteins	7 to 8	0	—
Urea	0.03	2	60
Creatinine	0.001	0.1	100
Ammonia	0.0001	0.05	500

Fig. 11.1 Relative concentrations of various dissolved substances in human blood plasma and in urine.

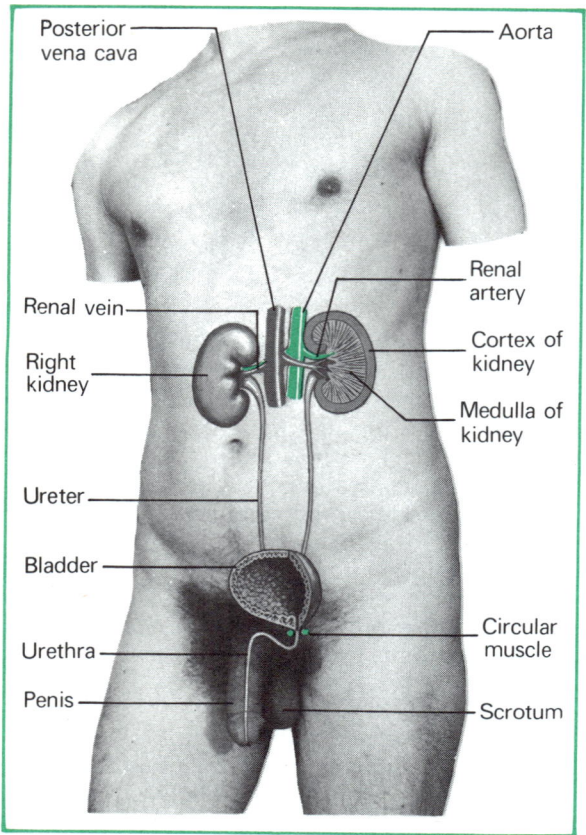

Fig. 11.2 Positions of the kidneys, ureters, and bladder, together with their blood supply.

needed to satisfy the demand for cell growth and replacement. Even in man, about 10 % of the protein consumed by the average Briton is de-aminated, and in Greenland, where the diet contains a higher proportion of protein, the figure is about 40 %.

Plants are holophytic. They make all their own proteins, fats, and carbohydrates and so are able to balance the supply of proteins according to their needs. The continuous deamination of large quantities of amino acids characteristic of animals has no counterpart in plants.

Disposal of ammonia   Ammonia is very soluble in water. Some animals, such as fresh water fish and invertebrates, have unlimited quantities of water available to them and the ammonia can be eliminated in large quantities of urine.

Urea and uric acid   Land-living animals, however, always conserve water and produce little urine. In mammals (including man), the amino groups of excess amino acids (amino group $= -NH_2$) is not converted into ammonia at all. Instead, liver cells turn it into urea (chemical formula $CO(NH_2)_2$). This is also soluble, is much less poisonous than ammonia, and is conveyed to the kidneys in solution in blood plasma. Figure 11.1 shows that urea is sixty times more concentrated in human urine

Fig. 11.3 Corrosion preparation of human kidney blood vessels and ureters.

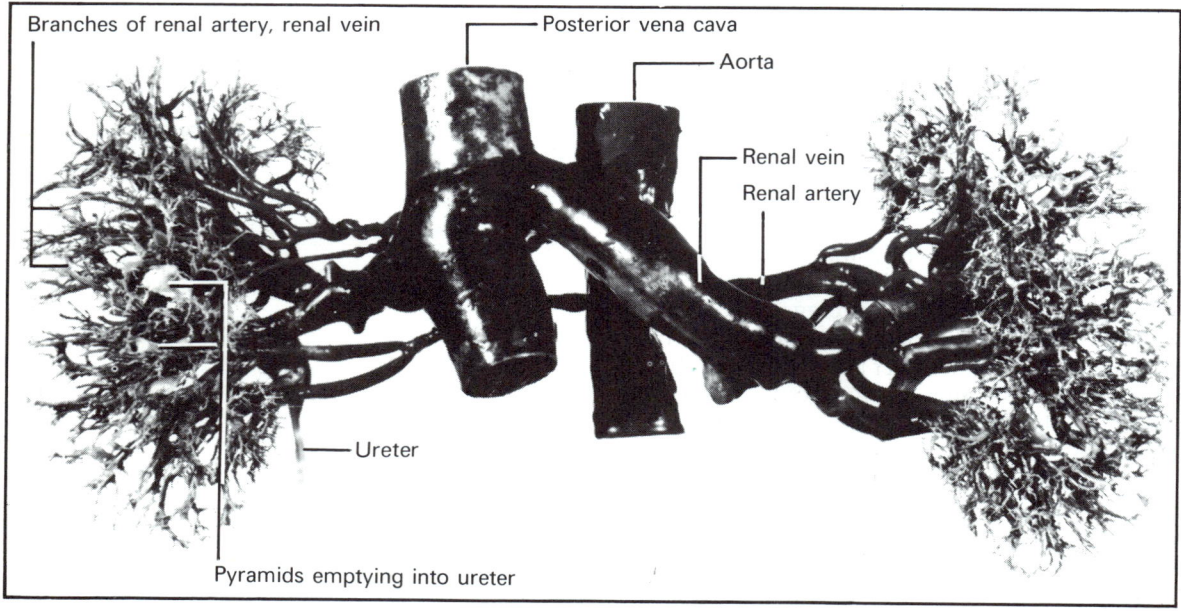

than in blood plasma.

An adult human consuming 70 g of protein per day produces about 30 g of urea per day, and it is the main dissolved component of human urine.

In reptiles, birds, and in insects, the amino groups of excess amino acids are converted into insoluble uric acid. This is a non-poisonous substance which gives bird droppings their characteristic white colour.

Functions of the kidneys

As just noted, one of the principal functions of the kidneys is to remove urea from blood plasma and to pass it into the urine. The kidneys also act as a water-regulator, eliminating excess water from the body. They also excrete excess ions.

Thus the kidneys monitor the composition of blood plasma and remove unwanted substances from it. Also dissolved in plasma are precious food substances, glucose and amino acids, and in producing urine the kidneys ensure that these are retained, not lost.

Positions of the kidneys and allied organs

Figure 11.2 shows the kidneys lying at the back of the abdomen. Their blood supply is the short **renal arteries** from the aorta and the short **renal veins** leading into the posterior vena cava. Passing from each kidney to the bladder is a muscular tube, the **ureter**. Urine is produced continuously and moves along the ureters by peristalsis.

The bladder has a muscular wall. When empty it lies within the bony hip basin, but as it fills its muscles relax. Its maximum capacity is about 500 cm^3, when it may reach as high as the navel.

Leading out of the bladder is the **urethra**, another muscular tube. This has a separate opening to the exterior in woman (Fig. 12.9(b)), but in man it joins the sperm ducts from the testes before passing through the **penis** (Fig. 12.4). At its junction with

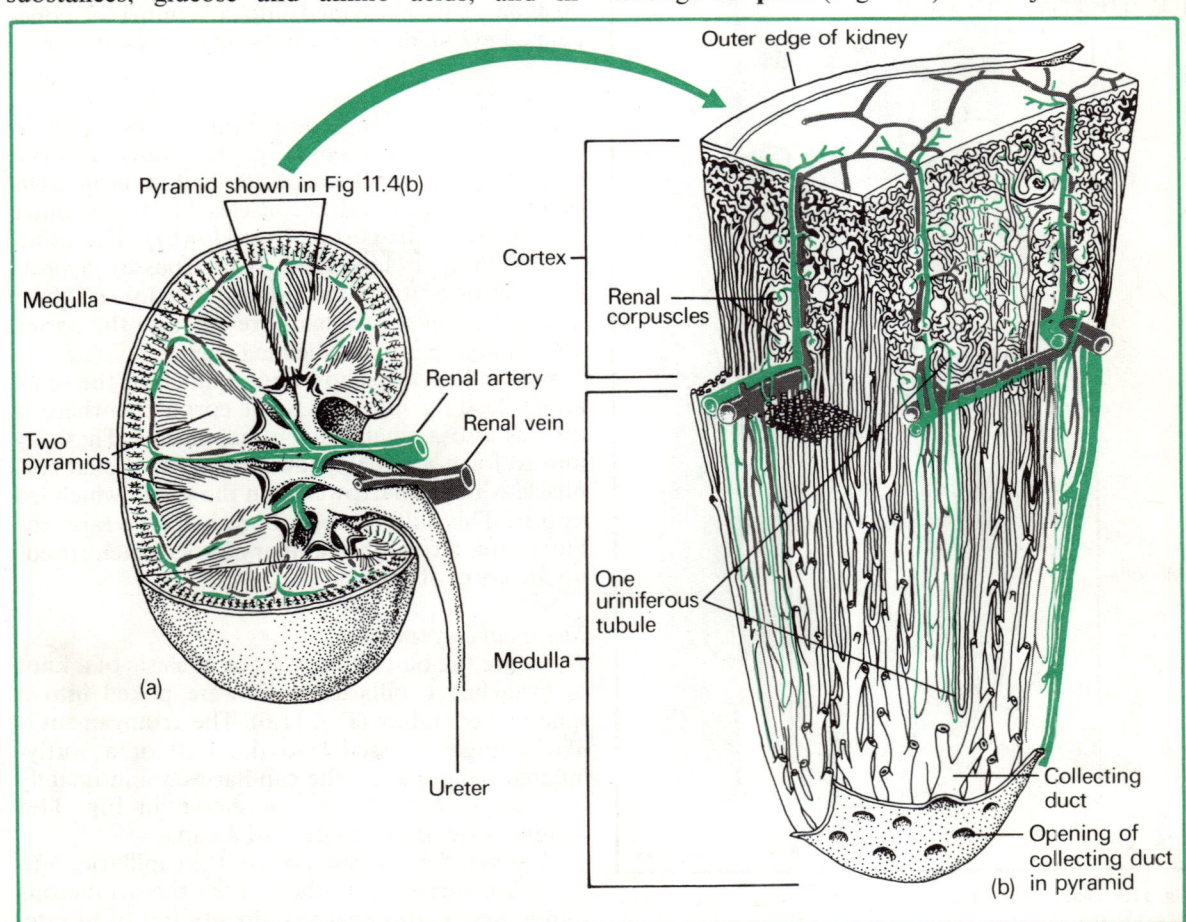

Fig. 11.4 Internal structure of the human kidney.

111

the bladder is a circular muscle. During urination this relaxes and the muscular wall of the bladder contracts, expelling urine.

Blood supply to the kidneys

As the kidneys are blood filters, it is to be expected that they will have an abundant blood supply. Figure 11.3 is a corrosion preparation of part of the human aorta, posterior vena cava, renal arteries, renal veins, and blood vessels in the kidneys. Why are the aorta and renal arteries so much narrower than the posterior vena cava and renal veins?

Internal structure of the kidney

Figure 11.4 shows that there are two distinct regions to the human kidney. The outer **cortex** forms a complete surface layer over the inner

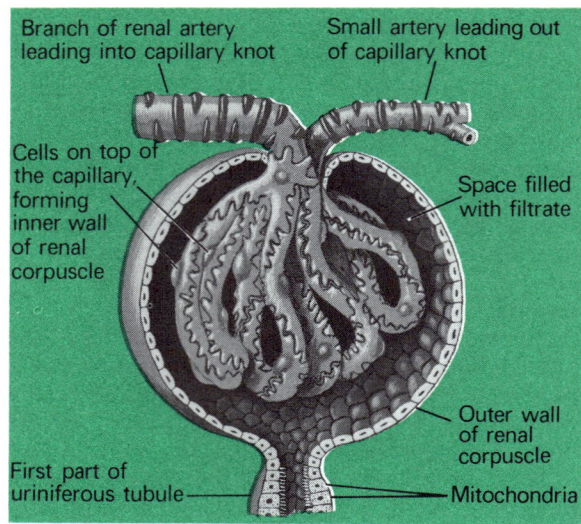

Fig. 11.6 A renal corpuscle.

medulla. This is divided into a number of cone-shaped **pyramids**, each emptying into the ureter.

The nephron

Each human kidney contains about one million blood filters or **nephrons** (Fig. 11.5). Each nephron consists of a **renal corpuscle** and a **uriniferous tubule**. All the renal corpuscles lie in the outer kidney layer (the cortex, Fig. 11.4(b)). The uriniferous tubule is U-shaped. It first passes inwards towards the centre of the kidney, then does a U turn and passes outwards again, re-entering the cortex and emptying into a **collecting duct**.

Figure 11.5 also shows one branch of the renal artery leading into the renal corpuscle where it divides into a small knot of capillaries. These rejoin to form a vessel leading out of the renal corpuscle which is narrower than the vessel which led into it. This difference in bore helps to raise the blood pressure in the capillary knot, which speeds up filtration of plasma.

The renal corpuscles

These are the blood filters. Each consists of a knot of branched capillaries as it were poked into a spherical container (Fig. 11.6). The arrangement is like a finger pressed into the wall of a softly-inflated balloon, and the capillaries are intimately covered by cells which are shown in Fig. 11.5 merely as the inside margin of a cup.

Plasma is force-filtered out of the capillaries into the space surrounding them. Like the uriniferous tubule below, this space is already full of filtrate. Experiments have been performed in which a micro-

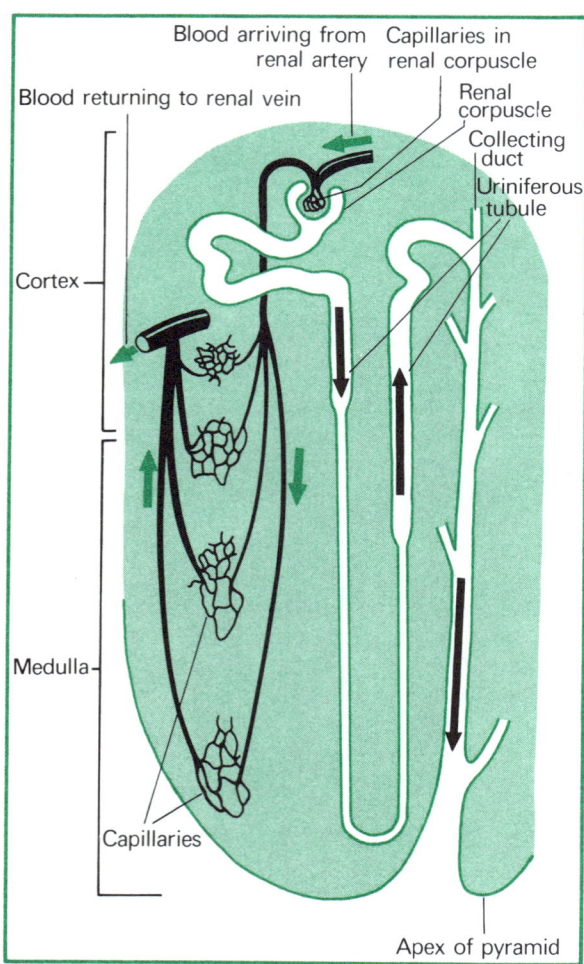

Fig. 11.5 Diagram illustrating the portions of a nephron and its associated blood vessels lying in the cortex and medulla of the kidney.

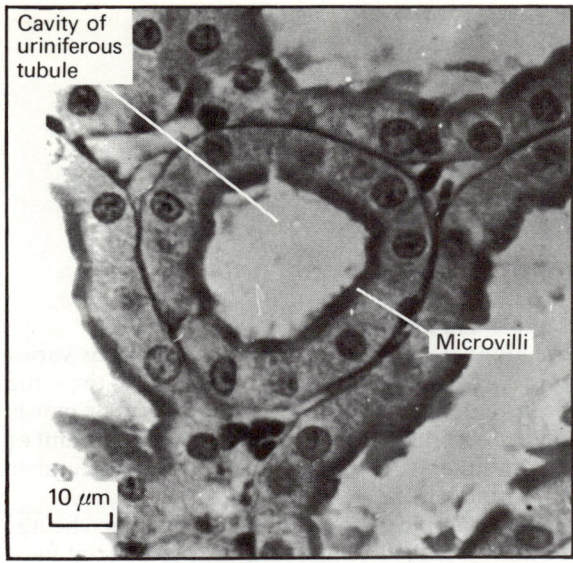

Fig. 11.7 Transverse section of the first portion of a uriniferous tubule (light microscope).

pipette was inserted into the renal corpuscle enabling a sample of freshly-formed filtrate to be extracted and analyzed. Unlike the blood plasma in the capillary knot, this did not contain blood cells or plasma proteins. As in the formation of tissue fluid (Fig. 8.6), these had been retained in the blood.

When a person is sitting at rest about one quarter of the output of his heart passes through his kidneys, and in 24 hours 170 litres of filtrate are produced (120 cm³ per minute).

Of course there are only 5–6 litres of blood in an adult man, and this contains about 3 litres of water. To make 170 litres of filtrate the blood has to be passed many times through the kidneys. Since an average of 1.5 litres of urine are passed per day, 99 % of the water that passes out of the renal corpuscles as filtrate must be re-absorbed into the blood stream, and this takes place in the uriniferous tubules.

The uriniferous tubules

These receive the freshly-formed filtrate, remove from it dissolved foods, such as glucose and amino acids, and most of its water, and allow dissolved wastes such as urea to pass on. In other words they convert filtered blood plasma into urine.

Figure 11.1 shows that glucose and amino acids, present in plasma, are absent from urine altogether. Urea and creatinine, however, are 60 and 100 times more concentrated respectively in urine than in blood plasma. The kidneys have not only removed

them from the blood, they have also concentrated them in the urine.

Much is known about how the uriniferous tubules perform these tasks, though the mechanisms are beyond the scope of this book. We should, however, note two features of many of the cells which make up the tubules. First, the face which lines the bore of the tubule is thrown up into microvilli (Figs. 11.7, 11.8). These increase the surface area of the cell considerably, and since dissolved materials are absorbed from the filtrate across the plasma membrane into the cell, the larger its surface area the more efficient this process will be. A similar consideration applied to the outermost cells of the villi in the small intestine (Fig. 6.21(e)).

Secondly, these cells contain abundant mitochondria (Fig. 11.8). We saw on p. 95 that selective uptake of ions from soil solution by root hairs required a supply of chemical energy. The same applies here, and the chemical energy (in the form of ATP) is generated in the mitochondria.

Fig. 11.8 Transverse section of a part of one cell from the first portion of a uriniferous tubule (electron micrograph).

12 Asexual and Sexual Reproduction. Human Reproduction

Reproduction means producing new organisms of a kind similar to the parent(s) in the basic features of structure, function and behaviour. Being able to reproduce is the fundamental distinction between living and non-living things.

Sometimes it is carried out by one individual (this is particularly true of unicellular organisms such as bacteria), but often it requires the cooperation of two individuals, male and female, and reproduction is better thought of in terms of a species than of an individual. Individuals have short life spans and die. But because of reproduction, species have much longer life spans—sometimes hundreds of millions of years.

To understand why reproduction sometimes involves only one individual and sometimes two we must briefly consider the structure of the cell nucleus, though this is dealt with in full in chapter 14.

Behaviour of the nucleus in reproduction

The cell theory (page 1) states that 'All cells arise from previously-existing cells'. Reproduction therefore involves the division of cells into smaller cells which then grow and differentiate. When cell division is studied it is found to centre around division of the nucleus. This contains instructions in the form of a chemical code which 'tells' the newly-formed cell how to make proteins, how to grow, and how to become a type of cell appropriate to its position and function in the developing organism.

By making extracts of tissues which contain large and plentiful nuclei, such as fish sperm, the nuclear material can be isolated and chemically analyzed. It is found to consist of **nucleoprotein**, a special form of protein in which protein molecules are associated with **deoxyribonucleic acid** (**DNA**). It is the DNA which embodies the **genetic code** or hereditary instructions.

DNA is located within the nucleus in organelles called **chromosomes** (Fig. 14.11). These form the **genetic material**.

Haploid and diploid cells

The number of chromosomes in the nucleus varies from species to species but all members of the same species possess the same number. In other words all members of a species possess the same amount of genetic material distributed between a fixed number of chromosomes.

For example, the cells of all human beings contain 46 chromosomes; the cells of all fruit flies, *Drosophila melanogaster*, contain 8 chromosomes. Within a particular individual human being every set of 46 chromosomes is exactly the same as every other set of 46: every cell contains exactly the same genetic material.

Between individuals, portions of the genetic material differ. A blue-eyed person contains slightly different genetic material from a brown-eyed person, though both persons possess 46 chromosomes in all of their cells.

But there is an important exception to this. The **gametes** or **sex cells** (**sperms** and **ova**) contain only *half* as many chromosomes as all the other cells. In human beings, the sperms and ova contain 23 chromosomes each; in *Drosophila melanogaster*, four each. Sperms and ova are said to contain the **haploid** number of chromosomes. All the other cells of the body contain the **diploid** number of chromosomes (Fig. 12.1).

If this were not the case and sperms and ova contained the same diploid number of chromosomes as all the other cells, then when human gametes joined at fertilization to produce a zygote, this would contain 92 chromosomes, twice as much genetic material as the cells of its parents.

Sexual reproduction

The *essential feature* of **sexual** reproduction is that two haploid gametes unite (**fertilization**) to form a diploid zygote from which a new individual develops. How the gametes are produced and how they are brought together varies enormously throughout the living world. Taking two examples, in human beings sperms are produced in one

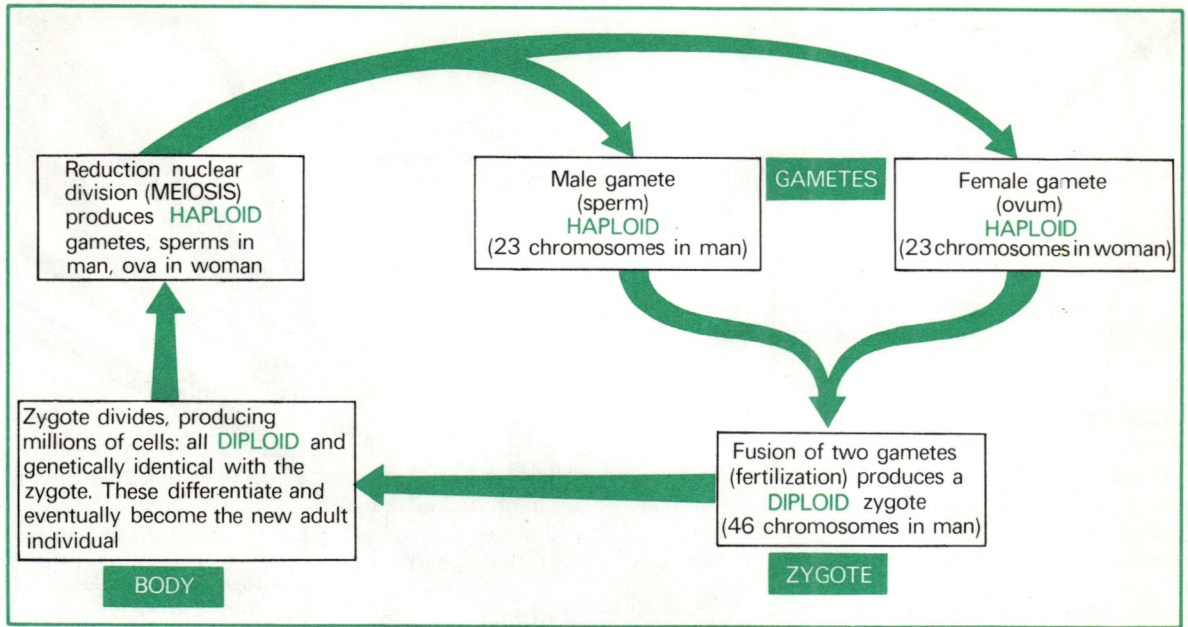

Fig. 12.1 The behaviour of the nucleus in the cycle of sexual reproduction.

individual (the man) and ova in another (the woman). That is, humans, like most animals, are **unisexual**. For the gametes to be brought together sperms are transferred from the man's body into the woman's body by the act of copulation.

By contrast, most flowering plants are not unisexual but **hermaphrodite**, that is one individual plant produces both male and female gametes. These are brought together by transfer of the male gamete (contained in the pollen grain) to the female gamete (contained in the carpel) by wind, water, insects, or other animals. But both humans and flowering plants share the same fundamental pattern of sexual reproduction, the production of haploid gametes by diploid individual(s) and the joining together of these at fertilization to give a diploid zygote from which the new organism develops.

With very few exceptions all living organisms of the most diverse kinds follow this pattern. It is one of the basic features of biological organization. This means that it must have evolved very early in the history of life and has been retained as life radiated over vast periods of time into the present diversity of organisms. This has occurred because sexual reproduction confers a major advantage on organisms possessing it. This is that the offspring are *not genetically identical with their parents*. You, for example, though you are broadly similar in basic features of structure, function, and behaviour

to your parents, differ from them in many details, and this is because the genetic material in your cells is derived partly from your father and partly from your mother.

We shall see in chapters 14 and 20 why this is a major biological advantage.

Asexual reproduction

However, reproduction can occur without the production and fusion of gametes, when it is said to be **asexual** (without sex). It involves the formation of a part of an individual which can become separated from the 'parent' and start an independent existence. (The word 'parent' is placed in inverted commas because no gametes are involved.)

All the cells produced are, like those of the parent, diploid. The biological advantage conferred by halving the number of chromosomes in the formation of gametes and restoring it in the zygote at fertilization is absent. New individuals produced by asexual reproduction are genetically identical with the parent.

Examples of asexual reproduction

Flowering plants Several examples of asexual reproduction were illustrated in chapter 3—rhizome of Solomon's Seal, potato tuber, tulip bulb, and crocus corm (pages 27–29).

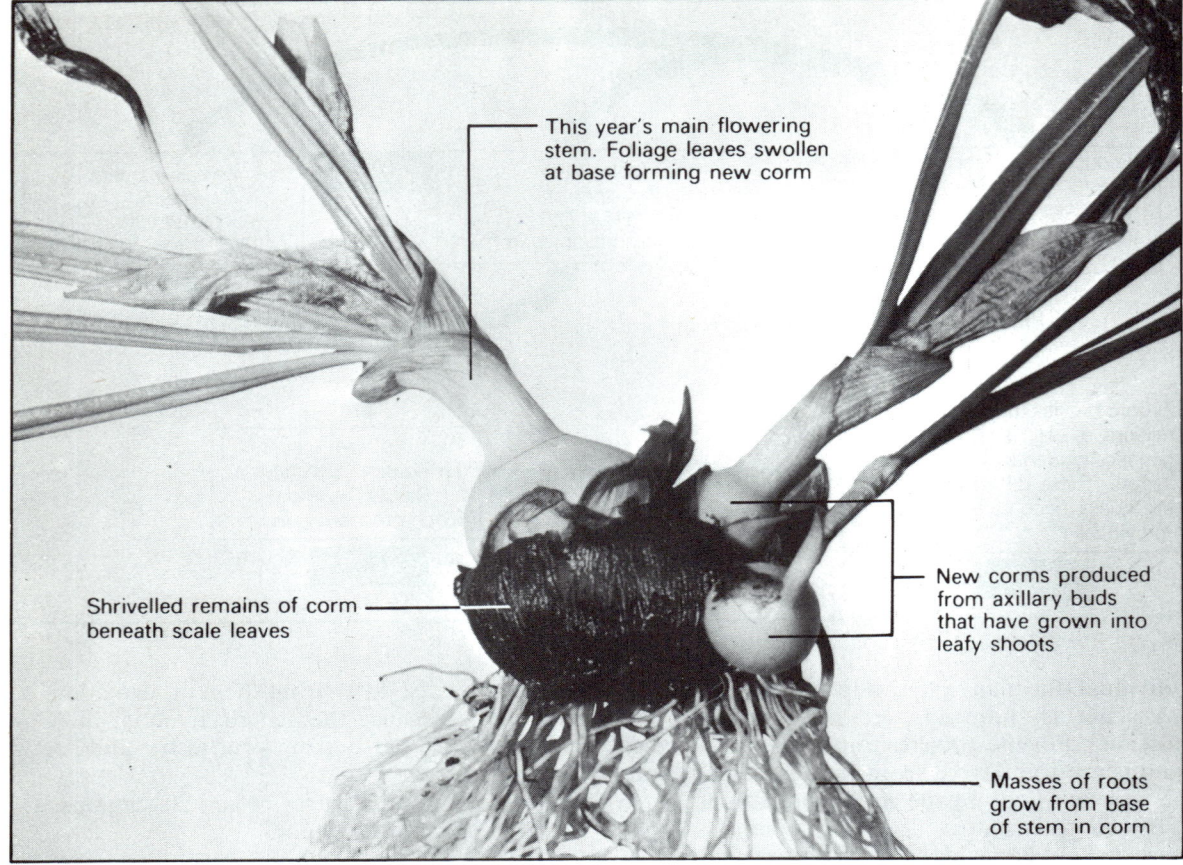

This year's main flowering stem. Foliage leaves swollen at base forming new corm

New corms produced from axillary buds that have grown into leafy shoots

Shrivelled remains of corm beneath scale leaves

Masses of roots grow from base of stem in corm

Fig. 12.2 Crocus: three new corms have been produced during this season's growth.

Compare the crocus corm before planting (Fig. 3.15) with a similar corm towards the end of the growing season (Fig. 12.2). The shrivelled remains of the spring corm lie beneath the scale leaves. At the base of the foliage leaves the new season's stem has swollen to produce a new corm, and two others have developed from axillary buds which also grew out in the new season. The old corm has reproduced itself, forming three new corms which can be separated and will grow independently of each other.

Animals Plants have a more restricted range of cell and tissue types than most animals, and a detached part can often develop the tissues and organs it lacks. But most animals are highly differentiated and a detached part of an animal does not usually contain all the tissues and organs necesary for life.

For example, a detached human limb lacks a blood supply from the heart, and so receives no oxygen from the lungs and no food from the gut and liver. It could not, therefore, grow and produce a new man. Only animals with a simple body structure and little or no organ differentiation are capable of asexual reproduction. It is common among unicellular animals and also in coelenterates (sea anemonies, corals, hydra) and echinoderms (starfish). Figure 12.3 shows hydra with an asexually-produced 'bud' which would eventually have become detached from the parent and led a separate existence.

Oogamy in sexual reproduction

The essential feature of sexual reproduction is the production of haploid gametes which join in fertilization to form a diploid zygote. From this the new organism develops (Fig. 12.1).

In some species male and female gametes are produced in equal numbers and look much alike. But the most common arrangement is for the male to produce very large numbers of small, mobile gametes (sperms), consisting of little more than a swimming nucleus, and the female to produce only a few relatively large and immobile gametes (ova) containing food reserves on which the developing zygote can draw. For example, a man releases about 400 million sperms each time he copulates, but a woman releases only one ovum each month.

This pattern occurs in the algae, mosses and liverworts, ferns, almost all invertebrates, all vertebrates, and, in a modified form, in conifers and flowering plants. It is called **oogamy** and is evidently both ancient and successful, having spread from remote ancestors to all the main divisions of the plant and animal kingdoms.

Oogamy and life on the land

Oogamous organisms like hydra which live in water can achieve fertilization merely by releasing their gametes. The mobile sperms may then encounter the immobile ova by chance. In many organisms one of the gametes secretes a chemical substance which attracts the other, making fertilization more likely.

Such effortless transfer of male gametes to female gametes cannot apply to land-living organisms. When life emerged from the water to the land various mechanisms were evolved to overcome this difficulty. Many vertebrates (and all mammals,

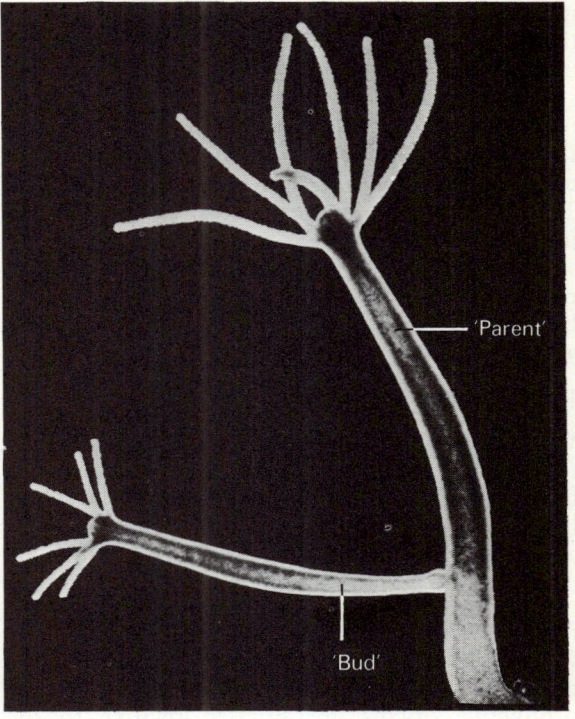

Fig. 12.3 *Hydra* budding (asexual reproduction).

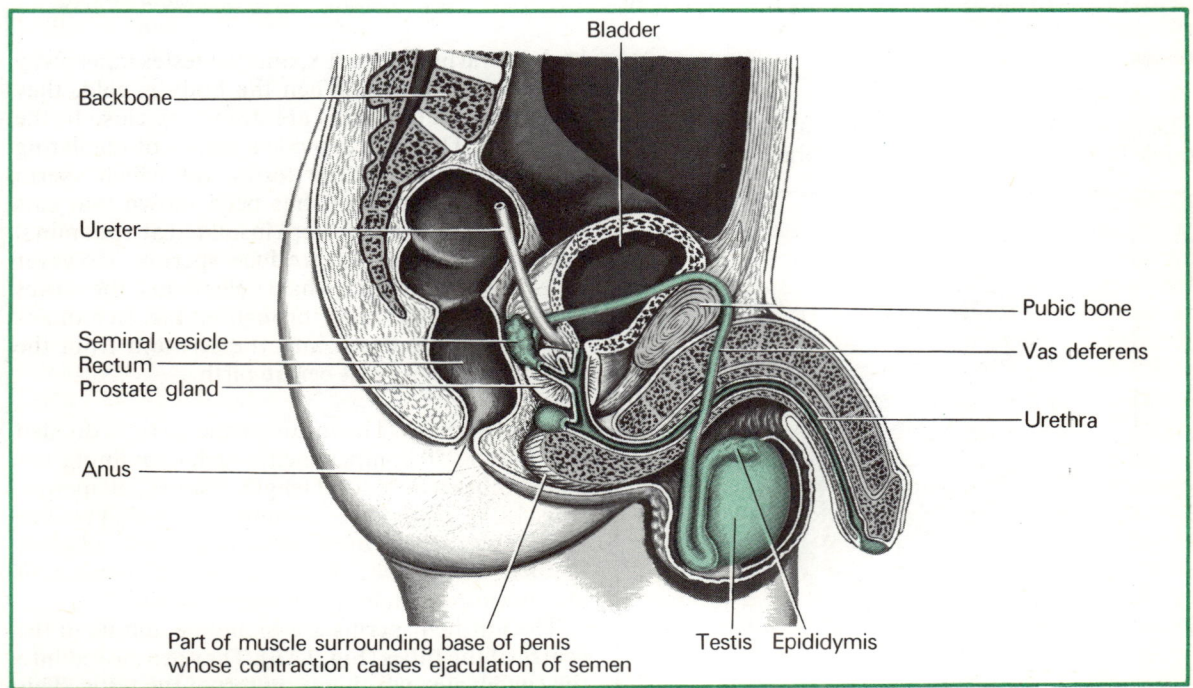

Fig. 12.4 Reproductive organs of a man, showing their relationship to other organs.

including man) copulate. This means that part of the body of the male animal, the penis, is inserted into the body of the female and liquid containing sperms (**semen**) is discharged close to the place inside the female body where the ovum is released. In this way the gametes are not exposed to the environment, where they would dry out and die. They are passed from the inside of one body to the inside of another.

Like many other animal activities, copulation requires active movement, and flowering plants have found a very different solution to this problem which we shall examine on page 134.

Reproduction in mammals

We will consider human beings as our example, for the obvious reason of natural interest and because more is known about human reproduction than about that of any other mammal. But the anatomy, the principles, and much of the behaviour described are common to all mammals.

Man

Figure 12.4 is a drawing of the lower abdomen of a man seen in section and showing how the reproductive organs are situated relative to the bladder, the rectum, and the backbone. Look back to Fig. 2.1 and observe the external genitalia, the penis (here **circumcised** and showing the opening of the urethra), and the **scrotum** containing the two **testes**.

The testes

The skin of the scrotum is unusual in that it contains large numbers of muscle fibres. When the

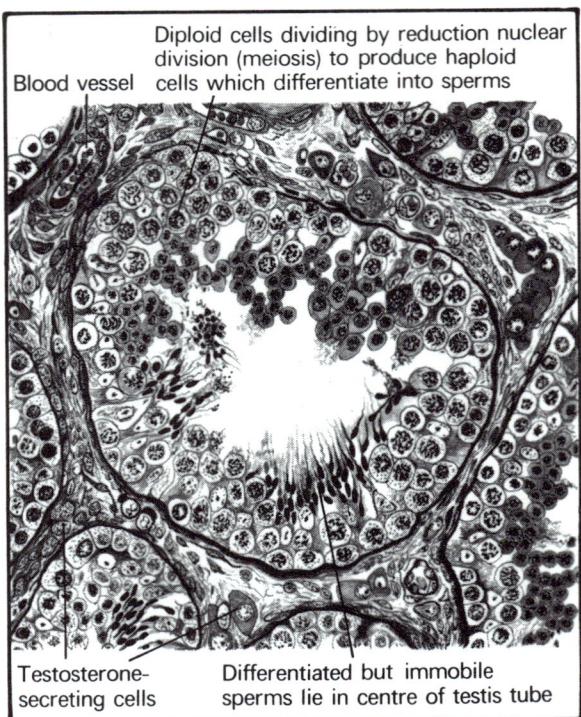

Fig. 12.6 Section of human testis. Sperm-producing cells line the walls of the tubes. Mature sperms lie in the centres of the tubes.

Labels on figure 12.6:
- Blood vessel
- Diploid cells dividing by reduction nuclear division (meiosis) to produce haploid cells which differentiate into sperms
- Testosterone-secreting cells
- Differentiated but immobile sperms lie in centre of testis tube

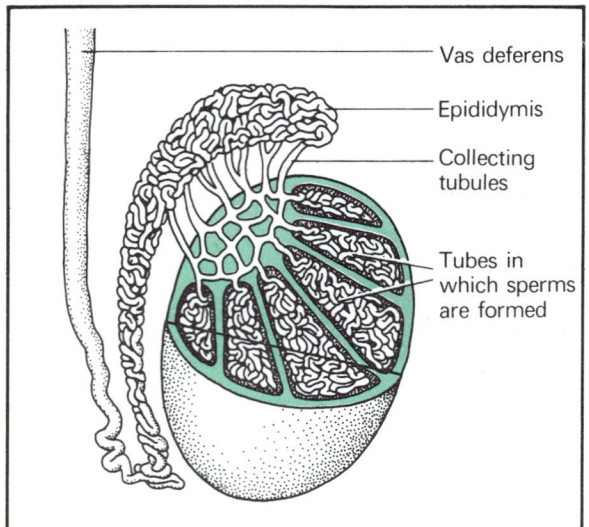

Fig. 12.5 Testis, epididymis, and vas deferens of a man.

Labels on figure 12.5:
- Vas deferens
- Epididymis
- Collecting tubules
- Tubes in which sperms are formed

body is warm these relax, and the testes hang away from the abdomen. When the body is cold, they contract, and the testes are drawn up close to the abdomen. This may provide a means of regulating the temperature of the testes, on which sperm production depends. It has been shown that cats whose testes have been kept insulated at abdominal temperature cease to produce sperms. However in a few mammals (whale, elephant) the testes remain in the abdomen throughout life. In humans they are pulled down into the scrotum from the abdominal cavity just before birth.

Inside the testis   The inside of the testis is divided into about 250 compartments each containing 1–4 coiled tubes whose total length is some 250 metres. (Fig. 12.5). Diploid cells lining their walls produce sperms by a special nuclear division called *reduction division* or *meiosis*. This halves the number of chromosomes, making the sperms haploid.

The finished sperms are immobile and lie in the centre of the tubes (Fig. 12.6). They are moved into the **epididymis**, which lies on top of the testis. This is another much-coiled tube and here the sperms are stored prior to ejaculation.

When a man reaches orgasm the sperms are moved by peristalsis along the **vas deferens** to a point at the start of the urethra (Fig. 12.4). Here the vas deferens is joined by the vas deferens from the other testis and by the two **seminal vesicles**. Surrounding this junction is the **prostate gland**. The seminal vesicles and the prostate gland secrete liquid containing nutrients which is added to the sperms, making semen. Then 8–10 powerful and violent contractions occur in the muscles surrounding the base of the penis. These squeeze the semen and shoot it out of the urethra through the erect penis and into the top of the vagina (Fig. 12.7). The onset of orgasm can to some extent be controlled voluntarily, but once it has commenced it is uncontrollable.

Hormonal functions of the testis The prime function of the testis is to produce sperms, but between its tubes are islands of cells (Fig. 12.6) which secrete the hormone **testosterone**. This has a widespread effect on the growth and development of boys. It causes increase in muscularity, growth of body hair (including the beard), growth of the larynx and so the breaking of the voice, and growth of the penis.

These are all **secondary sexual characteristics**, and if a boy is castrated they do not develop and he becomes a eunuch. Castration is commonly practiced on horses (which develop into geldings) and bulls (which develop into bullocks) in order to reduce their aggressive behaviour.

The **primary sexual characteristic** is the possession of either ovaries or testes.

The penis
The penis is one of the few parts of the male body whose skin bears no hair. Its tip (**glans**) is covered by a sheath of skin, the **foreskin**, which retracts when the organ is erect (Fig. 12.7). The foreskin is often removed shortly after birth in a small operation called circumcision.

A transverse section of the penis (Fig. 12.8) shows that it contains three shafts of spongy tissue. The bottom one is nearest the body when the penis is not erect and encircles the urethra, which is a common duct for both semen and urine. This shaft swells out at its end into the mushroom-shaped glans.

The penis is richly supplied with blood vessels and its skin (especially the glans) contains large numbers of touch-sensitive nerve endings. When a man is stimulated by touch or by psychic stimuli, blood is delivered to his penis faster than it can be drained away. This causes erection and the organ increases very considerably in length and girth.

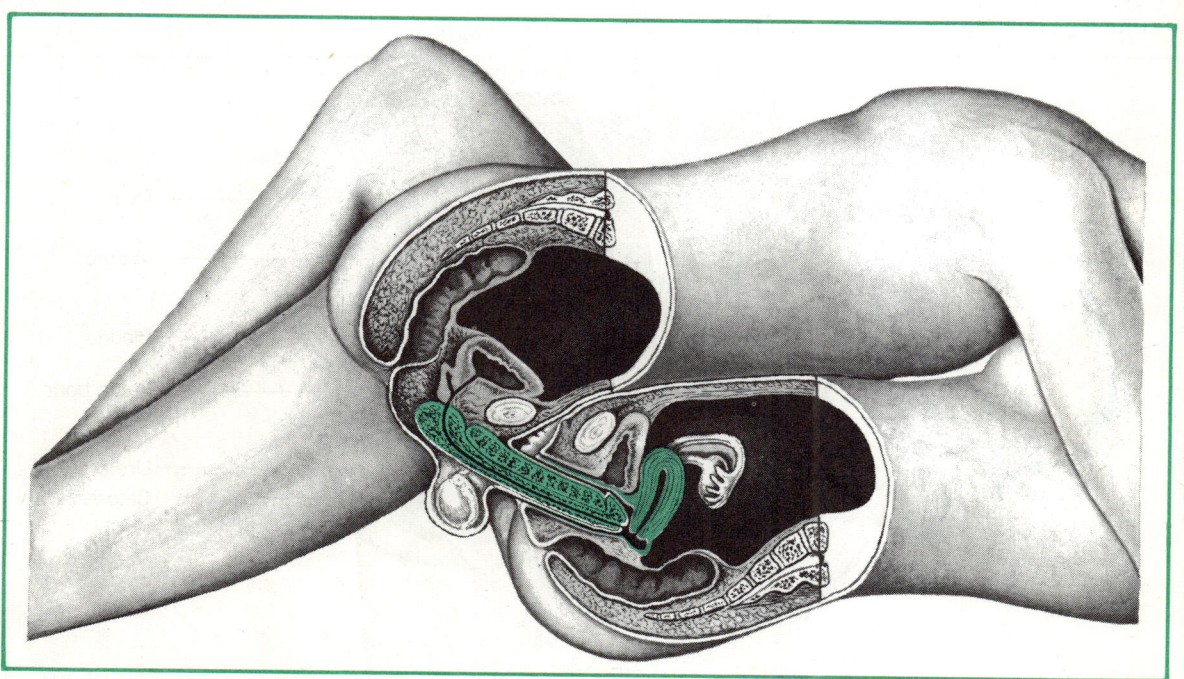

Fig. 12.7 Copulation. Note how the angle made by the erect penis with the man's body is complementary to the downward inclination of the vagina.

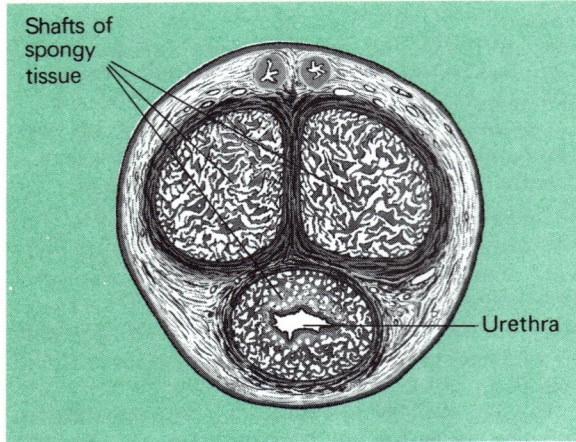

Shafts of spongy tissue

Urethra

Fig. 12.8 Transverse section of the penis of a man.

It now points upwards at an angle away from the man's body and in the most frequently adopted position in human sexual intercourse (Fig. 12.7) this is complementary to the downward and inward solpe of the vagina. The spongy tissue extends backwards beneath the pubic bone (Fig. 12.4), and the erect penis cannot be bent downwards. Neither is it possible for a man to pass urine when it is in this condition.

Woman

Figure 12.9 shows how the reproductive organs of a woman lie relative to the skeleton of her hip basin,

and to her bladder and rectum.

Her ovaries lie at the back of her body. At birth each already contains about 500 000 haploid ova, and no more are made subsequently. Between the

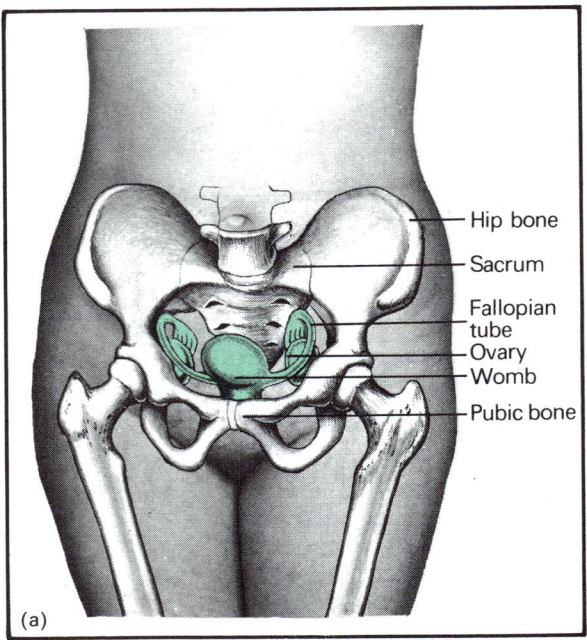

Hip bone
Sacrum
Fallopian tube
Ovary
Womb
Pubic bone

(a)

Fig. 12.9(a) Positions of the womb and the ovaries in a woman. **(b)** Reproductive organs of a woman, showing their relationship to other organs.

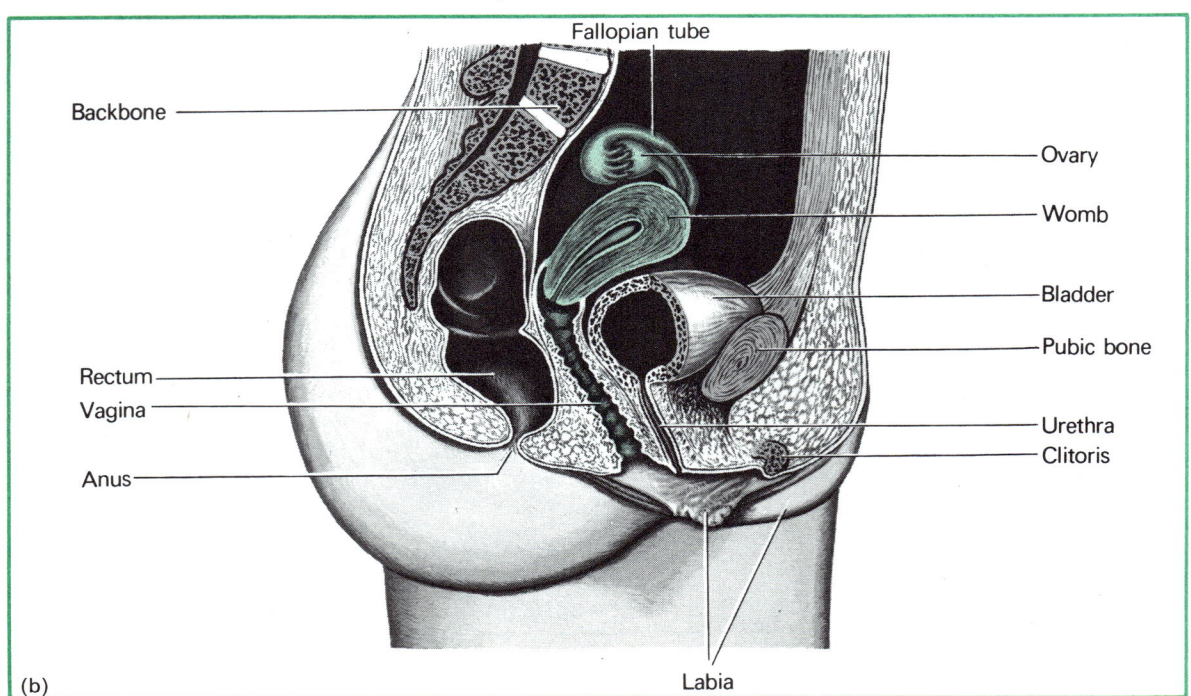

Fallopian tube

Backbone

Ovary

Womb

Bladder

Pubic bone

Rectum

Vagina

Urethra

Clitoris

Anus

Labia

(b)

ages of about 13 (the age of **puberty**) and about 45 (the **menopause**, or change of life), a woman liberates one ovum per month, a total of about 500; the rest degenerate. The ovum is shed into the abdominal cavity, but at the moment that this happens the frilled end of the **Fallopian tube** (Fig. 12.10) moves and clasps the ovary, thus ensuring that the ovum passes into it.

The tube is filled with liquid and the cells lining its inner wall bear tiny hairs, like large microvilli, which move and waft the liquid, and the ovum which is suspended in it, towards the womb. If the woman has had sexual intercourse at about the time the ovum is shed, sperms may meet it and one may penetrate it (fertilization). Sperms live and swim in the Fallopian tube for about 24 hours and the ovum is capable of being fertilized for up to 36 hours after it has left the ovary.

Leading downwards from the neck of the womb is the **vagina** or **birth canal** into which the penis is inserted during sexual intercourse. It opens to the exterior at the **vulva**. This is obscured by pubic hair in Fig. 2.1 but can be seen in Fig. 12.26(c) because it has been shaved for childbirth. The vulva is flanked by the inner and outer lips or **labia**.

The urethra is a separate duct and also has a separate opening from that of the vagina (Fig. 12.9(b)). Lying above this opening is a small organ, the **clitoris**, which corresponds in its development to the penis. During sexual intercourse it too becomes erect and its stimulation is essential for the woman's satisfaction. It contains the highest concentration of sensory nerve endings of any part of the body, male or female.

Menstruation

The lining of the womb is called the **endometrium** and can be seen in Figs. 12.10 and 12.14(b). It is about 2–3 mm thick, has a spongy texture, and contains many blood capillaries. Each month after puberty it is freshly prepared for the possible reception of a developing zygote, though this occurs only a very few times in a woman's life, perhaps never.

If no developing zygote arrives, the endometrium becomes detached from the womb and passes out of the body through the vagina. This process is called **menstruation** or **monthly period**. An adult woman loses between 50 and 500 cm³ of blood at each period, which may last for four or five days.

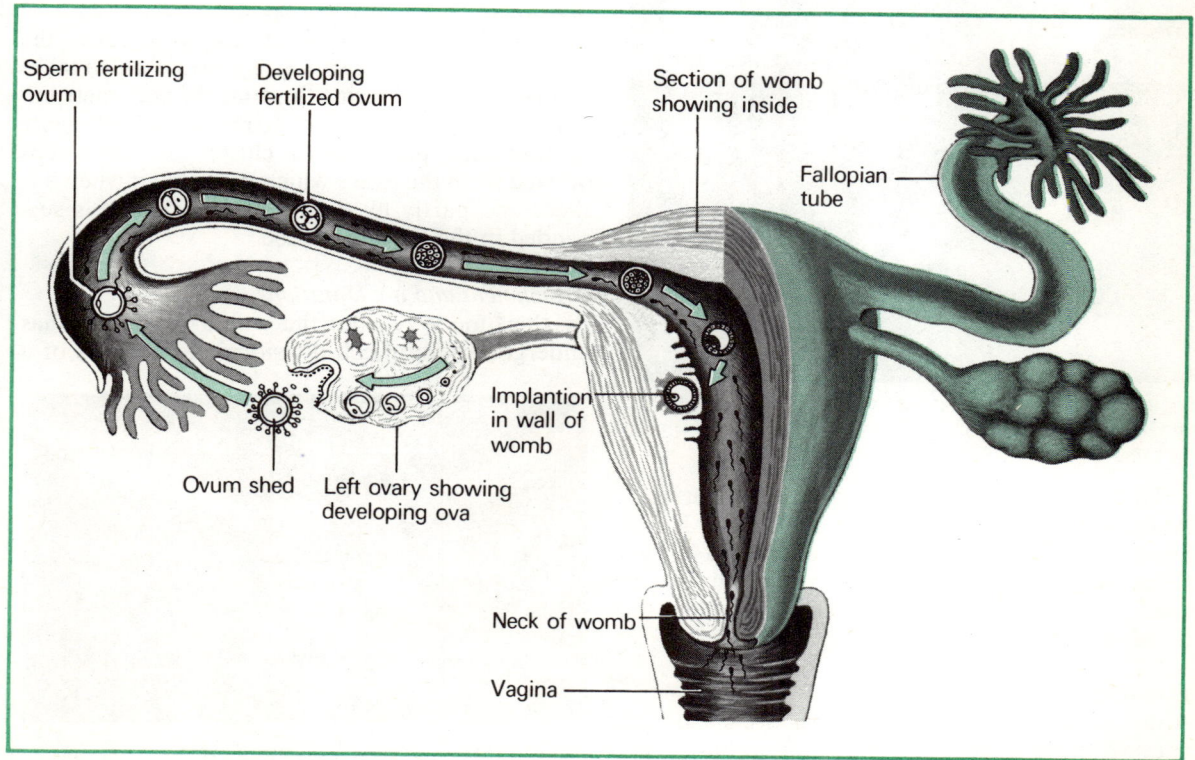

Fig. 12.10 Ovulation, fertilization, early development of the embryo, and implantation.

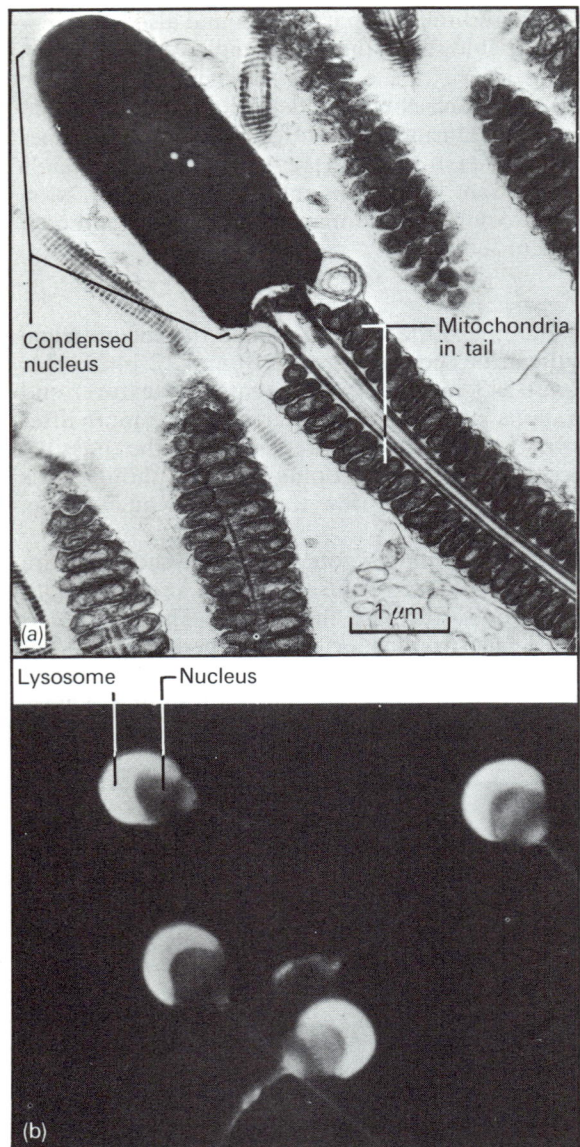

Fig. 12.11(a) Section of a bat sperm (electron micrograph). **(b)** Guinea pig sperm stained to show the lysosomes which cap their nuclei (light microscope).

The age when periods commence is very variable. About 100 years ago in northern Europe it was about 17, but today the average age is 14. This seems to be one of the features of twentieth century civilization, and it is not possible to say when the trend towards lowering the age will stop.

Fertilization
Although some 400 million sperms are ejaculated at one human orgasm, it is thought that only a few thousand gain access to the womb and swim up the Fallopian tube (Fig. 12.10). Full details of human fertilization are unknown for it cannot be observed in a living person. Only one sperm penetrates and so fertilizes the ovum, but the presence of a large number is necessary in order for one to penetrate, and if a man produces less than about 35 million sperms in his ejaculate he is likely to be infertile.

Figure 12.11(a) is a longitudinal section through the sperm of a bat and shows how the cell consists of little more than a condensed nucleus propelled by a tail in which there are large numbers of mitochondria. These provide the ATP necessary for movement, but since the sperm contains no food reserve it depends for chemical energy on sugar contained in the semen.

Mammalian sperms have their nuclei capped by by lysosomes containing digestive enzymes which dissolve a hole in the membrane of the ovum, permitting the sperm to enter. The lysosome can be seen if Fig. 12.11(a) is carefully examined. In Fig. 12.11(b) the lysosomes have been specially stained.

Figure 12.12 shows a human sperm about to penetrate a human ovum. When the haploid sperm nucleus has joined with the haploid ovum nucleus, the cell will be a diploid zygote, from which all the cells of the adult body will be derived by cell divisions.

The zygote is wafted on its way down the Fallopian tube to the womb, a journey which takes from 4 to 7 days. During this time the endometrium of the womb thickens ready to receive it. The zygote is nourished partly by a cluster of small cells derived from the ovary which surround it when it is shed, and partly by the liquid in which it is suspended in the Fallopian tube.

Attachment and implantation
By the time it reaches the womb the zygote has undergone several cell divisions and consists of a

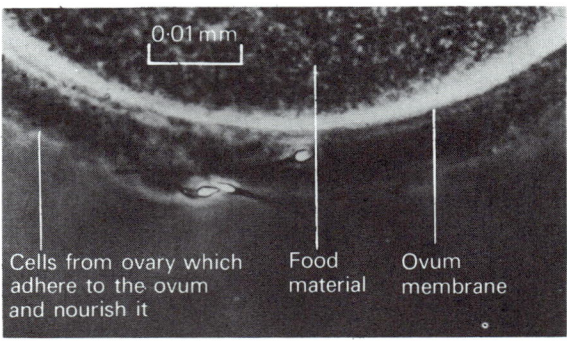

Fig. 12.12 Sperm penetrating a human ovum.

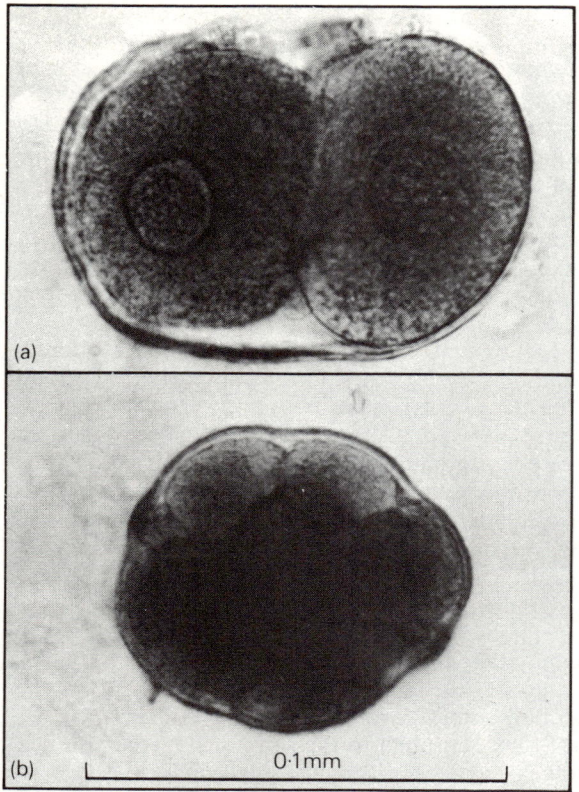

Fig. 12.13 Early stages in development of a human zygote. (a) two cell stage; (b) 16 cell stage.

hollow ball called the **blastocyst**. Figure 12.12 shows that between the ovum and the small cells which cluster round it is a thick membrane. The cell divisions which form the blastocyst take place within this, and it can still be seen in Fig. 12.13. When the blastocyst adheres to the endometrium it is no bigger than the ovum.

It rapidly becomes embedded in the endometrium, a process called **implantation**. First the ovum wall is shed (Fig. 12.14(b)), then it sinks inwards, perhaps digesting its way by secreting enzymes which dissolve endometrial cells. The surface layer of the endometrium rapidly grows over it and covers it. The woman is then said to be **pregnant**.

Figure 12.15 is a section through an implanted blastocyst. Note that the endometrium has grown over it. It is already differentiated into an **inner cell mass**, which will give rise to the baby and the **umbilical cord**, and a **wall**, which will produce the **chorion**. This is a shaggy outer coat which, on the side towards the wall of the womb, develops into the placenta.

Twins

It is usual for a woman to bear one baby at a time, but in Europe about one childbirth in 85 produces a pair of twins. One quarter of these pairs are **identical** and three quarters **non-identical**, and these may even be of opposite sexes.

Non-identical twins result from the woman discharging two ova at the same time, both of which

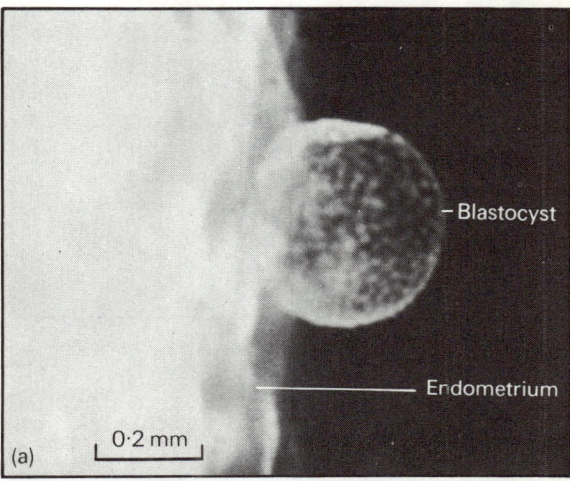

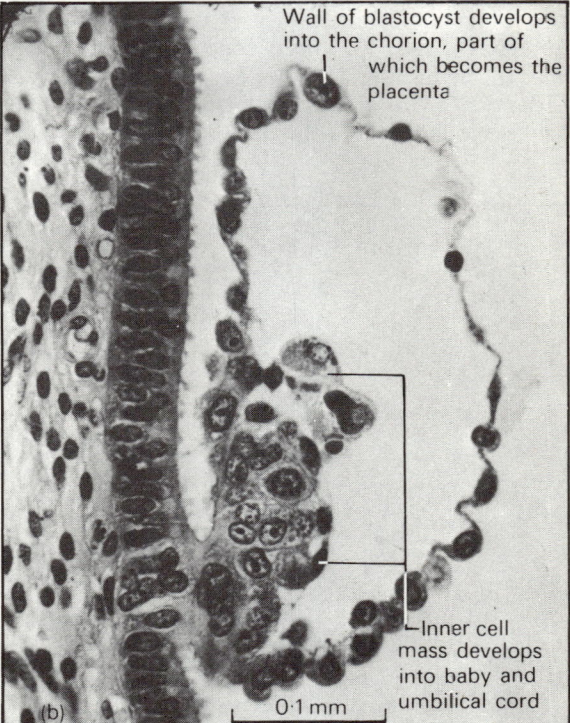

Fig. 12.14(a) Human blastocyst attached to the endometrium, seen from the side. **(b)** As (a), but seen in section.

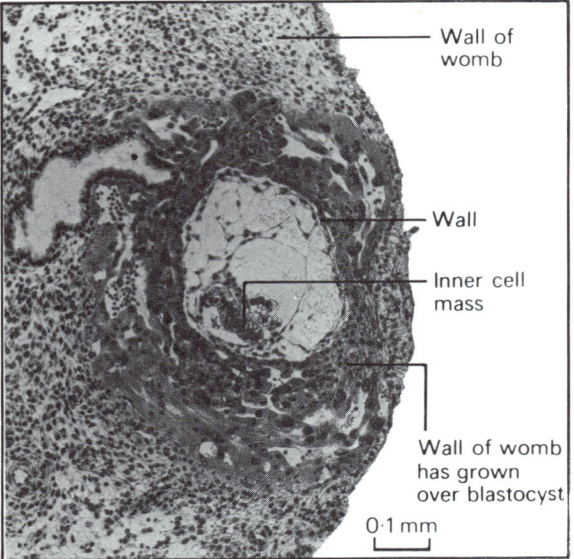

Fig. 12.15 Section through an implanted human blastocyst.

become fertilized and implanted. Each develops its own amnion, chorion, and placenta. Because all ova and all sperms are genetically different from each other, the two zygotes resulting from this double act of fertilization are as genetically dissimilar as if they had been conceived at different times in the mother's life and were not twins.

Identical twins arise from a single zygote. At an early stage the cells of the blastocyst separate into

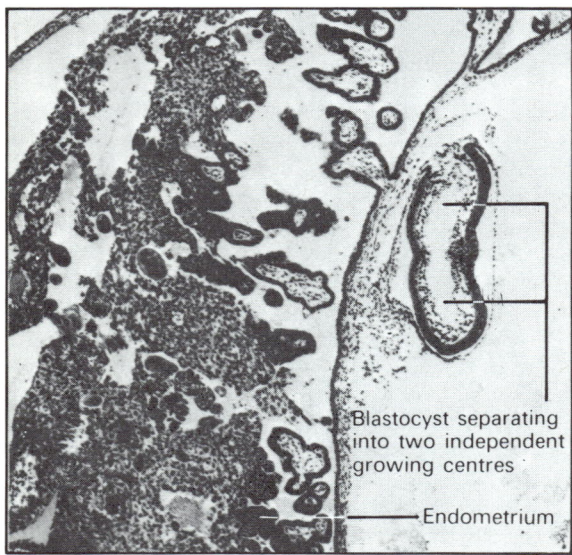

Fig. 12.16 Implantation of a twin human blastocyst.

two groups, each of which acts as an independent growing centre. It is impossible to observe this separation in a living person, but it probably occurs at the two-cell stage (Fig. 12.13(a)). Siamese twins, permanently-joined pairs of individuals, probably arise from an incomplete separation of the blastocyst at a later stage. Figure 12.16 is a remarkable photograph of a section through a developing human blastocyst which has already separated into two growing centres and would probably have given rise to twins. It corresponds to the stage seen in Fig. 12.14(b).

Identical twins share a single placenta, though each develops its own umbilical cord and amnion (Fig. 12.17). All the cells in both growing centres, being derived from the same zygote, contain exactly the same 46 chromosomes, that is they carry the same genetic information and are genetically identical. Any differences between identical twins therefore result from differences of environment, whether these be in the womb or in the outside world.

Later stages of development
Some of the later stages of development of the human embryo are shown in Fig. 12.18. They are not photographed to the same scale—the younger stages have been magnified because they are so minute, The sizes stated are from the crown to the rump.

Make a careful study of this figure and write an account of the visible differences stage by stage, refering to the following features: gill pouches (see page 18); somites (the serially-arranged blocks of tissue at the back of the embryo); limbs, including fingers and toes; eyes; ears; nose; mouth; head; chin; neck; brain; umbilical cord; rump; tail.

Amnion, chorion, and placenta
Figure 12.19 is a life-size drawing of a human embryo in the womb three weeks after fertilization. The inner cell mass has given rise to the **embryo** and the beginnings of the umbilical cord. The wall of the blastocyst has given rise to the chorion, a protective bag still covered by the endometrium. Its walls bear large numbers of capillary tufts, giving it a shaggy appearance.

As development proceeds the part of the chorion in contact with the womb becomes the placenta. Here the blood vessels develop further, but in the rest of the chorion they degenerate (Fig. 12.20). The whole structure projects into the cavity of the womb. The embryo is also surrounded by an inner

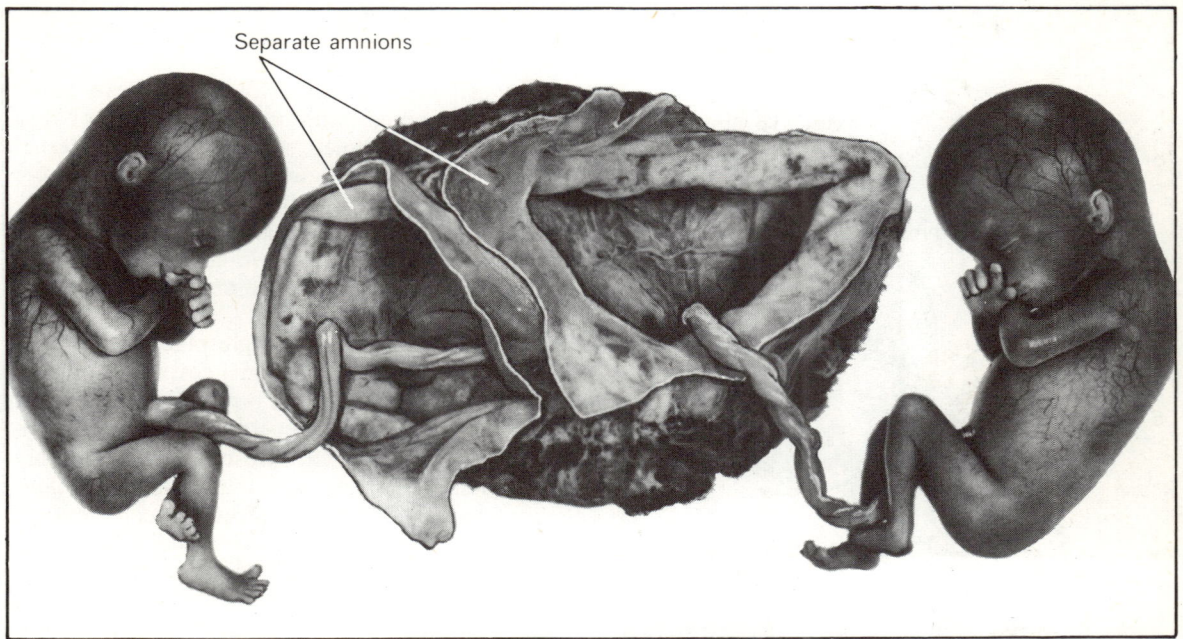

Fig. 12.17 Identical human twins attached to one placenta by separate umbilical cords.

membrane, the **amnion** or **caul**, lying within the chorion. This is full of liquid which buffers the embryo from physical shocks and, by acting as a thermal insulator, keeps it at an even temperature.

Figure 12.21 is a life-size photograph of a 7 week old human embryo lying inside its amnion. The chorion on the observer's side has been removed, but on the far side, lying behind the embryo, it can be seen developing into the placenta with its masses of shaggy blood vessels.

Respiration, nutrition, and excretion of the embryo
All the baby's organs are formed by the end of the seventh month of pregnancy, but it neither breathes nor feeds nor excretes. The alveoli of its lungs are in any case crumpled and do not expand until the first breath is taken; and the baby makes no breathing movements with its chest or diaphragm. Respiration, nutrition, and excretion are carried out on the baby's behalf by the mother via the placenta. She breathes for it, eats for it, and excretes for it.

By the end of the second month the embryo has its own beating heart, blood, and blood vessels. Its blood passes to the placenta in the two **umbilical arteries** and is returned in the **umbilical vein** (Fig. 12.22). The fully-grown human placenta is about the size of a dinner plate. The many branches of its arteries and veins can be seen in Fig. 12.23. They

give rise to tree-like tufts of capillaries. These lie in spaces surrounded by the mother's blood which drains into the spaces from her arteries and out of them from her veins.

The volume of the placenta is about 500 cm^3 of which about 350 cm^3 is occupied by the capillary tufts on the baby's side. The total length of the capillary tufts is estimated at about 50 km, and their total surface area at about 14 m^2. These figures emphasize the large surface area available for exchange of dissolved substances by diffusion from the baby's to the mother's blood and vice versa.

The mother's blood arrives at the placenta directly from her aorta. It contains high concentrations of dissolved foods and oxygen and low concentrations of urea and carbon dioxide. The baby's blood arrives at the placenta depleted of food and oxygen, which have been consumed by its active cells, and loaded with urea and carbon dioxide.

Food and oxygen diffuse from the mother's blood in the blood spaces into the baby's blood in the capillary tufts. Urea and carbon dioxide diffuse the other way. All the baby's blood (400 cm^3) passes through the placenta each minute, as does 500 cm^3 of the mother's blood, which under resting conditions is one tenth of the output of her heart.

Childbirth
The average **gestation period** (time spent in the

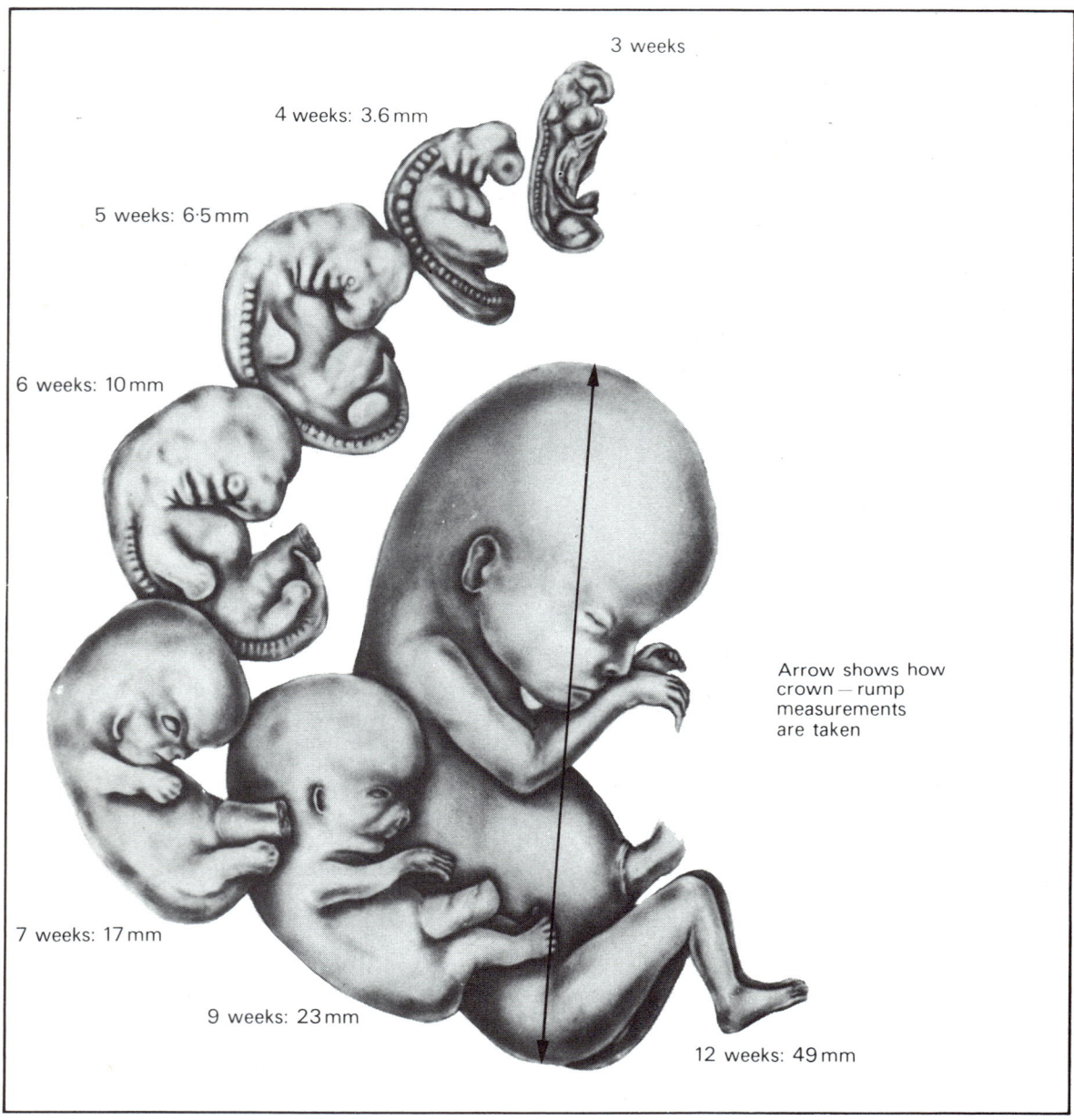

3 weeks

4 weeks: 3.6 mm

5 weeks: 6·5 mm

6 weeks: 10 mm

Arrow shows how
crown — rump
measurements
are taken

7 weeks: 17 mm

9 weeks: 23 mm

12 weeks: 49 mm

Fig. 12.18 Seven human embryos.

womb) for a human being is 267 days, that is 38 weeks or nine calendar months. A baby born prematurely at the end of seven months can usually survive if kept warm in an incubator and fed on drops of milk. If it is born before this it dies, an event called **miscarriage**.

In the later stages of pregnancy the normal position occupied by the baby is upside down with its head resting in the mother's hip basin. Figure 12.24 shows the skeleton of the hip basin. It is bounded at the back by the **sacrum** and **caudal vertebrae** (not shown in this view), and at the sides and front by the bones of the pelvic girdle, meeting at the junction of the **pubic symphysis**. Passing through this cavity, from front to back, are the urethra, the vagina, and the rectum (Fig. 12.9(b)).

The baby has to pass through the vagina, which is muscular and can stretch. The skeleton of the

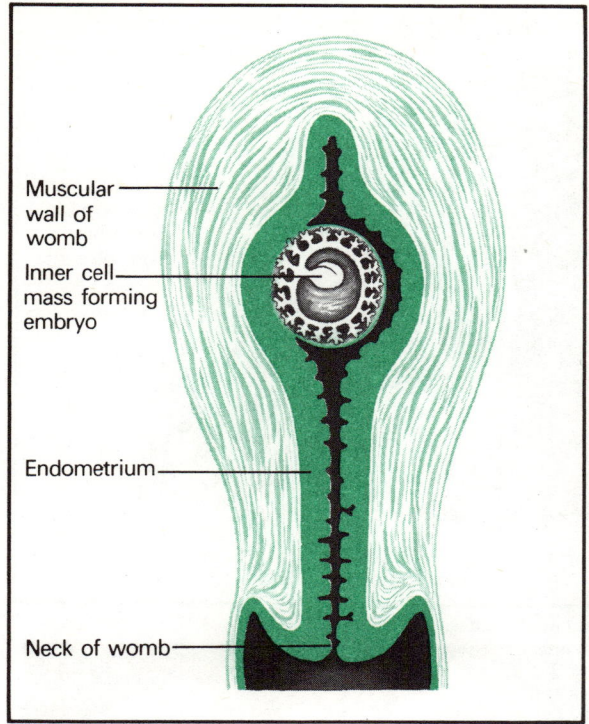

Fig. 12.19 Three week human embryo in the womb, life size.

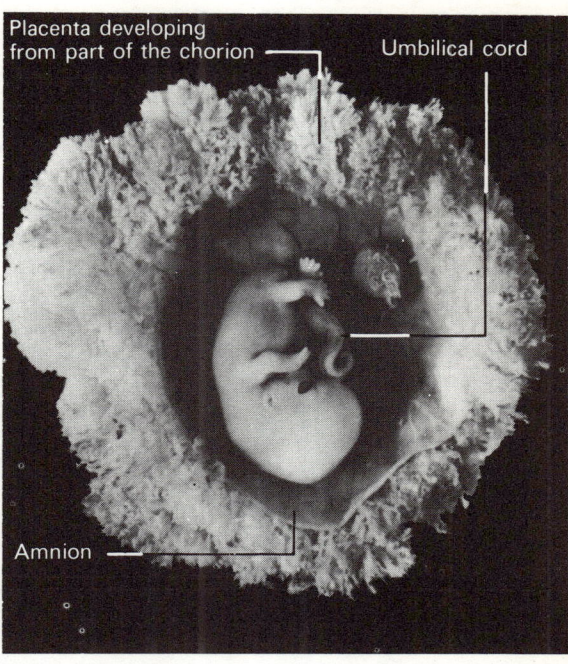

Fig. 12.21 Seven week human embryo. Amnion and placenta are clearly seen. The chorion on the observer's side has been removed, but on the far side the capillary tufts are seen. These are the developing placenta.

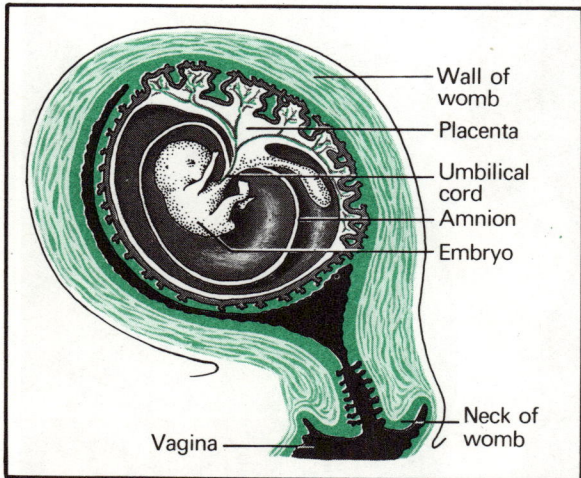

Fig. 12.20 Seven week human embryo in position in the womb.

hip basin cannot stretch, but the caudal vertebrae are pulled into a less curved position during childbirth, allowing the baby a little more room to pass (Fig. 12.25(c)).

The first stage of **labour** lasts 18–24 hours for a woman's first baby and 6–12 hours for subsequent ones. The muscular wall of the womb contracts, pressing the baby against the neck of the womb, which is thus stretched open (Figs. 12.25(a) (b)). The baby's head, by far the largest part of its body, can be moulded somewhat to fit the birth canal because the bony plates which make up the skull (see Fig. 6.12) are able to overlap slightly without damaging the brain.

The second stage of labour is shorter. Stronger contractions of the womb press the baby into the birth canal. The baby's shoulder turns so that it can slip more easily through the hip basin (Fig. 12.25(c)). The amnion breaks, releasing the fluid within it. Once the baby's head is born it needs only a few more pushes to expel the rest of its body.

Figure 12.26(c) shows the umbilical cord entering the baby at its navel and passing into the mother's vulva. The nurse is clamping the cord in two places, after which ties are made to prevent loss of blood. The cord is cut between them, thus sep-separating the baby from its mother. The blood vessels in the cord contract strongly when damaged, and this is the mechanism that prevents loss of blood in other mammals, when the mother bites the cord in two. The baby is born covered with a whitish cream and a little blood. After it has been bathed the cord is dusted and bandaged to its

127

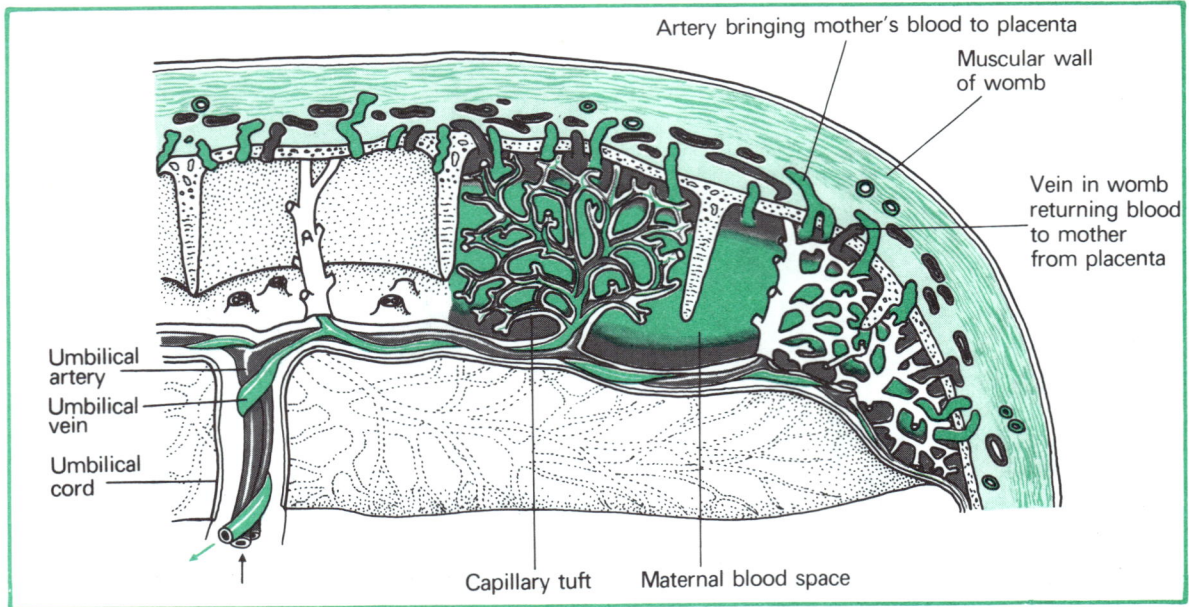

Fig. 12.22 Part of the human womb, showing the capillary tufts lying in the maternal blood spaces.

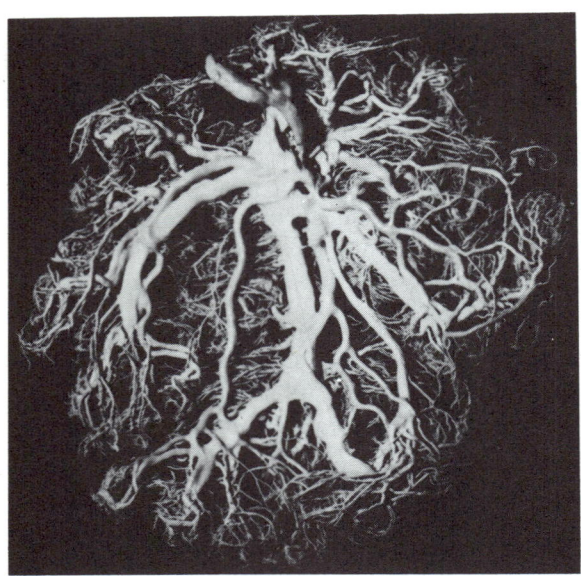

Fig. 12.23 Arteries and veins of the human placenta.

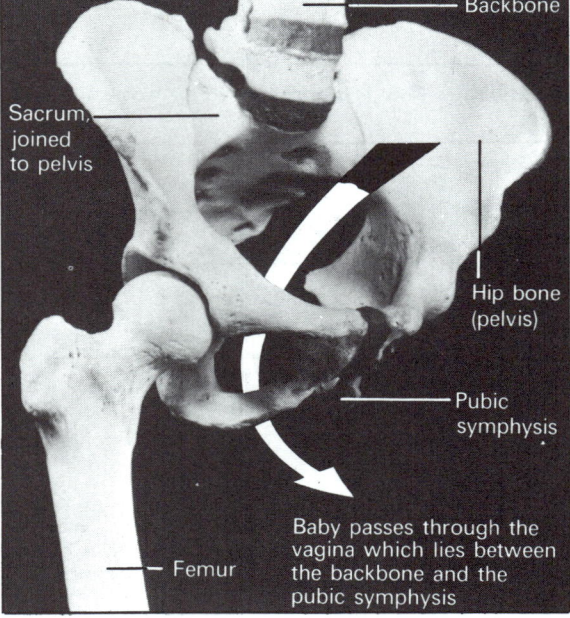

Fig. 12.24 Skeleton of the human pelvis.

body. It dries up and falls away from the navel after a few days. Other mammals lick the baby clean and eat the cord and the afterbirth.

In the last stage of labour, about half an hour after the baby has been born, the placenta becomes detached from the womb and is expelled through the vagina as the **afterbirth** (Fig. 12.27).

Mammalian reproduction contrasted with that of other vertebrates

The human ovum, like that of other mammals, is very small. It is only 10 μm in diameter, a completely different order of size from birds' eggs. Because the mammalian embryo lives as a parasite and is nourished by the mother from her placenta, the

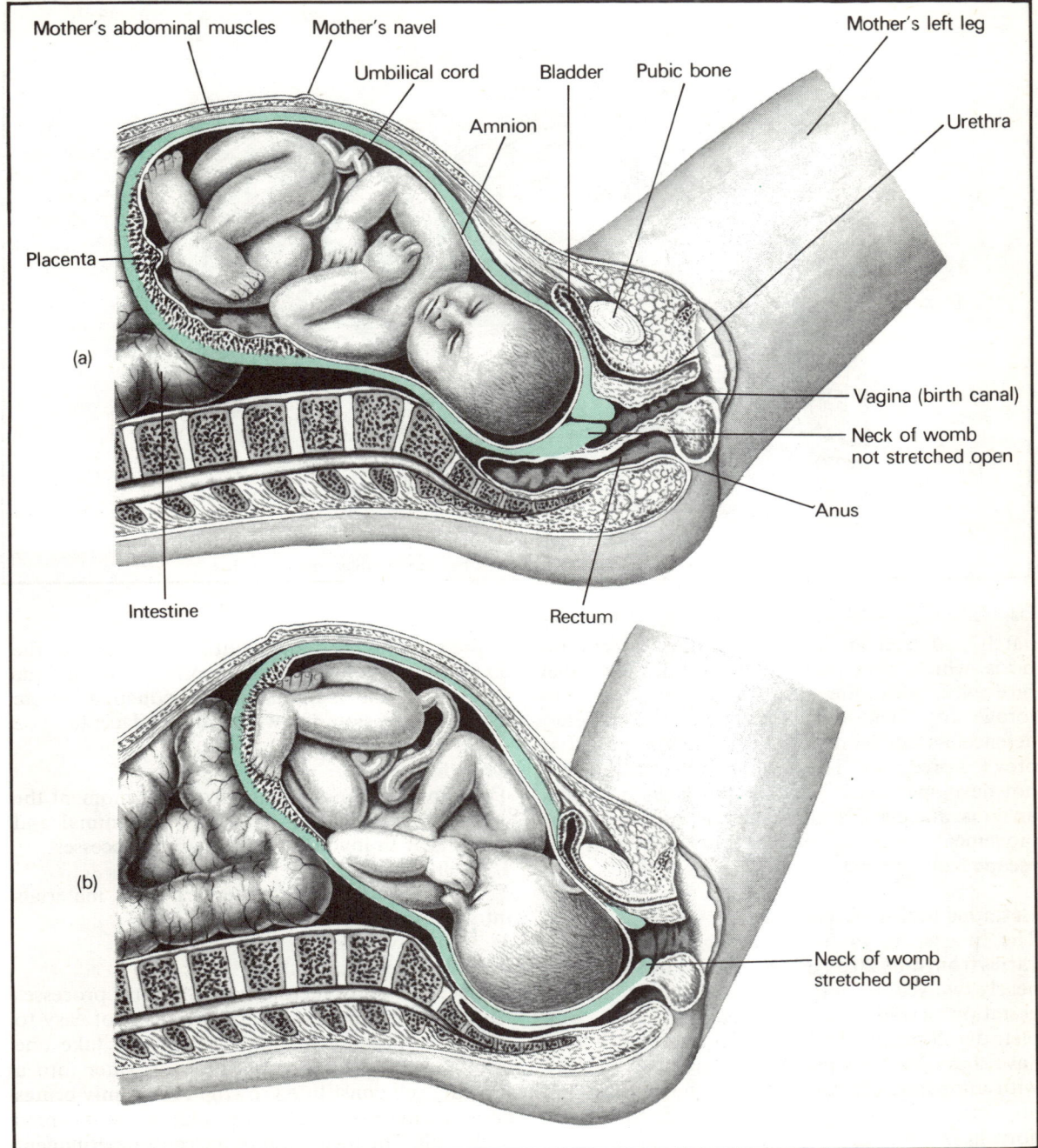

Fig. 12.25(a) The baby in position just before childbirth. **(b)** The first stage of labour: stretching of the neck of the womb.

ovum does not need a large yolk or a large 'white'. Its food supply is virtually limitless, and the young animal is born in an advanced state of development in which all its organs are already formed and functioning.

Fish, amphibian, reptilian, and bird embryos, by contrast, make their growth entirely at the expense of the food contained in the egg yolk and the water contained in the albumen or 'white' of the egg. When these sources are exhausted the animal must

129

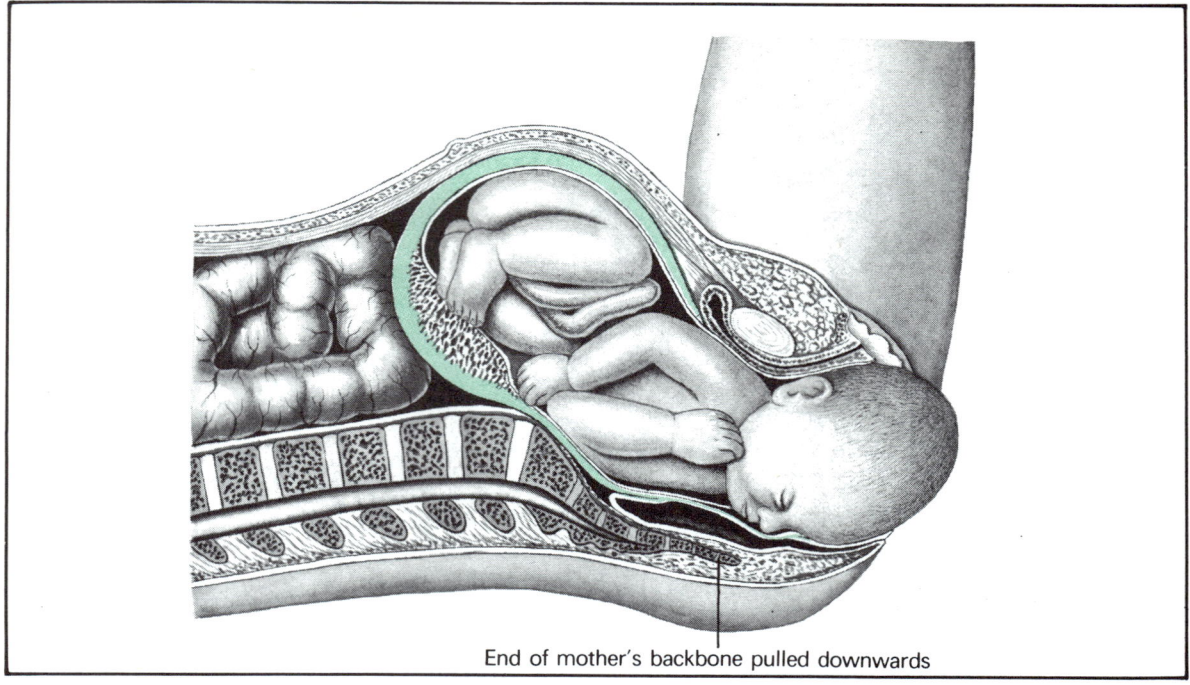

End of mother's backbone pulled downwards

Fig. 12.25(c) The second stage of labour: birth of the baby's head.

hatch and lead an independent life. Except for birds, which show parental care, and a few other rare exceptions, the newly-hatched young must forage for their own food. Being completely defenceless and almost helpless they are an easy prey for predators. The newly-hatched tadpole has not developed all its organs and lacks a mouth, an anus, and eyes. It can make only tiny squirming movements and can neither hunt for food nor escape from enemies.

Gestation period of mammals

The time spent in the womb (gestation period) varies from only 20 days in the mouse to 660 days—nearly two years—in the African elephant. Longer gestation periods are associated with more complete development at birth. Rodents and carnivores have short gestation periods and are born hairless, with unopened eyes and needing the warmth and protection of a nest. On the other hand, newly-born horses and deer can run within a few hours of birth.

Viviparity

The young of mammals are always born alive, and this is described as **viviparity**. Most other vertebrates lay eggs and so are not viviparous, though some fish and snakes are. Here the egg is retained in the body of the female until it hatches, but the embryo is not fed by the mother. The unique features of the mammalian reproductive system are the possession of a womb and a placenta.

Growth

The baby has been growing from the moment the egg was fertilized. In all organisms, animal and plant, growth involves the following processes:
1 Cell division.
2 The assimilation of chemical raw materials into new cells.
3 Cell growth.
4 Cell differentiation.

Since so many factors and so many processes enter into the process of growth, it is not easy to define or measure in simple terms. To take one example, does the incorporation of water into a growing cell constitute growth? It certainly brings about an increase in both the volume and the mass of the cell. The formation of the central permanent vacuole is the principal means by which plant cells enlarge.

Growth rate and its measurement
Growth rate is the *proportional change in growth over a period of time*. If growth is measured in terms of increase in mass, growth rate is the proportional

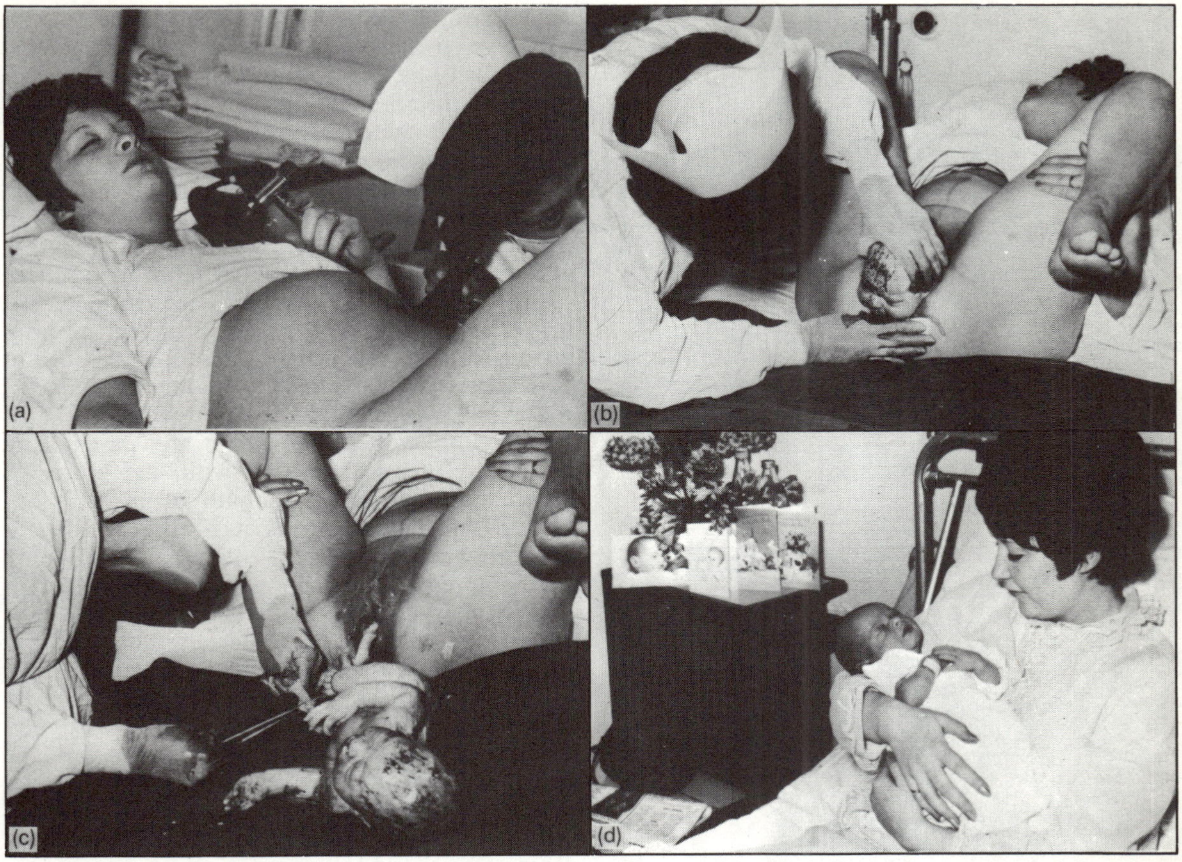

Fig. 12.26(a) The nurse is listening to the baby's heartbeats. Note the size of the mother's abdomen. She is holding a 'gas-and-air' mask for relieving the pain of childbirth.
(b) The baby's head is born.
(c) Clamping the cord. The baby is born covered with a whitish cream and a little blood. Note the size of the mother's abdomen, and the labia of her vulva.

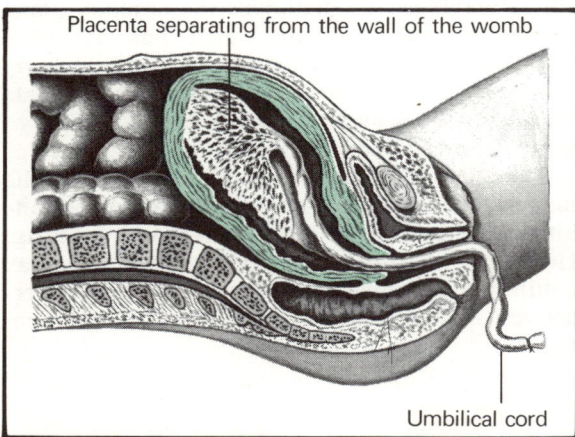

Placenta separating from the wall of the womb

Umbilical cord

Fig. 12.27 Separation of the afterbirth.

change in mass over a period of time (Fig. 12.28).

Measuring growth in terms of increase in mass involves the complication of body fluid. This can vary from day to day, and a well-watered plant, for example, will weigh more than it would if it were wilting. This complication can be avoided by killing the organism and drying it to a constant mass, but then, of course, growth rate can only be measured on samples of individuals from a population. This method is often used for plant growth. If animal growth rate is measured as increase in mass, care must be taken to weigh the animals at the same time each day and to feed and water them at the same time each day.

Genetical and environmental factors
Every species has its characteristic adult size. This is determined (a) by the **genetic information** carried in the chromosomes of the zygote and every cell derived from it by mitosis (b) by **environmental factors** of which a major one is food supply. Suppose a scientist wants to investigate the effect of varying food supply or temperature on growth of a par-

131

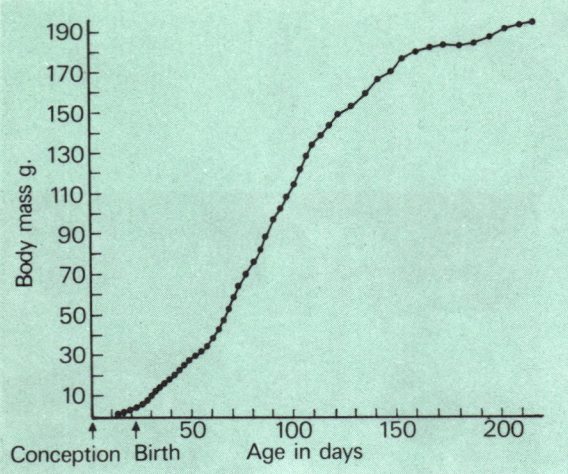

Fig. 12.28 Growth of the Norwegian white rat.

Fig. 12.29 A normal sized man, a pituitary giant, and a pituitary dwarf.

ticular strain of laboratory animal. Some animals will be given one treatment, others another, and the results will be compared. Only if the animals receiving the different treatments are genetically identical can the differences obtained be ascribed to the differences in food supply etc..

Such experiments must therefore be conducted on populations of genetically identical animals. In the case of plants, a genetically identical population can often be raised by asexual reproduction (see p. 115). Mammals do not reproduce asexually, but it is possible to raise a genetically identical population by **inbreeding**, that is repeated matings of closely-related individuals. Although sexual reproduction gives rise to offspring which are genetically unlike their parents, if the parents are genetically identical to start with, the source of difference is absent. Examples of inbred populations are pedigree dogs such as Pekinese and West Highland terriers.

An inbred animal that is extensively used in biological experiments is the Norwegian white rat. The growth rate of this animal is shown in Fig. 12.28, and provided the same sex animals are used and all experimental conditions, such as space, warmth, food supply, and water supply, are always the same, this experiment should be repeatable with the same results. Growth is shown measured as mass, and it is slow between conception and birth. It then speeds up and declines again after about 150 days, reaching a maximum at about 200 days. Insect growth follows a different pattern, and is described on page 233.

Hormones and growth

Animal and plant growth are also affected by internal factors, notably by chemical secretions called **hormones**. In mammals the **pituitary gland** secretes **growth hormone** which has profound effects on the rate of protein synthesis in the body and so on the rate of growth. Experimental rats injected with growth hormone can be made to grow to twice their normal size.

The effect in humans is seen in Fig. 12.29. The man in the middle was a pituitary giant whose pituitary gland was overactive when he was growing. His height was 278 cm. The man on the right was a pituitary dwarf whose pituitary gland was under-active when he was growing. He was only 90 cm tall.

13 Sexual Reproduction in Flowering Plants

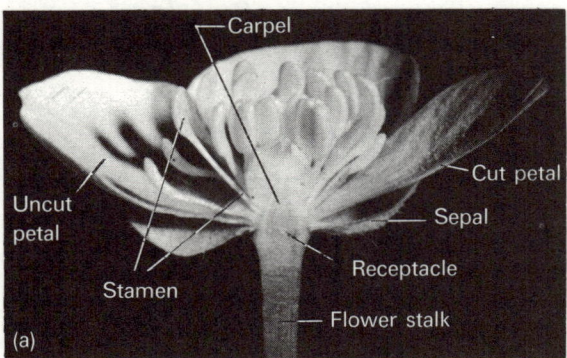

(a)

Stigmas of enlarging carpels covered with pollen grains

Withered stamens

Withered petal

(b)

Fig. 13.1(a) Half flower of buttercup.
(b) Pollinated flower of buttercup developing fruits.
(c) Fruits of buttercup. Sepals, petals, and stamens have all dropped off.

The most abundant and successful group of plants in the world today is the flowering plants. They occupy a wide range of habitats from the most arid deserts to total submersion in water. Although flowering plants are oogamous (see p. 116), they do not produce sperms. Their most distinctive feature is their method of sexual reproduction, based on the flower, which gives the group its name.

Structure of the flower

The structures of flowers are very varied. Some, such as orchids, rhododendrons, and lilies, are large, showy, and colourful. Others are so inconspicuous that the layman would not recognise them as flowers or even notice them at all, for example oak flowers (Fig. 13.9).

The structures of all flowers are based on a similar ground plan. This indicates that they are derived from a common ancestor by evolution. All produce either male gametes only or female gametes only (i.e., are unisexual) or produce both (i.e., are hermaphrodite). We will start by studying a fairly large, unspecialized flower, the buttercup (Fig. 13.1).

There are four distinct layers of parts borne one above the other and all attached to the swollen end of the flower stalk, the **receptacle**. The lowermost layer consists of the **sepals**. These are green and fairly leaf-like; they enclose and protect the flower when it is a bud. In buttercup flowers there are five sepals.

(c)

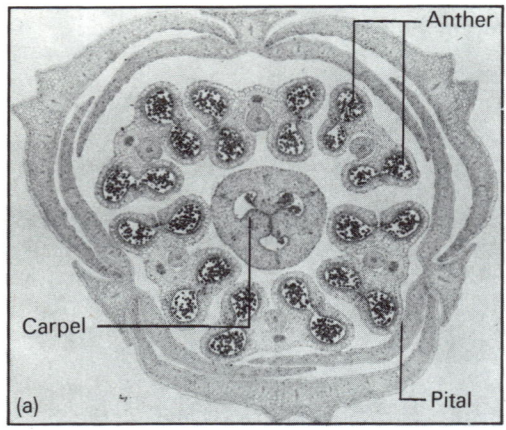

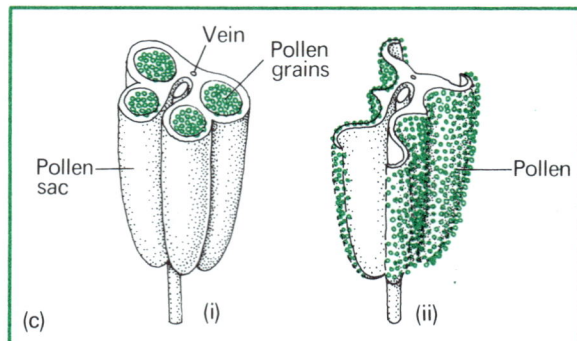

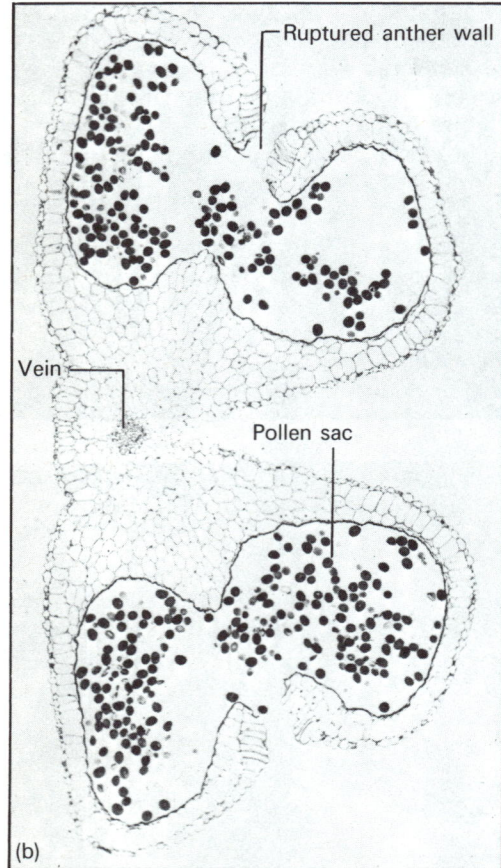

Fig. 13.2(a) Transverse section of lily flower bud (Courtesy Carolina Biological Supply Co.).
(b) Transverse section of a mature, burst lily anther (light microscope). (Courtesy Carolina Biological Supply Co..)
(c) Structure of a typical anther (i) unripe (ii) burst. The top half is drawn as cut off in each case.

Above them are five **petals**, which have narrower bases than the sepals. They are attached to the receptacle above the place where the edges of the sepals meet. The petals are also fairly leaf-like, except that they are not green but bright yellow. Their function is to attract insects to the flower.

Above the petals lies a group of **stamens**. Each has a stalk which bears at its top an **anther**, which produces pollen. In Fig. 13.1(a) the outer anthers are evidently older and more mature than the inner ones, for they have shed their pollen on to the upper surfaces of the petals.

Anther and pollen
Figure 13.2(a) shows a transverse section of a flower bud of lily. Six unburst anthers are shown. Figure 13.2(b) is a transverse section of one lily anther made after it had burst. The stalk of the stamen and the vein supply the developing **pollen grains** with food. The most obvious features, however, are the four **pollen sacs** filled with pollen grains. Figure 13.2(c) shows these in the form of a stereogram both before and after the anther has burst.

All the tissues of the young anther are diploid. In the formation of pollen grains, reduction nuclear division (meiosis) takes place, and the pollen grain nuclei contain the haploid number of chromosomes. This is a very similar arrangement to the formation of sperms in the testis tubes (see page 118), and seems to indicate that pollen grains are gametes (Fig. 12.1).

They are small, produced in large numbers and are carried, by wind or by insects, to the female gametes. Each of these is larger than the pollen grain and situated in the **carpel** of the same or another flower. Thus flowering plants are oogamous. Pollen grains differ from sperms, however, in that they cannot swim. Each contains two nuclei, both haploid, one of which acts as the male gamete (Fig. 13.3).

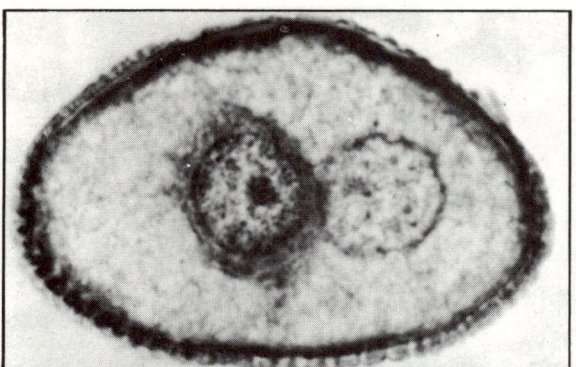

Fig. 13.3 Transverse section of a lily pollen grain. Note the two nuclei, one of which serves as the male gamete, and the thick wall. (Courtesy Carolina Biological Supply Co..)

The carpels

Above the stamens are the carpels. Each contains one haploid female gamete. All the tissues of the carpel except for the female gamete are diploid, and the production of the female gamete involves reduction nuclear division or meiosis. (This is a somewhat simplified version of the detailed facts, but it is essentially true.)

Thus the buttercup, like many other flowering plants, is hermaphrodite. It bears male gametes in its anthers and female gametes in its carpels. It would seem to be an easy matter for the plant to fertilize itself—all that is necessary is for pollen to be transferred from the anther to the receptive part of the carpel called the **stigma** (Fig. 13.4). Such an individual plant would be **self-fertile**; it need not

provide for the transfer of pollen from one plant to another, and would produce seeds (embryos) by itself.

But as we saw on page 132, such **inbreeding** leads to lack of variety in the offspring, and in the long term this is a strong evolutionary disadvantage. Many flowering plants, therefore, although hermaphrodite, possess mechanisms to prevent self-fertilisation, and Fig. 13.1(a) shows us one of them. Some of the buttercup anthers seen here are mature and have burst, but the carpels are so small and immature that we can barely make out their structure. They are incapable of becoming pollinated and fertilized at this stage of their development and this flower cannot, therefore, fertilize itself.

In Fig. 13.1(b) the flower is slightly older. The sepals, petals, and stamens have withered, but the carpels are now mature and their receptive stigmas can be seen covered with pollen grains which have been brought by insects from other buttercup plants. Beneath the stigma is a short stalk, the **style**, which crowns the large base of the carpel, the **ovary**. Fig. 13.4 is a diagrammatic longitudinal section of one such carpel.

Inside the ovary

Inside the ovary are one or many **ovules**. Each contains one female gamete and each will, after fertilization, develop into a seed. Figure 13.4 shows the wall of the ovary, the space between the ovary and the ovule, the ovule itself, and the female gamete within it. Piercing the wall of the ovule and leading directly to the female gamete is a tiny hole called the **micropyle**. This plays an important part in fertilization and can be seen in the mature seed. Since Fig. 13.4 is drawn as a mid-line section, this hole appears as a canal.

Pollination

Pollination is the *transfer of pollen grains, each containing a male gamete, to the receptive surfaces, the stigmas, of the carpels.*

Insect pollination

Flowering plants can be divided into those species that are pollinated by wind, which blows pollen grains from the anthers of one flower to the stigmas of another, and those that are pollinated by insects.

Insects visit flowers for food. They eat pollen and also gather **nectar**, a sugary solution, often scented, which flowers make in **nectaries**. Pollen grains adhere to various parts of the insect's body when it visits one flower and may later become

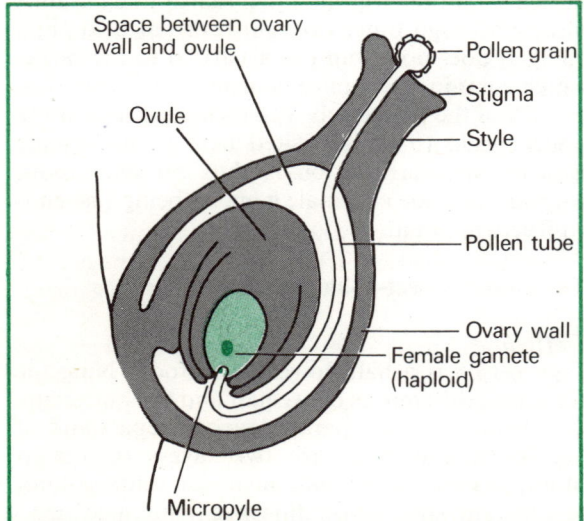

Fig. 13.4 Longitudinal section through a buttercup carpel. The pollen tube has entered the micropyle.

scraped off on the stigma of another. The pollen grains of insect-pollinated flowers are large and sticky and adhere easily to insect hairs and to stigmas.

Insects are an ancient group which appeared first in the Devonian period 400 million years ago (Fig. 20.9). They underwent a great evolutionary spurt at the time when the flowering plants appeared in the early Cretaceous period, 130 million years ago. Insects and flowering plants **radiated** together, that is they both developed a wide range of new species at the same time. As a result many flowering plants have highly specialized pollination mechanisms that can only be operated by one particular species of insect.

The buttercup flower does not show very specialized adaptation to particular insect visitors. It has large, brightly coloured, shiny flowers and produces nectar at the bases of its petals and an abundance of pollen. Most of this is probably destined to be eaten by insects. The flower is strong, and large flying insects can alight on it. While the insect reaches down for nectar, the underside of its body becomes liberally dusted with pollen, which may be wiped off on to the stigmas of the next flower that is visited.

Pollen tube and fertilization
Once on the stigma the pollen grain may germinate and form a **pollen tube** which grows downwards into the tissues of the style, by which it is nourished (Fig. 13.4). It is not known what stimulus causes this growth, but it is directed towards the micropyle, which the pollen tube seeks out and penetrates. The end of the tube then bursts and the male gamete, one of the nuclei within the pollen grain, is deposited next to the female gamete in the ovule. The gametes fuse (fertilization) producing a diploid zygote nucleus which starts to divide and produce the diploid tissues of the embryo or seed.

Figure 12.1 shows the behaviour of the nucleus in the cycle of sexual reproduction in man. It applies equally well to any other vertebrate. Draw a similar cycle for the life history of the flowering plant, and include the following terms: *diploid plant body; pollen grain containing male gamete; ovule containing female gamete; fertilization; division of diploid zygote nucleus to produce the seed.*

Biological significance of pollination
On page 117 it was pointed out that a major problem facing land-living organisms is the provision of water in which the male gametes can swim to find the female gametes. Most land-living animals have

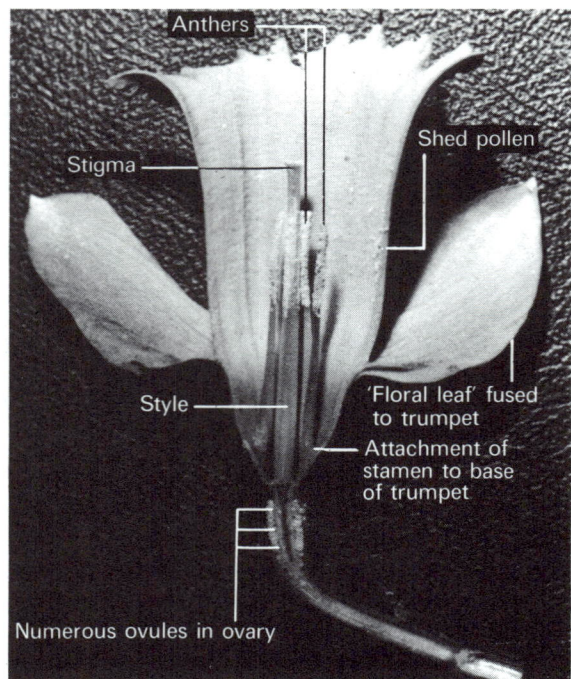

Fig. 13.5 Half flower of daffodil.

overcome this difficulty by transferring the male gametes, suspended in liquid (semen), into the female body by copulation. This may involve a penis, as in mammals.

Flowering plants have evolved a completely different method. The male gamete is a nucleus contained in the pollen grain. This is hardy and durable, having a tough outer drought-resistant coat (Fig. 13.3). It does not swim but is carried to the carpel which contains the female gamete in the ovule. The growth of the pollen tube causes the nucleus of the pollen grain to be deposited next to the female gamete, and fertilization is thus brought about without the male or female gametes being suspended freely in liquid.

Three other insect-pollinated flowers

Daffodil
Figure 13.5 is a half flower of daffodil. Note the following differences between it and the buttercup.
1 There is not a separate layer of sepals and of petals. Instead there are two layers of similar 'floral leaves', which all have the same colour, texture, and size. (Check this on a living specimen.)
2 There is a trumpet-shaped outgrowth from the floral leaves.

3 There is a single carpel and its ovary is situated *beneath* the floral leaves and the stamens. In buttercup flowers there are several carpels all situated above the petals and the stamens.

4 The ovary contains a large number of ovules.

5 There is only one style, which is long, and one stigma, held above the level of the anthers.

6 There are only six stamens (how many are visible in this photograph?), and they are fixed at their bases not to the receptacle but to the trumpet.

Try to observe insects pollinating these flowers. What kinds of insects are involved? How do they bring about pollination?

(N.B. Daffodils also reproduce asexually by means of bulbs.)

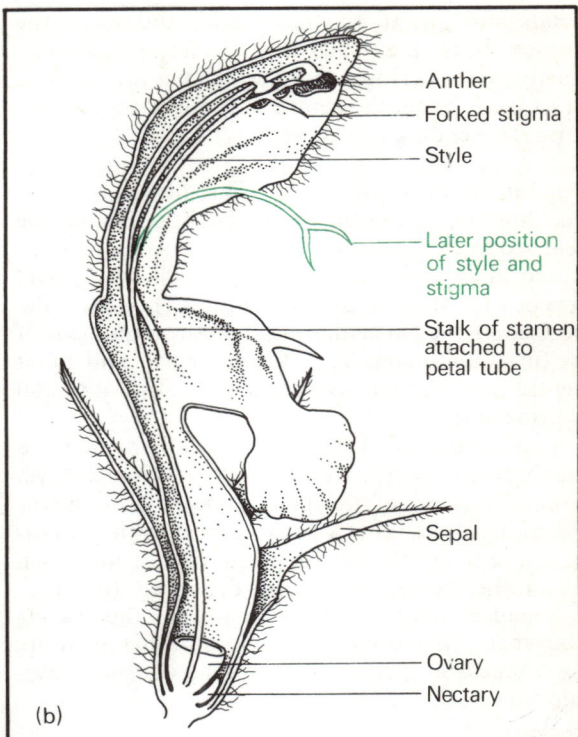

(b)

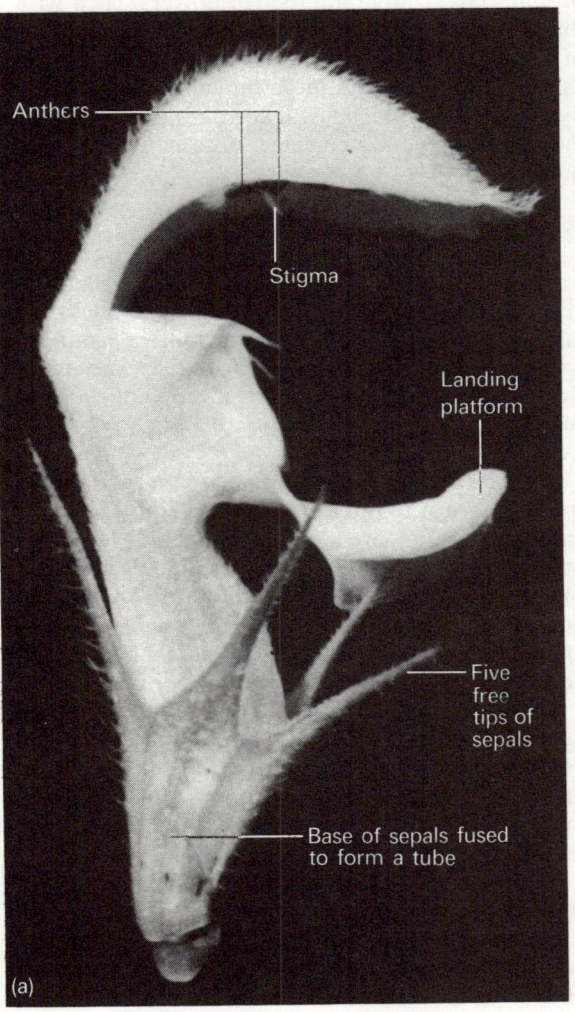

Fig. 13.6(a) Whole flower of white dead nettle photographed from the side.
(b) Half flower of white dead nettle.

White dead nettle (Lamium album)
Flower symmetry Unlike buttercup and daffodil, the flower of white dead nettle is symmetrical about only one plane. This means that there is only one way in which the flower can be cut into two longitudinally so as to give two identical halves. Find out from Fig. 13.6 which plane this is.

Longitudinal sections along any diameter of buttercup and daffodil flowers produce two identical halves. In general, flowers which have only one plane of symmetry are adapted in precise and specialized ways to pollination by particular insects.

Although the general appearance of dead nettle flower is markedly different from that of a buttercup, the same four layers of parts appear in the same sequence. First, five sepals, this time all joined together at their bases, forming a tube but free at their tips. Then the petals—perhaps we should say petal, for there is no obvious division into any particular number, though the usual practice is to say that the flower has five petals which are 'fused'.

Next come four stamens, two long and two short, The arrangement of these is such that two anthers are held above the other two giving a plate of anthers protected from rain by the hood of the petals. The stalks of the stamens are attached to the petals and not to the receptacle, and when the anthers burst they shed their pollen downwards. Lastly there is a single carpel which has four ovaries at its base. From the centre of these arises a single long style ending in a forked stigma.

Pollination of dead nettle The flower secretes nectar at the base of the petal tube. This can only be reached by long-tongued insects such as bees. When a bee alights on the lower lip of the petal and reaches forward to find the nectar, pollen is deposited in one particular place on the upper side of its furry abdomen. The stamens ripen and burst before the stigma is mature, and so self-pollination is prevented.

Later in the life of the flower, the stigma matures and becomes receptive to pollen. It now hangs down from the hood of the petal tube (Fig. 13.6(b)). When an insect visits a flower that is in this condition, the blob of pollen deposited on its abdomen by another flower touches the drooping stigma.

Because pollen is exactly located on the insect's abdomen, this pollination mechanisms is more economical of pollen than is that of the buttercup flower.

Dandelion (Taraxacum officinale)
Dandelion is a member of the largest family of flowering plants, the Compositae. Close inspection shows that a dandelion 'flower' is made up of a large number of small flowers. Fig. 13.7(a) is a

Disc at apex of flower bears a large number of small flowers (a)

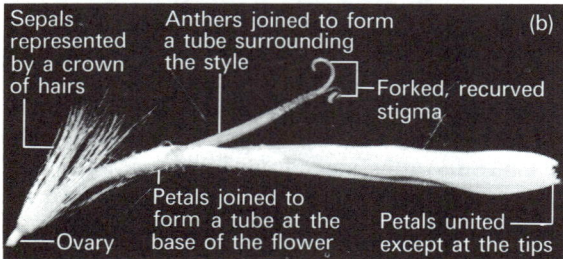

Sepals represented by a crown of hairs (b)

Anthers joined to form a tube surrounding the style

Forked, recurved stigma

Petals joined to form a tube at the base of the flower

Petals united except at the tips

Ovary

Fig. 13.7(a) Flower head of dandelion, half section. **(b)** Single flower of dandelion.

longitudinal section through the whole group of flowers, and Fig. 13.7(b) is a photograph of one intact flower.

The sepals are represented by a crown of hairs, which later develop into a parachute device for dispersing the seed. The five petals are fused to form a tube at their bases, but higher up this becomes a single strap-shaped 'petal' with five free tips. The *anthers* are fused together making a tube which surrounds the style, though the stalks of the stamens are free from each other. There is a single ovary situated, as in daffodil, beneath the level of all the other parts. From it arises a single style whose end is forked.

The Composite 'flower' displays great economy of material. A large number of flowers, each with the potential of producing a seed, share the labour of display and attracting insect visitors. The 'flower' can be pollinated by a large range of crawling and flying insects which, when poking among the individual flowers for nectar, become dusted with pollen.

Wind pollination
As pollen is drought-resistant, pollination can be brought about if it is blown from one plant to another. This method is far more hit and miss than insect pollination, and most of the pollen produced is wasted. Nevertheless it is obviously successful since all the grasses, which cover much of the land surface of the earth, use it. So do a wide variety of trees.

The flowers of wind-pollinated plants are always small so as not to interfere with the flight of pollen from them. They are also inconspicuous (greenish or brownish), unscented, and produce no nectar. These features are related to the fact that they do not attract insect visitors. Sometimes wind pollinated flowers are little more than a cluster of anthers on a tiny floral leaf (e.g. oak, Fig. 13.9(b)).

Small, dry pollen grains are produced in vast quantities. This contrasts strongly with insect pollinated flowers which produce smaller quantities of sticky pollen. The stigmas are often large and branched or feathery so as to offer the best chance of catching the wind-blown pollen.

Wind-pollinated trees bear their flowers in clusters of catkins which swing freely in the breeze. The flowers open and pollination is complete before the tree comes into leaf, so the flight of pollen from the tree is not impeded by the foliage.

Rye grass (Lolium perenne)
Figure 13.8(a) is a photograph of rye grass, whose structure is typical of all grasses.

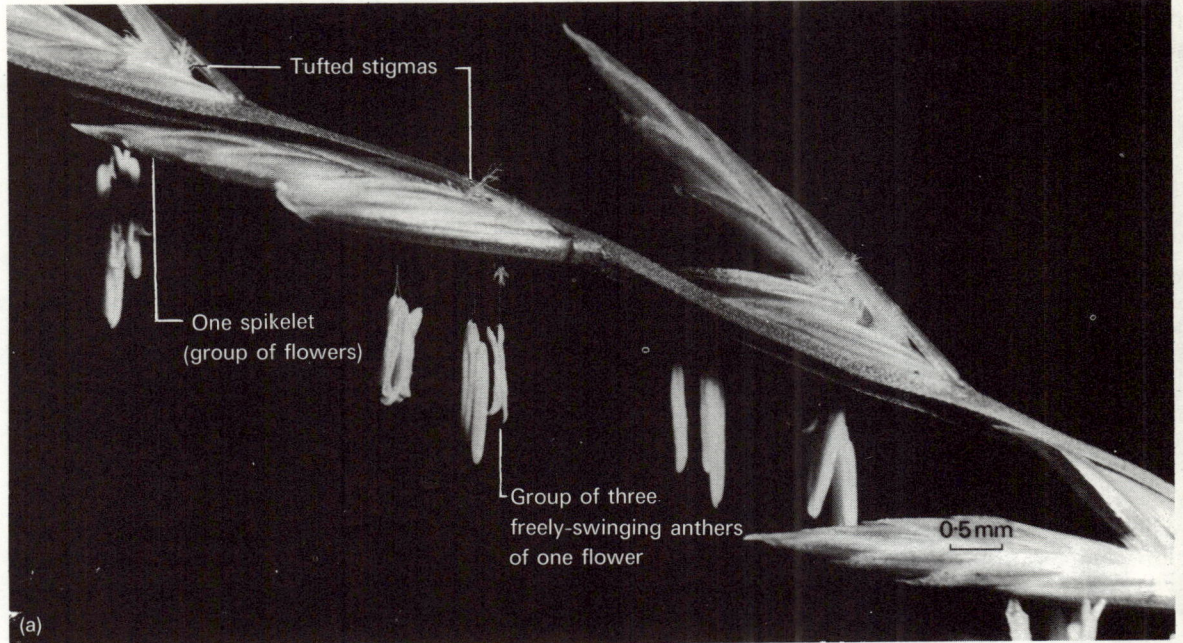

Tufted stigmas

One spikelet
(group of flowers)

Group of three
freely-swinging anthers
of one flower

0·5mm

(a)

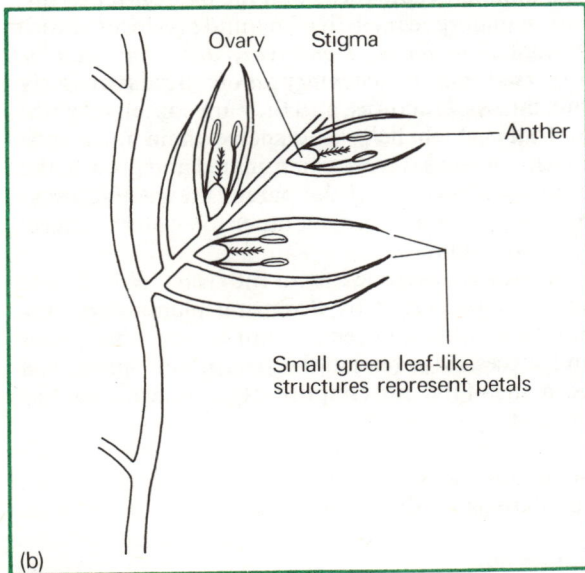

Ovary   Stigma

Anther

Small green leaf-like
structures represent petals

(b)

Fig. 13.8(a) Flowering stem of rye grass.
(b) Diagrammatic longitudinal section through rye grass flower head, showing the structure of the flowers.

The anthers occur in groups of three, hanging freely away from the plant on thread-like stalks. These are quite unlike the sturdy stalks which support the anthers in buttercup, daffodil, and dead-nettle flowers. The stigmas are prominent and tufted, pushed out from the plain green flowers to catch pollen. Figure 13.8(b) shows the structure of the flower in which the sepals and petals are represented by a number of small green leaf-like structures.

Oak (Quercus *spp.*)
Figure 13.9(a) shows a group of male catkins on a small twig of oak. The total number produced by the whole tree is huge. This is related to the wastage of wind-blown pollen. Figure 13.9(b) shows detail of one catkin. The flowers consist of almost nothing but anthers.

Female oak catkins (Fig. 13.9(c)) bear only a few flowers and are very inconspicuous. Their stigmas are prominent and branched, but otherwise the flower contains little more than an ovary surrounded by a group of tiny green leaves which will give rise to the acorn cup (Fig. 13.10).

Hermaphrodite and unisexual flowers

Oak flowers are unlike those we have looked at so far because they are unisexual, not hermaphrodite. An individual oak tree bears both male and female catkins, but some species of flowering plants produce individual plants which bear either male flowers only or female flowers only. Self-pollination is then impossible.

This situation is like that in animals, and such **outbreeding** has a very strong long-term evolutionary advantage because the offspring produced are genetically unlike their parents.

139

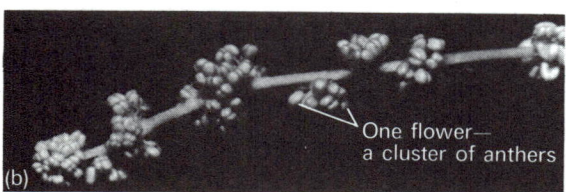

One flower— a cluster of anthers

Branched style

Green scales, will give rise to acorn cup

One female flower

Fig. 13.9(a) Oak twig with male catkins. Note the size of the young leaves.
(b) Male oak catkin.
(c) Oak twig with female catkin.

We might ask therefore why it is that all plant species do not produce unisexual flowers on different individual plants. The answer lies in the fact that plants, unlike animals, cannot move and seek out a mate. Cross-pollination, whether by insect or by wind, is a passive process over which the plant has no direct control. Seed production in plant species which have separate male and female plants is therefore by no means a certainty.

Most plants strike a balance between outbreeding and inbreeding. The short-term advantage of being sure to produce seed is ensured by self-pollination and self-fertilization. The long-term evolutionary advantage of producing genetically variable offspring is ensured by cross pollination. Thus although many hermaphrodite plants possess devices to prevent self-pollination (see page 135), these are usually not perfect, leaving open the possibility of self-pollination if cross-pollination does not occur.

Fruits

The diploid zygote formed in the ovule after fertilization undergoes cell divisions and produces a seed. A **seed** is *a dormant embryo plant which can be dispersed from its sedentary parent.* Seeds are fairly substantial structures, and they are contained within **fruits** which help to disperse them in a wide variety of ways. If seeds germinated and grew in the immediate vicinity of the parent they would compete with it and with each other for water, mineral salts, and light.

A fruit is developed from the ovary wall. Fruits are only produced by flowering plants and they have undoubtedly been a major factor in the rapid and successful spread of the group. This spread has been one of the most important biological events in the history of life on earth.

Some fruits and their adaptations for dispersing seeds

Buttercup
The simplest type of fruit is one in which the wall of the ovary grows to keep pace with the seed developing inside it, eventually becoming dry and woody. Such are the fruits of buttercup (Fig. 13.1(c)). They look like enlarged buttercup carpels (compare with Fig. 13.1(b)).

Oak
The woody part of the acorn is derived from the enlarged and hardened ovary wall. Compare Fig.

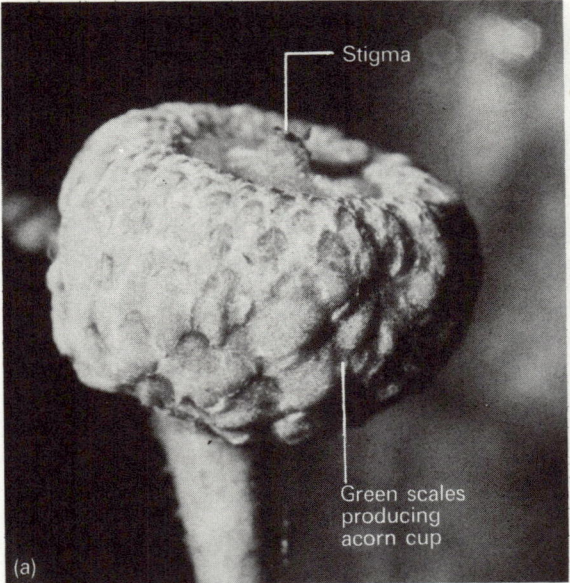

Stigma

Green scales
producing
acorn cup

(a)

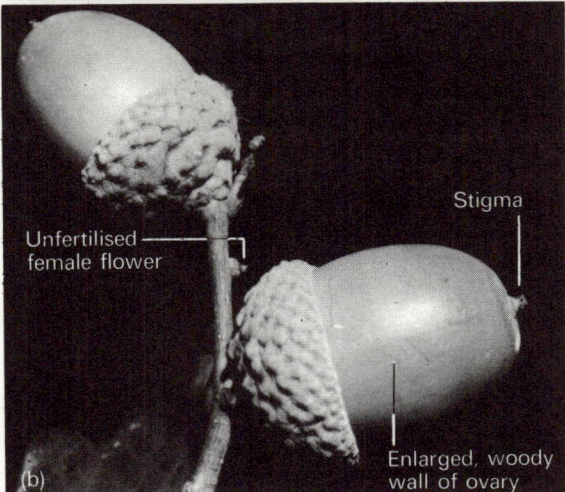

Unfertilised
female flower

Stigma

Enlarged, woody
wall of ovary

(b)

Fig. 13.10(a) Developing acorn.
(b) Mature acorns.

Wall of fruit covered
with spines

Fig. 13.11 Fruits of goosegrass.

13.10 with a female oak catkin (Fig. 13.9(c)). The remains of the stigmas are still visible, and the acorn cup is developing from the scaly green leaves which surround the base of the female flower.

Neither of these fruits has any mechanical device to aid dispersal of the seed each one contains. Buttercup fruits may be picked up in the mud on animal's hooves or men's feet or on vehicles. Acorns may roll a little distance or be carried away by squirrels and other animals, but most rot or germinate where they are.

Oak seed dispersal may therefore appear to be inefficient, but before drawing this conclusion we must remember that oaks are very long-lived trees. Specimens 300 years old are a common sight in England. Much of the country was covered by oak woodland until recent times, and if the land were not cultivated, this community would re-establish itself. Oaks spread 'slowly but surely'.

Goosegrass (Galium aparine)
This plant also has a simple fruit. The enlarged ovary wall develops very efficient hooked spines which adhere to human clothing and to the fur of animals (Fig. 13.11). How might this fruit, once detached from the plant in this way, find its way into the soil?

Poppy (Papaver spp)
Figure 13.12 shows the urn-like fruit with the withered remains of the stamens at its base. The curious flat stigmas form a cap at its apex. The hard, dry wall shrinks away from the cap leaving the familiar holes from which the small dry seeds shower when the long poppy stem is shaken in the wind.

Shepherd's Purse (Capsella bursa-pastoris)
This small plant is an abundant weed in most of Britain and is common on roadside verges (Fig. 13.13(a)). The seeds within the tiny fruits are not free but are attached to a septum. When ripe, the hard fruit wall blows off. Half of it has blown off in Fig. 13.13(b), and the minute seeds need only a small shake to make them spring away from the septum.

Laburnum and other legumes
Figure 13.14 shows a burst pod of *Laburnum*. Its wall is dry and hard and has two seams or lines of

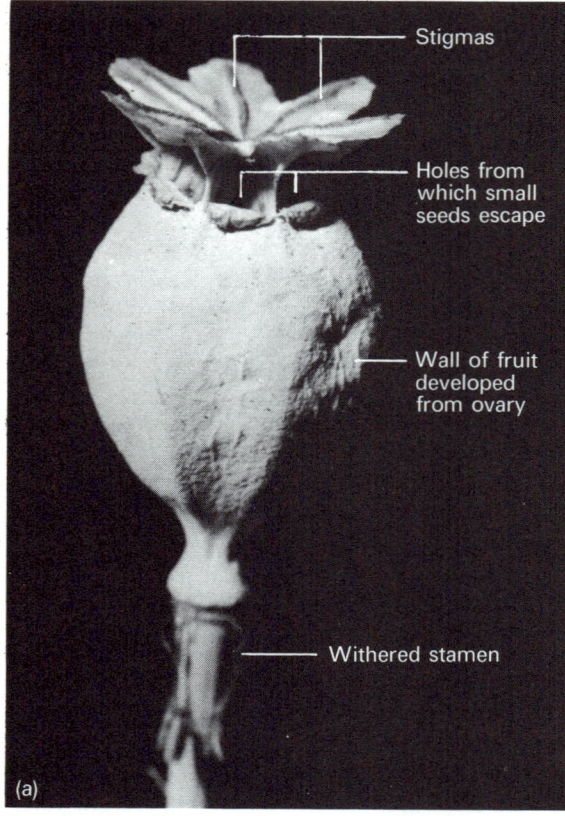

Fig. 13.12(a) Fruit of poppy.
(b) Poppy fruit sectioned horizontally.

weakness. These parted as the fruit dried.

In some legumes the splitting progress process is slow, but in others it is sudden, even explosive, as in gorse. The seeds may then be flicked as far as a metre from the parent plant. Note how the seeds are attached alternately to one or the other side of the pod.

Willowherb (Chamaenerion spp).
This common weed has a long thin dry fruit which splits open to reveal a multitude of tiny seeds (Fig.

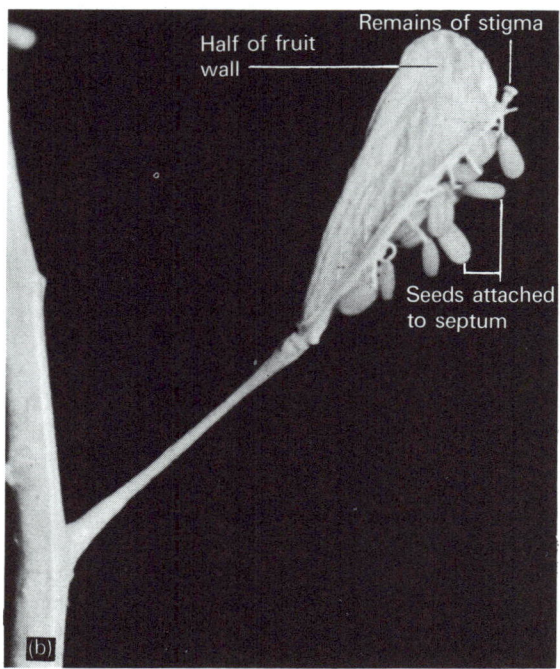

Fig. 13.13(a) Fruiting head of Shepherd's purse.
(b) Single fruit of Shepherd's purse with half of the fruit wall detached, revealing the seeds.

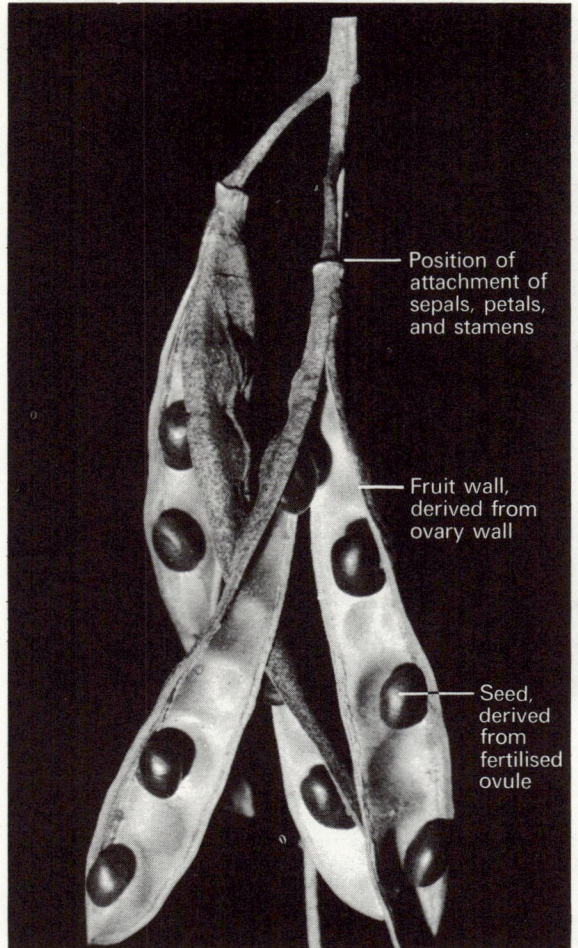

Fig. 13.14 Laburnum pods burst.

Labels on figure:
Position of attachment of sepals, petals, and stamens

Fruit wall, derived from ovary wall

Seed, derived from fertilised ovule

manure in which to grow. Birds scatter such seeds over long distances.

Seeds

Seeds are small dormant embryo plants. By being carried away from the immobile parent they bring about dispersal of the species *in space*. Because they may not germinate for a considerable time after they have been scattered, they may bring about dispersal *in time*.

Each seed contains an embryo root (the **radicle**), an embryo shoot (the **plumule**), and embryo leaves (the **cotyledons**). The whole structure is enclosed in a hard, dry seed coat, the **testa**.

Often the cotyledons are large and make up the bulk of the seed, for example in lettuce, Fig. 13.17(a). Their cells are packed with insoluble food reserves, proteins, fats, or starches, which have been made by the parent plant. Other seeds, such as cress (Fig. 13.17(b)) and the cereals, possess a dead tissue called the **endosperm**, which forms the food reserve.

Germination of seeds

Provided conditions are right, seeds germinate and grow into young plants. Figure 13.18 shows some of the stages in germination of broad bean. The testa swells and wrinkles, then tightens again as the underyling cotyledons swell. The testa cracks and the radicle emerges and grows downwards. This develops root hairs in the root hair zone. Now the plumule emerges, hooked over to protect the young leaves as they push through the rough soil.

Conditions necessary to bring about germination

Water supply
The water content of an ungerminated seed may be as little as 2.5 % of its total mass whereas growing plant tissue contains about 75 % water.

When the insoluble proteins, fats, and starches of the food stores have been converted into smaller, soluble molecules, these must dissolve in water in order to move to the growing points of the embryo. Also the bulk of a living plant cell consists of the central permanent vacuole, filled with cell sap. Water is therefore needed to fill the vacuoles in newly-formed cells.

The need for water can be shown by setting up two covered dishes one lined with wet cotton wool, the other with dry cotton wool. Similar seeds are placed in both and both are kept at the same room temperature. The seeds on the wet cotton wool swell and germinate, the others remain as they are. The only difference between the two treatments is

13.15). Each seed is crowned by a tuft of fine silky hairs which form a parachute to carry it away from the parent plant.

Bramble (Rubus fruticosus)
The berry consists of a collection of small fruits each developed from a single carpel. Figure 13.16 shows the remains of the sepals and stamens, and these indicate that the berry, although it contains several fruits, has developed from a single flower.

The longitudinal section shows this more clearly. Each individual fruit contains a seed surrounded by a fruit wall. The wall is divided into three distinct layers, an outer skin, a middle juicy pulp, and an inner woody seed covering (the pip). This protects the seed from digestion when the fruit is passing through the gut of an animal. When passed out in an animal's droppings, the seed is provided with

Fig. 13.15 Seeds of *Chamaenerion sp.*

Individual fruits

Remains
of stamens

Remains of sepals

(a)

Fig. 13.16(a) Fruits of blackberry.
(b) Longitudinal section through a fruit of blackberry.

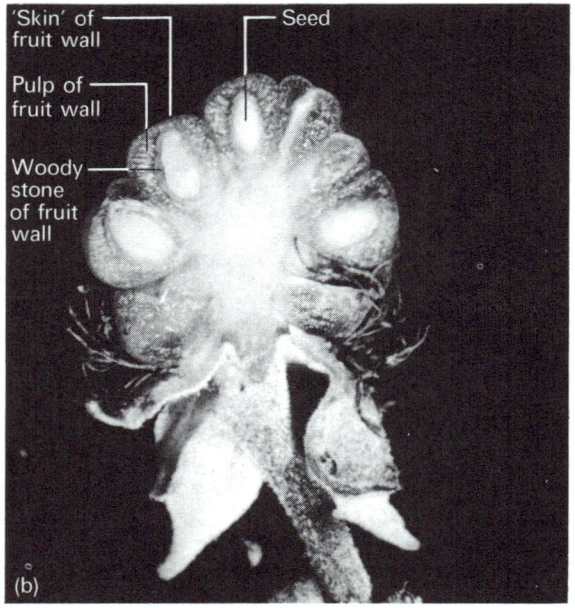

'Skin' of
fruit wall

Pulp of
fruit wall

Woody
stone
of fruit
wall

Seed

(b)

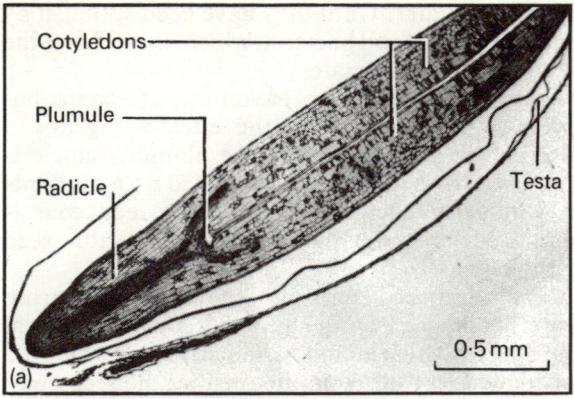

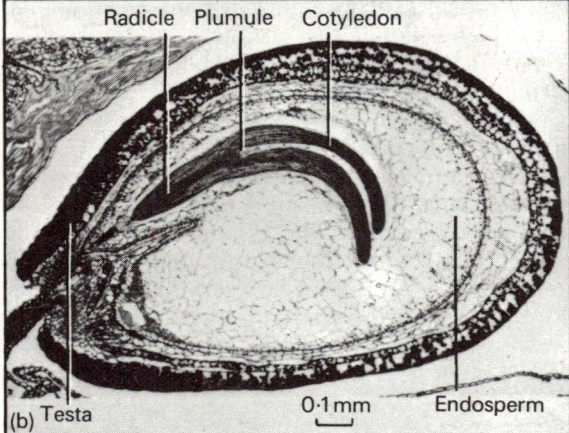

Fig. 13.17 Longitudinal section of a lettuce seed (large cotyledons) (a), and of a cress seed (small cotyledons, large endosperm), (b).

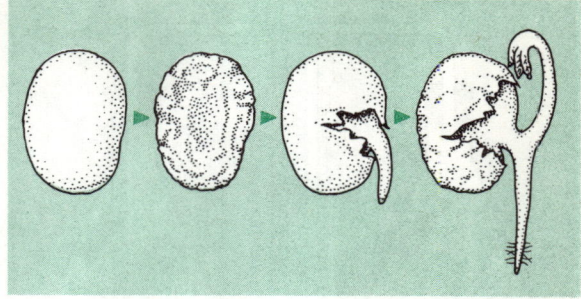

Fig. 13.18 Stages in the germination of broad bean seeds.

the presence or absence of water, that is to say the
dry seed experiment acts as a control for the wet
seed experiment.

Oxygen

Germinating seeds are making new cells rapidly.
This requires not only the chemical raw materials,
derived from the cotyledons or the endosperm, but
also chemical energy. ATP is the source of this
energy, and as ATP is generated by tissue respiration (p. 67), it follows that germinating seeds need a
good oxygen supply.

The apparatus shown in Fig. 13.19 shows that
the presence of oxygen is essential for germination
to take place. The solution of alkaline pyrogallol
absorbs oxygen, and it is safe to assume that the air
in the left hand flask is oxygen-free. After a few days
at room temperature, seeds in the right hand flask
germinate, but those in the left hand flask do not.

The apparatus shown in Fig. 13.20 can be used to
show that germinating seeds absorb oxygen. After

two days at room temperature, the water level has
risen in the left hand apparatus. Try to explain why
this is so, remembering:

1 Both sets of apparatus are equally affected by
changes in room temperature.

2 Soda lime absorbs carbon dioxide.

The output of heat by germinating seeds was
demonstrated in the experiment described in Fig.
7.2.

Temperature

Temperature has a strong effect on the activities of
enzymes (Fig. 4.4) and so will affect germination.
Since proteins are denatured at temperatures
higher than 40°C we should expect to find that
germination is slow at low temperatures, increases
to a maximum as the temperature rises, and falls to
zero at temperatures above 45°C, when the seeds
are killed. This is easily confirmed by experiment.

Dormancy

Seeds of some plants may not germinate even
though the environmental conditions of water
supply, oxygen supply, and temperature are favourable. For example, freshly-harvested wheat and
barley seeds will germinate only after an interval of
several weeks. During this period the seeds are

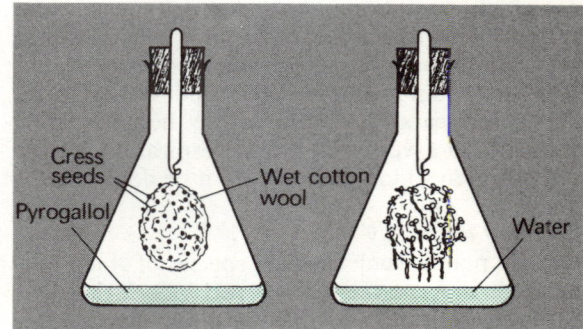

Fig. 13.19 In the left hand flask, alkaline pyrogallol removes
oxygen from the air, preventing germination of the cress seeds.

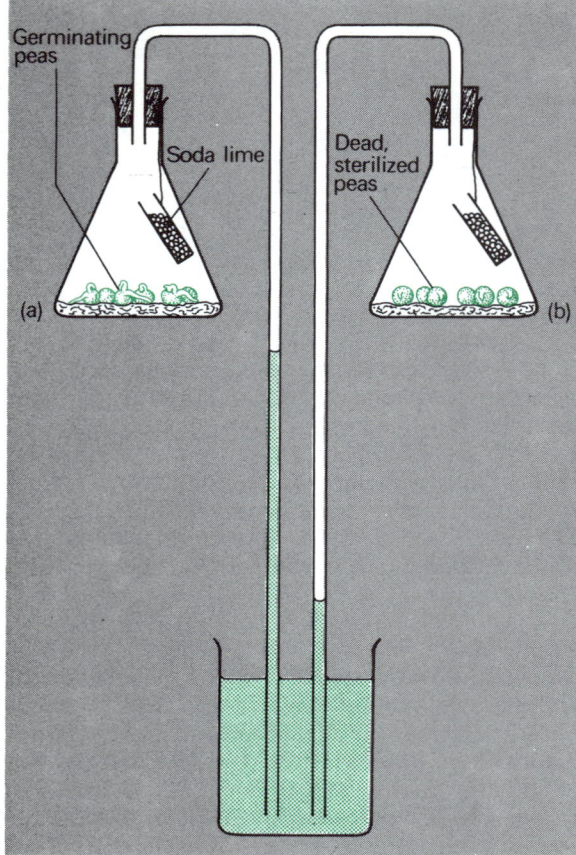

Fig. 13.20 Respiring pea seeds consume oxygen and give out carbon dioxide, which is absorbed by the soda lime.

said to be **dormant**, and they will not, therefore, germinate until the spring, when the young seedling will stand a better chance of establishing itself successfully.

Some seeds can remain dormant for years, even for decades, thus avoiding long periods of unfavourable conditions. Arid deserts, such as the Arizona desert and the red centre of Australia, suddenly become carpeted with flowering plants after heavy rain fall which may occur only once in every seven years. During dormancy the activities of seeds fall to a very low level. Respiration in dry seeds is often so low that it cannot be detected.

Causes of dormancy

In a few plants, such as some species of *Anemone*, the seeds are not fully developed before they are scattered and time must pass to allow this process to be completed. Some seeds (e.g. lupin) have very thick, waterproof seed coats which prevent the

uptake of water. Until they have been softened by the activities of soil bacteria, which decay them, the seeds cannot germinate.

Most seeds, however, have thin seed coats, but even so these are often the effective agents in preventing germination. If the plumule/radicle is removed from the seed and supplied with nutrients it will germinate, and even if the seed coat is damaged by pricking or scratching, the seed germinates.

Such thin seed coats may prevent oxygen from entering the seed. Or the seed may contain chemical substances which inhibit germination and which are prevented from leaving the seed by the seed coat. This is the cause of dormancy in many desert plants. After high rainfall the chemical inhibitors contained in the seeds are washed out, allowing germination to take place.

Other plants possess the same dormancy mechanism. In rye grass (*Lolium perenne*, Fig. 13.8), a plant hormone called **abscisic acid** is known to inhibit germination. When it is washed out of the seed, germination occurs. Figure 13.21 shows the effect that increased concentrations of this substance had in preventing the germination of non-dormant rye grass seeds.

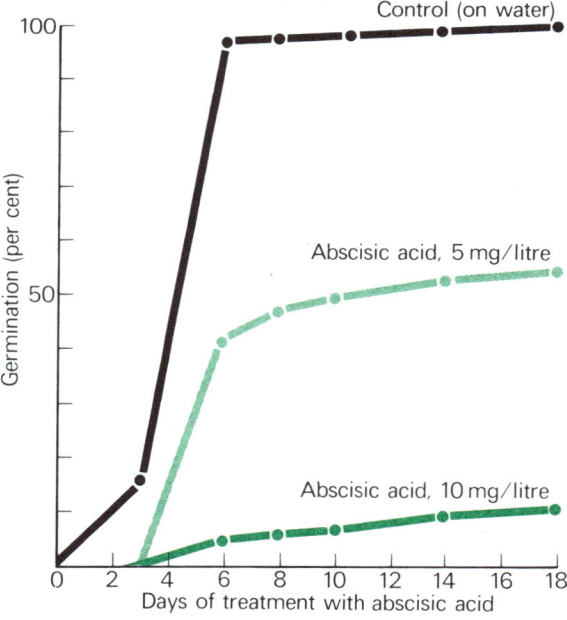

Fig. 13.21 Graph showing the inhibitory effect that the plant hormone abscisic acid has on the germination of rye grass seed. The degree of inhibition is proportional to the concentration of abscisic acid.

14 The Study of Inheritance

The need for death and the need for inheritance

The longest individual life spans are very short in terms of the age of the earth, which is about 4500 million years. But organisms reproduce, and offspring inherit the structure, function, and behaviour of their parents. Because of inheritance species are longer lived than individuals, though eventually, as the fossil record shows, they too become extinct. Their places are taken by new species that have arisen from old species by natural selection.

The word 'selection' implies variety in organisms from which selection can be made and this is provided by the death of individuals and their replacement by others which are slightly different from them. Death of the individual is therefore an essential feature of evolution.

Differences between organisms that are maintained by inheritance

We have encountered in various parts of this book some of the extents to which organisms differ from one another. One fundamental division is between the Prokaryota and the Eukaryota (p. 9). Another is the distinction between single-celled and multi-celled organisms, and another the difference between plants and animals, based on the fact that plants can photosynthesize.

All living organisms *resemble* one another in basic features of cell organisation. These are:

1 All cells are separated from their environment by a plasma membrane.
2 All cells are made of similar chemical materials (mainly proteins, fats, and carbohydrates) that are not found outside the living world.
3 All cells obtain chemical energy in a similar way.
4 All cells contain hereditary instructions in the form of a similar chemical code.

At the same time, organisms *differ* from one another in coarser and in finer ways—for example, the difference between an elephant and a man on the one hand and between two Englishmen on the other. For these similarities and differences to be maintained from generation to generation, the hereditary instructions must possess two opposite characteristics. They must be *fairly stable*, but also be *capable of change* so as to bring about the variety that we see among the offspring of sexually reproducing organisms.

The site of the hereditary instructions (genetic material)

In the sexual cycle of reproduction (Fig. 12.1) the individual is reduced to a single-celled zygote. It must follow that the gametes which give rise to the zygote at fertilization contain the hereditary instructions, or genetic material, telling the zygote how to develop into an organism similar to but not identical with its parents. As the male gamete is little more than a swimming nucleus Fig. 12.11(a)), it is likely that the genetic material is contained in the nucleus.

Mendel's study of inheritance

In the early nineteenth century many attempts were made to discover whether there were any rules governing inheritance, both in laboratory animals and in plants. Nobody, however, made any progress in the matter until the remarkable investigations of Gregor Mendel, an amateur scientist (he was an Austrian monk) whose keen interest in mathematical puzzles may have helped him to solve a problem that had been too complex for his predecessors. He carried out his investigations in the years 1856–65 and published the results in 1865 in a paper that, with Charles Darwin's *Origin of Species*, forms one of the few great landmarks in the history of biological thought. At one stroke it lays the foundations of the science of genetics which all the vast amount of subsequent work in the field has endorsed.

Mendel's assumptions about the genetic material

Before stating these, we must remember the mid-nineteenth century biological background. In 1865, microscopy was in its infancy. Schwann had noted the presence of the nucleus in cells in 1839, but had mistakenly described it as a 'vesicle', that is some sort of fluid-filled space. Chromosomes, which we now know to form the major part of the nucleus, were not described until 1888, 23 years after the publication of Mendel's paper. There is no record of Mendel ever having looked down a microscope.

Nevertheless Mendel *assumed* that the genetic material must be located in the gametes, as these form the link between one generation and the next in sexually reproducing organisms. He stated that the genetic material must be:

(*Assumption 1*) **Structural**, that is, it must have some physical basis, which we now know to be part of a chromosome.

(*Assumption 2*) **Particulate**, that is, individual aspects of the organism must be governed by separate portions of the genetic material, which we now call **genes**. For example, whether a pea plant produces white flowers or violet flowers is governed by one particle of the genetic material, or by one gene.

(*Assumption 3*) **Equally contributed by both gametes.** Knowing as we do now that the sperm and the ovum both contain a nucleus, it is natural for us to make this assumption. But it was not obvious in 1865 when the structure of the nucleus, let alone its function, was not known.

Mendel's careful choice of experimental organism

Mendel chose to use for his investigations the garden pea (*Pisum sativum*), which was grown for food in the monastery gardens. But he states clearly the three biological characteristics that his experimental organism must have 'in order to avoid from the outset every risk of questionable results'. The garden pea satisfied all these requirements, which were as follows:

1 The varieties of plants used must possess differences which are permanent from generation to generation. For example, if he chose to use a strain of peas with violet coloured flowers, then when self-fertilized these plants must give rise to seeds all of which will produce violet-flowered plants. Similarly the seeds resulting from the self-fertilization of white-flowered plants must all give rise to white-flowered plants.

Another way of saying this is that each strain chosen for investigation must **breed true** when self-fertilized. True-breeding strains are called **pure lines**. Mendel chose fourteen strains which fall into seven groups of pairs, as follows:

(a)	(i)	plants with round seeds	(ii)	plants with wrinkled seeds
(b)	(i)	plants with yellow cotyledons	(ii)	plants with green cotyledons
(c)	(i)	plants with inflated pods	(ii)	plants with tight pods
(d)	(i)	plants with green pods	(ii)	plants with yellow pods
(e)	(i)	plants with violet flowers	(ii)	plants with white flowers
(f)	(i)	plants with terminal flowers	(ii)	plants with axillary flowers
(g)	(i)	plants with long stems	(ii)	plants with short stems

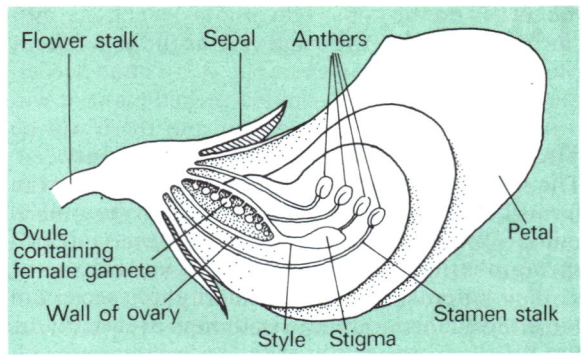

Fig. 14.1 Half flower of garden pea.

His first experiments were to prove that these fourteen strains were in fact true breeding. He grew each strain, self-pollinated its flowers, and gathered the resulting seeds. The next year he planted each of the fourteen sets of seeds and confirmed that each gave rise to plants exactly the same as the parent for that particular characteristic.

No previous investigator of inheritance had taken the precaution of using pure lines, though this has been an essential feature of all genetical experiments performed since 1865.

2 When a cross had been made between parent plants of strain Y and strain Z, and the resulting first generation (called the F_1) seed planted out the following year, the pollination of the F_1 plants must be easily controllable. No other pollen carried by chance insect visitors should be allowed to bring about pollination of these F_1 plants or the resulting second generation (F_2) plants would be of unknown male parentage.

Figure 14.1 is a longitudinal section of a garden pea flower, which is hermaphrodite. To ensure that a particular pea flower pollinates itself, all that is necessary is to tie a muslin bag round it. The anthers will then ripen and burst, depositing their pollen on the stigma.

Alternatively, to prevent a particular flower from pollinating itself, it can be opened while it is still a bud and long before its anthers are ripe. The anthers can be cut off, making the flower into a unisexual female flower.

3 The offspring of matings should be as fertile as the parents were so that further matings can be carried out with them.

A further important practical point can be seen by inspecting the table opposite. The first four pairs of characteristics concern the appearance of the seeds and the pods. The results of such matings can be obtained in the same year as the experiment was carried out. For example, if a round-seeded plant is mated with a wrinkled-seeded plant, it will produce seeds in the same year and these will be either round or wrinkled or something in between. The remaining three pairs of characters concern the flowers and the length of the stem, and the inheritance of these characters cannot be observed until the year after the mating (or **cross**) was made. Taking the cross: tall plant x short plant, seeds will be produced in the autumn that the cross was made, but it will not be possible to tell whether these will give rise to tall plants, short plants, or something between. Only after planting these seeds next spring can a result be obtained.

Mendel's careful planning of his experiments
Another essential ingredient of the success of Mendel's experiments was his realization that meaningful results would come only from matings (crosses) which involved small, clear-cut differences between the parents, such as those shown on page 148. He also thought it vital to count the numbers of offspring which inherited any particular characteristic. Mendel states that he intends:
a To observe the different kinds of offspring that result from the crosses.
b To ensure that he records the various plants produced in each generation accurately, and not to get observations from one generation muddled with those from another.
c To count the numbers of each type of offspring from the crosses and to determine their numerical relationships to one another.

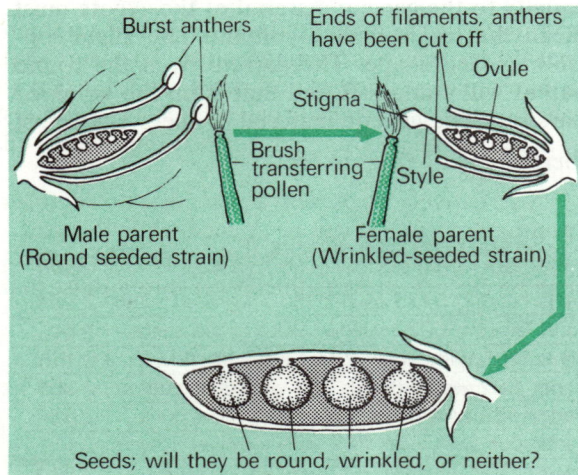

Fig. 14.2 Method used by Mendel for crossing two strains of pea plants.

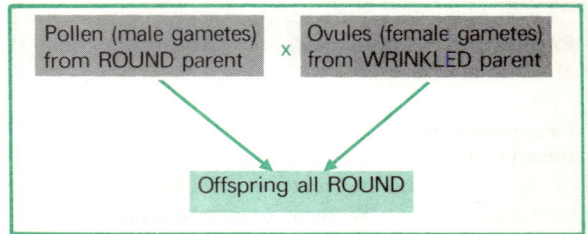

Fig. 14.3 One of Mendel's crosses.

Nobody had thought of this approach before, though it has been the basis of all genetical investigations since. It should be contrasted with the mid-nineteenth century attitude to inheritance which was that if two organisms of the same species were crossed, their offspring would show a vague blend or mixture of the parental characteristics in which no order could be made out.

Mendel's experiments
Mendel crossed each of the seven pairs of true-breeding (pure-line) plants shown in the table on page 148.

To make clear what this means, consider the first pair, plants which produce round seeds crossed with plants which produce wrinkled seeds (Fig. 14.2). A group of plants of each type was raised to the flowering condition. Before the anthers of the wrinkled-seeded strain were ripe, they were all removed by pulling them off with forceps. Thus these flowers could not pollinate themselves. Then the stigmas of these flowers were dusted with pollen gathered from the anthers of the round-seeded strain, using a water colour brush. The flowers were then enclosed in muslin bags so that no insects could visit them, bringing with them chance pollen. They set seed.

Mendel inspected the seeds to see whether they were round (like those of the pollen parent strain) or wrinkled (like those of the ovule parent strain). All the seeds were round, none was wrinkled (Fig. 14.3). Similar results were obtained from the other six crosses. In every case one of the parental characters was shown in the offspring and not the other. Mendel called this the **dominant** character, the character which did not appear in the F_1 offspring being **recessive**.

Blending inheritance proved false
If he had gone no further than this, Mendel would have made a major advance in genetical understanding. The view of his contemporaries (including Charles Darwin) was that when such crosses

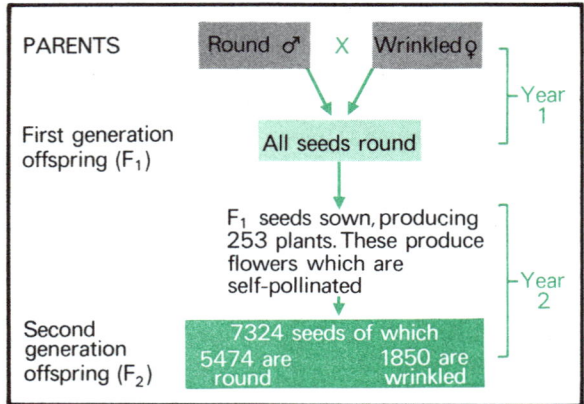

Fig. 14.4 Mendel's results in crossing round seeded x wrinkled seeded pea plants and self-fertilizing the F_1 offspring.

We see here how important it is that the offspring of matings should be as fertile as the parents that produce them (point 3, page 148). The recessive character can be seen in the second generation (F_2) offspring, but not in the first (F_1).

Numbers of offspring
True to his determination to count the offspring showing distinct characters (point (c), page 149), Mendel observed that 5474 of the F_2 seeds were round and 1850 wrinkled, a ratio of 2.96 to 1. The results from all seven crosses are shown in Fig. 14.5. It is obvious that there must be some significance in the ratio 3 dominant : 1 recessive offspring, which appears in the second generation (F_2) of all the experiments.

Mendel's explanation of the 3:1 F_2 ratio
In order to understand Mendel's explanation of this ratio, we must first define some more genetical terms.

Genes and alleles
Mendel assumed that the genetic material is structural and particulate (page 148). In modern terms, one part of the genetic material, a gene, affects (say) seed shape. *This gene can exist in two forms, and these are called* **alleles** *of the gene*. We can represent the dominant allele for roundness of seed as **R** and the recessive allele for wrinkledness of seed as **r**. **R** and **r** are alternative forms, or alleles, of one gene.

Genotype and phenotype
Mendel's assumption that both gametes contribute equally to the genetic material of the zygote must mean that each gamete contains one allele controlling seed shape. The gamete from the round parent will contain **R** and that from the wrinkled parent **r**. The genetic material of the zygote must therefore be $\frac{\mathbf{R}}{\mathbf{r}}$ (Fig. 14.6(c)).

were made the offspring would show a blend of the parental characteristics, such as partly wrinkled seeds, or pale violet flowers. These experiments showed that **blending inheritance** was a false idea, and this had immense implications for the theory of the way in which evolution works (page 276 et seq.).

But Mendel's third assumption about the genetic material (page 148) was that it was equally contributed to the zygote by both gametes. In this case the ovule parent must have contributed genetic material for determining wrinkledness of seed to the zygote, even though this had not appeared in the offspring. It could be that this genetic material had somehow been eliminated from the zygote, but Mendel chose to think that it was still there. He then proved this by sowing the first generation (F_1) round seeds in the following spring and ensuring that the flowers they produced pollinated themselves. The pods of the resulting second generation (F_2) plants contained both round and wrinkled seeds. This means that the F_1 seeds which produced them must have contained genetic information about wrinkled-seededness, even though they themselves were round (Fig. 14.4).

Cross		First generation offspring (F_1)	Second generation offspring (F_2) obtained from self-fertilization of (F_1)	Ratio
1	Round x wrinkled seeds	All round	253 plants bearing 5475 round: 1850 wrinkled seeds	2.96:1
2	Yellow x green cotyledons	All yellow	258 plants bearing 6022 yellow: 2001 green seeds.	3.01:1
3	Violet x white flowers	All violet	929 plants, 705 with violet: 224 with white flowers	3.15:1
4	Inflated x tight pods	All inflated	1181 plants, 882 with inflated: 229 with tight pods	2.95:1
5	Green x yellow pods	All green	580 plants, 428 with green: 182 with yellow pods	2.82:1
6	Axial x terminal flowers	All axial	858 plants, 651 with axial: 207 with terminal flowers	3.14:1
7	Tall x short stems	All tall	1064 plants, 787 with tall: 277 with short stems	2.84:1

Fig. 14.5 The results of Mendel's experiments.

This zygote is the offspring of a cross between a round-seeded parent and a wrinkled-seeded parent. How, then, should we represent the genetic material of these parents?

Since the round-seeded parent came from the self-fertilization of a round-seeded plant, each of the gametes contributing to the zygote from which it grew must have contained the allele **R**. The zygote they produced was therefore $\frac{R}{R}$ (Fig. 14.6(a)). Similarly, the wrinkled-seeded parent must be represented as $\frac{r}{r}$, for the gametes which gave rise to it both contained the allele **r** (Fig. 14.6(b)).

The genetic constitution of an organism is called its **genotype**. This term need not refer to all the genes the organism possesses, and here we can think of the genotypes of the parents as $\frac{R}{R}$ and $\frac{r}{r}$, and of the F_1 offspring as $\frac{R}{r}$.

The different visible and measurable features of an organism, such as seed shape, are collectively known as its **phenotype**. These seeds are phenotypically round or phenotypically wrinkled.

Homozygotes and heterozygotes

The round-seeded parent $\frac{R}{R}$ arises from a zygote which possesses two identical alleles of the gene **R**. It is therefore said to be **homozygous** for **R**. Similarly the wrinkled seeded parent **r** is homozygous for **r**. In contrast, the zygote giving rise to the F_1 offspring has two different alleles of the same gene, and its genotype is $\frac{R}{r}$. It is said to be **heterozygous** for **R**.

The 3 : 1 ratio

Figure 14.6(c) shows the parents $\frac{R}{R}$ and $\frac{r}{r}$ giving rise to gametes each containing one allele, **R** or **r** respectively. These join in the zygote producing an organism with phenotypically round seeds but heterozygous and with the genotype $\frac{R}{r}$.

Following the same logic, we must assume that when this organism produces gametes, they too will carry one allele of the gene, and will be genotypically either **R** or **r**.

Mendel now made two further assumptions neither of which he could prove by direct observation. They were:

Assumption 4 When the anther, whose genotype is $\frac{R}{r}$, produces gametes (pollen see p. 134) whose genotypes are **R** and **r**, then equal numbers of **R** and of **r** gametes are formed. The same is true for the female gametes.

Assumption 5 When pollination and fertilization take place, all male and female gametes will stand the same chance of meeting each other. An **R** male gamete will stand an equal chance of fuzing with an **R** or an **r** female gamete. Nothing directs it preferentially to any particular female gamete. This is referred to as **random fertilization**.

Figure 14.7 shows that if assumptions 4 and 5 are made, the F_2 offspring will be produced in the ratio of 1 dominant homozygote $\frac{R}{R}$: 2 heterozygotes $\frac{R}{r}$: 1 recessive homozygote $\frac{r}{r}$; that is a phenotypic ratio of three round to one wrinkled.

Since Mendel obtained a very close fit to a 3 : 1 ratio with all the pairs of phenotypic characteristics he investigated (Fig. 14.5), it is reasonable to conclude that all five assumptions he made (pages 148, 151) were correct.

Summary of Mendel's findings

The experiments so far described are often summarized in a statement called **Mendel's first law**, though he did not make any such statement himself. This is that *alleles of the same gene contributed to the zygote by the male and the female gametes do not fuze or join or blend in the zygote, but remain separate and can subsequently part from one another.*

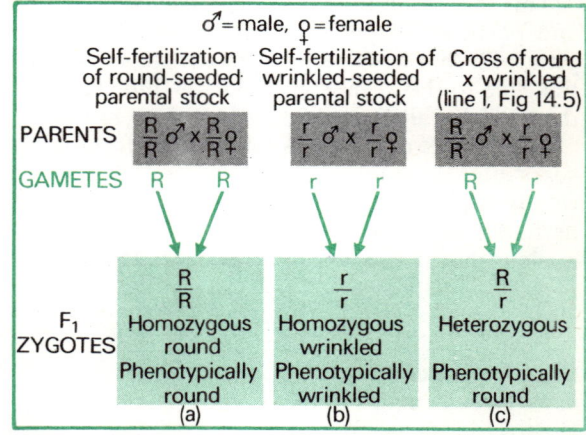

Fig. 14.6 Genotypes of parents, gametes, and F_1 zygotes in true-breeding self fertilizations (a, b), and in the cross round seeded x wrinkled seeded.

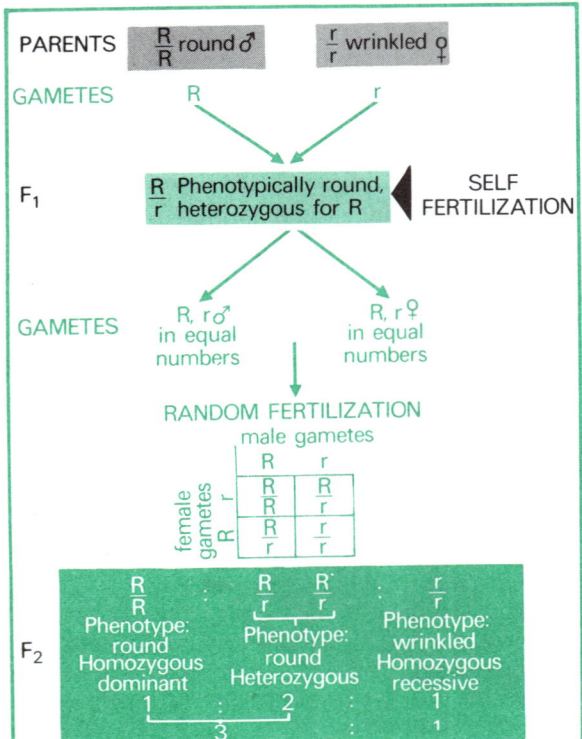

PARENTS $\frac{R}{R}$ round ♂ $\frac{r}{r}$ wrinkled ♀

GAMETES   R   r

F₁ $\frac{R}{r}$ Phenotypically round, heterozygous for R   ◀ SELF FERTILIZATION

GAMETES   R, r♂ in equal numbers   R, r♀ in equal numbers

RANDOM FERTILIZATION
male gametes

Fig. 14.7 Formation of **R** and **r** male and female gametes in equal numbers followed by random fertilization results in a 3:1 ratio of round seeds to wrinkled seeds.

When, for example, the zygote $\frac{R}{r}$ is produced (Fig. 14.7), **R** and **r** remain separate in it. When this zygote has given rise to a new plant, the alleles **R** and **r** pass separately into its gametes.

This law assumes (1) the existence of structural and particulate genetic material (i.e. alleles of a gene); (2) that both parents contribute an allele of the gene to the zygote; (3) that the dominant and the recessive alleles exist together in the zygote.

All these assumptions were verified fifty or more years after the publication of Mendel's paper by a totally different approach to the problem of inheritance—examination of the behaviour of the nucleus during cell division (see pages 164, 165).

Investigation of colour inheritance in maize fruits

Figure 14.8 shows four views of a maize cob. Each of the 'seeds' is in fact a dry fruit which has resulted from the pollination and fertilization of one female flower.

Since the fruits are retained on the cob it is easy to count the numbers of different phenotypes, in this case blue fruits (which appear black in the photograph) and yellow fruits (appear white). The rows have been numbered and the four photographs show the cob in successively turned positions so that you can count the full number of blue and yellow fruits on it.

The female parent was heterozygous for fruit colour. Its genotype was $\frac{B}{b}$, where **B** is the allele producing blue fruits and **b** the allele producing yellow fruits. The male parent which produced the pollen to pollinate and fertilize the female flowers was also heterozygous, $\frac{B}{b}$.

1 Count the numbers of the two types of fruits.
2 Calculate the ratio of the two types.
3 Explain how the result you obtain agrees with Mendel's experiments with peas.

Recessive backcross

Homozygous dominant individuals (e.g. $\frac{R}{R}$) and heterozygotes (e.g. $\frac{R}{r}$), have the same phenotypic appearance. They can be distinguished from each other by means of the **recessive backcross**.

If a homozygous dominant individual is crossed with a homozygous recessive individual, all the offspring are heterozygotes and all display the dominant characteristic (Fig. 14.9(a)). However, if a heterozygous individual is crossed with a homozygous recessive, half the offspring will be heterozygotes displaying the dominant characteristic, and half recessive homozygotes displaying the recessive characteristic (Fig. 14.9(b)).

This prediction involves assumptions 4 and 5 (page 151), namely that the $\frac{R}{r}$ heterozygotes will form **R** and **r** male and female gametes in equal numbers, and random fertilization.

Recessive backcross in maize

Figure 14.10 shows three views of a maize cob resulting from a recessive backcross.
1 Count the numbers of blue and yellow fruits on the cob.
2 Calculate the ratio of the two types.
3 Explain what the genotypes of the male and female parents must have been in order to produce this result.

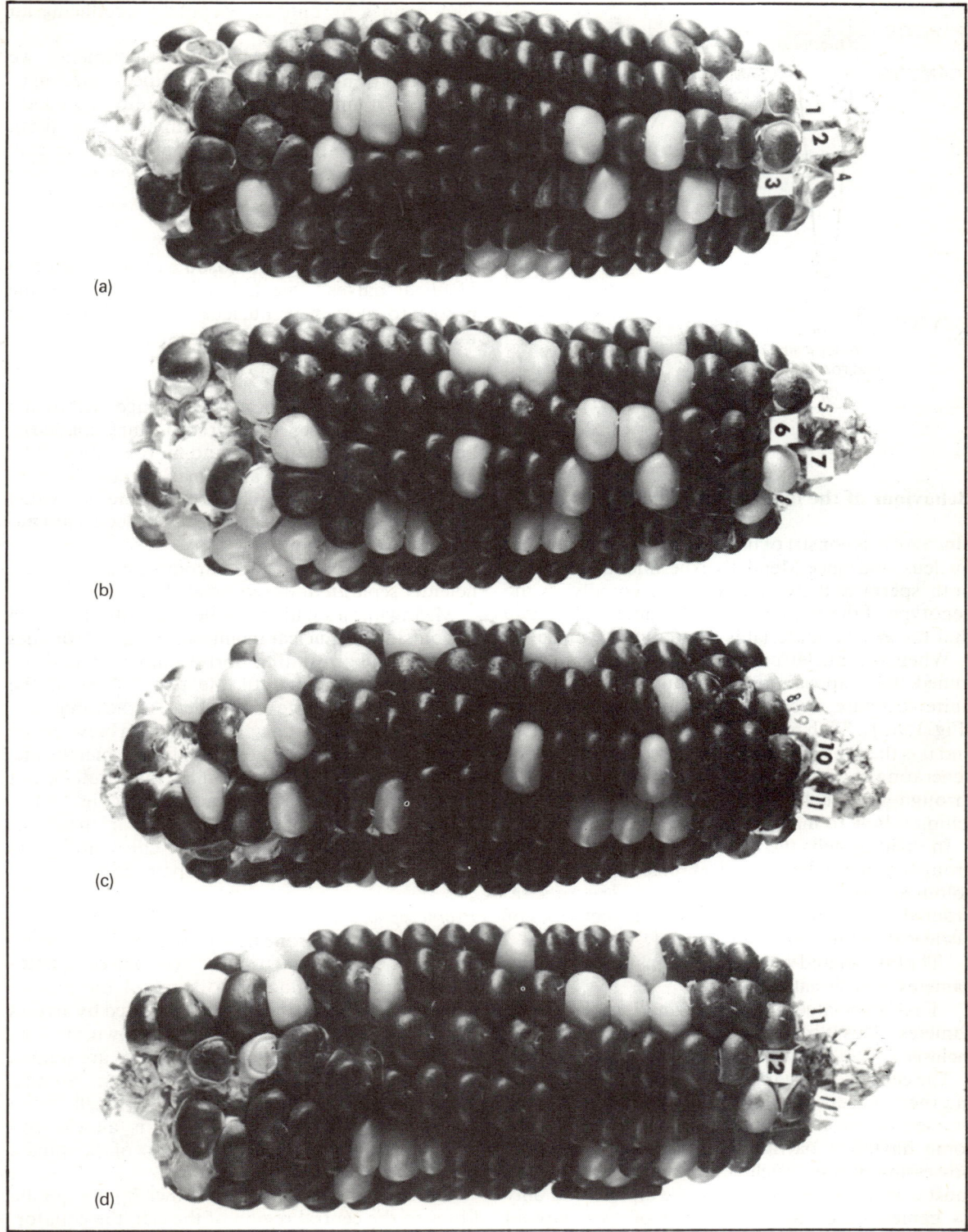

Fig. 14.8 Maize cob derived from the self-fertilization of a plant heterozygous for fruit colour $\frac{B}{b}$.

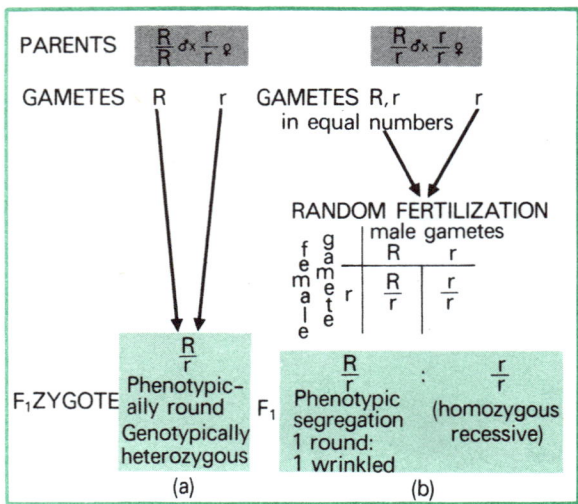

PARENTS

GAMETES   R   r   GAMETES R, r   r
                     in equal numbers

RANDOM FERTILIZATION

Fig. 14.9 Distinguishing between homozygotes and heterozygotes by the recessive backcross.

Behaviour of the nucleus during cell division

Since sperms consist of little more than a swimming nucleus, and since Mendel's experiments show that both sperm and ovum contribute equally to the genotype of the zygote, it is reasonable to conclude that the genetic material is located in the nucleus.

When the nuclei of *non-dividing* cells are examined, they appear granular but show little definite structure, except for the nuclear membrane (Fig. 1.9(a)). Such nuclei are said to be in **interphase** and it is then that they are carrying out their task of generating coded chemical instructions which pass through the nuclear pores into the cytoplasm, telling it how to make proteins from amino acids.

In dividing cells the nucleus is seen to contain a group of organelles called **chromosomes** (Greek, coloured body). These behave in very precise, ordered, and elaborate ways. Two distinct types of nuclear division can be recognized;

1 That associated with cell divisions not producing gametes. This is called **mitosis**.
2 That associated with the formation of haploid gametes.. This is called **reduction nuclear division**, or **meiosis**.

The cells in an organism are of two types regarding their nuclei: *diploid* cells, each of which bears two sets of chromosomes, each individual chromosome having a partner; and *haploid* cells, each possessing one set of chromosomes (Fig. 12.1). In most organisms the body cells are diploid and only its gametes are haploid. Fusion of two haploid gametes at fertilization produces a diploid zygote.

This divides by mitosis, eventually producing an adult organism.

When considering Mendel's experiments we have been thinking all the time of the formation of gametes. It might be logical, therefore, to look first at the reduction nuclear division (meiosis). However, since this is more elaborate and lengthy than mitosis, it is easier to start by considering mitosis.

Mitosis

In this process each chromosome divides into two identical halves. One half of each chromosome passes to each new nucleus. The two resulting nuclei are therefore genetically identical.

Prophase

The first stage, **prophase**, takes place within the nuclear membrane. The chromosomes condense, becoming shorter, thicker, and more easily visible. Two stages of prophase are seen in Fig. 14.11(a) and (b). Both show that each chromosome is divided lengthways into two strands, each called a **chromatid**.

In stage 14.11(b), the chromosomes are sufficiently separate to be counted. There are six.

Meanwhile a small organelle, the **centriole**, which lies outside the nuclear membrane (Fig. 1.7) divides into two. One resulting portion moves round the nuclear membrane until the two halves of the centriole are opposite each other. At the same time fibres of protein, the **spindle fibres**, form between the two centrioles, outside the nuclear membrane. Other protein fibres of a similar kind radiate out from the centrioles in the form of a star (Fig. 14.12). The spindle fibres play an important part in mitosis for they become attached to the chromatid pairs and draw the members of each pair apart.

Prometaphase

The nuclear membrane now dissolves and a special region of each chromatid, the **centromere** (kinetochore), becomes attached to some of the protein spindle fibres. Centromeres are indicated by arrows in Fig. 14.11. Prometaphase is shown in Fig. 14.11(c), though only the chromosomes are stained and the spindle fibres cannot be seen. Spindle fibres can be seen in the unstained, living cell shown in Fig. 14.13(a). Figure 14.13(b) is an electron micrograph showing the attachments of the spindle fibres to the centromeres.

The chromosomes are now drawn by the spindle fibres to the central region of the cell, the equator. Here they all lie in one plane. This is **metaphase**, and

(a)

(b)

(c)

Fig. 14.10 Recessive backcross in maize.

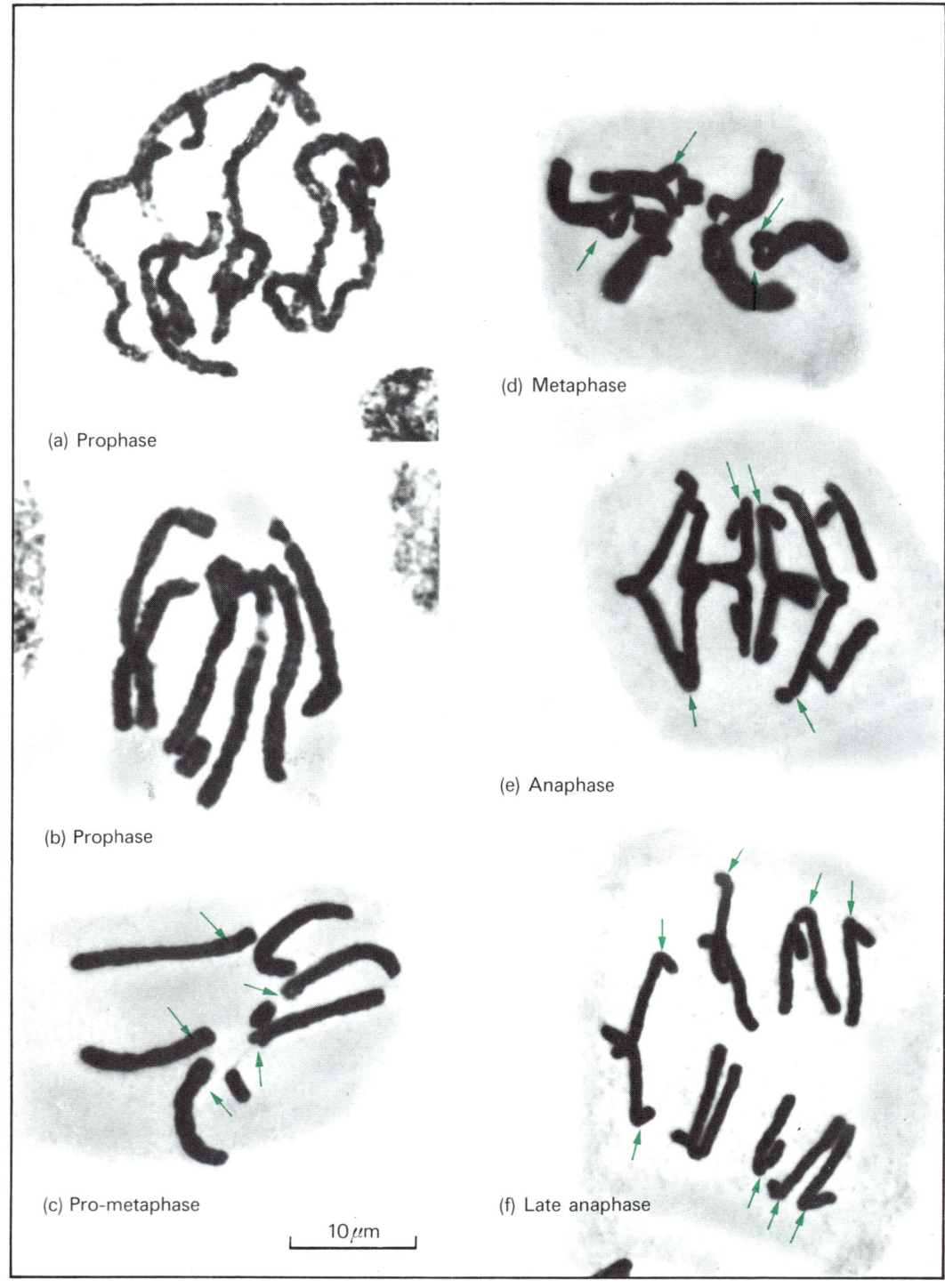

(a) Prophase

(b) Prophase

(c) Pro-metaphase

(d) Metaphase

(e) Anaphase

(f) Late anaphase

10 μm

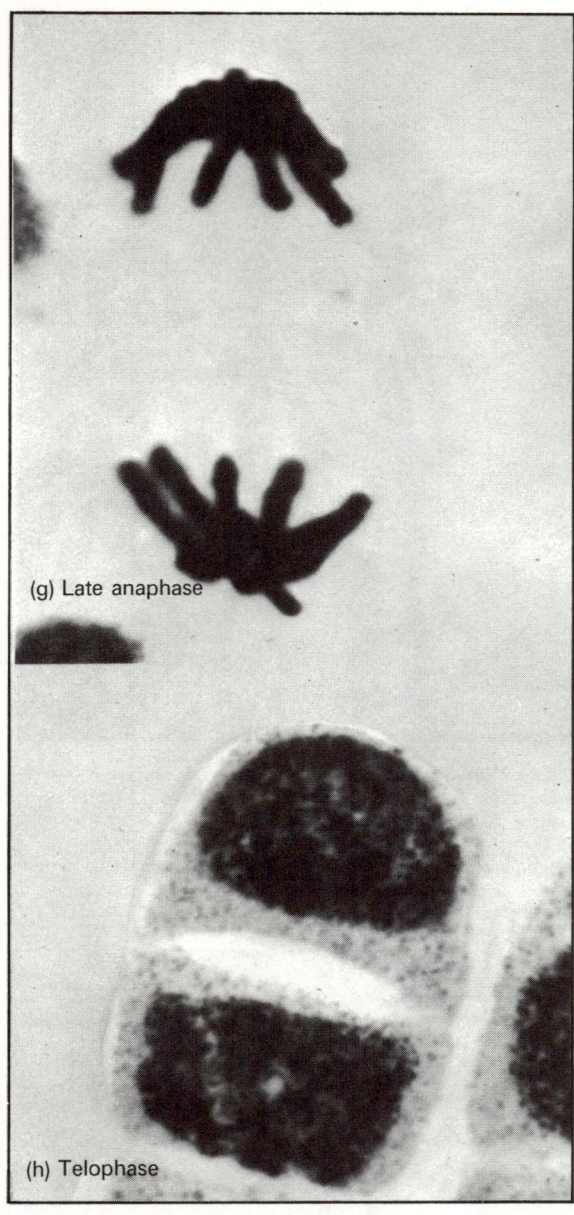

(g) Late anaphase

(h) Telophase

it is seen in side view (with the equator at right angles to the plane of the paper) in Figs. 14.11(d) and 14.13(a).

Anaphase

The centromeres in each pair of chromatids are now drawn apart by the spindle fibres to which they are attached, peeling apart the identical chromatid partners of each chromosome. This is shown in Figs. 14.11(e) and (f).

Two groups of chromatids are now moving towards the opposite poles of the cell. Each is a single-stranded structure, but in the next interphase it will duplicate, forming a double strand. We must therefore cease to think of it as a chromatid which was part of the old cell. It is now a chromosome which will be part of one of the two new daughter cells.

Telophase

In Fig. 14.11(g) the chromosomes in the two groups are seen close to one another. The chromosomes now lengthen, resuming their diffuse interphase condition (Fig. 14.11(h)).

A nuclear membrane forms round each nucleus. In animal tissues the cytoplasm between the two nuclei constricts, producing two cells. In plant tissues a cellulose wall develops separating the two daughter cells.

Mitosis produces genetically identical cells

The exact division of the chromosomes into two equal sets in mitosis means that each daughter cell possesses exactly the same genetic information as the parent cell, *provided that the sister chromatids in each chromosome are genetically identical.* There is much evidence that they are. Genes are situated on the chromosomes in a linear order like a string of beads. Each chromatid in a chromatid pair is an identical string of genes.

It is difficult to say how many genes an organism possesses, but in man it is probably of the order of

Fig. 14.11 *Mitosis in Crocus balansae* (diploid number of chromosomes, 2n = 6). (a), (b), stages in *prophase.* The chromosomes are shortening and appear as paired threads or chromatids (arrows = centromeres). In (b) it is clear that there are six chromosomes. The nuclear membrane is still present, though not visible in these photographs, and the spindle has not yet appeared. (c) *Pro-metaphase.* The nuclear membrane has dissolved. The spindle has appeared, though not visible in this photograph, and the chromosomes (paired chromatids) have become attached to some of its threads by their centromeres (arrows). (d) *Metaphase* the chromosomes have moved to the middle zone of the cell, the equator. They have become still shorter and thicker, and are seen here from the side, i.e. the equator of the cell is perpendicular to the page. (e)

Anaphase. One chromatid from each pair is passing to one pole of the cell (oriented north/south), being led by its centromere along the spindle fibre. The corresponding chromatid is passing to the opposite pole. The chromatids are still associated along part of their length. (f), (g) Late stages of anaphase showing two nuclei each consisting of exactly the same amount of the same kind of genetic material: each contains one chromatid from each of the original chromosomes. At this stage we must stop calling the chromatids chromatids, for although single, they are the *chromosomes* of the daughter nuclei. (h) *Telophase.* The chromosomes have uncoiled and are indistinguishable. The cytoplasm of the original cell has divided into two, producing two cells each with an identical nucleus.

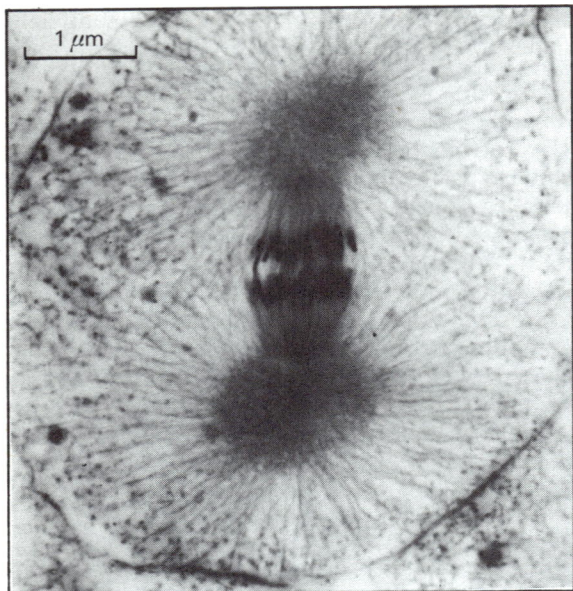

Fig. 14.12 Spindle fibres and stars of fibres radiating from the centrioles. Anaphase of mitosis in the white-fish *Coregonus clupeoides*.

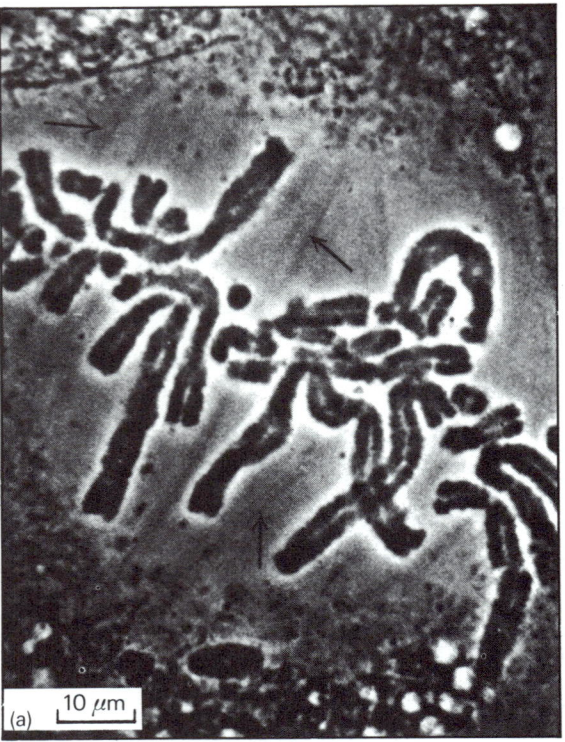

Fig. 14.13(a) Spindle fibres (arrows). Chromosomes are attached to some of them by their centromeres. Metaphase of mitosis in the plant *Haemanthus katherinae*. This is a living, unstained preparation. (Light microscope.)
(b) Chromosomes attached to spindle fibres (arrows) at their centromeres. Electron micrograph of mitotic metaphase in the plant *Haemanthus katherinae*.

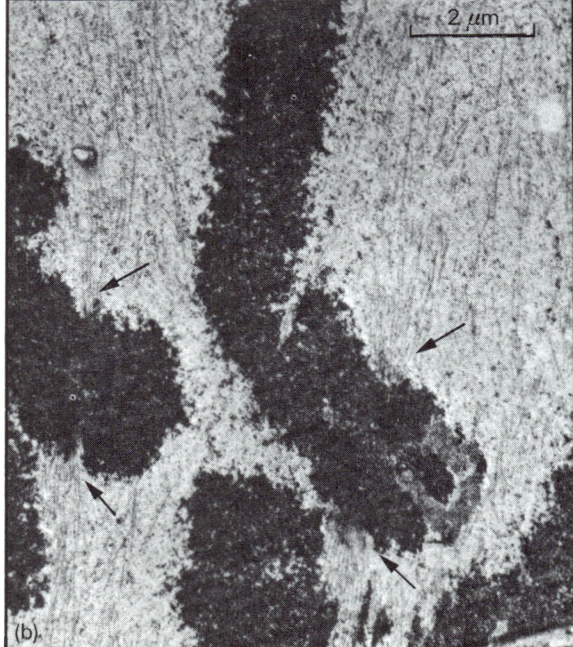

15 000 distributed between 46 chromosomes. Genes are too small to be seen with the light microscope, but in Figs. 14.1(a) and (b) we can see that corresponding regions of sister chromatids appear similar in size, shape, and amount of staining.

Clones

Descendents produced from a single 'parent' by asexual reproduction form a **clone**. Members of a clone are genetically identical.

Asexual reproduction in higher plants and animals (see page 115) also involves only mitotic cell divisions. The cells of such offspring are therefore genetically identical with those of their parents. Provided the environmental conditions under which parent and offspring are reared are the same, the phenotype of the offspring will be identical with that of its parent. This is the basis of horticultural propagation of prize varieties by corms, bulbs, tubers, rhizomes, and of grafting techniques.

A field of potatoes is an example of a clone, and a pair of human identical twins is another. These are formed after mitotic divisions of the zygote have given rise to the blastocyst which separates into two growing centres (see p. 123). If you know a pair of identical twins you will have a measure of the extent to which variation in the environment can influence the development of two genetically identical organisms.

Clones and evolution

If a species is well adapted to a stable environment there is no immediate advantage in its individuals being genetically different from each other. Under these conditions asexual reproduction, producing a clone of genetically identical individuals, is an advantage. Clones are more possible in plants than in animals, and clones of fescue grass (*Festuca spp*) in hill pastures of Britain are known to be up to 1000 years in age.

Such clones are at a *long-term* disadvantage however if they cannot reproduce sexually as well. Their offspring are unable to adjust to long-term changes of climate, such as the ice ages of the recent geological past. If attacked by virus or other diseases, they have no genetically different individuals who might be immune and might enable the species to survive.

Reduction nuclear division (meiosis): the formation of gametes

Halving the diploid number of chromosomes present in the body cells is an essential feature of sexual reproduction (Fig. 12.1). The special nuclear division that brings this about is called reduction nuclear division or meiosis. It results in haploid gametes. When two of these unite in fertilization they form a diploid zygote which divides by mitosis to produce the body cells of the new individual.

Major differences between mitosis and meiosis

Before describing meiosis it is worth noting the major differences between mitosis and meiosis. These are:

1 Whereas mitosis ensures that the daughter cells contain the same amount of the same kind of genetic material, meiosis ensures the exact opposite, namely that *every gamete* produced by it is *genetically unique*.

2 A cell undergoing meiosis passes through *two* distinct phases of nuclear division with an interphase between them.

3 Although in mitosis the chromosomes behave independently of each other, in meiosis chromosomes associate intimately in pairs.

Meiosis in male grasshopper

Our photographs (Fig. 14.14) show meiosis in the formation of sperms in the grasshopper *Chorthippus brunneus*. The diploid number of chromosomes in this animal's body cells is 16. The overall result of this meiosis is to produce four haploid sperms each containing eight chromo-

somes from one diploid sperm-mother cell containing sixteen chromosomes.

The first phase (meiosis I)

The **prophase** of the meiosis I is longer and more elaborate than the prophase of mitosis. It, too, takes place inside the nuclear membrane. The centriole divides and produces the spindle at the same time as prophase is proceeding. Prophase of meiosis I is illustrated in Fig. 14.14(a)–(f). The following features should be noted:

1 The 16 chromosomes of the diploid cell attract each other in pairs. In Fig. 14.14(c) we see eight *pairs* of **partner chromosomes.**

2 Each member of the pair duplicates. Each of the associated pairs of chromosomes in Fig. 14.14(c) now becomes four associated chromatids (Fig. 14.14(d)).

Metaphase (Fig. 14.14(g)) and anaphase (Fig. 14.14(h)) follow prophase. You will see in Fig. 14.14(h) that *pairs of chromatids, that is whole chromosomes* are moving to the opposite poles of the cell. In this photograph the poles are north/south. Where the centromere is at the centre of a chromosome, four arms can be seen trailing behind it. Where the centromere is at the end of a chromosome, two arms (two chromatids) trail behind it. In this stage *partner chromosomes are separated from each other*. Each resulting nucleus contains only one member of each pair of partner chromosomes.

This stage should be compared with anaphase of mitosis (Figs. 14.11(e) and (f)), where the centromeres lead *single* chromatids to opposite poles of the cell. In meiosis, however, two haploid nuclei are going to be produced at the end of meiosis I, each containing only eight chromosomes. The diploid sperm-mother cell nucleus that started this meiotic division possessed sixteen chromosomes.

A telophase and interphase conclude the first phase of meiosis (meiosis I) (Figs. 14.14(i) (j)). A nuclear membrane forms round each of the two haploid nuclei, and the cytoplasm of the cell may divide into two.

The second phase (meiosis II)

Meiosis I has produced two haploid nuclei each containing eight chromosomes. Each of these now undergoes a further division in the second phase of meiosis, meiosis II. This produces a total of four haploid nuclei from the original diploid sperm mother cell.

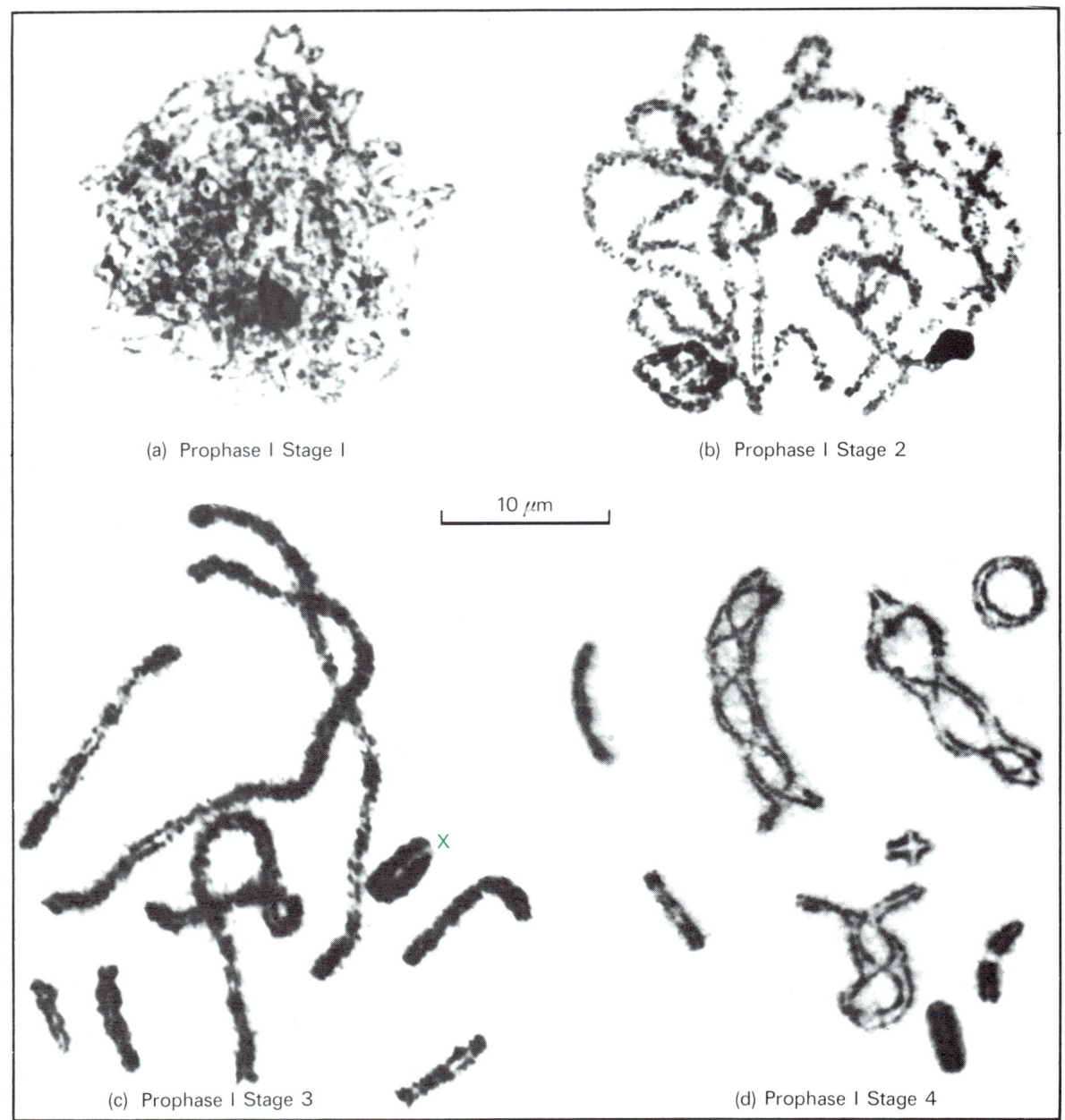

(a) Prophase I Stage I

(b) Prophase I Stage 2

10 μm

(c) Prophase I Stage 3

(d) Prophase I Stage 4

Fig. 14.14 *Meiosis in male grasshopper (Chorthippus brunneus)* (diploid number of chromosomes, 2n, = 16). (a) Prophase I stage 1: chromosomes are shortening and appear as single threads. (b) Prophase I stage 2: partner chromosomes have associated along their whole lengths producing paired threads. (c) Prophase I stage 3: further shortening reveals eight pairs of chromosomes plus the X chromosome. (d) Prophase I stage 4: each chromosome has duplicated into two chromatids and associations are therefore now of groups of four threads. The close association has loosened and chiasmata are abundant. (e) Prophase I stage 5: further shortening marks the end of prophase I. The spindle has now appeared, though not visible in this photograph. (f) Pro-metaphase I: each chromatid possesses a centromere, so each chromosome pair possesses four. These become attached to spindle fibres which draw the chromosomes to the equator of the cell. (g) Metaphase I: the chromosome pairs all lie in the equatorial region, and are seen here from the side with the equator at right angles to the page. (X chromosome has not passed to the equator.) (h) Anaphase I: separation of chromosomes, each consisting of two chromatids, into two sets of eight each. Poles of the cell are north/south on this page. The X chromosome has passed to the south pole.

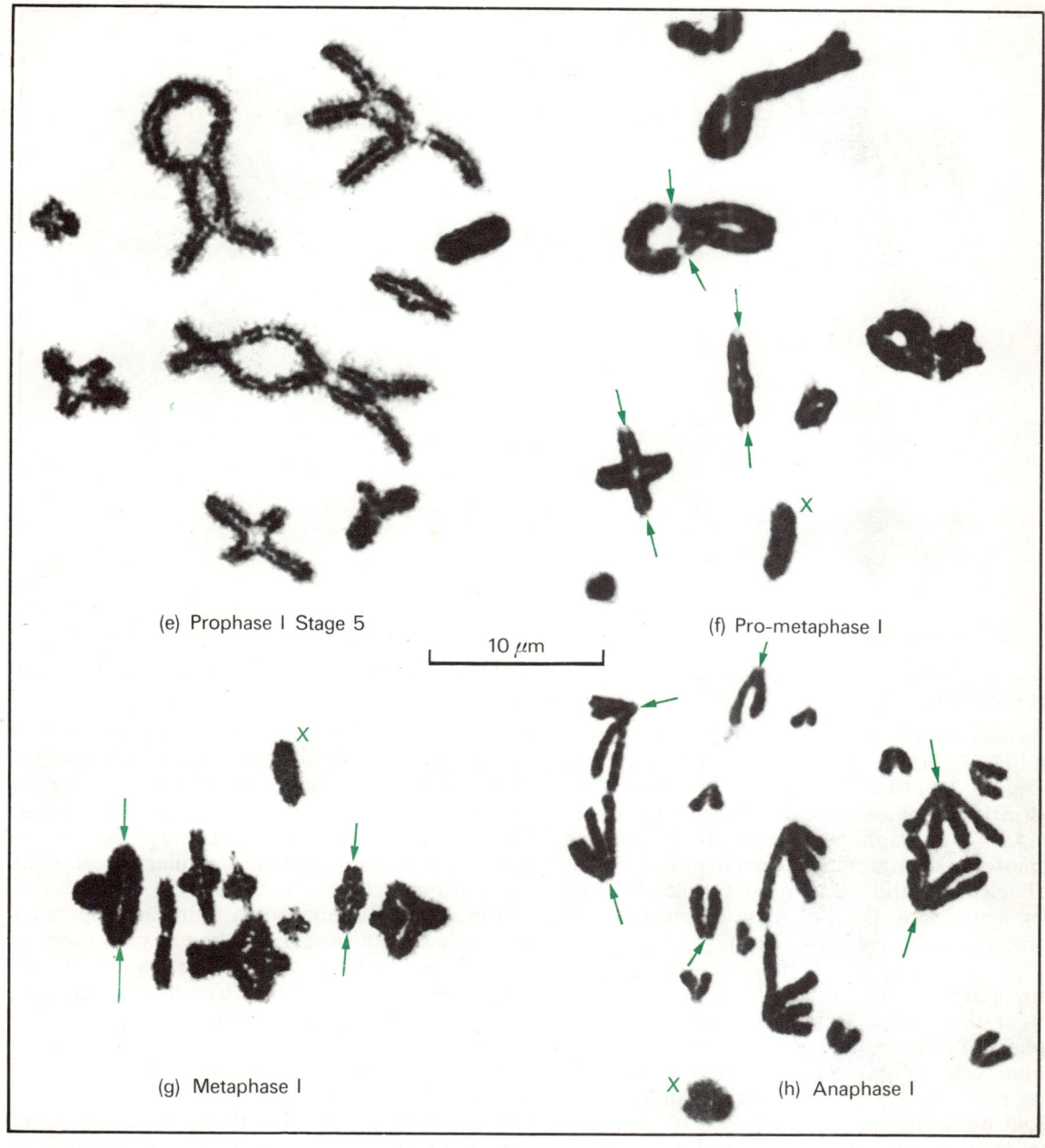

(e) Prophase I Stage 5

10 μm

(f) Pro-metaphase I

(g) Metaphase I

(h) Anaphase I

(i) Late anaphase I: the chromosomes are losing their individual identity. (j) Interphase: the chromosomes have uncoiled and a nuclear membrane has formed, though it is not visible in this photograph. (k) and (l), prophase meiosis II: chromosomes have shortened by coiling. Each consists of a pair of chromatids. Nucleus (k) does not contain the X chromosome, and is derived from the nucleus at the north pole of (h); nucleus (l) does contain the X chromosome and is derived from the nucleus at the south pole of (h). (m) and (n), metaphase meiosis II, seen from above so the chromosomes are arranged at the circumference of a circle formed by the spindle fibres to which they are attached by their centromeres. (o) and (p), anaphase II: the chromatids in each chromosome have parted and are being led to opposite poles by their centromeres. Thus four haploid nuclei, each containing eight chromatids have been produced from the original diploid sperm mother cell. This contained 16 chromosomes. Each of the eight chromatids in the four anaphase II nuclei will become a chromosome of a sperm. (N.B. the nuclei in (p) contain nine chromosomes, eight plus the X chromosome).

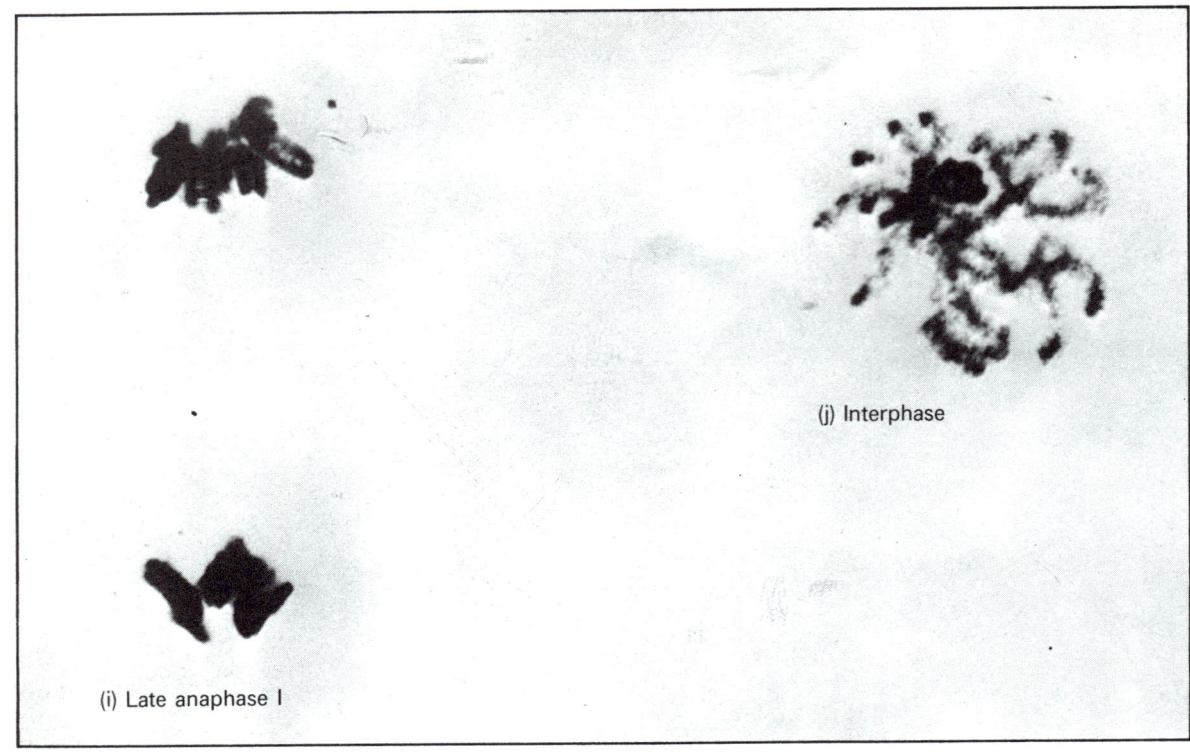

(j) Interphase

(i) Late anaphase I

Fig. 14.14 (i) and (ii).

The meiosis II division of the north pole nucleus from Fig. 14.14(h) is shown in Figs. 14.14(k, m, and o). Figure 14.14(k) is prophase of meiosis II; 14.14(m) is metaphase of meiosis II seen from above; and 14.14(o) is anaphase of meiosis II. In this each chromosome separates into two daughter chromatids. Each chromatid is drawn to one pole of the cell by its centromere. Each of these chromatids will become a chromosome of a sperm.

Differentiation of gametes

Each of the four cells produced at the end of anaphase II differentiates into a sperm, much like the ones seen in Fig. 12.11. In the case of flowering plants, the diploid pollen mother cell produces four haploid nuclei by meiosis, and each of these becomes the nucleus of a pollen grain (Fig. 14.15).

In the production of haploid female gametes (ova), the process of meiosis is the same as that described for the production of sperms. Far fewer ova are produced than sperms or pollen grains, and it is usual for only one of the four haploid nuclei resulting from meiosis to differentiate into the female gamete. The other three disintegrate.

Fertilization gives a diploid zygote

The four haploid nuclei produced from the above meiosis each contain the haploid number (8) of chromosomes. Two of these nuclei will develop from the north pole and two from the south pole sets of chromosomes seen in the anaphase I of Fig. 14.14(h). In an exactly similar way, the ova of grasshoppers each contain a similar set of eight chromosomes.

The partner chromosome pairs are separated at anaphase of meiosis I but the pairs are restored by fertilization (Fig. 14.16). The zygote and all the body cells derived from it by mitosis are diploid and contain sixteen chromosomes.

Independent movement of chromosomes at anaphase I of meiosis

On page 159 it was said that meiosis produces genetically unique gametes. This is its fundamental biological importance. Evolution takes place by natural selection, and meiosis and sexual reproduction provide the genetically variable offspring from which selection can be made (see page 268).

A major factor in the formation of genetically unique gametes is the independent behaviour of partner chromosomes at anaphase of meiosis I, which we must therefore consider in more detail.

Figure 14.14(h) shows that the partner chromosomes of each pair have separated and one member

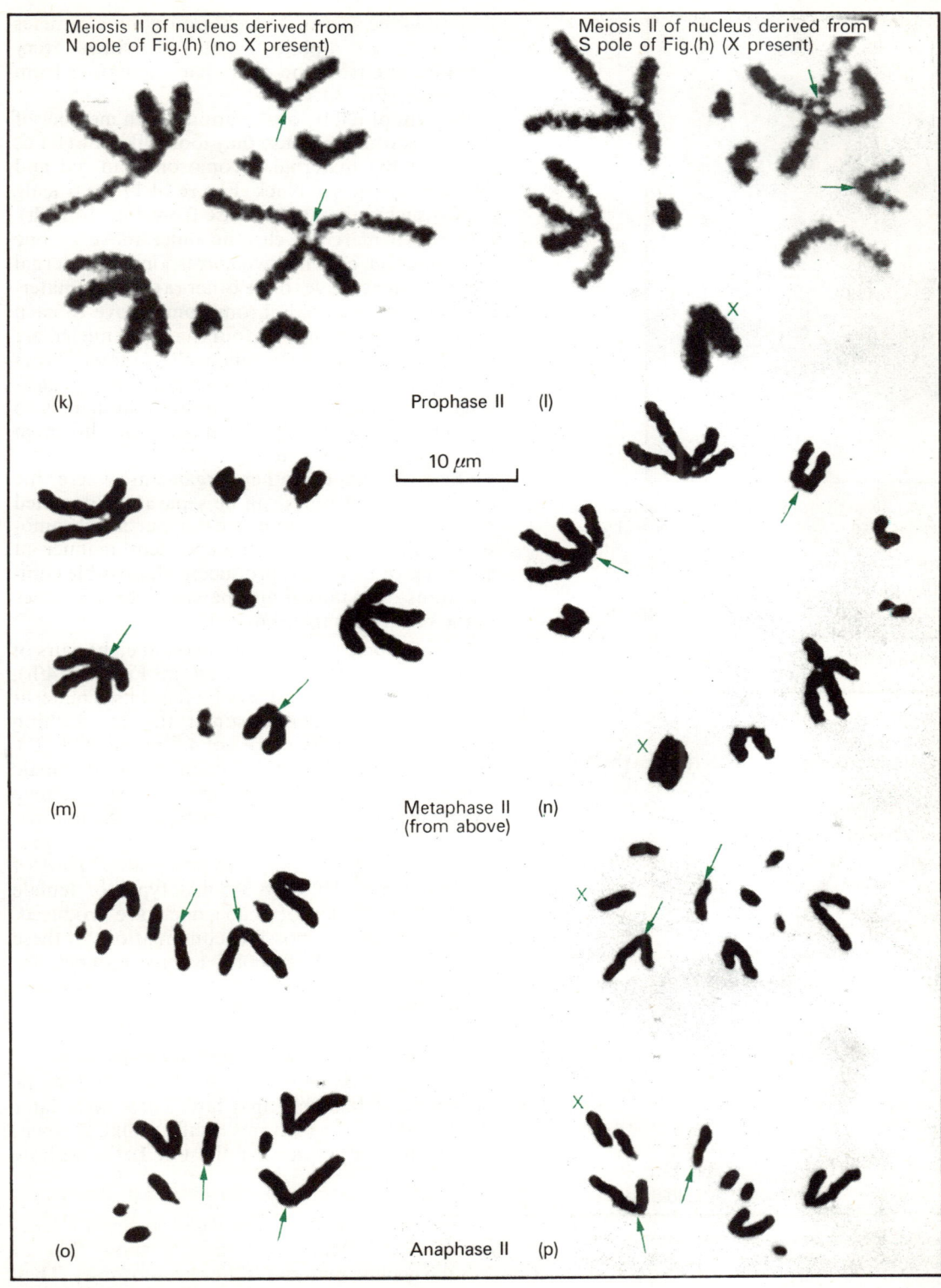

Meiosis II of nucleus derived from
N pole of Fig.(h) (no X present)

Meiosis II of nucleus derived from
S pole of Fig.(h) (X present)

(k)   Prophase II   (l)

10 μm

(m)   Metaphase II   (n)
             (from above)

(o)   Anaphase II   (p)

Fig. 14.14 (k)–(p).

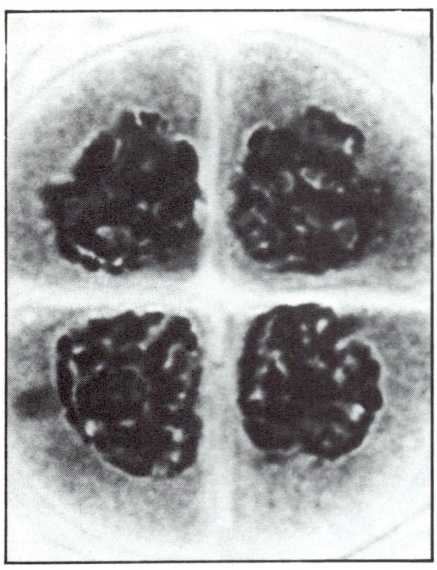

Fig. 14.15 Meiosis in the formation of pollen grains. Cell walls are forming between the four nuclei at telophase II. (Rye.)

is passing to the north pole of the cell, the other to the south pole. One member of each of these pairs was derived from the maternal haploid set of chromosomes present in the ovum. The other member was derived from the paternal haploid set

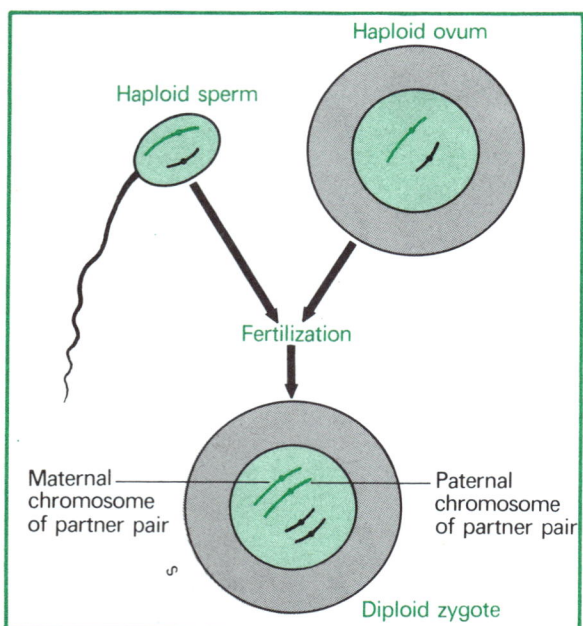

Fig. 14.16 The haploid sperm and the haploid ovum each contain one member of each pair of partner chromosomes. Fertilization produces a diploid zygote in whose nucleus both members of every partner pair are present.

present in the sperm. At fertilization these gametes produced the diploid zygote in which every chromosome from the ovum had its partner from the sperm (Fig. 14.16).

We cannot tell by observation which member of the pair is which because they look alike. But let us represent the maternal chromosomes in red and the paternal ones in black. Figure 14.17 is a tracing of two such pairs at anaphase from Fig. 14.15(h). Either two maternal chromosomes move to one pole, in which case the two corresponding paternal chromosomes move to the other (a), or one maternal and one paternal chromosome move to each pole (b). The resulting four haploid nuclei are therefore of four, or 2^2 genetically different types *provided the two chromosome pairs move independently of each other*. (This means that there is no reason why (say) only the movement shown in Fig. 14.17(a) is possible.)

Observations on other organisms where the partner chromosomes can be separately identified by observation, has confirmed that partner chromosomes do move in this independent manner at anaphase of meiosis I, producing all possible combinations of maternal and paternal chromosomes in the resulting haploid nuclei.

But of course there are not two but eight pairs of partner chromosomes moving in Fig. 14.14(h), and the number of possible combinations of maternal and paternal chromosomes in the resulting haploid nuclei is therefore not 4 (2^2), but 128 (2^8).

In other words, 128 different types of female gamete and 128 different types of male gamete will be produced by grasshoppers, and the number of possible combinations of these at fertilization is 128^2, or 16 384. In man, who possesses 23 pairs of chromosomes, 2^{23} or 8 388 608 types of female and 8 388 608 types of male gamete are produced, and the number of possible combinations of these at fertilization is 8 388 608^2, a huge number, far greater than the world human population which is about 4000 million (in 1976).

Genes and chromosomes behave in similar ways

On page 148 we said that three assumptions underpinned Mendel's first law. These were later confirmed by investigations totally unlike his own, namely the study of nuclear division. Let us see how this is so.

1 The existence of structural and particulate genetic material. Many experiments show that in fact the genetic material *is* the chromosomes. They provide the structure. The particles are genes,

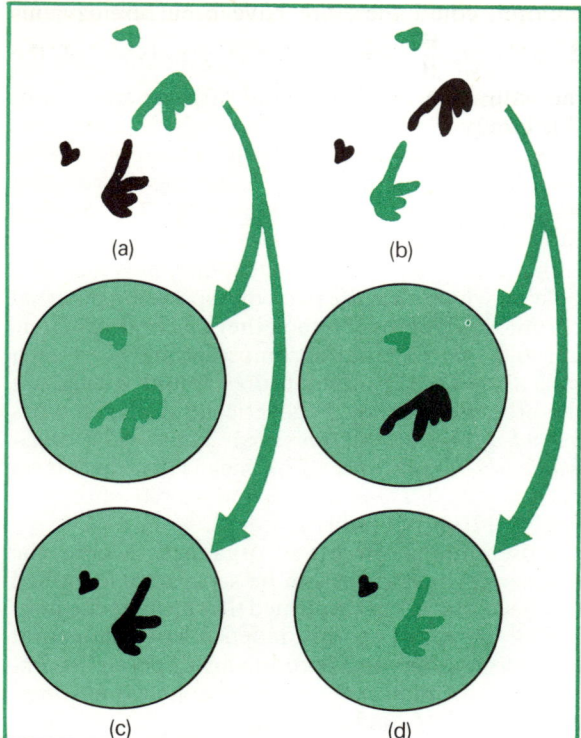

Fig. 14.17 Independent migration of two pairs of partner chromosomes at anaphase I ((a), (b)) produces four (2^2) haploid interphase nuclei. These are all genetically different from each other. Two contain either maternal or paternal chromosomes only (c), the other two contain both maternal and paternal chromosomes (d).

which are arranged in the chromosomes in a linear sequence like a string of beads.

2 Both parents contribute an allele of a gene to the zygote. In Fig. 14.7, each F_2 zygote combines an allele, which may be **R** or **r**, from the male gamete, with an allele, also **R** or **r**, from the female gamete.

This is paralleled by the behaviour of the chromosomes, for each gamete contains one member of a partner chromosome pair which it contributes to the zygote at fertilization (Fig. 14.16).

3 That the dominant and recessive alleles exist together in the zygote. If partner chromosomes bear the same genes this too is explicable by chromosome behaviour. For example, assume for the sake of argument that the chromosomes seen in Fig. 14.17 belong not to grasshopper but to the garden pea and that this cell is a zygote. Let the larger chromosome be the site of the gene whose

alleles are **R** and **r**. Then the maternal chromosome (red) may bear the allele **R** and its paternal partner (black) the allele **r**. The two alleles coexist in the zygote on the separate members of a partner pair of chromosomes.

Chromosomes and alleles segregate in similar ways

In explaining the 3:1 F_2 ratios in his experiments, Mendel assumed that when a heterozygote such as $\dfrac{R}{r}$ forms gametes, half will contain the allele **R** and half the allele **r** (assumption 4, p. 151).

Meiosis provides the physical basis for this. If the large red chromosome in Fig. 14.17 bears the allele **R** and the black one the allele **r**, then since these chromosomes move to opposite poles of the cell at anaphase I, one of the alleles passes into one haploid nucleus, the other into the other. This will occur in all the meioses giving rise to all the gametes. That is to say, 50% of the gametes will contain the allele **R** and the other 50% the allele **r**.

The origin of genetic differences: gene mutation

So far we have considered the reassortment of the genetic material in meiosis, but not the origin of genetic differences. What causes the differences between alleles of the same gene, such as **R** and **r**?

Chromosomes, and the genes they contain, are duplicated in the interphase of mitosis and in prophase I of meiosis. Very occasionally something goes wrong with the duplication process and a slightly different form or allele of a gene is produced. This is called **gene mutation** (Latin mutare, to change).

Frequency of mutation

We would expect mutation to be a rare event, for otherwise the genetic material would be unstable. We noted on page 147 that the genetic material must be stable so that similarities can be passed on from generation to generation. The essential variety among offspring which provides the basis of natural selection could not be due to gene mutation, for the genetic material would then be too unstable to maintain these similarites. Genetic variety in the offspring is provided by the reassortment of the genetic material in meiosis.

The frequency of mutation is best investigated in rapidly-reproducing organisms such as bacteria. Here it has been found that about one copy of a particular gene in every 50 million is mutant. In

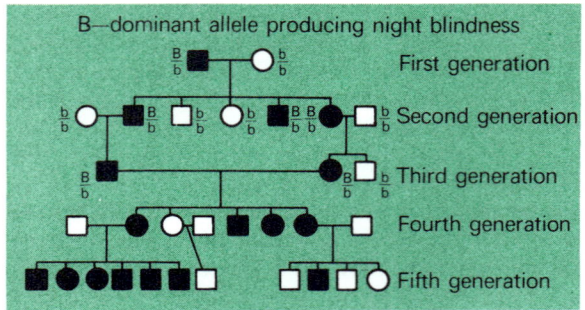

Fig. 14.18 Pedigree showing the inheritance of night blindness in five generations of a family.

children could therefore have been homozygous dominants, $\frac{B}{B}$. Work out the genotypes of the individuals in the fourth and fifth generations of this family.

the fruit fly *Drosophila melanogaster* most genes tested show one mutant for every 30 000 copies made. Although rare, mutation is the ultimate source of genetic variation.

Another important feature of mutation revealed by laboratory investigations is that it is reversible. Mistakes can be made in the gene copying process both ways, for example from **R** to **r** or from **r** to **R**.

Heritable mutations
If a mutation takes place in a body cell, the genotype of that cell will become different from those surrounding it, but the genetic change will not be passed on to succeeding generations. For mutation to be inherited it must take place in the nuclei of the sperm or ova mother cells, in the sperm or ova themselves, or in the zygote.

Figure 14.18 shows the pedigree of inheritance of a dominant mutation causing night blindness in humans. Males are shown as squares, females as circles. Night-blind individuals are shown solid black, normal-sighted individuals white.

The man in the first generation was the first in this family to show the phenotype of night blindness. None of his ancestors had been affected, and the mutation must therefore have taken place in the nuclei of the gamete mother cells which gave the gametes which formed him, or in the nuclei of the gametes themselves, or in the nucleus of the zygote from which he arose. That he was heterozygous for the gene is shown by the fact that only three of his five children by a normal woman inherited night blindness. These were likewise heterozygotes and when two of them married normal partners, their affected children were also heterozygotes.

In the third generation there was a marriage between first cousins, some of whose affected

15 Sensitivity and Coordination

Coordination in unicellular organisms

Coordination means the working together of parts to make a harmonious whole.

To take an example, when a single-celled protozoan, such as *Amoeba*, feeds, its cytoplasm moves towards the food particle and then surrounds it. It will not move towards inorganic particles, such as glass, of similar size. The cell is able to detect the presence of food. This is followed by movement of the plasma membrane and the cytoplasm to form a food vacuole (Fig. 1.15, route 1).

Next the food vacuole moves towards the Golgi apparatus and fuses with a lysosome, whose digestive enzymes dissolve the food. After digestion is complete and the soluble products have been absorbed into the cytoplasm, the food vacuole moves to the surface of the animal again. Its membrane fuses with the plasma membrane and so discharges the indigestible remains. All these events happen in the correct sequence, that is to say they are **coordinated**.

Coordination in multicellular organisms

The cells of multicellular organisms show many similar examples of coordinated activities, perhaps the most spectacular being the events of mitosis and meiosis. But multicellular organisms have an additional need for coordination, for they contain many specialized types of cells, none of which is capable of functioning alone. To take one example, a muscle cell carries out the special task of contraction, producing movement. But it has to be supplied with glucose and oxygen; and its waste products, carbon dioxide and lactic acid, have to be removed.

The mammalian body contains organs which can best be studied as systems, such as the circulatory, respiratory, and digestive systems. But the workings of the systems are coordinated to produce the organism, and this is the level of coordination that this chapter is concerned with.

Examples of coordinated activities in the mammalian body

The demand for energy

The demand for energy varies with the activity of the body. The demand is satisfied by increasing the supplies of glucose and oxygen in the blood and by increasing the rate at which the blood circulates.

The concentration of glucose in the blood is kept very constant at about 0.1%, however hard the body may be exercising. When it is being consumed more rapidly (by the tissues) it is released more rapidly into the blood from the glycogen store in the liver (see page 61).

Demand for extra oxygen is partially met by increasing the rate and the depth of breathing. An adult man breathes 15–20 times per minute, taking in about 0.5 litres of air at each breath, a total of about 8 litres of air per minute. In severe exercise this can rise to as much as 70 litres per minute, but even this supply of oxygen is often inadequate, and an oxygen debt may follow (p. 78).

The rate of heartbeat in an adult man at rest is about 75 per minute. Each ventricular systole expels 60 cm^3 of blood, which gives an output of about 4.5 litres per minute. This rises with increasing activity both by increasing the rate of beat and the strength of each beat (the degree of contraction of the heart muscle). In an Olympic athlete, the rate can rise to 180 beats/min (by reducing the period of diastasis, p. 90). At the same time increased strength of beat expels up to 160 cm^3 of blood from each ventricle at each beat, and the total output from each ventricle is now 29 litres/min.

These coordinated rises in the output of the heart and in the rate and depth of breathing take place automatically. They are not controlled consciously. Presumably, therefore, there are structures in the body which check on the need for glucose and oxygen and which are then able to influence the heart and the muscles of the chest and diaphragm.

Experiments have shown that two sets of nerves lead to the heart. One causes an increase in the rate and strength of beat; the other has the opposite effect, rather like the accelerator and brake of a car. These nerves arise from a particular part of the brain which responds to pressure changes in the heart and blood vessels. It is also sensitive to the concentration of carbon dioxide in the blood. The higher this is, the more rapidly and strongly the heart is made to beat. A similar arrangement controls the rate and depth of breathing.

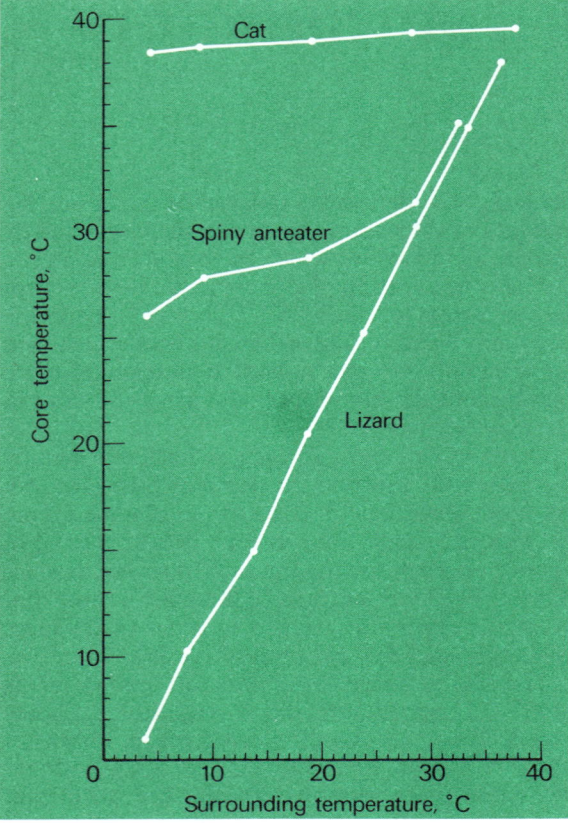

Fig. 15.1 Relationship between the core temperature of a lizard, a spiny anteater, and a cat, and the temperature of the surroundings.

remain at about 35°C no matter whether the surrounding temperature is higher or lower.

Figure 15.1 shows how the internal body temperatures of a lizard (a reptile), a spiny anteater (a primitive egg-laying mammal from Australia), and a cat, vary with external temperature. That of the lizard is directly proportional to the surrounding temperature. Since all body activities depend on temperature-sensitive enzyme-controlled chemical changes, this means that in a cold season, or at night, its activity is low, and it forms sluggish prey for an animal that is not restricted in this way. The spiny anteater has better temperature control than the lizard, but it is not as good as that of the cat. The cat's body temperature hardly varies over the range of 5°–40°C in the surrounding temperature.

Obviously the temperature of every part of the mammalian body cannot be the same, for external parts are exposed to the air and have a low blood supply compared with, say, the liver. Figure 15.2 shows the results of eleven experiments in each of which a man lay naked in a temperature-controlled room for three hours, after which the temperatures of his head, feet, and rectum were measured. In the first experiment the surrounding temperature was 22°C, in the second, 23°C, etc.. While his internal (core) temperature remained fairly constant at 37°C, that of his head was less well controlled, and that of his feet showed almost no control. Most mammals have much more hair than man, and the temperatures of their body surfaces are correspondingly better controlled.

Redistribution of blood during exercise

During exercise the blood supply to the human gut is very much reduced while that to the muscles is increased. This is possible because the walls of the arteries are muscular and can contract, thus reducing the bore of the vessel (p. 91).

Similar redistribution of blood is very marked in diving mammals such as whales, seals, and porpoises. The Weddell seal has been recorded as remining submerged for 70 min, during which time it was actively swimming but took no breaths. Its rate of heartbeat fell to about 10 % of the resting surface value and the circulation of its blood was almost entirely confined to its heart, lungs, and brain. Its muscles were supplied with oxygen from a special reserve stored as oxymyoglobin, a substance similar to oxyhaemoglobin.

Mammalian temperature control

One of the distinguishing features of mammals is their warm-bloodedness. Their body temperatures

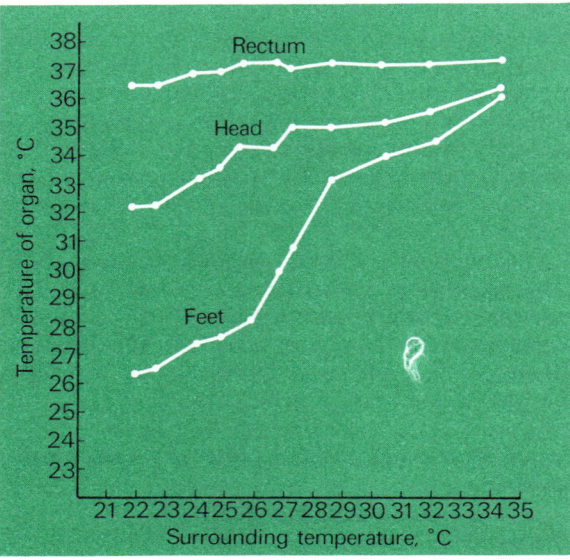

Fig. 15.2 Relationship between temperatures of the human feet, head, and rectum, and the surrounding temperature.

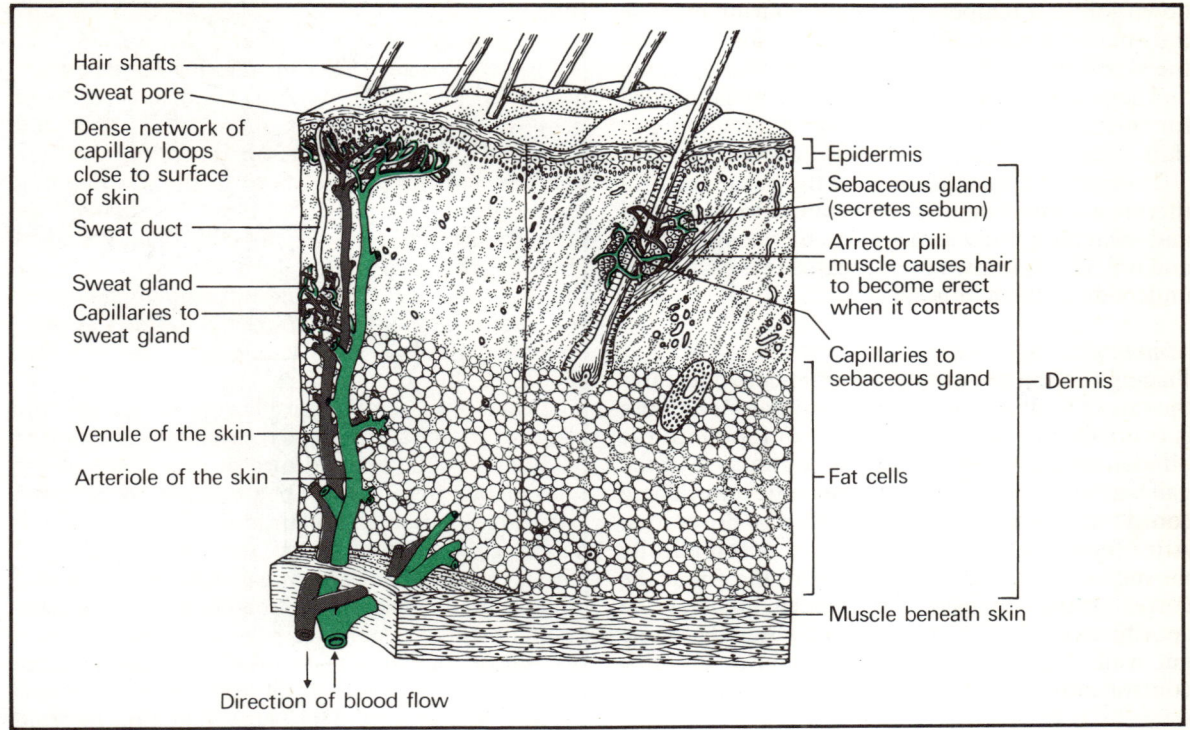

Hair shafts

Sweat pore

Dense network of capillary loops close to surface of skin

Sweat duct

Sweat gland

Capillaries to sweat gland

Venule of the skin

Arteriole of the skin

Epidermis

Sebaceous gland (secretes sebum)

Arrector pili muscle causes hair to become erect when it contracts

Capillaries to sebaceous gland

Dermis

Fat cells

Muscle beneath skin

Direction of blood flow

Fig. 15.3 Stereogram of human skin. Note the various temperature-controlling structures.

Heat generation The steady body temperature is maintained by a balance of heat-generating, heat-conserving, and heat-losing mechanisms. 30–40 % of the energy derived by every cell from aerobic respiration appears as heat (see page 66), and this is the heat-generating mechanism. Internal organs are prevented from over-heating by the circulation of the blood through them which evens out internal temperature in a similar way to that in which the water-cooling system of a car prevents the engine from over-heating.

Heat conservation Some heat is lost on the warm breath, but most of the heat-conserving and heat-losing mechanisms lie in the skin (Fig. 15.3).

The hairs can be erected by the **arrector pili muscles**, which are attached at one end to the connective tissue sheath surrounding the base of the hair and at the other to the surface of the dermis. Erect hairs create a layer of still, warm air round the animal which is difficult to blow away. Air is a poor conductor of heat, so this layer of still air prevents heat loss. (The best conductor of heat is silver, and taking its heat conductivity as 100, that of air is only 0.005.) The effectiveness of this

mechanism is shown by the ability of husky dogs to sleep comfortably in sub-zero temperatures. Exactly the same principle applies to human clothing.

The layer of fat cells in the dermis minimises heat loss from the underlying tissues because fat is also a poor conductor of heat.

Heat loss Heat loss is a response to over-heating. It is brought about in two ways.

First, immediately beneath the epidermis of the skin are dense loops of small capillaries. The bores of the arteries leading to them can be varied by nervous control of the muscular artery walls. We are all familiar with the flushing of the face after exercise, and conversely with its blanching on a cold day. The blood circulating in these capillaries loses heat by conduction through the epidermis and by convection and radiation through the hairs. When the animal is hot, these lie flat against the skin surface.

Secondly, the dermis contains sweat glands. There are some 2.5 million in the adult human skin. About 500 cm^3 of sweat, which is mainly water, is lost per day by a man employed in a sitting down job in a temperate climate. During exercise the amount produced increases, forming drops on the skin surface; but at all times the skin is damp and cool. The sweat evaporates and the heat needed for this is obtained from the blood circulating in skin capillaries. The blood is thus cooled.

Temperature control implies the existence of a thermostat checking the temperature of the blood and switching the various parts of the system on and off. This is known to be located in a part of the underside of the **mid-brain** (Fig. 15.22).

Mammalian digestion

Placing food in the mouth automatically increases the rate of flow of saliva. This coordinated response is controlled by nerves. The entry of food into the duodenum automatically causes contraction of the gall bladder and secretion of pancreatic juice. These coordinated responses are controlled by hormones. After food has been swallowed it is automatically moved along the various portions of the gut at the correct speeds by peristalsis, which involves the coordinated activity of the circular muscle of the gut wall (Fig. 6.16(c)). These responses are also controlled by nerves.

Regulation of body fluid

The human body contains 49 litres of water disposed between plasma, tissue fluid, and cell fluid (Fig. 8.3). Variable amounts of water may be drunk. Water is lost as sweat, the amount depending on the type of activity as this is one of the body's cooling mechanisms. The sweat glands are supplied with nerves which cause them to secrete more or less sweat. In extreme conditions a man may lose as much as one litre of sweat per hour, though this cannot continue for long without the blood becoming concentrated and the circulation failing.

The kidney is the main water regulator of the body because a variable amount of water can be reabsorbed from the filtrate produced by the renal corpuscles. Cells in the mid-brain detect changes in the osmotic pressure of the blood and secrete a hormone (called ADH) which controls the amount of water reabsorbed by the uriniferous tubules (page 113).

Sensory and effector cells

These examples show that the body contains **sensory** cells of many different kinds which check on the body's internal state. They cannot themselves cause the necessary adjustments to be made. These are brought about by **effector** cells, such as the muscle cells of the heart, the secretory cells of

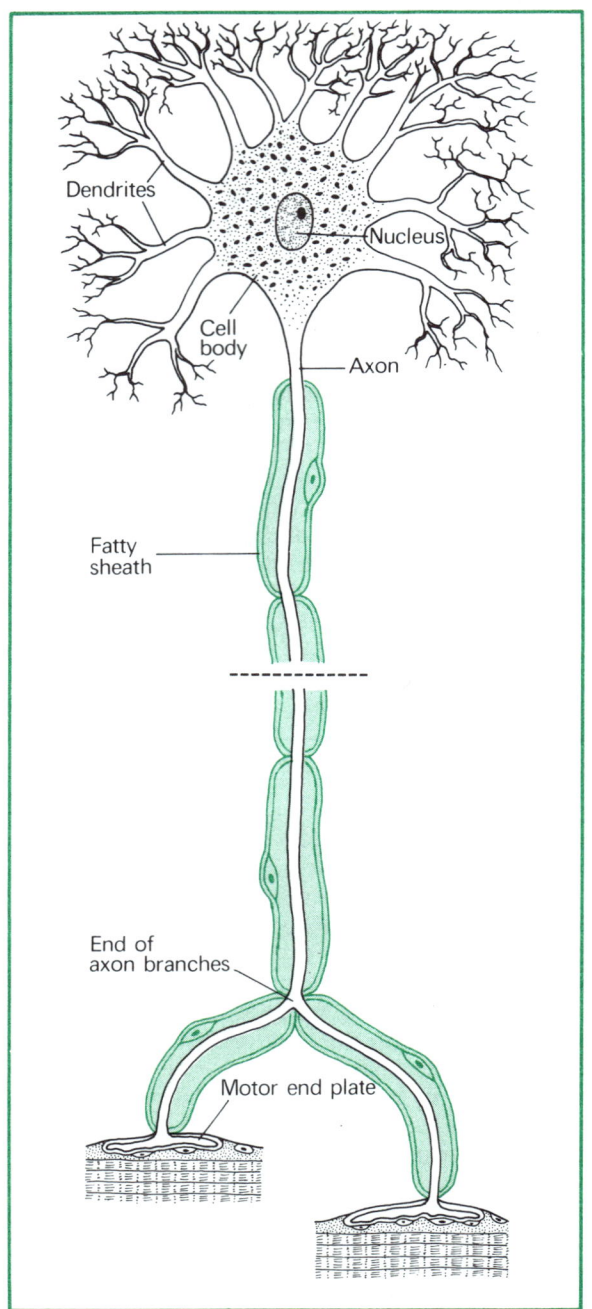

Fig. 15.4 A vertebrate motor nerve cell.

the sweat glands, or the water-absorbing cells of the uriniferous tubules.

The link between sensory and effector cells is sometimes provided by nerves, sometimes by chemical messengers (hormones) carried in solution in the blood plasma.

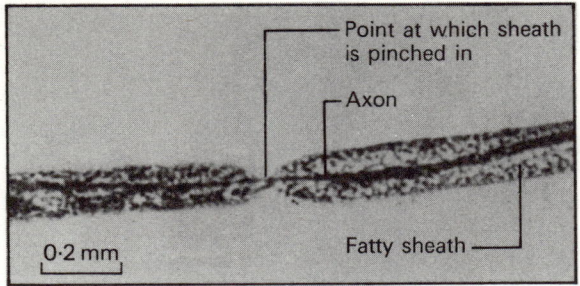

Fig. 15.5 Fatty sheath surrounding part of a nerve axon, showing the pinched-in node.

Internal and external stimuli

A **stimulus** is a change in the local environment of a cell or an organism. The stimuli considered so far have all arisen within the body.

But external stimuli such as light, sound, chemical substances in solution, and pressure and temperature changes, also fall on the body, which has specialized organs for detecting them. In vertebrates these are the eyes, the ears, the sensory cells of the tongue and nasal cavity, and the pressure and temperature-sensitive cells of the skin. Many of these organs are situated in the head, which is the first part of the body of a fish or a tetrapod to encounter changes in the environment.

The vertebrate nervous system

Stimuli affect sensory cells. These connect with **sensory nerve cells** leading to the spinal cord and to the brain. An effector action, called a **reflex action**, may automatically follow. But the incoming information may instead be combined with other information that is either being received at the same time or has been received on previous occasions and is stored in the memory. This combination may result in no action being taken.

Information which leaves the brain and spinal cord for the effector organs, that is the muscles and glands, passes along **motor nerve cells**.

Structure of a nerve cell

Figure 15.4 is a drawing of a vertebrate nerve cell. One end is enlarged to form the **cell body** which is drawn out into tapering structures called **dendrites**. These enlarge the surface area of the cell body. Within the cell body is the nucleus.

Most nerve cell bodies are situated in or near the brain or spinal cord. Leading out of each is a **nerve fibre** whose length can be a few millimetres or as much as 70 cm in man. Only the central part of the fibre, the **axon**, is continuous with the cell body. Surrounding the axon is a fatty sheath which is

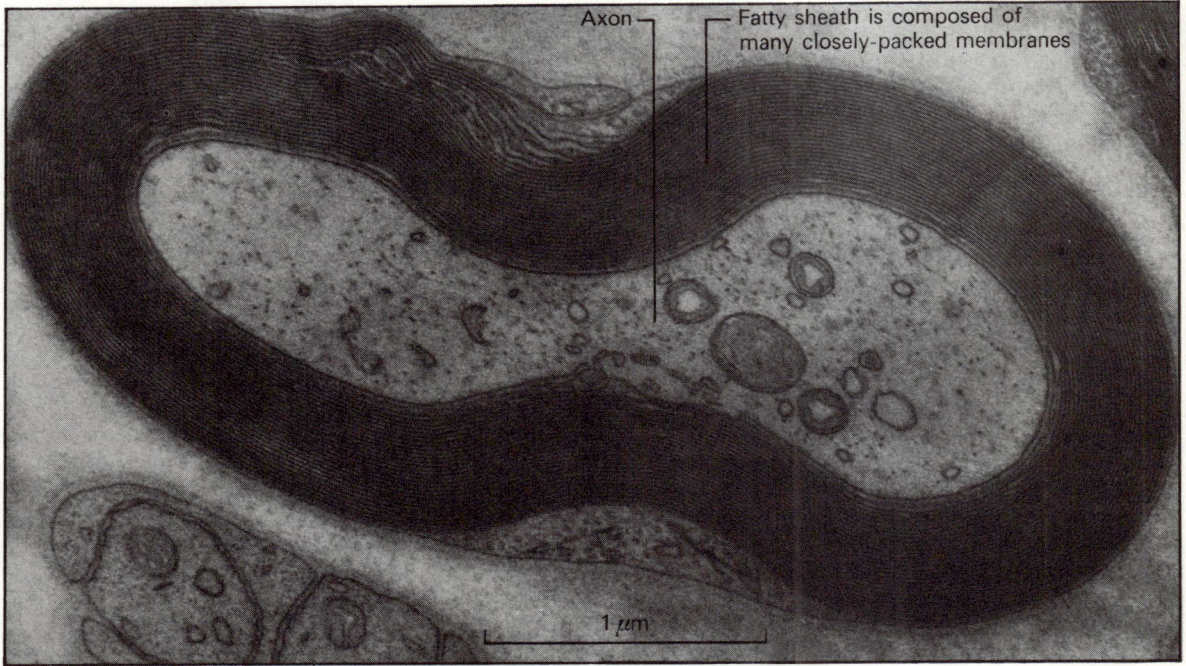

Fig. 15.6 Transverse section of a nerve axon and its surrounding fatty sheath, composed of many membranes (specimen somewhat flattened) (electron micrograph).

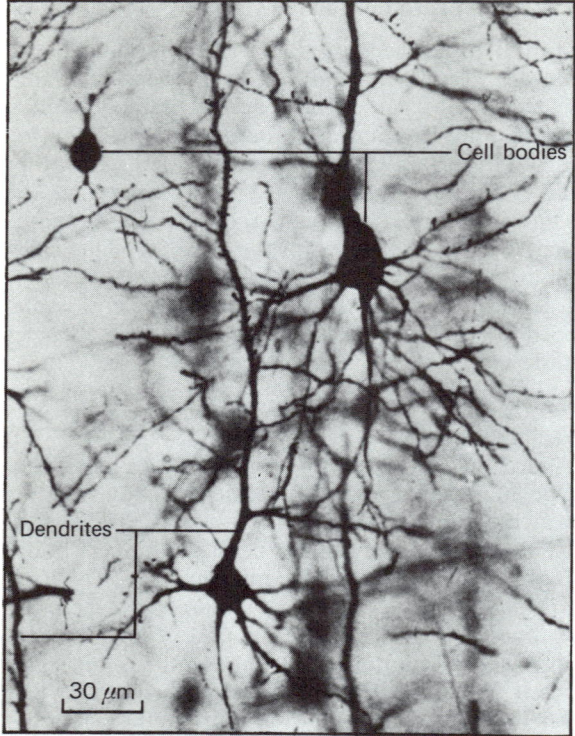

Fig. 15.7 Nerve cells in the cerebral cortex of a cat. Each has many dendrites and no fatty sheath.

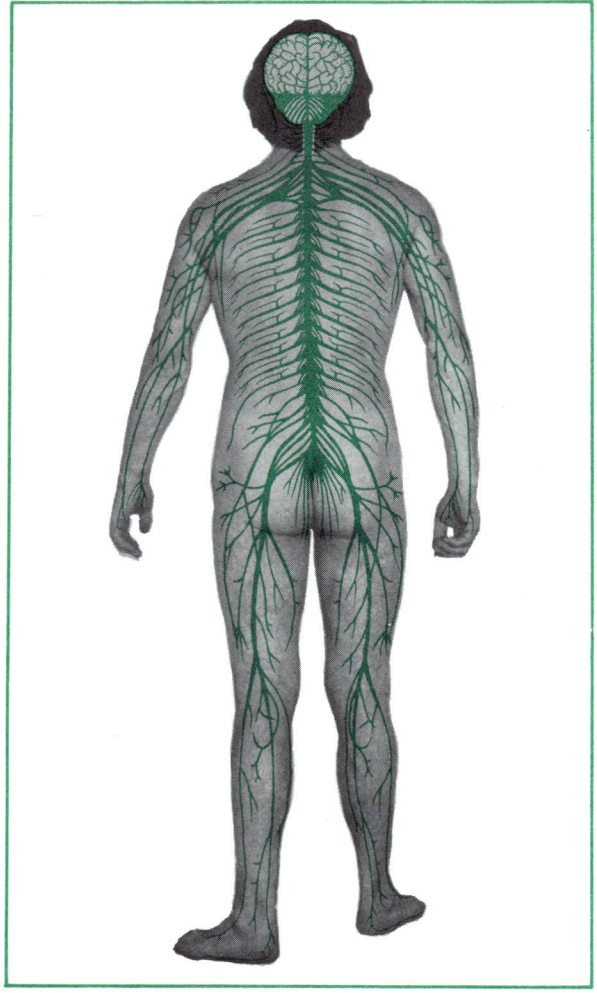

Fig. 15.8 Brain, spinal cord, and main nerves arising from the spinal cord.

pinched in at intervals of about 1.5 mm (Fig. 15.5). Figure 15.6 is an electron micrograph of a transverse section across axon and sheath. The sheath can be seen to be composed of many closely-packed membranes, and it forms an effective electrical insulator round the axon.

At the end farthest from the cell body the nerve fibre branches. Each branch ends either in a **synaptic bulb** or a **motor end plate**. Synaptic bulbs contact the plasma membrane of another nerve cell body, the junction being called a **synapse** (Fig. 15.10). Motor end plates end in contact with a muscle fibre (Fig. 15.12).

Nerve cells in the brain
There are some 10^{10} nerve cells in the human brain—a huge number. Most of them do not possess fatty sheaths. Figure 15.7 shows a few nerve cells in the brain of a cat; the dendrites are stained clearly and can be seen making many synapses with each other.

Nerves in the body
We must distinguish between the terms **nerve** and **nerve cell**. Figure 15.8 shows the main nerves

arising from the human spinal cord. Each consists of a bundle of nerve cell axons bound together and seen in Fig. 15.9(a) in transverse section. In this photo the fatty sheaths have been stained black. The structure is much like a multi-core electric cable.

The nerve impulse
The **message** or **impulse** passing along a nerve fibre is electrical but it differs in speed and in form from an electric current passing along a metallic wire. Electric current travels at a speed of 300 million metres/sec, but the highest recorded speed for transmission of a nerve impulse is 120 metres/sec. Electric current travels in the entire solid substance of a wire, but a nerve impulse travels only along the plasma membrane surrounding the axon.

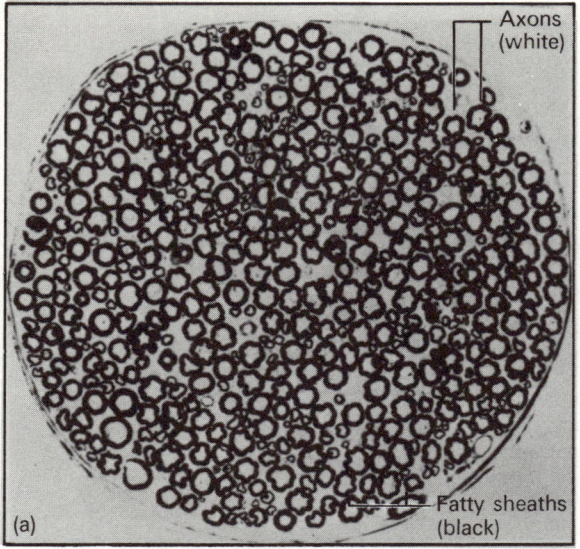

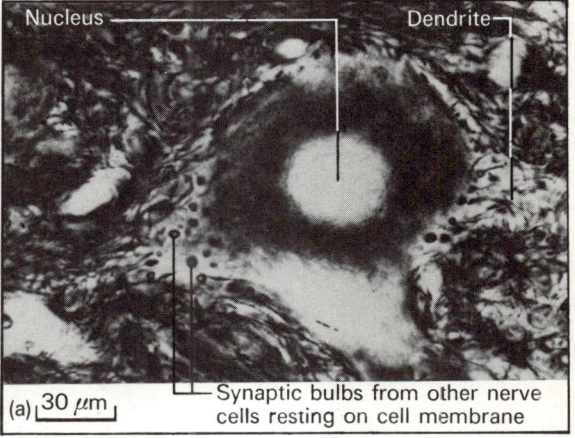

(a) 30 μm

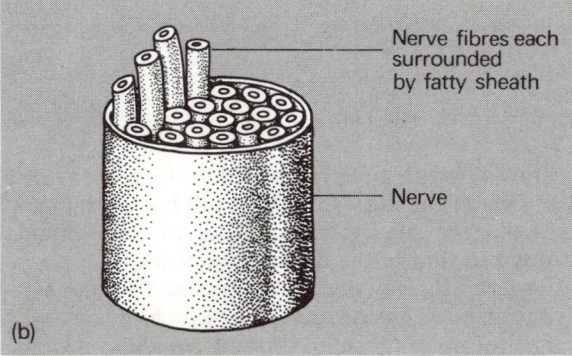

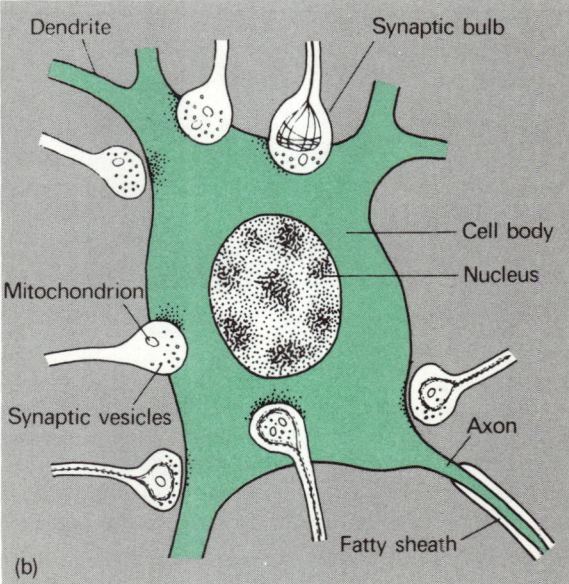

Fig. 15.9(a) Transverse section of a human nerve showing individual nerve fibres (light microscope).
(b) Stereogram of a nerve.

Fig. 15.10(a) Cell body of a nerve cell from the spinal cord of a cat. Note the synaptic bulbs from other nerve cells which make contact with it (light microscope).
(b) Explanatory sketch for (a).

The nerve impulse is an anonymous code

Impulses passing along all nerve fibres, whether sensory or motor, are all of the same kind and strength. All sensory cells generate the same anonymous electrical impulse. Impulses arising in the sensitive cells of the retina of the eye as a response to the stimulus of light are exactly the same as those arising in the taste buds of the tongue as a response to the presence of chemical substances in the food.

The various receptor centres in the brain all receive the same kinds of impulse though they interpret them differently, producing the sensations of sight, taste, and so on. If it were possible to re-wire the optic nerve from the eye so that its impulses stimulated cells in the taste centre of the brain, messages from the retina would be interpreted as tastes.

Nerve cell circuits: synapses

The strength of the nerve impulse does not fade as it passes along the axon. It might seem, therefore, that when the impulse arrives at the synaptic bulb of one nerve cell there is nothing to prevent its passing straight on to the cell body of the next. Figure 15.10 shows the cell body of a nerve cell. On the surface of its plasma membrane are hundreds of synaptic bulbs from other nerve cells.

If the situation was this simple a stimulus applied to one nerve cell might be conveyed to all the other nerve cells in the body. Such an arrangement would not allow the existence of definite nerve cell circuits

173

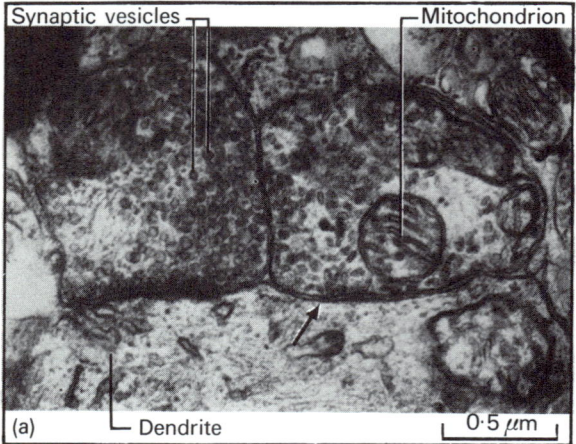

(a)

Synaptic vesicles — Mitochondrion

Dendrite 0·5 μm

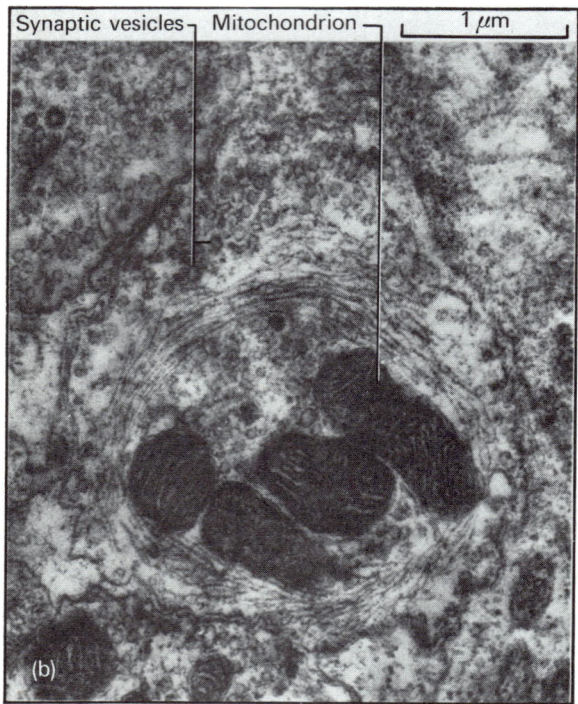

Synaptic vesicles — Mitochondrion 1 μm

(b)

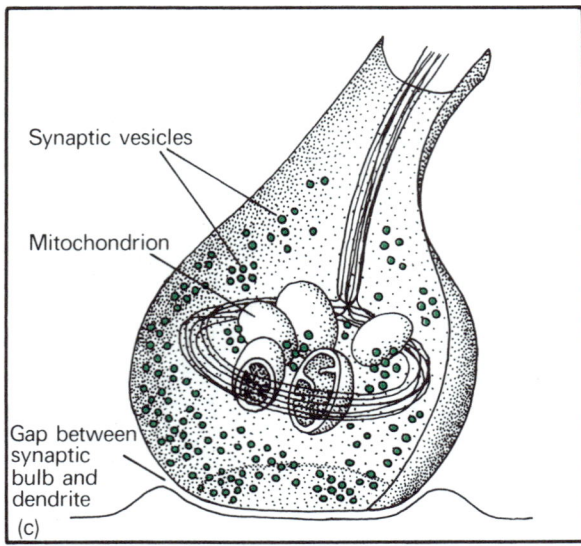

Synaptic vesicles

Mitochondrion

Gap between synaptic bulb and dendrite

(c)

Fig. 15.11(a) Section through two synaptic bulbs synapsing with a dendrite. Arrow shows gap between synaptic bulb and dendrite (electron micrograph).
(b) Transverse section through a synaptic bulb (electron micrograph).
(c) Diagram to explain (b).

section through two synaptic bulbs forming synapses with a dendrite. Note the small but distinct gap between the plasma membranes of the synaptic bulbs and that of the dendrite (arrow).

Figure 15.11(b) is a transverse section across a synaptic bulb. Inside there are mitochondria and a large number of small, round **synaptic vesicles**. Each is a minute package of a hormone. When a nerve impulse arrives at the synaptic bulb, it causes one or more synaptic vesicles to move to the surface of the synapse and to discharge its hormone into the gap between the two plasma membranes. It is the hormone which starts off the nerve impulse

along which impulses can be a passed to predetermined targets, which is the situation that really applies. Nerve circuits are made possible by the existence of **synapses** which may or may not allow the impulse to pass on. The impulse can, in any case, only pass across the synapse in one direction, namely onwards. So each synapse forms a valve; and because of synapses impulses can only pass along individual nerve circuits in one direction.

Structure of the synapse
Figure 15.11(a) is an electron micrograph of a

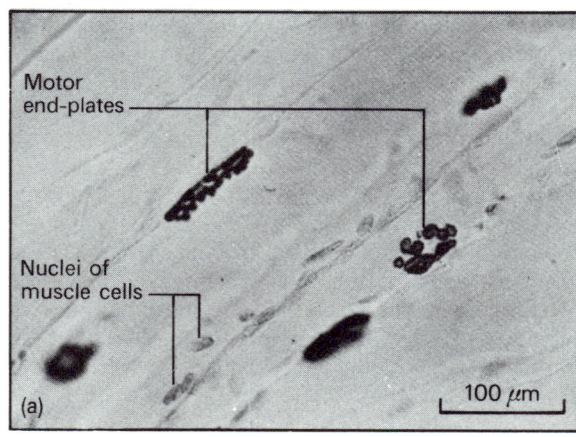

Motor end-plates

Nuclei of muscle cells

(a) 100 μm

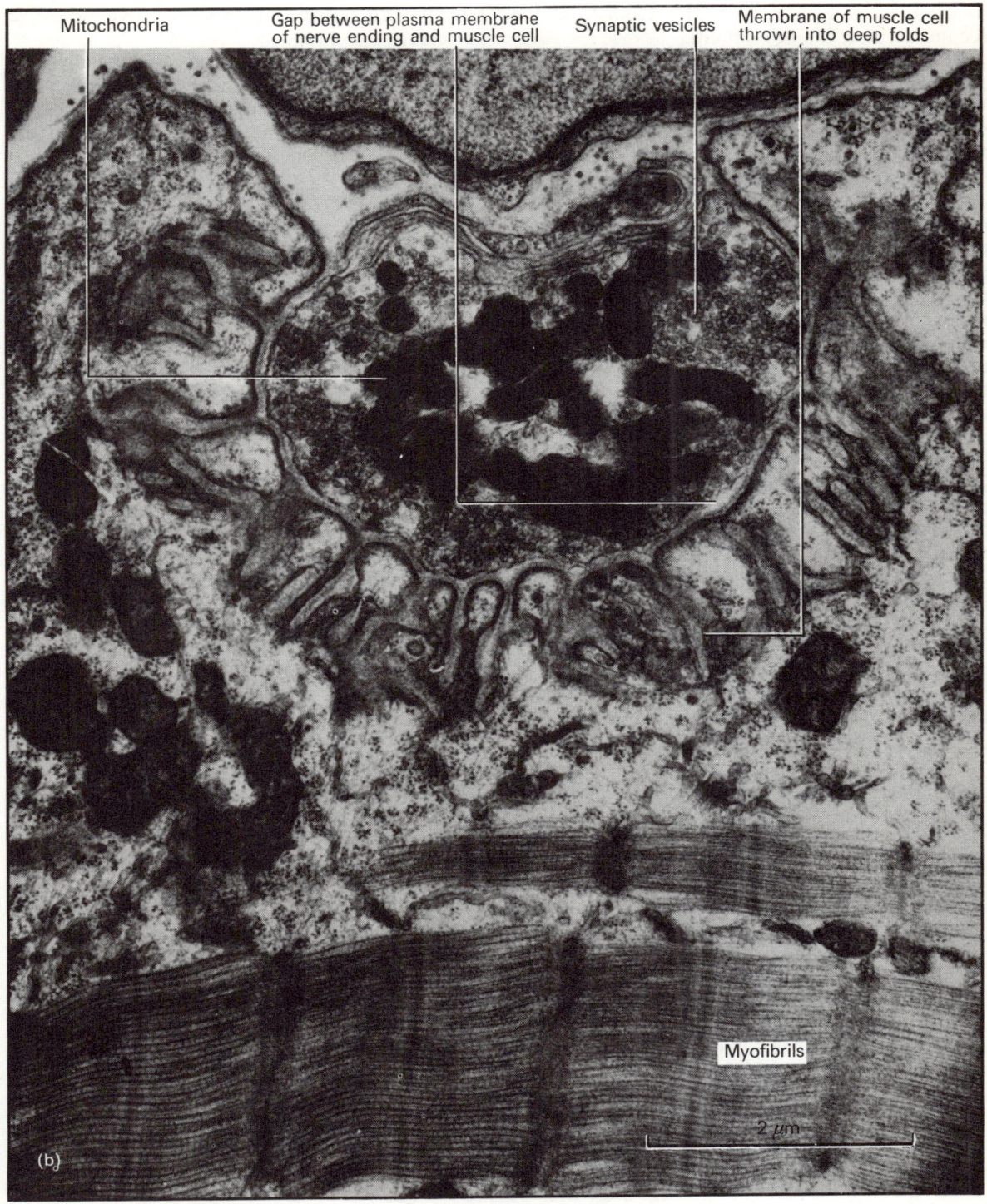

Mitochondria | Gap between plasma membrane of nerve ending and muscle cell | Synaptic vesicles | Membrane of muscle cell thrown into deep folds

Myofibrils

2 μm

(b)

Fig. 15.12(a) Motor end plates on voluntary (striped) muscle cells (light microscope).
(b) Electron micrograph of a vertical section through a motor end plate and part of the voluntary muscle cell that it serves.

in the plasma membrane of the dendrite of the second nerve cell in this circuit.

Motor end plates

When the end branch of a nerve cell makes contact with a muscle cell it forms a motor end plate (Fig. 15.12(a)).

The electron microscope shows that the structure of the motor end plate is essentially the same as that of the synapse, and it functions in a similar way (Fig. 15.12(b)). At the junction the surface of the muscle cell is thrown into deep folds. There is a distinct gap between the plasma membranes of the nerve ending and the muscle cell. Hormone is discharged into this by the synaptic vesicles when the nerve inpulse arrives at the motor end plate. The discharge of hormone causes the muscle cell to contract.

The spinal cord

The spinal cord is continuous with the brain. It passes downwards through the **foramen magnum** of the skull into the bony canal formed by the **neural canals** of successive vertebrae (Fig. 17.7(b)). Observe half a pig or sheep carcass in a butcher's shop, showing the neural canals and the spinal cord lying inside them.

Arising from the cord are 31 pairs of spinal nerves. Each has two origins on the cord; a dorsal root, which bears a swelling, and a ventral root. These join to form a single nerve which emerges from the **intervertebral notch** between two successive vertebrae (Fig. 17.9).

Figure 15.13 shows a section of the spinal cord in position in the neural canal. From the side (Fig. 15.14) the attachments of the dorsal and ventral roots are seen to extend over some length of the cord.

Inside the spinal cord

Figure 15.15 is a photograph of a stained transverse section of a human spinal cord. It shows two distinct areas; an inner H-shaped **grey matter**, which contains nerve cell bodies, and surrounding it the **white matter**. This contains only axons leading into and out of the grey matter. The fatty sheaths surrounding the axons produce the white appearance.

Reflex actions and the reflex arc

Much of the nervous coordination of the vertebrate body does not involve the brain and takes place via the spinal cord. This was first demonstrated in the eighteenth century by Stephen Hales who observed that if the hind leg of a live frog was dipped into warm water, the animal automatically withdrew it. When a cut was made at the base of the brain, severing it from the spinal cord, the animal still withdrew its hind limb from warm water. But when the spinal cord was destroyed by running a wire up and down it the hind limb was not withdrawn from warm water.

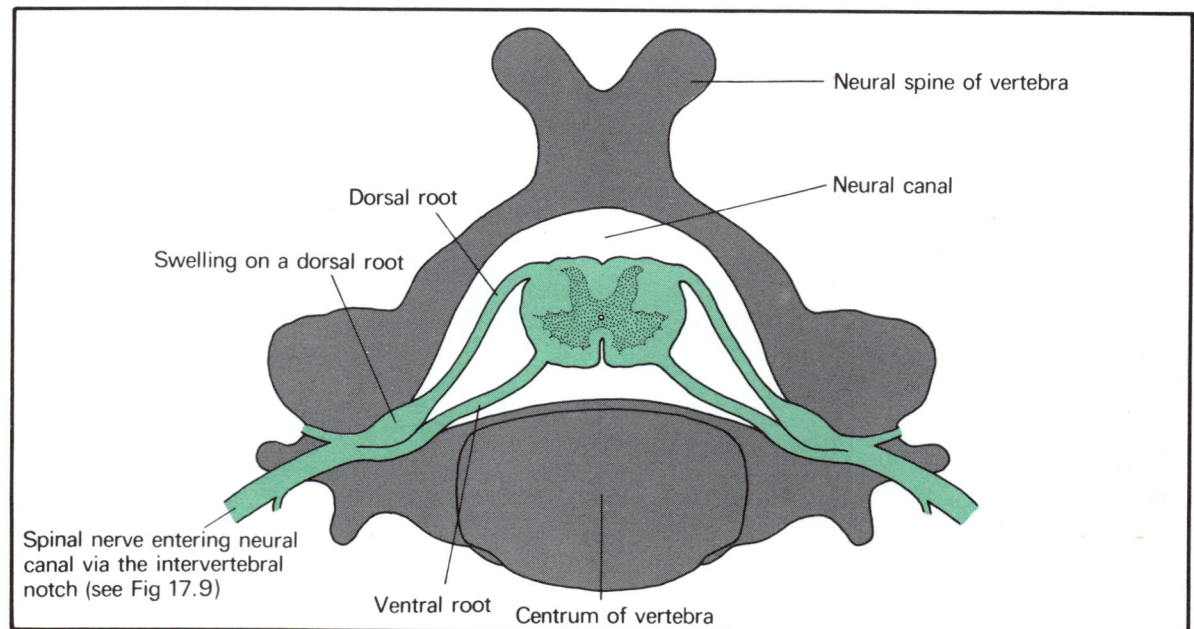

Fig. 15.13 Section through a cervical vertebra showing the spinal cord in place, suspended in liquid in the neural canal.

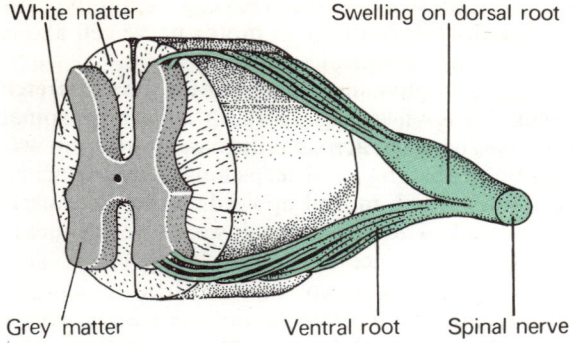

Fig. 15.14 Spinal nerve branching into dorsal and ventral roots and joining the spinal cord.

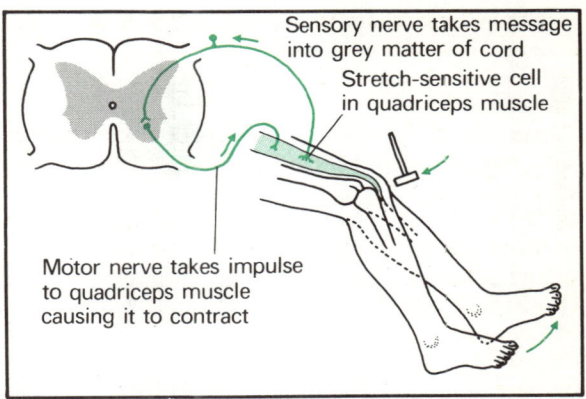

Fig. 15.16 Nerve circuit in the knee jerk reflex.

This shows that contraction of the muscles of the hind limb in response to heat stimulus detected by the skin does not require the cooperation of the brain, but it does require an intact spinal cord. The nerve pathways coordinating this activity must be located in the spinal cord.

Such **involuntary** responses are called **reflex actions**. Other examples are blinking, coughing, secreting saliva, secreting tears, the automatic adjustment of the iris of the eye to light of different intensities, and the automatic adjustment of the thickness of the lens in accommodation (see page 191). Many activities carried out regularly become automatic and reflex, for example walking, or even driving a car.

The knee jerk reflex

Demonstrate the stretch reflex in your own quadriceps muscle by crossing your right leg over your left and getting a partner to give a light tap with the edge of a book under your knee cap.

In this reflex action, tapping the tendon attached to the quadriceps muscle stimulates a stretch-sensitive sensory cell in the muscle (Fig. 15.16). This stimulates a sensory nerve whose axon takes the impulse into the grey matter of the spinal cord.

Its path is shown as nerve circuit 1 in Fig. 15.17. Its cell body lies in the swelling on the dorsal root of the spinal nerve. Its axon enters the cord via the dorsal root. Inside the grey matter this nerve cell synapses with the cell body of a motor nerve whose axon leaves the cord in the ventral root. This passes via the spinal nerve to the quadriceps muscle which it stimulates.

This is the simplest reflex nerve circuit or **reflex**

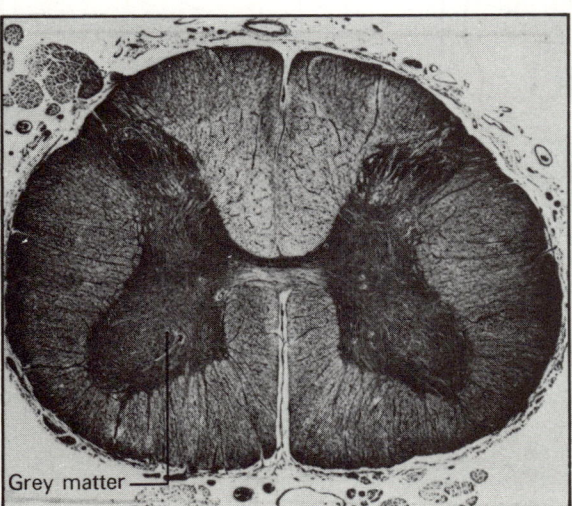

Fig. 15.15 Transverse section of human spinal cord. Note nerve cell bodies in the grey matter.

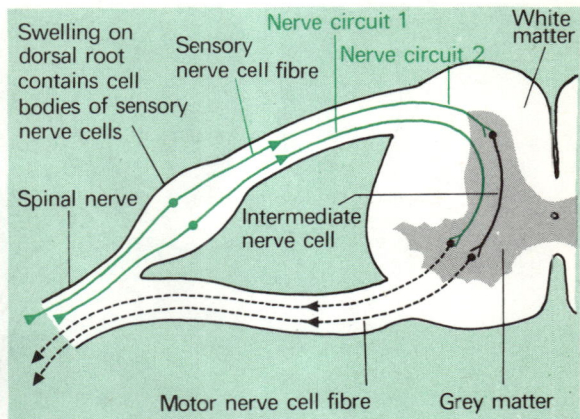

Fig. 15.17 Nerve circuits of two reflexes. Circuit 1: reflex involving two nerve cells, sensory and motor. This is the arrangement in the knee jerk reflex (Fig. 15.16), and in reflexes associated with muscle tone. Circuit 2: reflex involving sensory, intermediate, and motor nerve cells. The intermediate nerve cell allows the sensory input to be diffused to other parts of the cord, and to the brain (see Fig. 15.19).

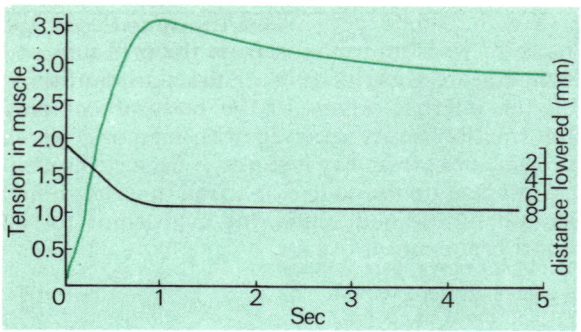

Fig. 15.18 Rapid lowering through a distance of 8 mm of a cat whose spinal cord had been severed just behind the brain causes an immediate reflex contraction of its quadriceps muscle. The peak of contraction coincides with maximum stretch.

arc. It consists of (i) a sensory cell, (ii) a sensory nerve cell entering the cord and synapsing with (iii) a motor nerve cell which leaves the cord and stimulates (iv) the motor (muscle) cell at the motor end plate.

All spinal nerves contain axons of both sensory and motor nerves and so conduct impulses to and from the cord. But there is a strict division in the circuit when the spinal nerve forks before joining the cord. All sensory nerve cell axons enter the cord via the dorsal root and all motor nerve cell axons leave it via the ventral root.

A quantitative investigation of a similar **stretch reflex** is shown in Fig. 15.18. A cat whose spinal cord had been cut across at the base of its brain was laid on a table. Its quadriceps muscle was attached to a lever which moved upwards when the muscle contracted. The table was lowered very quickly through a distance of 8 mm (black line). This stretched the quadriceps muscle which underwent an immediate reflex contraction (red line), whose peak coincided with the maximum stretch.

When the spinal nerve serving the quadriceps muscle was cut there was no response if the table was lowered. This was also true when only the dorsal root of this nerve was cut, showing that the sensory flow into the cord takes place via the dorsal root.

More complex reflexes
If a barefooted man steps unexpectedly on to a very hot object he automatically withdraws his foot. This simple reflex operates through a reflex arc similar to the stretch reflex. But the response will

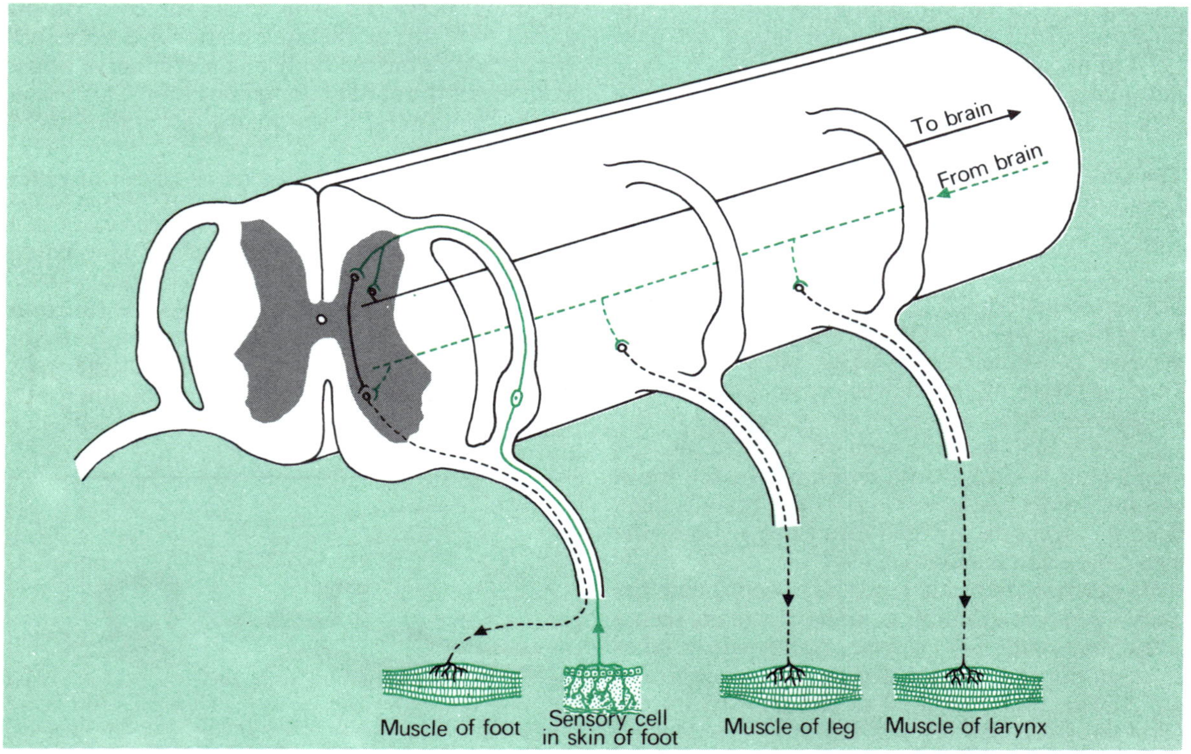

Fig. 15.19 Stereogram to show how the intermediate nerve cell synapses with a nerve cell passing to the brain. A motor nerve cell from the brain bearing instructions in response to this information synapses with motor nerve cells at various levels of the cord Thus one stimulus can cause a complex of responses from various parts of the body.

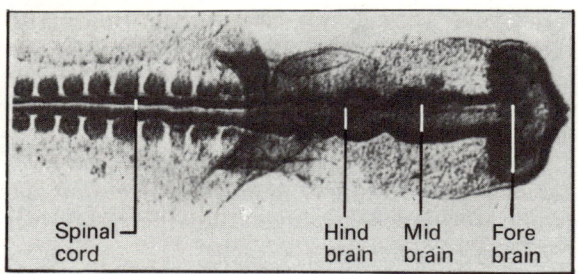

Fig. 15.20 Developing brain and spinal cord in a 33-hour-old chick embryo (Courtesy Carolina Biological Supply Company).

probably be more complex than withdrawing the foot. Other parts of the body may be moved—the trunk, arms, and neck; and the man may call out. This means that one sensory impulse in one spinal nerve has produced a **motor outflow** in many spinal nerves serving many parts of the body, and not only on the same side as the foot which touched the hot object.

Such complex coordination is made possible because some sensory nerve cells do not synapse in the grey matter of the spinal cord directly with motor nerve cells but with **intermediate** nerve cells (Fig. 15.17, nerve circuit 2). In Fig. 15.19, we see that the intermediate nerve cell can pass both up and down the length of the spinal cord, synapsing with motor nerve cells at different levels and so producing motor outflows along different spinal nerves. If the stimulus is strong enough it will pass up the cord to the brain where it will be felt as pain. Since large numbers of nerve cells pass from the brain to all the spinal nerves, the sensory impulse in one spinal nerve is potentially connected with a motor outflow along all the spinal nerves.

The vertebrate brain

During embryonic life the brain and spinal cord develop as a tube on the upper side of the body. These are two of the characteristic features that distinguish vertebrates from invertebrates (p. 18). The front end of the tube soon shows three enlargements (Fig. 15.20). These develop into the brain and the nerves that arise from it. The brain is thus, from its inception, divided into three regions, the **fore-**, **mid-**, and **hind-brains**.

Its function, in broad terms, is to coordinate the body's activities, though as we have already seen a considerable amount of reflex coordination takes place in the spinal cord without reference to the brain. The brain receives information from the main sense organs, the eyes, the ears (or balancing organs), and the sensitive linings of the nose and the tongue. All these, like the brain itself, are situated at the front, exploring end of the animal.

Via the spinal nerves and the spinal cord the brain receives information from the pressure- and temperature-sensitive cells of the skin, and from all the internal organs of the body. Combining information newly received with memory, which in some unknown way it stores, it decides moment by moment on the course of action that the animal should follow and sends out instruction to the appropriate motor organs.

Fish brain

The fish brain is relatively simple, and in longitudinal section (Fig. 15.21) we see that the three parts are developed more or less equally extensively. The forebrain is extended forwards as the **olfactory lobes** which receive and relay impulses from the olfactory (smell) organs, one of the fish's most important senses. Its roof, the **cerebrum**, is small, and the only observable effect on the fish of severing its forebrain is to abolish its sense of smell.

The midbrain bears a small protrusion on its upper surface, the **pineal body**, of unknown function. On its lower surface is the **pituitary body**, which produces many hormones. Behind the pineal body the midbrain forms the **optic lobes** which receive and relay impulses from the eyes.

The hindbrain is expanded on its upper surface to form the **cerebellum** which is responsible for maintaining balance and coordinating muscular activity. Its lower side is thickened to form the **medulla oblongata** which controls respiratory movements and heartbeat.

This brief account shows that different regions of the brain are specialized for receiving and relaying sensory information and for generating motor impulses. They contain specialized *centres*, groups of nerve cells associated with the function in question—for example the **respiratory centre** in the medulla. Something is known about the ways in which these are connected to one another, but the organization of the brain is very largely not understood.

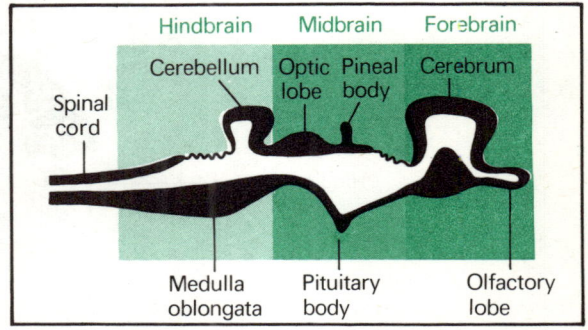

Fig. 15.21 Longitudinal section through a fish brain.

The mammalian brain

Figure 15.22(b) is a vertical section through the brain of a rat, a small and relatively unintelligent mammal most of whose activities are reflex. It has a limited capacity to learn and a very limited capacity for solving new problems.

Two notable differences between this brain and the fish brain are the relatively large cerebrum, which overhangs the midbrain, and the large cerebellum. In a section of a human brain (Fig. 15.22(a)) we see that the cerebrum is by far the largest part, and when viewed from the outside (Fig. 15.23) the cerebrum, and the smaller cerebellum, are the only parts visible.

The development of the cerebrum is the most distinguishing feature of man. Both cerebrum and cerebellum are deeply infolded, thus increasing their

Fig. 15.22(a) Vertical section through a rat brain. Grey matter tinted.
(b) Vertical section through a human brain. Grey matter tinted.

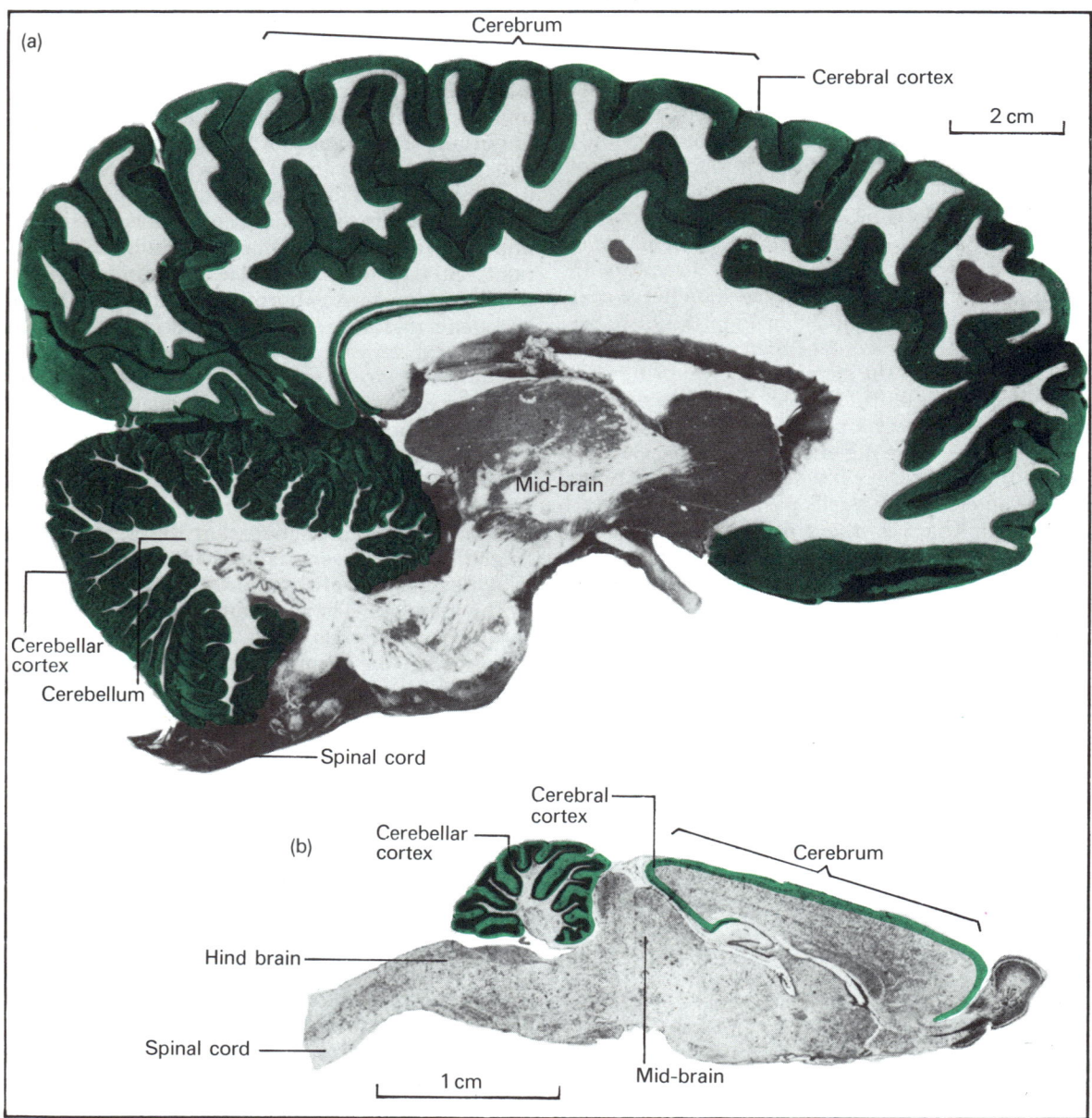

(a) Cerebrum
Cerebral cortex
2 cm
Mid-brain
Cerebellar cortex
Cerebellum
Spinal cord

(b) Cerebral cortex
Cerebellar cortex
Cerebrum
Hind brain
Spinal cord
1 cm
Mid-brain

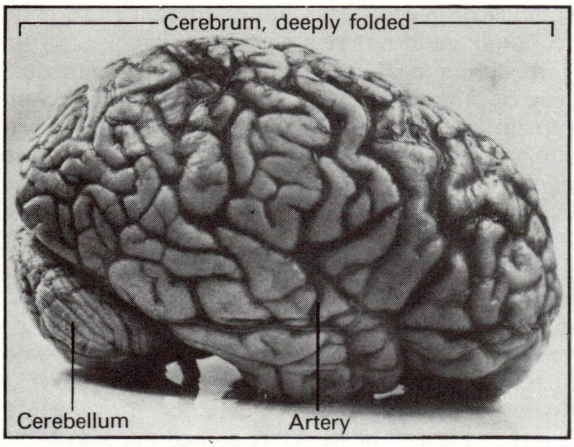

Fig. 15.23 Human brain from the side.

surface areas. The positions of the grey and white matteres are here reversed compared with the spinal cord. The grey matter surrounds the white. Because it is on the outside, the grey matter in these regions is called the **cerebral cortex** and the **cerebellar cortex**, and the deep infoldings are clearly an adaptation to enable more nerve cell bodies to be packed in. With some 10^{10} nerve cells, each making several thousand synaptic connections with others, the human brain is the most complex object in existence.

The cerebral cortex is the region where high level activities such as learning, memory, and thought, take place. Very little is known about the ways in which it works, though it has been possible to identify areas associated with particular sensory and motor functions (Fig. 15.24). The fish brain is not capable of higher mental activity, and its cerebrum is very poorly developed.

The cerebellum and cerebellar cortex, the ex-

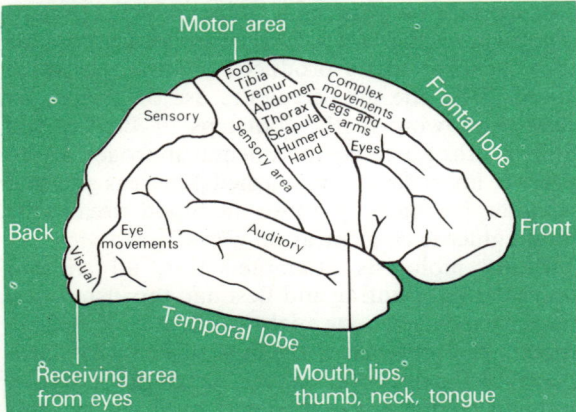

Fig. 15.24 Regions of the human cerebral cortex known to be associated with particular functions.

panded roof of the hindbrain, perform a similar function in both fish and man, the coordination of movement. Their greater relative development in man is related to his far more complex, varied, and delicate movements, such as my typing this book or, in perhaps the most breathtaking example, a concert violinist in action.

The lower part of the hindbrain, the medulla oblongata, contains centres which control respiration and heartbeat, as in the fishbrain. If a cat's brain is severed in front of the medulla it is no longer capable of voluntary movements because the outflow from the motor area of the cortex is cut off. But it continues to breathe and its circulation functions normally. Other reflex centres are also situated here, such as those controlling swallowing and salivation.

Hormonal coordination

In 1849 a German, Berthold, performed an experiment with cockerels which gave a clue to a second, apparently quite distinct, system of coordination in the vertebrate body. This is chemical coordination, brought about by chemical messengers or **hormones** (Greek, I arouse to activity). These are released from the cells which produce them into the blood. They are carried via the tissue fluid to all the body cells, some of which they affect.

The testes of birds are situated internally. Berthold castrated a young cockerel, removing both its testes. It developed into a capon, a bird in which the secondary sexual characteristics of cockerels, the possession of a comb and wattle, an aggressive attitude to other cockerels, display behaviour to the hen, and the cock-a-doodle-do cry, are absent. (N.B. the primary sexual characteristic is the possession of either testes or ovaries.) Four months later he killed the bird and found no testes inside it. They had not regenerated.

Castrating a second male bird, he replaced one testis among the intestines. It developed into a cockerel. When he killed and examined it two months later, Berthold found the testis attached to the wall of the intestines. It had developed connections with the animal's blood system, had grown, and contained sperms. Thus it seemed that as well as serving its prime function of producing sperms, the testis had a profound effect on development, since this proceeded in a different manner in its absence.

Berthold suggested that the testis had its effect by producing a chemical substance which was released into the blood stream. We now know that this is true not only for birds but for mammals, and on page 119 we noted the production of **testosterone** by the island cells of the human testis. The mammalian ovary also produces several hormones which, together with hormones from the pituitary body, coordinate ovulation, the preparation of the endometrium for implantation, menstruation, and childbirth.

Bayliss and Starling discover secretin

Pancreatic juice flows into the duodenum within two minutes of commencing a meal (p. 59). At the turn of the nineteenth century it was thought that this was a reflex action. The stimulus initiating it seemed to be the passage of acid stomach contents into the duodenum, for the introduction of hydrochloric acid into the duodenum of experimental animals caused a flow of pancreatic juice some two minutes later.

Bayliss and Starling cast serious doubt on this mechanism, however. They cut the nerve supply that connected the brain and spinal cord with the pancreas and duodenum of an experimental dog. When they injected 30 cm^3 of 0.4% hydrochloric acid into its duodenum, a flow of pancreatic juice still followed two minutes later. Apart from possible experimental error in not destroying the nerve

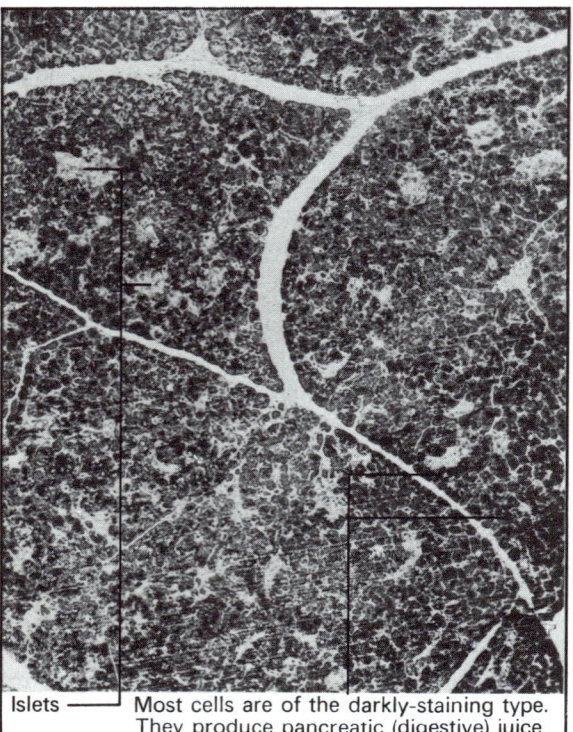

Islets —— Most cells are of the darkly-staining type. They produce pancreatic (digestive) juice

Fig. 15.25 Transverse section of human pancreas (light microscope).

supply completely, this indicated that pancreatic secretion could not be a nervous reflex.

Bayliss and Starling suggested that the acid reacted with cells of the lining of the duodenum causing them to produce or release a chemical substance. This was conveyed by the blood to the pancreas and caused secretion. They called the chemical substance **secretin**.

To test this possibility they scraped the lining off a small piece of duodenum, mashed it in a mortar with 0.4% hydrochloric acid, filtered the mash, and injected the filtrate directly into one of the dog's veins. A copious flow of pancreatic juice followed 70 sec later. This was taken as conclusive proof that no nervous reflex is involved in pancreatic secretion but that a hormone (Starling coined the word) is.

We must note that the experimenters did not attempt to isolate secretin or to investigate its chemical nature. But its effects as a coordinator are much more specific than those of the sex hormone whose production by the testis of cockerels was inferred by Berthold.

Banting and Best discover insulin

It had been known since 1889 that removal of the pancreas from a dog caused it to develop the disease **diabetes mellitus**, from which man also suffers. Its principal symptom is that the glucose concentration of the blood plasma rises from the normal 0.1% to 0.2–0.5%. This is accompanied by a running down of the glycogen reserves in the liver and muscles, and sugar is also made from proteins and fats, which causes wasting of the muscles. Poisonous chemicals are produced, and the disease is fatal.

Later it was established that the pale-staining **Islet tissue** (Fig. 15.25) is the part of the pancreas responsible for controlling blood glucose concentration. (The dark-staining tissue of the pancreas produces pancreatic juice, which contains digestive enzymes, see p. 59.) It was thought possible that the Islet tissue produced a hormone which could affect the uptake or the use of glucose by cells. Following in the footsteps of Bayliss and Starling, the obvious experimental approach was to prepare an extract from mashed pancreas and find whether, on injecting it into the blood stream, the blood glucose concentration was affected.

Many biologists attempted this, all without success, until Banting and Best had the inspiration that any such hormone might be a protein. If it were, then in the presence of the digestive enzymes produced by the dark-staining regions of the pancreas, it might be digested and so destroyed. To test this possibility they tied a thread round the pancreatic duct of an experimental dog (see Fig. 6.10). This

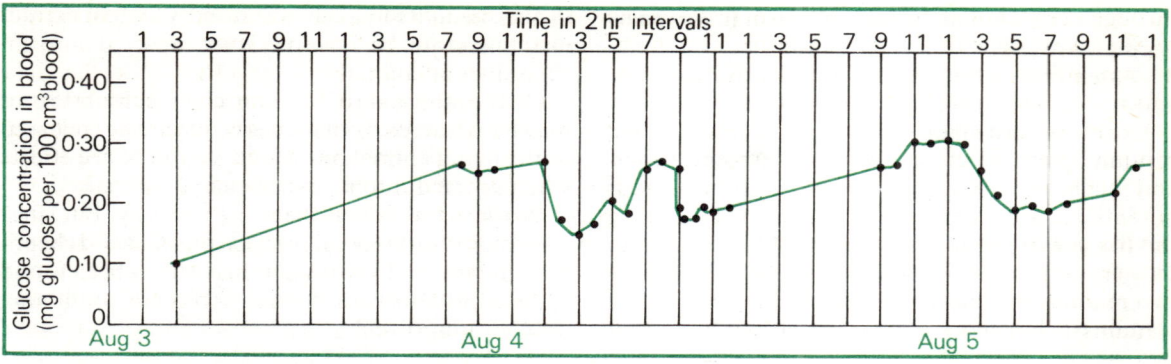

Fig. 15.26 Experimental results of Banting and Best on injecting extract from Islet tissue into a dog whose pancreas had been removed. For explanation see text.

had the convenient effect of causing all the dark-staining tissue to die, while leaving the Islet tissue unharmed. In this way they prepared an extract from pure Islet tissue, and Fig. 15.26 is a reproduction of the experimental results they obtained when they injected it into the bloodstream of an experimental dog on August 3, 1920.

The Y (vertical) axis shows blood glucose concentration. The normal level for the dog was 0.09 %, that is 90 mg of glucose per 100 cm³ of blood. The X axis shows time in two-hourly intervals. The dog's pancreas was removed at 3 p.m. on August 3, and in the next 17 hours its blood glucose concentration rose to 0.27 %, where it remained steady for the next five hours.

At 1 p.m. on August 4 they injected the dog with 5 cm³ of extract of pure Islet tissue. The rapid fall in its blood glucose concentration to 0.15 % two hours later has been called 'one of the most important events in medical history'. It opened the way to relief for millions of human diabetics.

The pancreatic extract did not restore the normal level of 0.09 % glucose, and the effects gradually wore off so that by 9 p.m. the blood glucose concentration was back at 0.26 %. Now a second injection of 5 cm³ of extract was given, with the same effect as the first. Again the treatment wore off, and the experimenters administered a third dose at 2 p.m. on August 5. The results endorsed their previous findings.

Clearly Banting and Best were right in thinking that the hormone, which they called **insulin**, is destroyed by digestive enzymes. They confirmed this by mixing 20 cm³ of pure Islet extract with 10 cm³ of active pancreatic juice obtained from a normal pancreas that had not been operated on. After incubating the mixture for 3 hours at body temperature they injected it into the experimental

dog. There was no fall in blood glucose concentration, indicating that the insulin had been destroyed.

Experimental investigation of hormones

These experiments illustrate the basis for establishing the existence of a hormone:

1 The tissue or organ suspected of producing the hormone is removed by surgery. The resulting symptoms are presumed to be due to lack of hormone.

2 An extract of the tissue is made and injected into

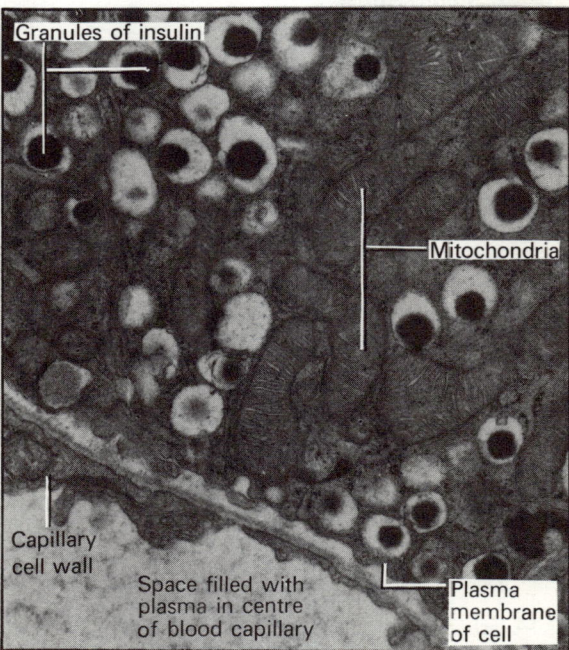

Fig. 15.27 Part of an insulin-secreting cell from mouse pancreas (electron micrograph).

the deficient animal, restoring normal, or near-normal, activity.

3 Attempts are then made to purify the hormone and to establish its chemical identity.

Hormones are made in very small quantities and produce their effects at very low concentrations, and such experiments are not without serious potential sources of experimental error. In making the tissue extract, the hormone will be exposed to oxygen and to cell chemicals which may destroy it, creating the false impression that the tissue contains no hormone. Hormone may not be present in a tissue in its final chemical form and an extract may therefore be inactive when injected into an experimental animal.

On the other side of the coin, chemical substances may be extracted from tissues which are not released in the living animal but which may produce effects when injected into an experimental animal.

Work in this field is therefore not easy. But since 1920 many hormones have been purified and chemically identified. In particular insulin has been found to be a protein with a small molecule containing only 51 amino acids.

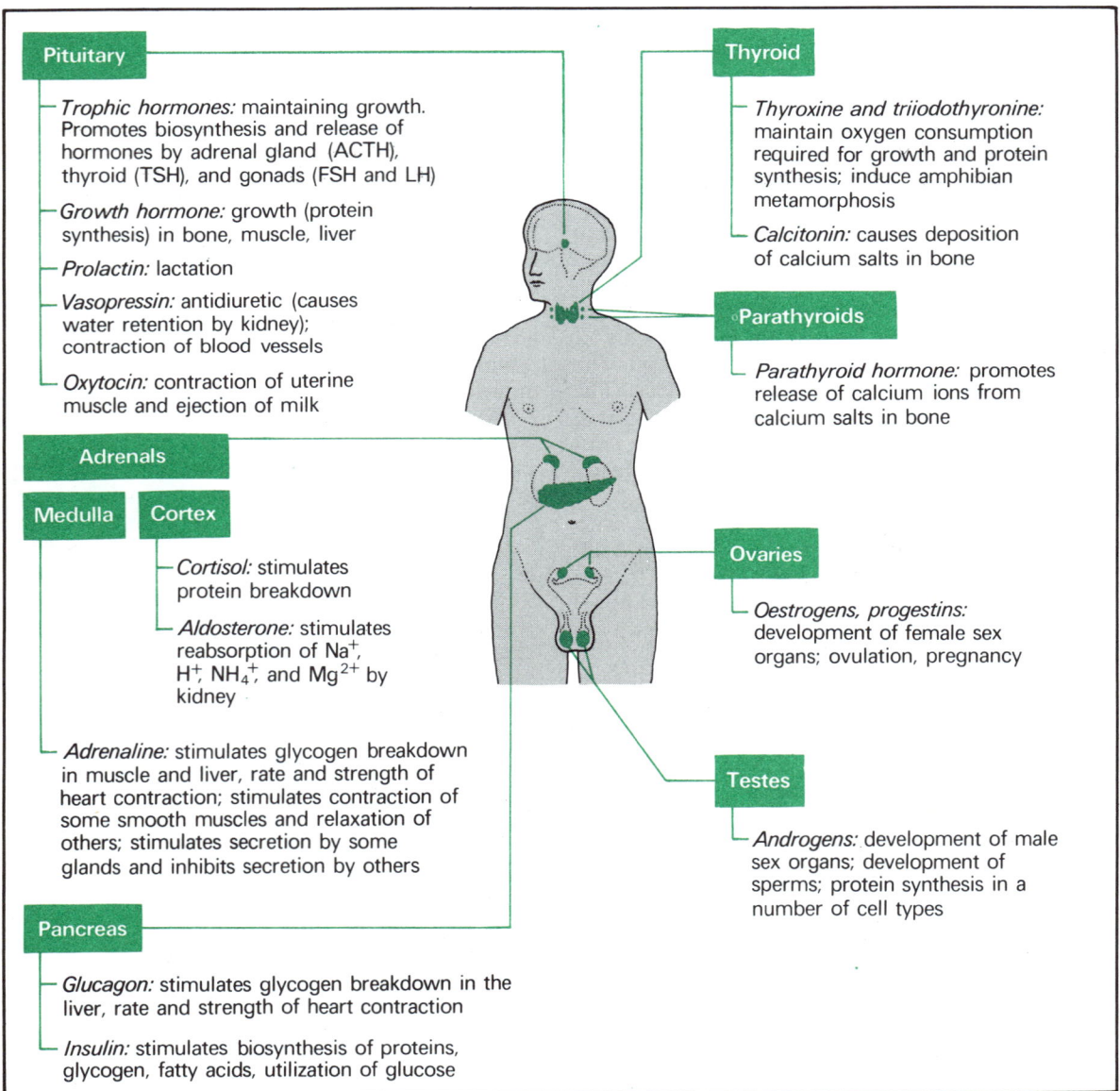

Pituitary

—*Trophic hormones:* maintaining growth. Promotes biosynthesis and release of hormones by adrenal gland (ACTH), thyroid (TSH), and gonads (FSH and LH)

—*Growth hormone:* growth (protein synthesis) in bone, muscle, liver

—*Prolactin:* lactation

—*Vasopressin:* antidiuretic (causes water retention by kidney); contraction of blood vessels

—*Oxytocin:* contraction of uterine muscle and ejection of milk

Adrenals

Medulla **Cortex**

—*Cortisol:* stimulates protein breakdown

—*Aldosterone:* stimulates reabsorption of Na^+, H^+, NH_4^+, and Mg^{2+} by kidney

—*Adrenaline:* stimulates glycogen breakdown in muscle and liver, rate and strength of heart contraction; stimulates contraction of some smooth muscles and relaxation of others; stimulates secretion by some glands and inhibits secretion by others

Pancreas

—*Glucagon:* stimulates glycogen breakdown in the liver, rate and strength of heart contraction

—*Insulin:* stimulates biosynthesis of proteins, glycogen, fatty acids, utilization of glucose

Thyroid

—*Thyroxine and triiodothyronine:* maintain oxygen consumption required for growth and protein synthesis; induce amphibian metamorphosis

—*Calcitonin:* causes deposition of calcium salts in bone

Parathyroids

—*Parathyroid hormone:* promotes release of calcium ions from calcium salts in bone

Ovaries

—*Oestrogens, progestins:* development of female sex organs; ovulation, pregnancy

Testes

—*Androgens:* development of male sex organs; development of sperms; protein synthesis in a number of cell types

Fig. 15.28 The endocrine glands of the human body, and some of the known functions of the hormones they secrete.

Figure 15.27 shows another aspect of advance in knowledge of insulin. It is an electron micrograph of an insulin-secreting cell from mouse pancreas showing the membrane-bounded sacs containing insulin. When blood glucose concentration rises these move to the surface of the cell and are discharged into the tissue fluid and so into the blood.

Some other hormones
Tissues and glands which produce hormones are called **endocrine glands** or ductless glands, because they release their secretions directly into the blood stream and not along some tube or duct such as the pancreatic duct. Figure 15.28 shows the known endocrine glands of the human body and the effects that some of their hormones produce.

Other hormones than insulin affect the body's use of sugar. Adrenaline from the adrenal gland stimulates the conversion of glycogen stored in the muscles and the liver into glucose. This has the opposite effect on blood sugar concentration from that of insulin. At least three other hormones are known to influence the use of glucose, and the maintenance of a steady 0.1 % concentration in the blood is achieved by a balance of sugar-mobilising and sugar-consuming processes. Both are governed by hormones.

Some hormones affect the activity of other endocrine glands. The pituitary body in particular produces hormones which stimulate the thyroid gland to secrete the hormone **thyroxine** and the adrenal gland to secrete the hormone **cortisone**.

Hormonal and nervous coordination compared
1 Hormones act by modifying the rates of particular chemical reactions within those cells that respond to them. They have steady, long-term effects throughout the life of the cell. By contrast, nervous stimulation is very short-term.
2 A variety of hormones affect various aspects of the chemical workings of the cell, but nervous stimulation produces a very limited range of effects, namely muscle contraction or glandular secretion.
3 The effects of hormones are often general. For example, insulin affects the glucose uptake of all the cells of the body. Other hormones however, are more particular, and the dark-staining cells of the pancreas are the only ones in the body that are known to respond to secretin.

Because they are present in the blood and tissue fluid, hormones come into contact with all cells. The fact that a particular hormone produces an effect on some cells and not on others is due to a property of the plasma membrane. This is known to

possess specific **receptor sites** which 'recognise' and react with particular hormones.

The action of nerves is quite different from this. Electrical messages, all of which are of the same kind, and pass along clearly-defined channels of communication to specific targets.
4 Hormones have to be delivered from the endocrine glands to their target cells by the blood. However rapid this is, some delay is involved. Here again there is a strong contrast with the nervous system, where messages travel at the rate of between one and 120 metres per second.

Inter-relationships of the hormonal and nervous systems
On page 181 it was suggested that hormonal coordination is quite distinct from nervous coordination, but reflection shows that this is not so. Some endocrine glands are stimulated to secrete their hormones by instructions from the brain delivered by nerves—for example the secretion of **adrenaline** by the adrenal gland. The rate at which blood is delivered to endocrine glands, and so the rate at which hormones are removed from them, is also governed by nerves, which control the bores of all arteries.

Synaptic vesicles contain hormones. Messages are carried along nerve axons as electrical impulses, but at every synapse transmission is governed by hormones.

Coordination in higher plants

Nervous coordination, aided by information from the senses, has much to do with muscular movement. Plants do not move freely, have no sense organs, and have nothing corresponding to a nervous system.

Plant hormones
The first indication of the existence of plant hormones came from some very simple experiments on oat seedlings in 1910. When oat seeds germinate the first green leaf is protected by a sheathing structure some 2 mm in diameter, called the **coleoptile**. This pushes up through the soil for a distance of 2–4 cm. It has a rounded end and when the leaf pierces it, it ceases to grow.

When the tip of a growing coleoptile was cut off, elongation at once slowed down. It resumed when the cut-off tip was replaced (Fig. 15.29(a)). Perhaps the tip was producing a hormone which diffused downwards and stimulated elongation. This hypothesis was tested by removing tips and standing them on small blocks of agar jelly. The blocks were then used to replace freshly-removed tips on other coleoptiles (Fig. 15.29(b),(c)). Elongation was again stimulated. The suggested explanation of these results was that a growth hormone (called **auxin**) had first diffused from the tip into the agar block and then from the agar block into the second, de-tipped coleoptile.

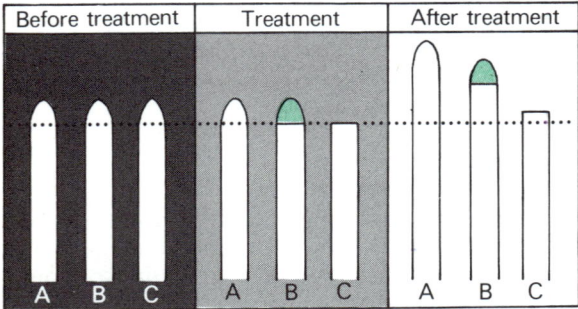

Fig. 15.29(a) Effect of oat coleoptile tip on growth of the coleoptile. (A) coleoptile untreated; (B) tip removed and replaced—growth almost normal; (C) tip removed, growth retarded.

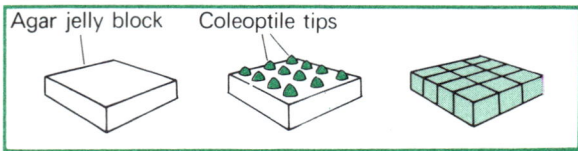

(b) Method of collecting auxin from coleoptile tips.

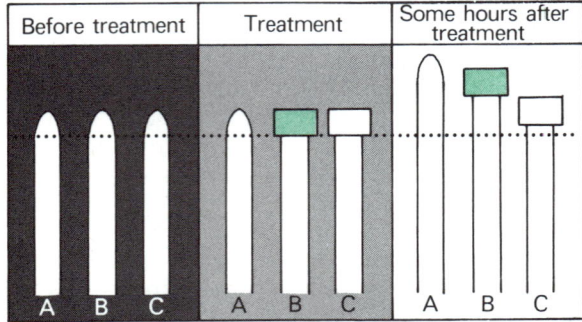

(c) Auxin collected from coleoptile tips has the same effect on growth of the coleoptile as the tips themselves. (A) untreated coleoptile; (B) tip removed and replaced with agar block containing auxin collected from tip—growth nearly normal; (C) control in which tip is removed and replaced by a plain agar block.

Care has been taken to speak of elongation of the coleoptile rather than its growth. Auxin has since been chemically isolated and identified, and it is known to affect the elongation of plant cells. This takes place by the taking in of water into the developing central permanent vacuole, a process that does not involve cell division and so may not be regarded as growth (see page 130).

Sensitivity in plants: tropisms
Though possessing no sense organs, or even sensory cells, parts of plants respond in definite ways to stimuli from the environment. When the stimulus is from a *particular direction* and the response to it is caused by *growth or elongation*, the response is called a **tropism**.

Phototropism
The stems of many plants grow towards the strongest source of light. This is called **positive phototropism**, and is obviously an advantage in enabling the plant to display its leaves for photosynthesis. The growth of roots is not affected by light. Positive phototropism can be demonstrated with mustard seedlings. Three lots are germinated; one is kept in complete darkness; the second is exposed to equal illumination on all sides (for example in a greenhouse); the third is kept in a dark box which has a window on only one side. Whereas the first two treatments produce straight, upward-growing seedlings, the third produces seedlings which are all bent towards the light.

Auxin and phototropism An experiment performed in 1935 showed that unequal distribution of auxin in a coleoptile stump produced much the same growth curvature as that in the positively-phototropic mustard seedlings.

Coleoptile tips were removed and stood on agar jelly. Auxin diffused down into the agar. Now fresh coleoptiles were de-tipped and the first leaf inside each was pulled up slightly to make a rest against which the small agar block containing auxin could be placed. The result a few hours later is shown in Fig. 15.30.

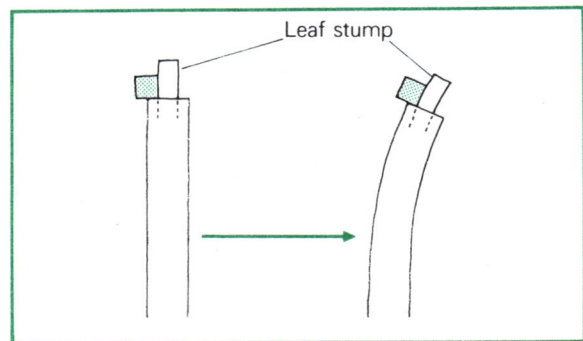

Fig. 15.30 Method used for attaching agar block containing auxin to one side of a coleoptile stump.

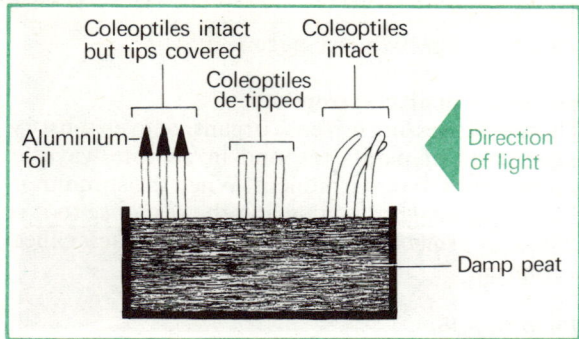

Fig. 15.31 Demonstration that the tip of oat coleoptile is the region sensitive to one-sided illumination. The region of response lies behind the tip.

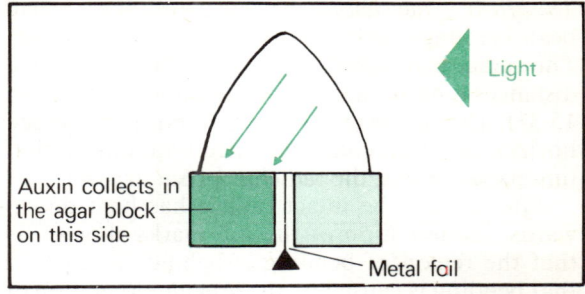

Fig. 15.32 One-sided illumination of oat coleoptile tip brings about redistribution of auxin in it.

The explanation given was that auxin had diffused downwards from the block and had not spread out in the coleoptile stump. It therefore caused a greater elongation of the cells on the same side of the coleoptile as the agar block.

The experiment illustrated in Fig. 15.31 confirms that the tip is the region of the coleoptile which detects light falling from one side, though the region which responds lies behind the tip. Three lots of oat seeds are germinated in damp peat. When the coleoptiles are about 3 cm high, they are subjected to the treatments shown. All three sets are placed in a black box illuminated from one side only. Both de-tipping and shading the tips prevents the positive phototropic response seen in the untreated, control coleoptiles.

Is auxin responsible for phototropism? Thus we know: (1) that the unequal distribution of auxin in oat coleoptiles can produce a growth curvature similar to that seen in positive phototropism; (2) the tip of the coleoptile, which is known to be the region which produces auxin, is the region which detects one-sided illumination.

Further experiments showed that when coleoptile tips that had been exposed to one-sided illumination were placed on de-tipped coleoptiles which were then kept in the dark, the coleoptiles bent. The side of the tip farthest from the one-sided illumination produced the bending.

This seems to indicate that one-sided illumination of the coleoptile tip brings about a redistribution of auxin in it. In an evenly-lit tip, auxin is distributed evenly, but in a tip lit from one side there is less auxin on the lit than the unlit side. This view is confirmed by the following experiment. Tips were taken from coleoptiles and stood on two agar blocks separated by metal foil to prevent diffusion of auxin (Fig. 15.32). They were then lit from

one side. The two resulting blocks were tested for auxin content by seeing how much bending they would produce in freshly de-tipped coleoptiles. The results showed more auxin in the blocks taken from the shaded sides of the coleoptiles.

However, we must be cautious in accepting auxin redistribution as an explanation of all phototropic behaviour. The experiments described were not conducted on shoot tips but on coleoptiles; and the intensity of illumination used was very much less than that encountered in natural situations.

Geotropism
Primary (main) roots grow downwards in the soil. The force of gravity acts in one plane, towards the centre of the earth, and primary roots are therefore **positively geotropic**. Branch roots are not positively geotropic and grow horizontally, perpendicular to the plane of gravity.

The main stems of most plants are **negatively geotropic** and grow perpendicular to the surface of the earth. Side branches are usually insensitive to gravity and grow parallel to the soil.

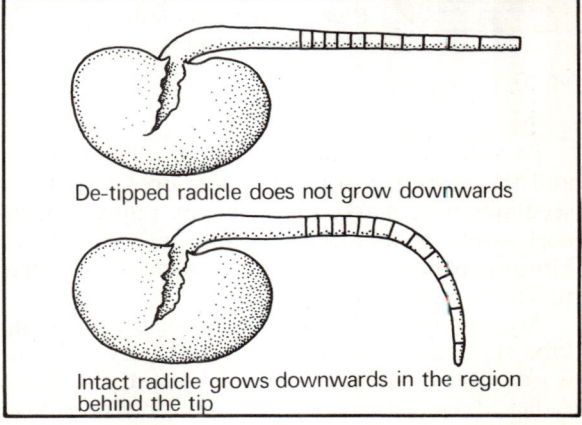

De-tipped radicle does not grow downwards

Intact radicle grows downwards in the region behind the tip

Fig. 15.33 Removal of root tip prevents the positive geotropic response.

The effect of the root tip in geotropism Two broad bean seedlings with straight radicles are selected. The radicles are marked with indian ink at equal distances both on their upper and lower sides (Fig. 15.33). The tip is cut off one, and both are pinned horizontally in a container lined with damp blotting paper so that the radicles do not dry out.

After a day, the intact radicle has bent downwards. Examination of the ink marks shows: (1) that the region of bending is behind the tip; (2) that bending is caused by the greater elongation of the upper side of the radicle compared with the lower side. The ink marks on the upper side separate more than those on the lower.

The de-tipped radicle does not bend. As with phototropism, therefore, the tip is sensitive to one-sided stimulus, this time of gravity, and the elongating region behind the tip is the region of response.

The klinostat Experiments on geotropism are limited by the fact that the force of gravity cannot be switched off, nor its plane of action altered. However, by using the **klinostat** (Fig. 15.34) in which plant material is slowly revolved, the one-sided effect of gravity can be eliminated as each side of the material is exposed to gravity in turn.

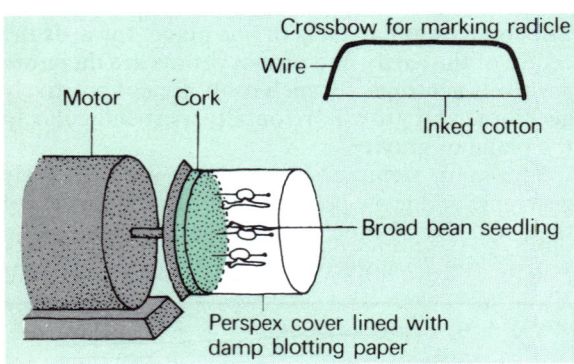

Fig. 15.34 The klinostat.

Two klinostats should be used, one which revolves and the second, a control, which does not. Bean seedlings with marked radicles are pinned to the cork inside the klinostat, which is lined with damp blotting paper. Some of the radicles have their tips removed, others are left intact.

After a day, none of the radicles on the revolving klinostat, intact or de-tipped, has grown downwards. But on the stationary klinostat, the intact radicles have grown down. The de-tipped ones have not. By revolving, all parts of the radicles are ex-

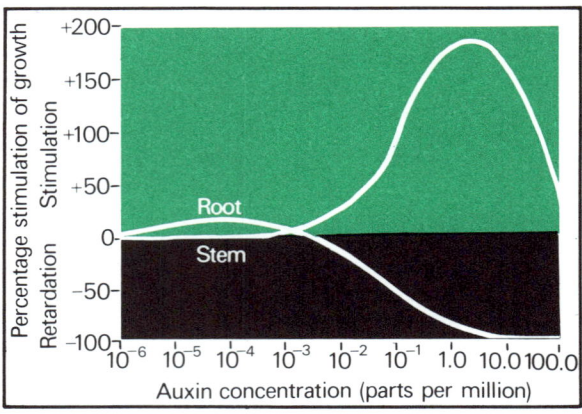

Fig. 15.35 Effects of auxin in stimulating or retarding growth of stem and root of oat seedlings. Over the range 10^{-6} to 10^{-3} parts auxin per million, root growth is stimulated but stem growth is unaffected. At higher concentrations root growth is retarded but stem growth is stimulated.

posed to the one-sided force of gravity in turn. When the radicles are still, gravity must produce some effect on their tips which then causes the growing region behind the tip to bend.

Is auxin responsible for geotropism? Small amounts of auxin are produced in root tips, and it is not difficult to explain positive geotropism in roots in terms of auxin.

First we must note that auxin has very different effects on the growth of shoots and roots of oat seedlings (Fig. 15.35). This graph shows the effects of different concentrations of auxin (on the X, horizontal axis) in stimulating *or retarding* stem and root growth. Results on the Y (vertical) axis are expressed as percentage stimulation or retardation of stem or root compared with untreated control material.

Over the range 10^{-6} to 10^{-3} parts of auxin per million the hormone stimulates root growth but has no effect at all on stem growth. At higher concentrations of up to 10^1 parts auxin per million, stem growth is stimulated but root growth is *retarded*.

So if the one-sided effect of gravity on the horizontal root tip were to cause auxin to be more concentrated on the lower than on the upper side, auxin would diffuse back into the root and cause the lower side to grow more slowly than the upper, and this would produce bending.

How gravity might cause such a distribution of a chemical substance dissolved in the cytoplasm of cells is quite unknown.

16 The Eye and the Ear

Positions of the eyes

The eyes are contained in the **orbits**, deep bony sockets in the skull which protect them (see Fig. 6.12). Most vertebrates have eyes on the sides of their heads. This gives a wide range of vision, especially if the animal does not possess a flexible neck (e.g. a frog).

Primates (e.g. monkeys, apes, man) have both eyes at the front of the head. This gives them stereoscopic (depth) vision and permits the close examination of objects held in the hands (Fig. 16.1). This ability is thought to have contributed a lot to the development of curiosity and intelligence in

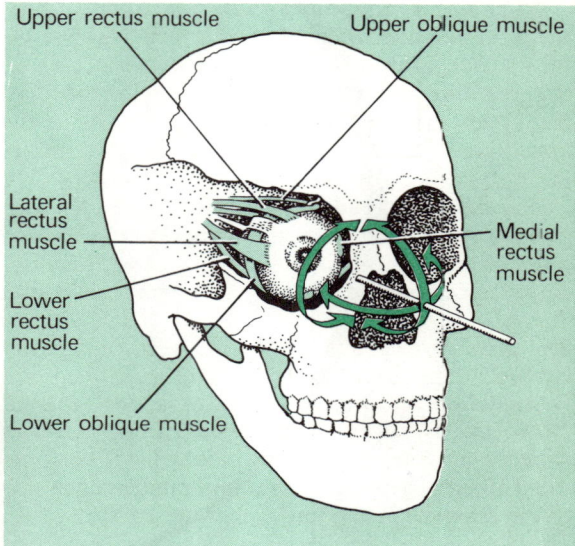

Fig. 16.2 The three pairs of muscles attached to the sclera and the eye movements that they produce.

Fig. 16.1 Frontal vision is associated with close examination of objects, and therefore with their delicate handling. Young baboon.

primates. In some mammals, and also in birds, the range of vision is augmented by head and neck movements.

Eye muscles and movement

Each eye is held in place by three pairs of muscles which are attached to its wall, or **sclera**. Fig. 16.2 shows how each pair produces movement in one of the three planes of space.

Structure of the eye

Most of the spherical wall of the human eye is white and tough. It is called the **sclera** (Figs. 16.3, 16.4). The liquid contents of the eye, the **aqueous** and **vitreous humours**, press against the sclera and so keep the eye in shape.

In front, the sclera is replaced by the transparent **cornea**. When light passes from air into the cornea it is strongly bent. It is the cornea, not the lens, which is the principal image-forming structure of the eye.

Behind it is the **lens**, attached to the **suspensory ligament** that is in turn attached to the muscles of the **ciliary body**. When light passes from the

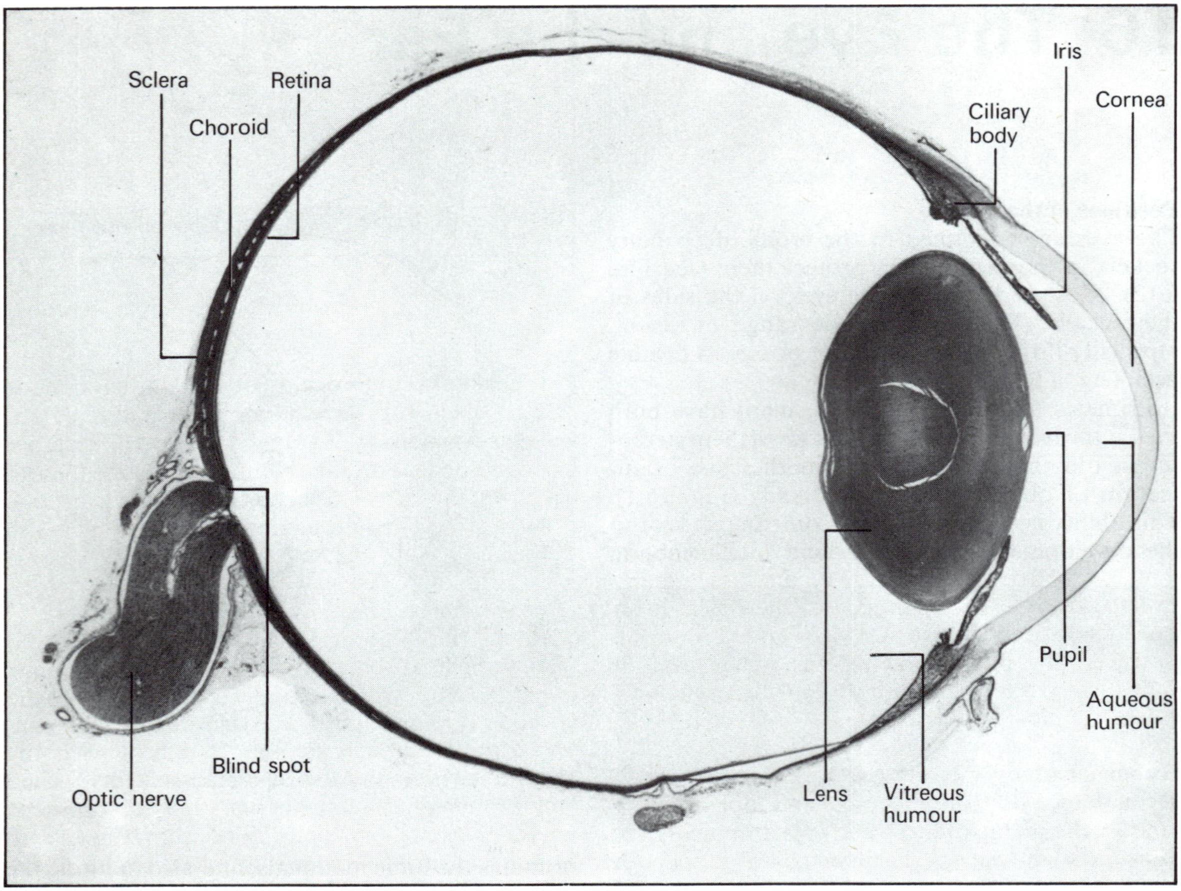

Fig. 16.3 Horizontal section through the eye of a rhesus monkey.

aqueous humour through the lens and into the vitreous humour, it is bent only slightly. The lens provides a fine focussing adjustment for the eye.

The coloured **iris** lies in front of the lens and provides it with a stop. The gap in its centre, the **pupil**, narrows when strong light falls on the eye, and broadens in dim light. Both these actions are nervous reflexes.

The light-sensitive layer is the **retina** and nerve axons pass directly from it into the **optic nerve** which is connected to the midbrain. Between the retina and the sclera lies the **choroid**. This layer contains black-pigmented cells which prevent light from entering the eye except from the front via the cornea. It also possesses a rich supply of blood capillaries (Fig. 16.5).

Formation of the image on the retina
The cornea and the lens bend light which passes through them, producing an image on the retina.

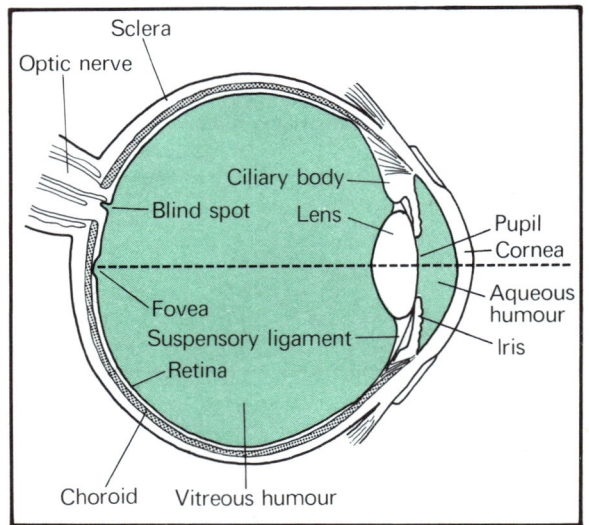

Fig. 16.4 Horizontal section through right human eye.

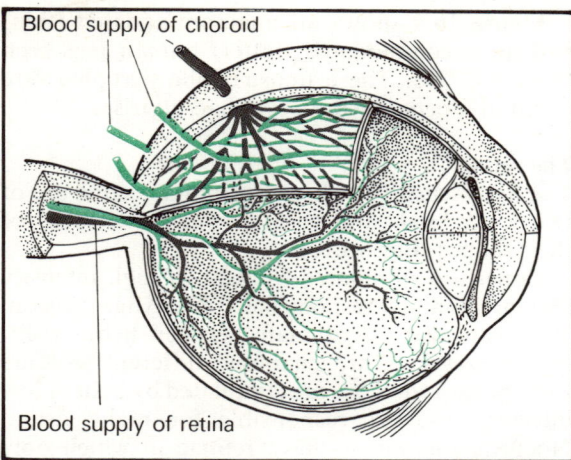

Blood supply of choroid

Blood supply of retina

Fig. 16.5 Stereogram of the blood vessels in the choroid

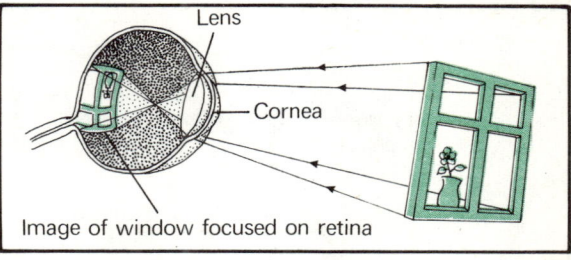

Lens

Cornea

Image of window focused on retina

Fig. 16.6 The cornea and the lens form an image which is both inverted and laterally inverted (right side of object on left side of image and vice versa).

The image is very small compared with the object being viewed, is upside down, and is laterally inverted (that is, the right side of the object is on the left side of the image and vice versa, Fig. 16.6). However, the brain **perceives** the image as if it were formed the right way up and the right way round.

Accommodation

If you glance up from this page and look out of a window the image of a distant object immediately comes into your view in sharp focus. To enable both near and distant objects to be focussed on the retina, the shape of the lens is automatically adjusted. This is a reflex nervous action and is called **accommodation.**

When looking at a distant object, the lens is made thin (Fig. 16.7(a)). When looking at a near object, the lens is made fatter (Fig. 16.7(b)). The distance between the lens and the retina is the same in both cases, but the fatter lens bends light passing through it more strongly, so the image that it forms is still in focus.

How the shape of the lens is changed

The lens has a natural elasticity and if removed from the eye takes on a much more spherical shape.

The sclera, which is under pressure from the vitreous humour, pulls the suspensory ligament, which pulls the lens, making it thin. This is the condition when looking at distant objects. When looking at near objects, the ciliary muscle contracts, drawing the ciliary body forwards (Fig. 16.8). The suspensory ligament is attached to the ciliary body, and this action causes the suspensory ligament to become slacker. The tension on the lens is thus reduced and it automatically assumes a more spherical shape.

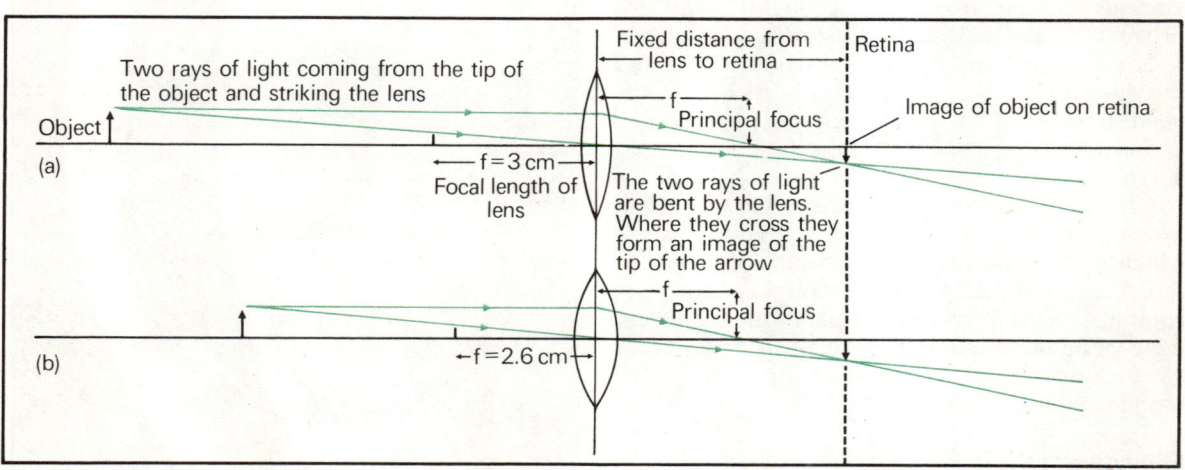

Fig. 16.7(a) Formation of an image on the retina by an object distant from the eye. Lens is unaccommodated and thin. **(b)** Formation of an image on the retina by an object close to the eye. Lens is accommodated and fat.

191

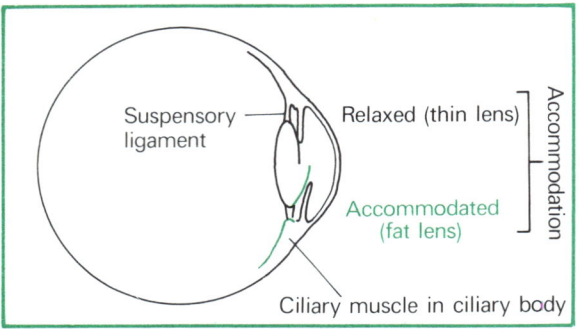

Suspensory ligament

Relaxed (thin lens)

Accommodation

Accommodated (fat lens)

Ciliary muscle in ciliary body

Fig. 16.8 Changes in positions and sizes of the ciliary body, lens, and pupil as the mammalian eye accommodates from distant (top) to near (bottom) vision.

Fig. 16.9 Change in the shape of human lens during accommodation from distant (a) to near (b) vision.

Figure 16.9 shows an unaccommodated (thin) and an accommodated (fatter) human lens seen from the side. These unusual photographs were taken of a person whose eyes had no irises.

The retina

The light-sensitive cells of the human retina are of two kinds, **rods** and **cones**, so called because of their shapes (Fig. 16.10).

Cones are stimulated by light of high intensity and are mainly responsible for daylight vision. They can also discriminate between light of different wavelengths, enabling different colours to be perceived. Rods are stimulated by light of low intensity and are responsible for night vision. Nocturnal mammals have retinae in which only rods are found. That rods cannot distinguish colours is a matter of common experience, for colours cannot be seen by moonlight.

The human **fovea** (Fig. 16.4), is the part of the retina with the highest resolving, or discriminating power. It contains only cones. Elsewhere rods predominate. The human retina possesses about six million cones and 120 million rods.

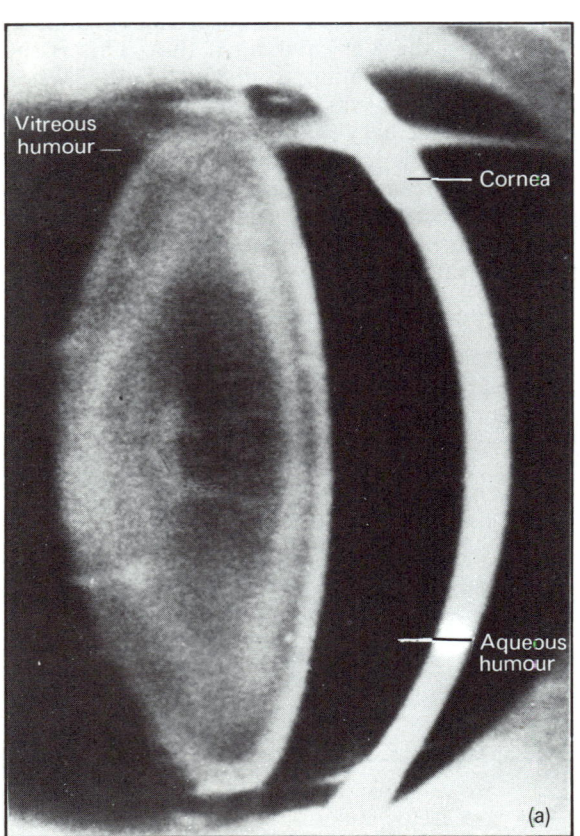

Vitreous humour

Cornea

Aqueous humour

(a)

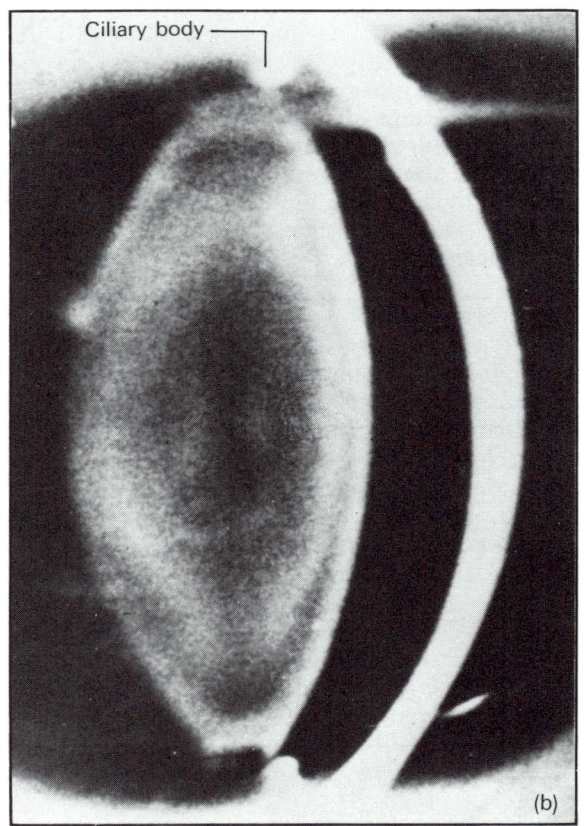

Ciliary body

(b)

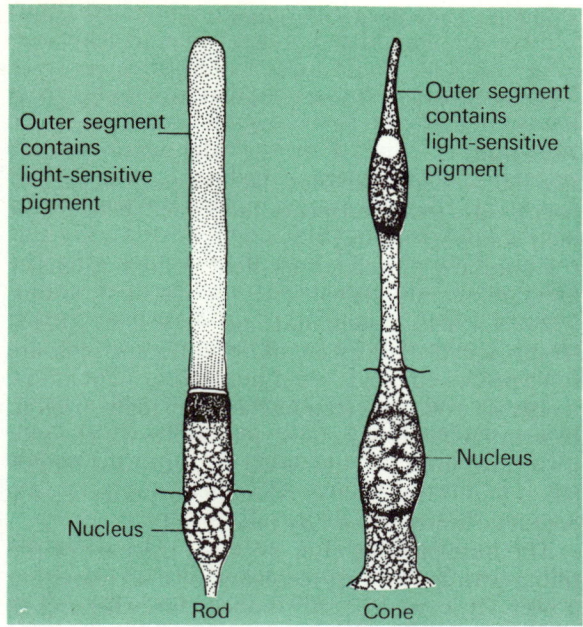

Fig. 16.10 A rod and a cone.

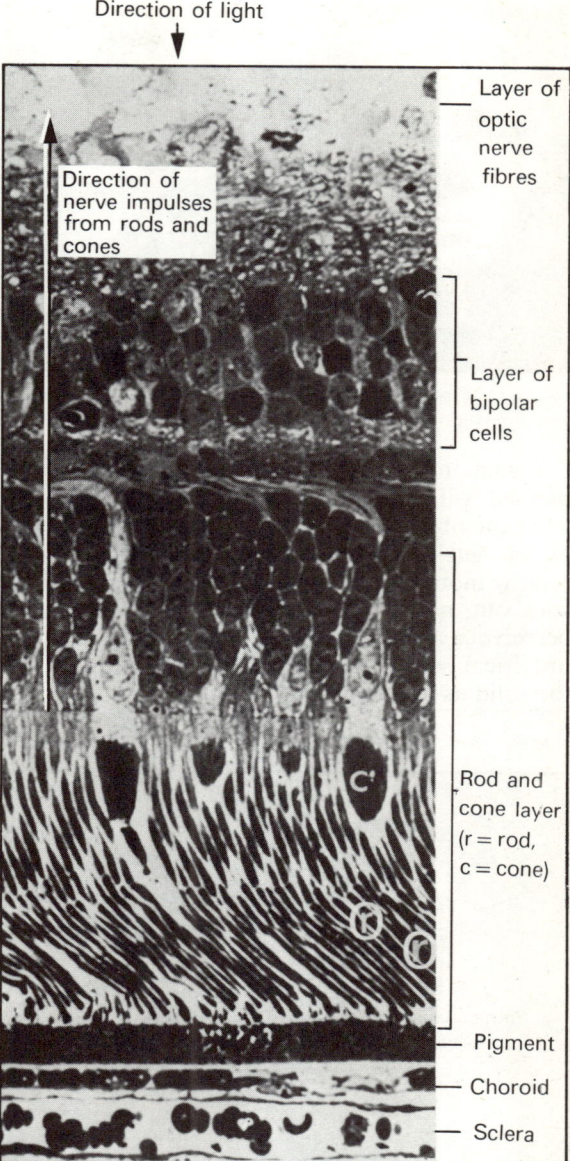

Fig. 16.11 Transverse section of the human retina showing the various layers. (Light microscope.)

Visual pigments

The outer segments of the rods and cones both contain light-sensitive pigments. These are chemically changed when light falls on them, initiating the nerve impulse which passes to the brain. The impulse is relayed first to bipolar cells, then to the optic nerve fibres. These cells can be seen in Fig. 16.11. The bipolar cells probably enable a great deal of the visual information falling on the retina to be discarded.

Visual perception

Visual perception is a different kind of phenomenon from photography because it goes beyond the available evidence presented to the brain by the retinae. For example, closing one eye eliminates stereoscopic vision, but it does not appear to do so. The brain compensates so well that it is even possible to play a game of table tennis with one eye shut. It is only by contrived conditions that we are aware of the loss of stereoscopic vision.

The distinction between the retinal image and the image perceived by the brain is illustrated by Fig. 16.12, a drawing of a wire cube which gives alternative perceptions when looked at for some time. As the retinal image does not change, this must be a central brain phenomenon.

Fig. 16.12 Drawing of a wire cube that gives alternative perceptions.

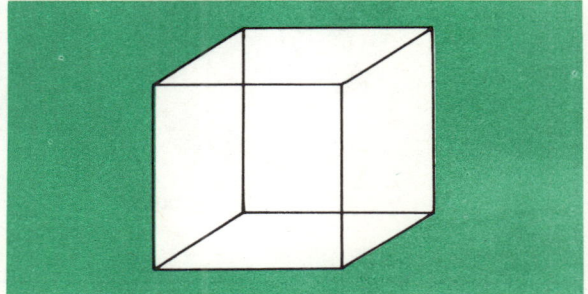

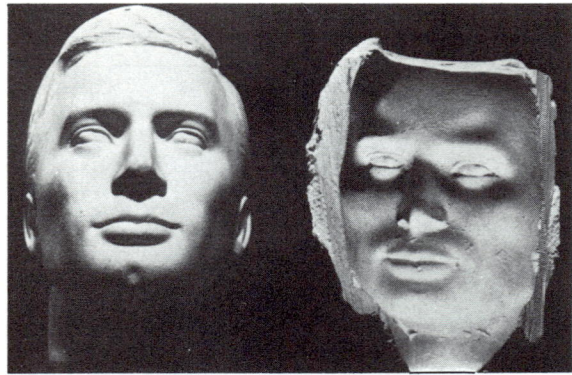

Fig. 16.13 A sculptured mask (left) and a hollow mould taken from it (right).

Visual information from the retina is supplemented with information from the brain about what the object is *expected* to be. Fig. 16.13 shows on the left a sculptured face and on the right a hollow mould cast from it. The mould appears to stick out, not in, for the brain does not expect to perceive a receding face. The lighting is quite uncritical, and the effect is just as pronounced in the solid as in this photograph.

Fig. 16.14 General structure of the human ear.

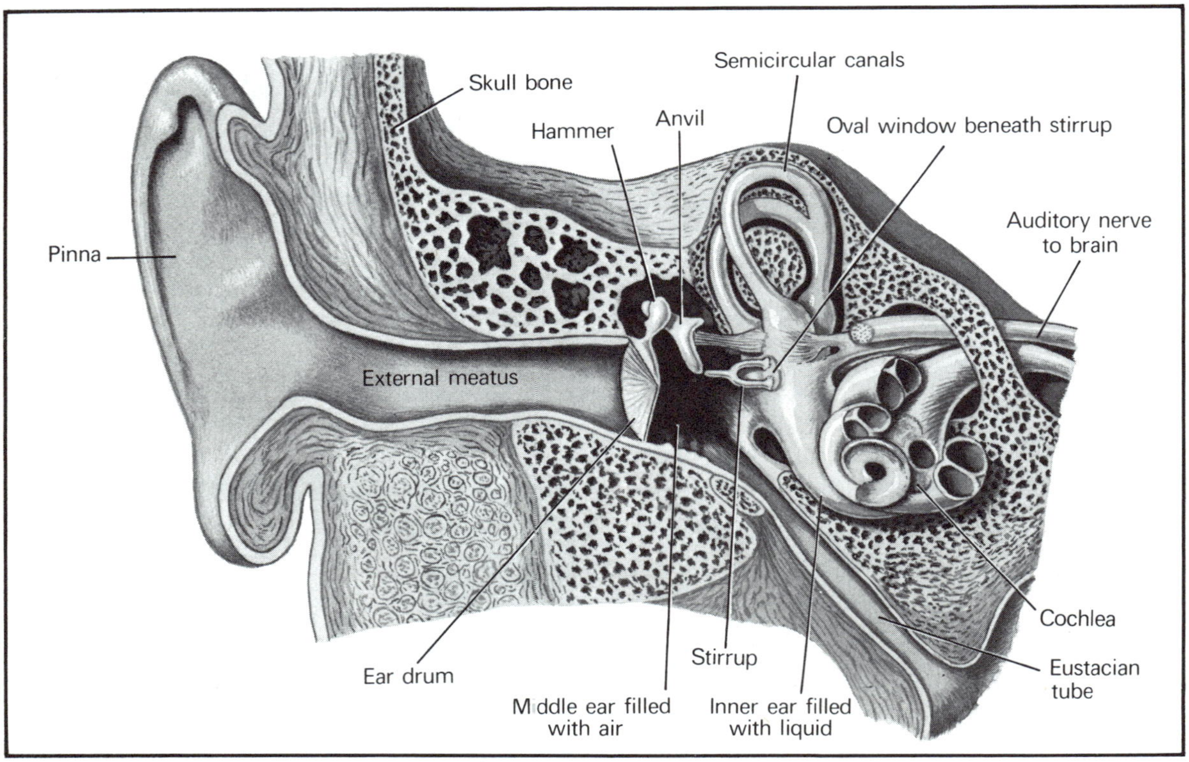

The ear: hearing and balance

The only parts of the mammalian ear visible from the outside are the **pinna**, a sound-catching device which can be 'pricked' by muscular action in some species, and the **external meatus** or ear hole (see Fig. 6.12). The rest is buried in the skull bone, which is very thick and contains many air spaces in this region. Figure 16.14 is a drawing showing the positions of the principal structures of this complicated organ. Figure 16.15 is a vertical section showing some of the parts of the human middle and inner ears.

The meatus leads to the **eardrum**, a conical membrane shaped like a radio loudspeaker. It completely separates the **external ear** from the **middle ear**. The meatus is lined by cells that secrete wax, keeping both it and the ear drum supple.

The middle ear is also air-filled. The **Eustacian tube** leads from it to the back of the throat and is sealed by a valve which is opened by the act of swallowing. This safety device keeps the air pressures on both sides of the ear drum equal.

Attached along the radius of the drum is one of three small **middle ear bones** (or **ossicles**), the **hammer** (Fig. 16.16). It makes a joint (articulates)

[Diagram labels: Skull bone, Hammer, Anvil, Semicircular canals, Oval window beneath stirrup, Auditory nerve to brain, Pinna, External meatus, Cochlea, Eustacian tube, Ear drum, Middle ear filled with air, Stirrup, Inner ear filled with liquid]

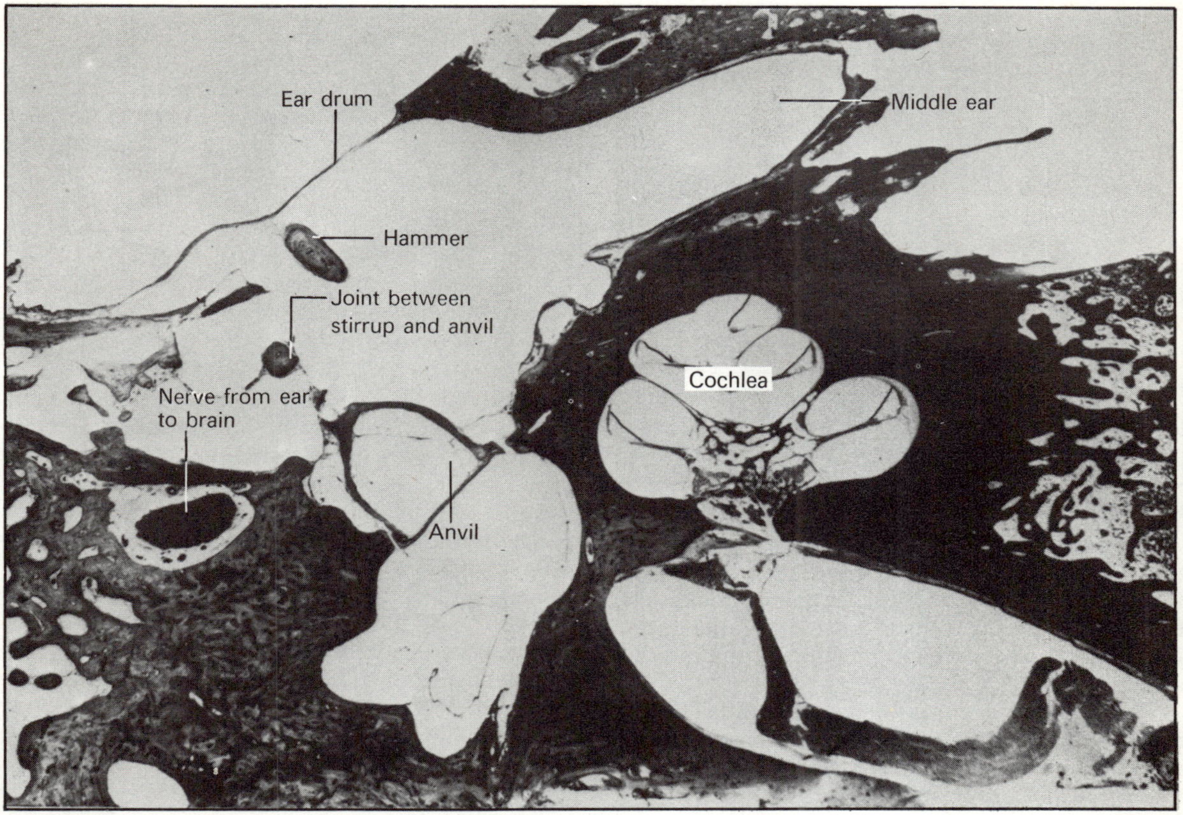

Fig. 16.15 Some of the parts of the human middle and inner ears embedded in the skull bone. Vertical section, light microscope.

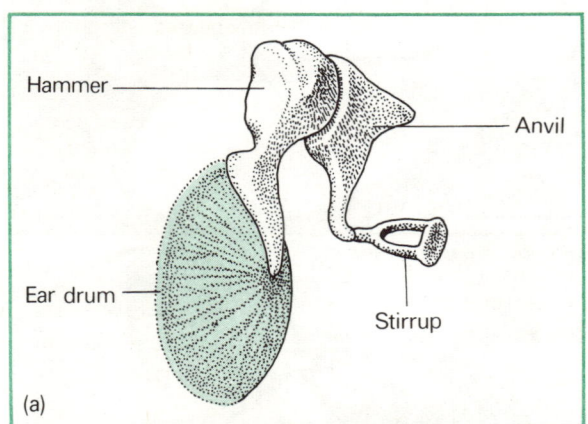

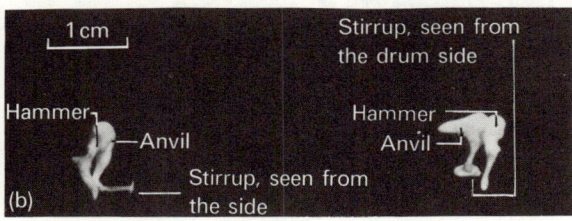

Fig. 16.16(a) The human ear drum and ossicles. Note the attachment of the hammer along the radius of the drum.
(b) Two views of the human ear ossicles.

with the anvil, which articulates with the **stirrup**. The flat side of the **anvil** presses against the fluid-filled **inner ear** at the **membranous oval window**.

When sound waves fall on the drum it vibrates. The vibrations are passed on to the oval window via the ossicles. Because the area of the drum is twenty two times that of the oval window, the pressure transmitted is magnified twenty two times. The drum has no natural resonance, which means that as soon as sound waves cease to fall on it, it ceases to vibrate. Not all sound waves reach the inner ear via the drum and ossicles. All the skull bones vibrate, as you can demonstrate by grinding your teeth.

The inner ear

The **inner ear** (Fig. 16.17) is a complicated structure only part of which, the **cochlea** (Greek, a snail) is concerned with hearing. This is a tube twisted

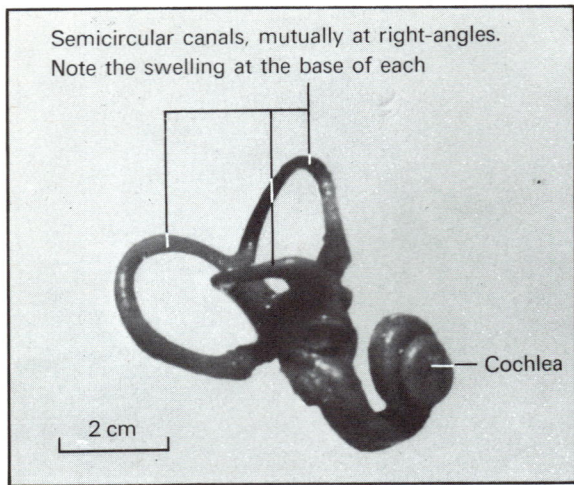

Fig. 16.17 Human inner ear—cochlea, and semi-circular canals.

through two and a half turns like a snail shell. It is sub-divided internally along its whole length by two membranes, producing three parallel tubes. Figure 16.18 is a stereogram showing the internal arrangement.

The whole inner ear is filled with liquid. The vibration-sensitive cells rest on the lower membrane with their tips embedded in a ribbon-like material which reaches only part way across the tube (Fig. 16.19).

Vibrations of the oval window cause vibrations in the liquid of the inner ear. The lower membrane vibrates, pulling the sensitive cells it supports

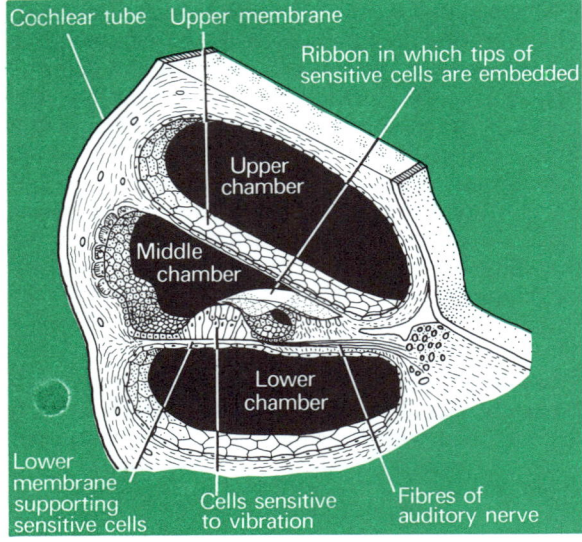

Fig. 16.18 Stereogram of part of the cochlear tube.

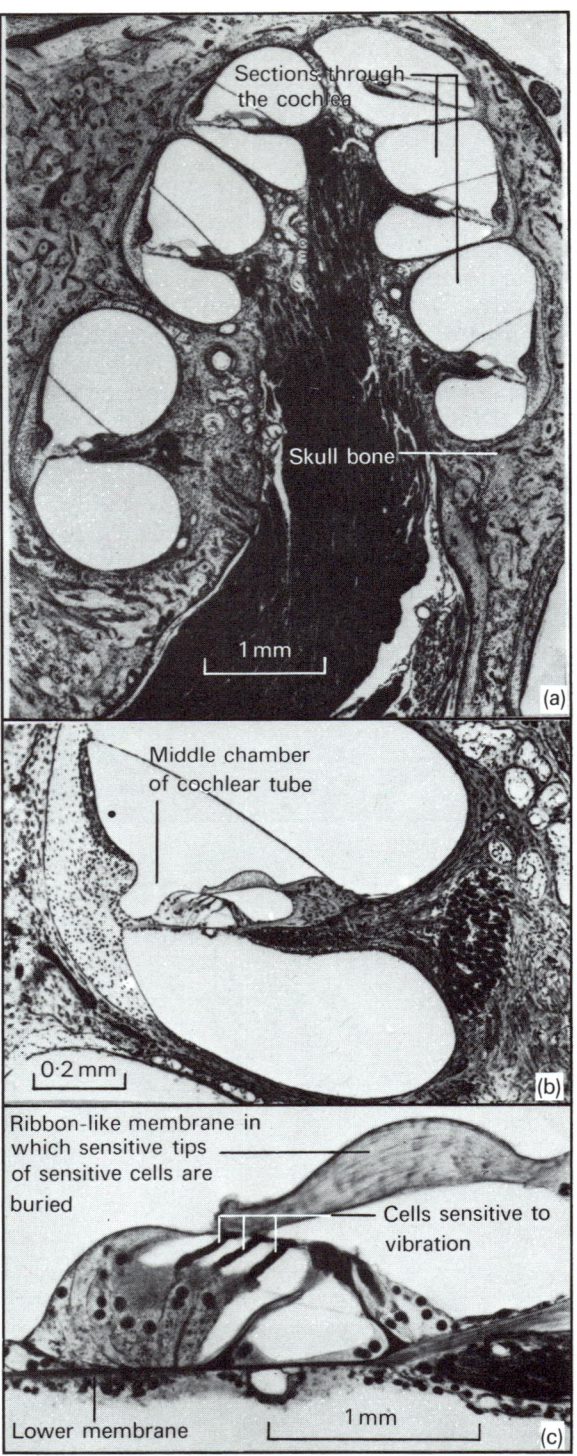

Fig. 16.19(a) Longitudinal section through the cochlea of a cat. (b) Part of (a) enlarged. (c) Part of (b) enlarged. (All light microscope.)

against the ribbon-like material in which their tips are embedded. This causes them to generate nerve impulses in the nerve fibres with which each one is associated. Different parts of the spiral cochlea respond to different frequencies.

Balance

Figure 16.17 also shows the **semi-circular canals** of the inner ear, each with a swelling at its base containing sensitive cells. The canals are placed mutually at right angles. They too are fluid-filled, and disturbance of the liquid within them stimulates the sensitive cells in the swellings. This initiates nerve impulses that are interpreted by the brain as information about the body's position in space. For example, nodding the head up and down causes movement of the fluid in the vertical canal that lies parallel to the front-back axis of the body.

Because the canals are mutually at right angles, the combined information from them provides for all positions of the head in space. Further information about balance comes from the eyes and from the sensory cells in the skin.

17 The Skeleton, Muscles and Movement

Bones belong to vertebrates

No invertebrate possesses bone. Bone is a characteristic vertebrate tissue which first appeared in the fossil record over 500 million years ago in the Ordovician period. In the first fish it did not form an internal skeleton but bony armour, protective plates (Fig. 17.1). Today, however, most fishes and all land vertebrates have bony internal skeletons. The sharks and rays are an exception, having gristle (**cartilage**) skeletons.

Bone as a calcium and phosphate store

Bone contains a large proportion of calcium phosphate, and it may be that its prime function in early fish was to act as a store of these ions. Calcium ions in the tissue fluid are essential for the proper functioning of muscles and nerves. Phosphate, in the form of ATP, is a necessary part of every chemical reaction in the cell which involves the consumption of chemical energy. Both calcium and phosphate ions are scarce in sea water.

Functions of bones

Bones serve:

1 As reservoirs of calcium and phosphate ions. These can be immediately released under the influence of hormones produced by the thyroid and parathyroid glands. The average daily exchange of calcium between bone and tissue fluid in a man is about 0.5 g Ca^{2+} ions out of a total of 1000 g.

2 As rigid levers on which contracting muscles can pull. This causes movement because at various places between bones there are moveable joints.

3 To protect soft, delicate organs. The chest skeleton protects the heart and lungs. The skull

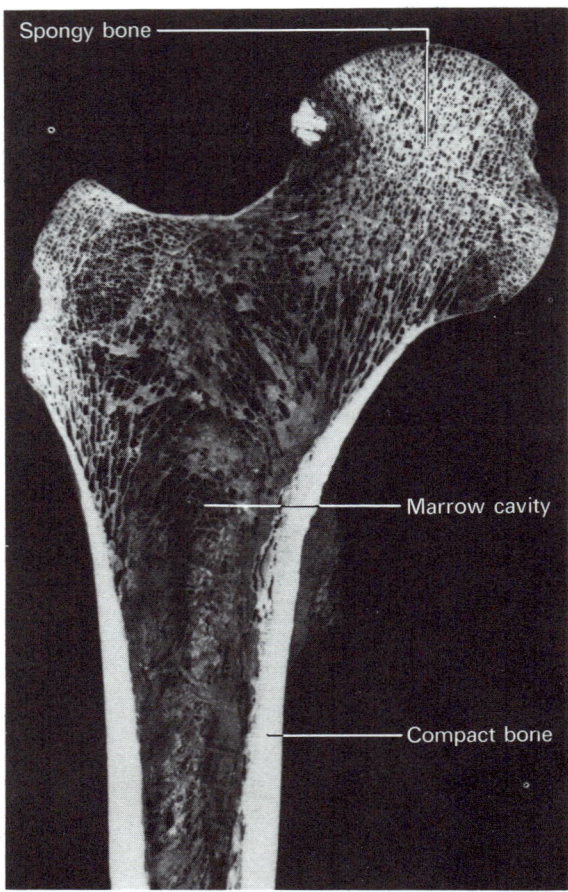

Fig. 17.2 Longitudinal section through upper part of human femur.

and backbone protect the brain and the spinal cord. The pelvic basin protects the bladder.

4 To house the red bone marrow which manufactures red and white blood cells.

Internal structure of bone

Figure 17.2 shows a longitudinal section through the upper part of a human thigh bone (the **femur**). The shaft looks solid (**compact bone**), but the upper end is composed of **spongy bone**. In life the hollow interior is filled with bone marrow. In the young animal this is red, producing red blood cells; in the adult it is yellow, being a store of fat cells.

Fig. 17.1 Fossil of *Hemicyclaspis*, one of the earliest fish. The front end was covered with a bony armour. There were no jaws. (Ordovician).

Because it is hollow, a limb bone is much lighter than if it were a solid rod. The heavier an animal, the more energy needed to move it. To gain this energy the animal has to eat more food. Thus natural selection will strongly favour lighter animals. But what of the strength of a bony cylinder compared with a bony rod?

Muscles are attached by tendons to the *sides* of limb bones. They therefore exert uneven **stresses** on the bones when they contract. Experiments with rods show that such **strains** produced by uneven stresses fall almost entirely at the edge of a rod, and a rod is just as efficient at resisting the strain if its centre is omitted.

Careful inspection of Fig. 17.2 shows that the spicules of bone are not arranged haphazardly but in definite lines. These are placed in the optimum positions to resist stress, thus combining maximum strength with minimum mass.

Flat bones, such as those of the skull, consist of plates of compact bone with spongy bone between; irregular bones like vertebrae are almost entirely spongy.

Compact bone

Compact bone is not really solid, as Fig. 17.3 shows. It is riddled with fine holes which contain blood vessels and nerves, and arranged in concentric layers round these are the **bone cells.**

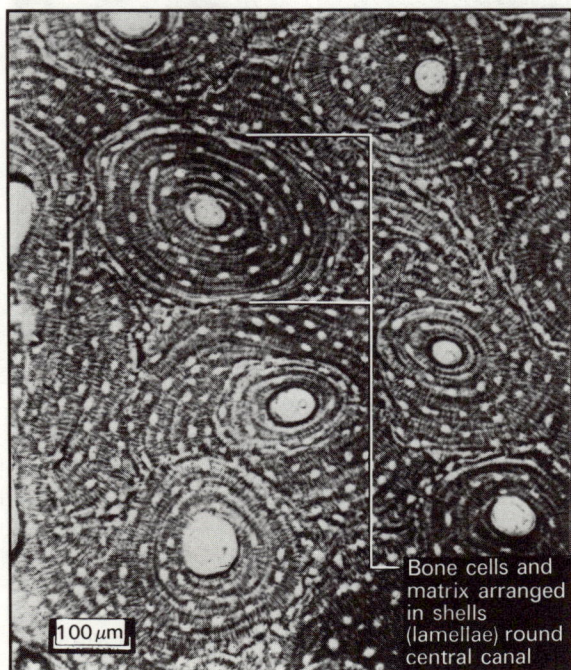

Fig. 17.3 Transverse section of compact bone (light microscope).

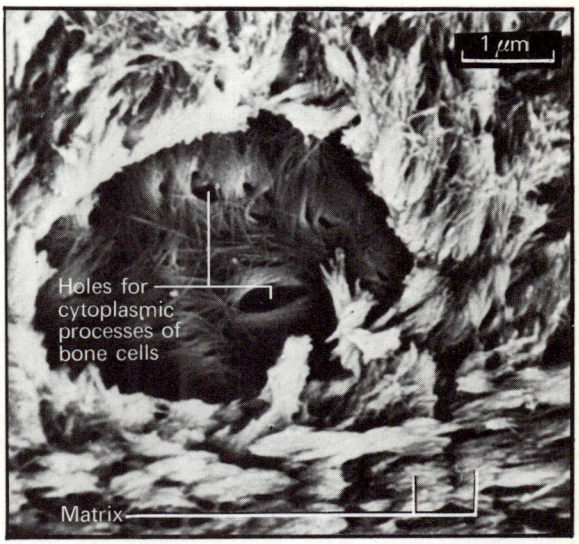

Fig. 17.4 Scanning electron micrograph of the space in the bone matrix in which a bone cell lies. Note the holes leading from the space into which the cytoplasmic processes of the bone cells pass.

Each bone cell lies in a space in the surrounding **matrix** (Fig. 17.4), and extending from it are very fine cytoplasmic processes, each lying in an equally fine canal (Fig. 17.5). These contact processes from other bone cells, all of which are thus in communication. The matrix consists of collagen fibres (see p. 30) impregnated with minute interlocking crystals of calcium phosphate.

Bone is an efficient building material

Bone has a very high breaking strength when compressed, especially when its low density (only twice that of water) is taken into account. In other words, bone is a very **stiff** material.

Bone offers almost as much resistance to stretching as to compression and is difficult to bend or twist. In other words, bone is a very **rigid** material. Stiffness and rigidity come from the intimate association of tough collagen fibres with the stiff, strong, hard, calcium phosphate crystals. Fibres and crystals are disposed in all directions in the matrix, which resembles reinforced concrete.

Bone is also very **tough**. Toughness is the converse of brittleness and means that cracks do not spread. For these reasons, bone is a superior building material to nearly all those employed by man.

Joints

When two bones meet they form a joint. Some of these are moveable (**synovial**) joints (Figs. 17.6,

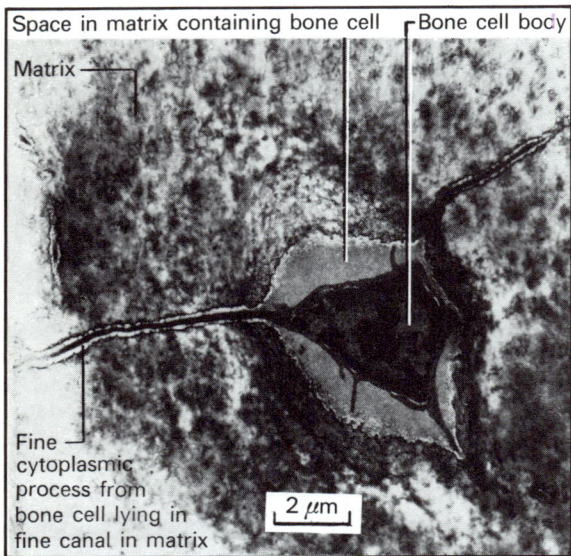

Fine cytoplasmic process from bone cell lying in fine canal in matrix

Fig. 17.5 Section showing a bone cell lying in its space in the bone matrix with radiating cytoplasmic processes (electron micrograph).

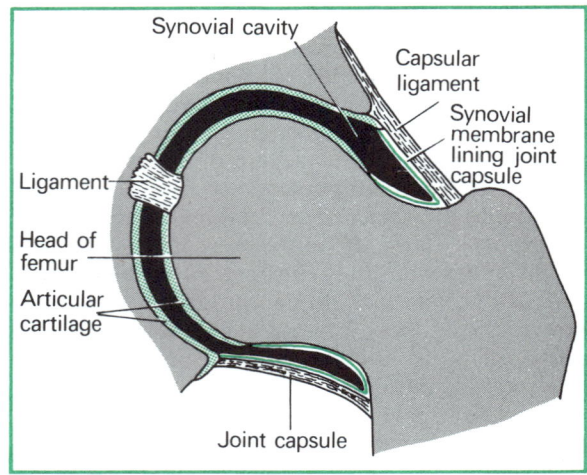

Fig. 17.6 Vertical section through the human hip joint, a synovial joint.

17.21). Here the end of each bone is covered by a layer of gristle (cartilage). These form the rubbing surface. The bones are held together by ligaments—bands of tough collage fibres forming the **joint capsule**. It is lined by a membrane which secretes sticky mucus into the **synovial cavity**. The joint is thus lubricated in much the same way as a mechanical machine is lubricated with oil.

Other joints are not moveable, such as the pubic symphysis (Fig. 12.24) and the joints between the sternum and the bony parts of the ribs (Fig. 7.7). Here a continuous bridge of cartilage joins the two bones.

The human skeleton

All vertebrate skeletons are based on the same basic pattern of a skull, a backbone, two limb girdles, and in amphibians, reptiles, birds, and mammals four pentadactyl limbs (Fig. 2.9). The pattern has become adapted in the course of evolution to many different needs. We will study the human skeleton.

Skull

The skull contains the brain and forms the jaws, embedded in which are the teeth. The mouth cavity is divided into an upper and a lower portion by the hard and the soft palates (Fig. 6.11). Figure 17.7(a) shows the **cranium** surrounding the

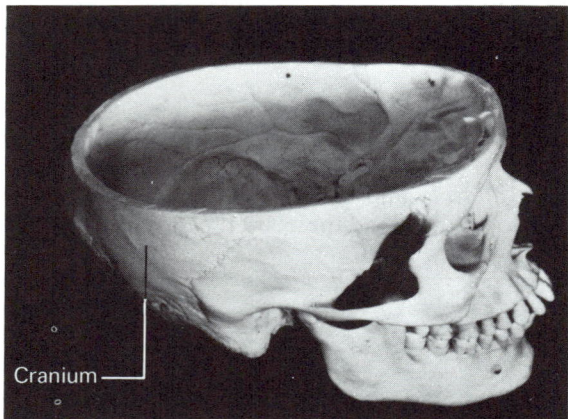

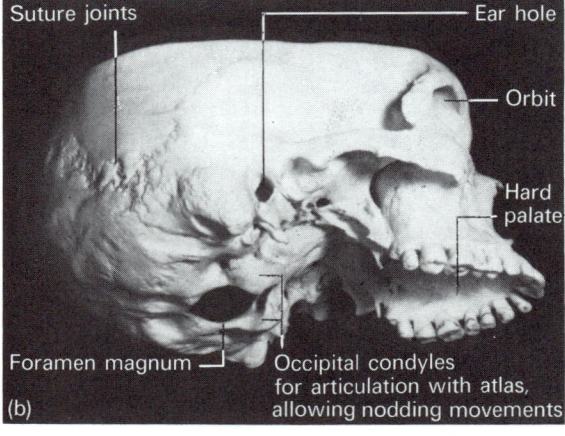

Fig. 17.7(a) Human skull, top half of cranium removed. **(b)** Human skull from below.

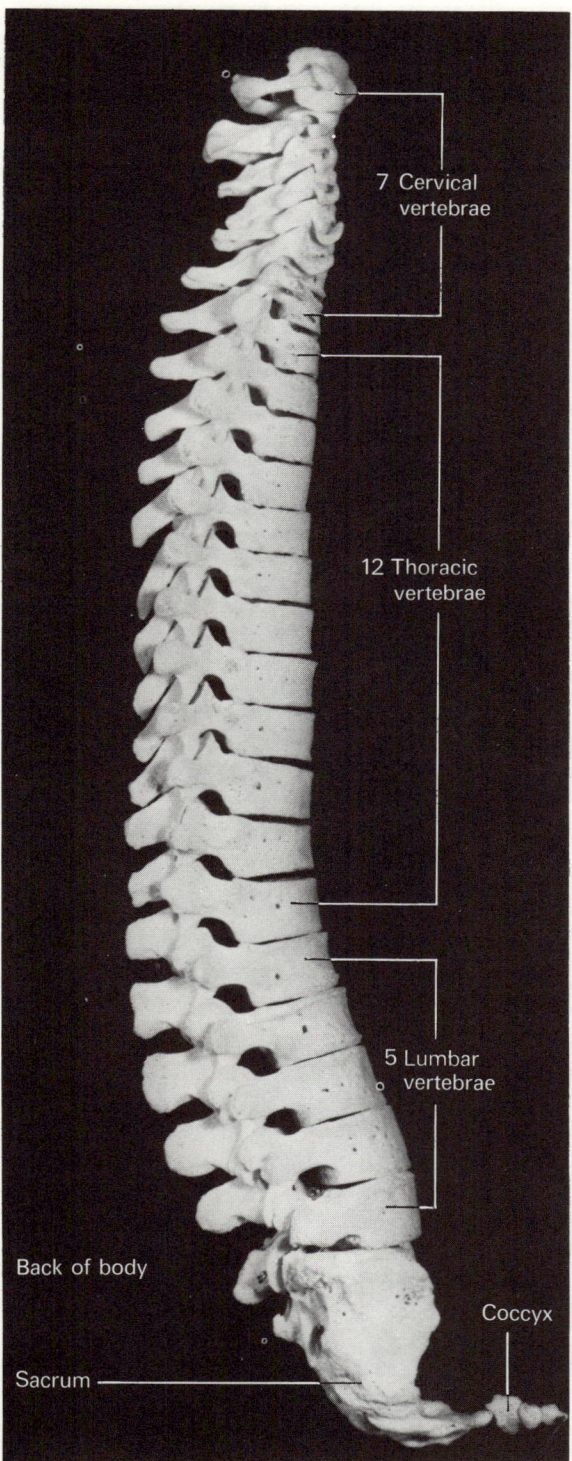

7 Cervical vertebrae

12 Thoracic vertebrae

5 Lumbar vertebrae

Back of body

Coccyx

Sacrum

Fig. 17.8 Human vertebral column from the right side.

brain, the eye sockets (orbits), and the joints between the skull bones.

Figure 17.7(b) has been taken from below. Note the large hole, the foramen magnum, through which the spinal cord passes to join the brain. On either side of it are the **occipital condyles**, two outwardly-curving bony protrusions which rest in the first vertebra forming a hinge joint which permits up and down (nodding) movements of the head.

The backbone (vertebral column)
This gives the vertebrates their name and consists in man of 33 individual vertebrae (Fig. 17.8). Five regions of the backbone can be distinguished, the **cervical**, **thoracic**, **lumbar**, **sacral**, and **caudal** regions, but all the vertebrae are built on the same structural ground plan. There is no abrupt change from one region to the next.

Common functions of the vertebrae
The vertebrae provide: (a) a tough, hard, but *flexible* pillar on which the body is built; (b) a bony canal (the **neural canal**) which houses and protects the spinal cord and spinal nerves.

The pillar is provided by the **centra** of the vertebrae. Figure 17.9 is a photograph of four lumbar vertebrae taken from the right hand side. Successive centra are joined together by the **intervertebral discs**, made of cartilage. Each disc is slightly compressible. They are thick in the cervical (neck) region and in the lumbar (back) region, for it is here that the spine bends. Between the thoracic (chest) vertebrae they are thin, and the five sacral vertebrae develop as a single bone without intervertebral discs.

Adaptations of different vertebrae

Lumbar vertebrae
Lumbar vertebrae provide a good introduction to the basic pattern of the vertebrae. Figure 17.10 shows a human lumbar vertebra from the right side and from above. The neural canal, whose sides and top consist of the **neural arches**, houses the spinal cord which floats in liquid (Fig. 15.13). A pair of spinal nerves emerges, one on each side, from between the **intervertebral notches** of successive vertebrae (Fig. 17.9).

Above the neural arch is the **neural spine**, and protruding from the sides of the arch, like two wings, are the **transverse processes**. These all form anchorages for muscles.

Finally, the vertebrae are prevented from rotating *freely* on each other by the **pre-** and **post-**

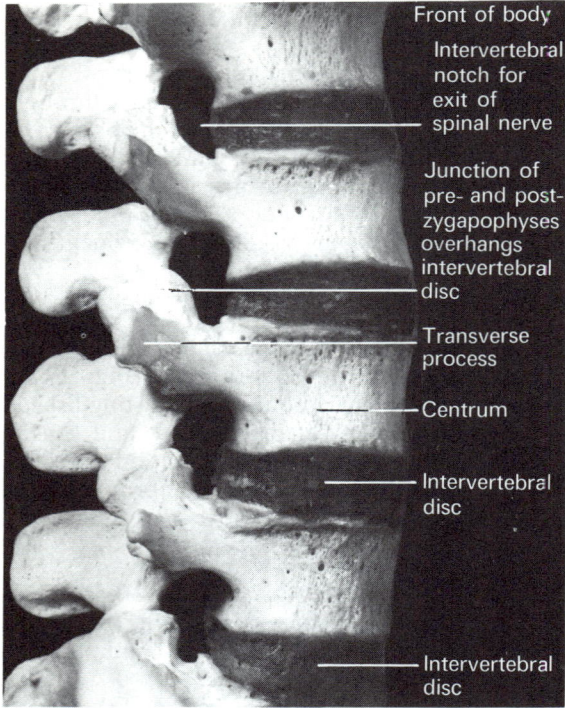

Front of body

Intervertebral notch for exit of spinal nerve

Junction of pre- and post-zygapophyses overhangs intervertebral disc

Transverse process

Centrum

Intervertebral disc

Intervertebral disc

Fig. 17.9 Three human lumbar vertebrae from the right side.

zygapophyses. In the lumbar region these are carried high on the neural arch. The prezygapophyses are curved and face inwards. The postzygapophyses are correspondingly curved and bowed outwards, like segments of a cylinder. The postzygapophyses of one vertebra articulate with the prezygapophyses of the next, making a joint which permits a certain amount of rotation. This should be contrasted with the zygapophyses in the thoracic region (Fig. 17.13) which are not curved but flat, making rotational movement between successive thoracic vertebrae very difficult.

Cervical vertebrae
These are the first seven vertebrae. They show special adaptations, permitting movements of the head forwards, backwards, sideways, and rotationally. These movements are particularly associated with sight.

Atlas The first is the atlas vertebra. Figure 17.11 shows that it is very much modified from the basic pattern. There is no centrum, no neural spine, no transverse processes, no zygapophyses. The atlas consists of little more than a bony ring divided into two compartments.

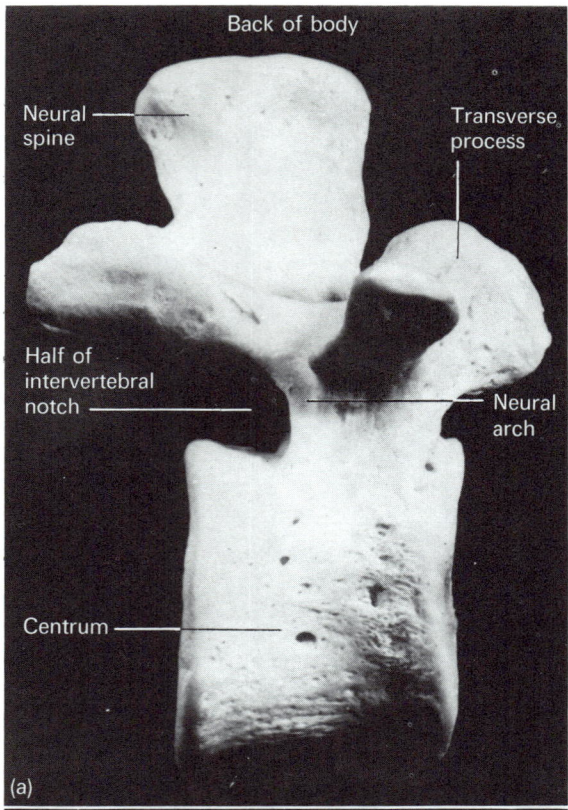

Back of body

Neural spine

Transverse process

Half of intervertebral notch

Neural arch

Centrum

(a)

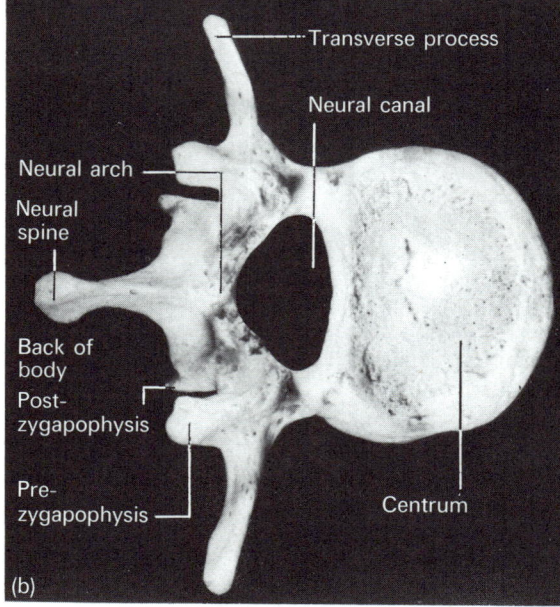

Transverse process

Neural canal

Neural arch

Neural spine

Back of body

Post-zygapophysis

Pre-zygapophysis

Centrum

(b)

Fig. 17.10(a) Human lumbar vertebra from the right side.
(b) Human lumbar vertebra seen from above.

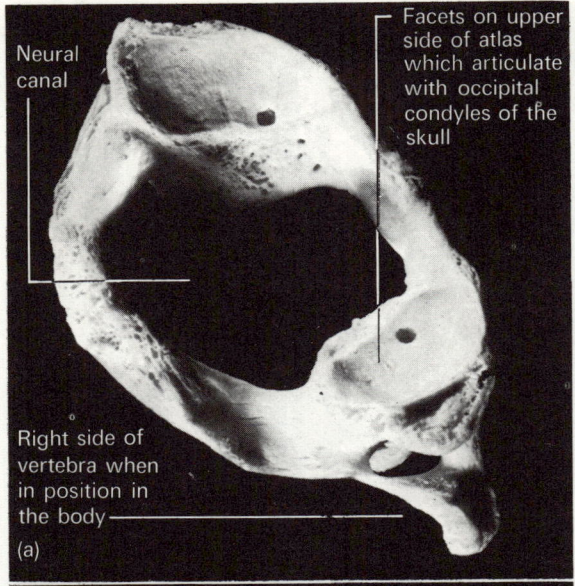

Neural canal

Facets on upper side of atlas which articulate with occipital condyles of the skull

Right side of vertebra when in position in the body

(a)

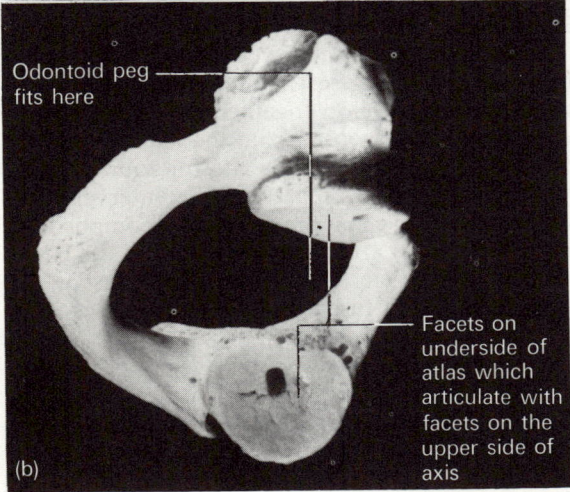

Odontoid peg fits here

Facets on underside of atlas which articulate with facets on the upper side of axis

(b)

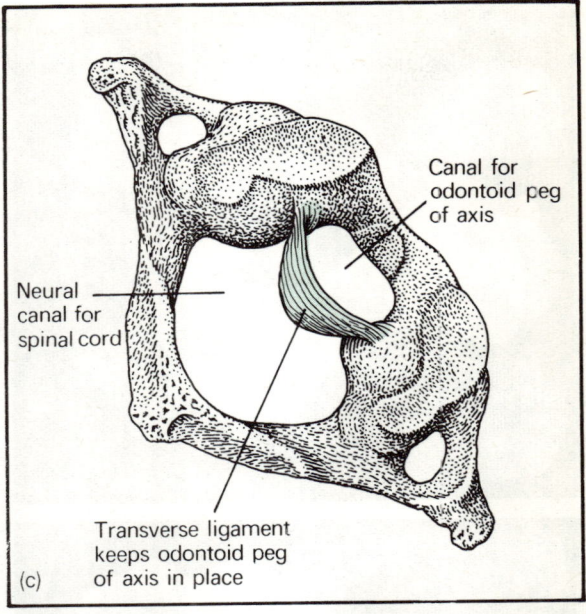

Canal for odontoid peg of axis

Neural canal for spinal cord

(c)

Transverse ligament keeps odontoid peg of axis in place

Fig. 17.11(a) Human atlas vertebra seen from above. **(b)** Human atlas vertebra seen from below. **(c)** Drawing of human atlas vertebra showing the transverse ligament.

Its upper side bears two concavities on which rest the occipital condyles of the skull, enabling movement in one plane only (nodding). Its lower side also bears two concavities. These articulate with the centrum of the second cervical vertebra (the axis). This joint allows very free side to side rotation of the atlas and so of the head.

Axis (Fig. 17.12) The centrum of the axis vertebra bears a forwardly-pointing **odontoid peg** which rests beneath a ligament in part of the neural canal of the atlas (see Fig. 17.11(c)). The joint between the atlas and axis is so free that but for the odontoid peg it might easily be dislocated. The axis has no prezygapophyses but does possess

a pair of postzygapophyses which articulate with the prezygapophyses of the third cervical vertebra. They can be seen in Fig. 17.12(b).

Note in Fig. 17.12 that both pre- and post-zygapophyses are flat and disc-shaped and are situated low on the neural arch. The pre-zygapophyses face upwards and forwards but not inwards. They meet the postzygapophyses which face downwards and backwards, but not inwards. This permits a gliding motion as the neck bends forwards, and also very free rotation.

Thoracic vertebrae

There are twelve thoracic vertebrae. Their special function is to anchor the ribs (Fig. 7.9). They move very little either forwards, backwards, sideways, or rotationally. The intervertebral discs are narrow, the neural spines are long and backwardly-pointing, restricting backwards movement.

Inspection of Fig. 17.13, two thoracic vertebrae photographed from the right side, shows in addition the flat zygapophyses. The prezygapophyses face upwards and forwards, the postzygapophyses face downwards and backwards. They are borne high on the neural arch, and these features all make forwards and backwards movement of the chest skeleton almost impossible.

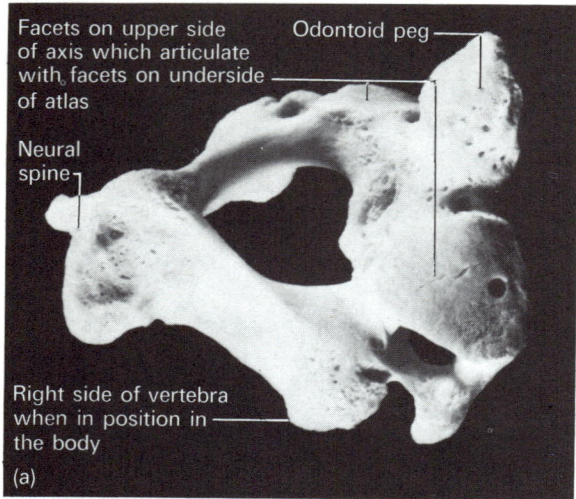

Facets on upper side of axis which articulate with facets on underside of atlas

Odontoid peg

Neural spine

Right side of vertebra when in position in the body

(a)

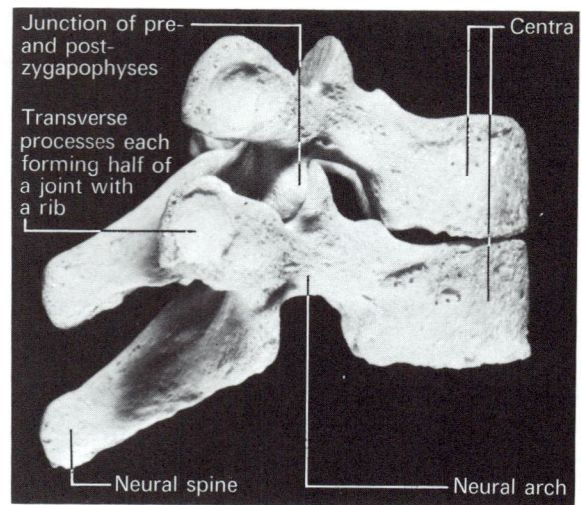

Junction of pre- and post-zygapophyses

Centra

Transverse processes each forming half of a joint with a rib

Neural spine

Neural arch

Fig. 17.13 Two human thoracic vertebrae seen from the right side.

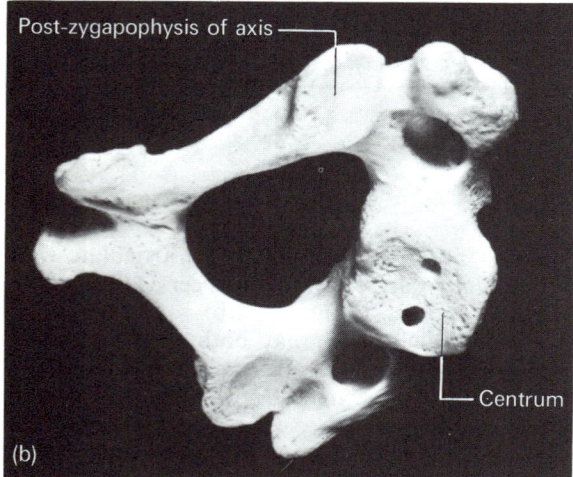

Post-zygapophysis of axis

Centrum

(b)

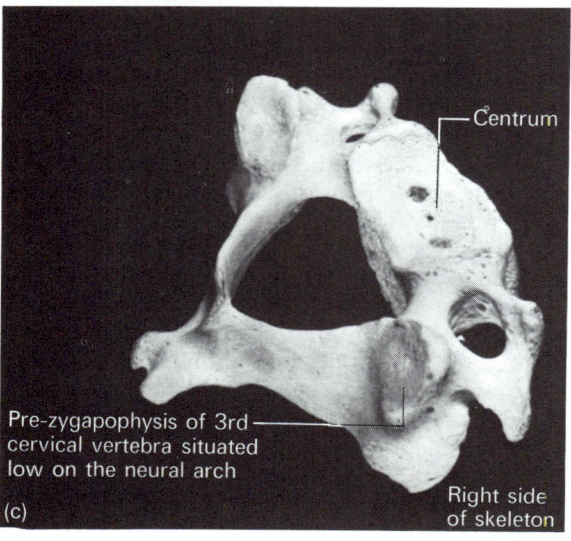

Centrum

Pre-zygapophysis of 3rd cervical vertebra situated low on the neural arch

Right side of skeleton

(c)

Sacrum

Following the five lumbar vertebrae is the **sacrum**. This single bone arises in the embryo from five initials which develop together as a unit. Figure 17.14 shows the sacrum from the front. The four pairs of holes provide the exits for four pairs of spinal nerves. They represent the intervertebral notches and prove that the sacrum is a composite of five vertebrae. Its function is to anchor the backbone to the pelvic (hip) girdle by means of the large, flared, and fused transverse processes (Fig. 17.16).

Man is one of the few vertebrates to possess no tail, though one makes a transient appearance in the embryo (Fig. 12.18). The four small **coccygeal vertebrae** (the **coccyx**) are the only trace of it in the adult.

Pectoral and pelvic girdles

The **pectoral girdle** in man is the shoulder blades (Fig. 17.15). It resembles the **pelvic (hip) girdle** (Fig. 17.16) in that both consist of two mirror-image halves and both provide the socket of the ball-and-socket joint with the limb bone.

They differ in two major ways. First, the shoulder blades are quite separate from one another (Fig. 2.4(b)), whereas the two halves of the pelvic girdle form an immovable joint at the **pubic symphysis**. Secondly, although the shoulder blades form a

Fig. 17.12(a) Human axis vertebra from the right side.
(b) The same from underneath.
(c) The third cervical vertebra from above.

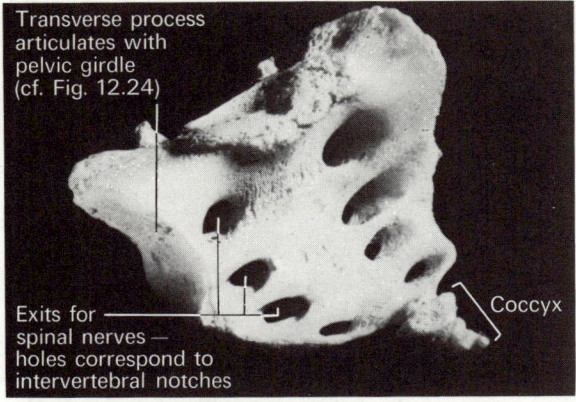

Fig. 17.14 Human sacrum from the front.

joint with the **collar bone** (Fig. 7.7), over nearly all their surfaces they lie embedded in the muscles of the back. Demonstrate the free movements of which they are capable by first making your shoulder blades touch and then moving them round to the front of your body as far as you can. This freedom contributes to the wide-ranging and precise movements of human arms compared with those of the legs. The pelvic girdle, by contrast, is firmly anchored to the base of the spine and is incapable of movement.

The collar bones

These are prominently developed in primates (e.g. monkeys, apes, man) compared with quadrupedal mammals such as cats and dogs, where they are very small. Nearly all primates are tree dwellers and the free arm movements associated with swinging from trees is one of their characteristic features. Many of the arm-moving muscles are attached to the collar bones, and if a man fractures his collar bone the arm on that side cannot be moved.

The limbs

Evolution of the pentadactyl limb

Amphibians, reptiles, birds, and mammals all have **pentadactyl limbs**. The basic pattern of these is: one upper limb bone, two lower limb bones, a set of wrist or ankle bones, a set of palm or foot bones, and a set of finger or toe bones (Fig. 2.9).

Fish, however, were the first vertebrates to evolve. They possess pectoral and pelvic girdles which support fins, but these do not show the pentadactyl arrangement of bones. In the Devonian period 370 million years ago certain fishes evolved lobed fins with narrow bases. It is thought that these were used for locomotion on the land from one pool to another during the prolonged seasonal droughts that are known to have occurred at that time. Most of these fish also possessed lungs. Figure 17.17 is a photograph of a fossil pelvic fin of one of these fishes, and the pentadactyl pattern of bones can be clearly discerned in it.

Having evolved, the pentadactyl limb underwent a great deal of adaptive radiation, producing limbs that are used for a variety of purposes, for

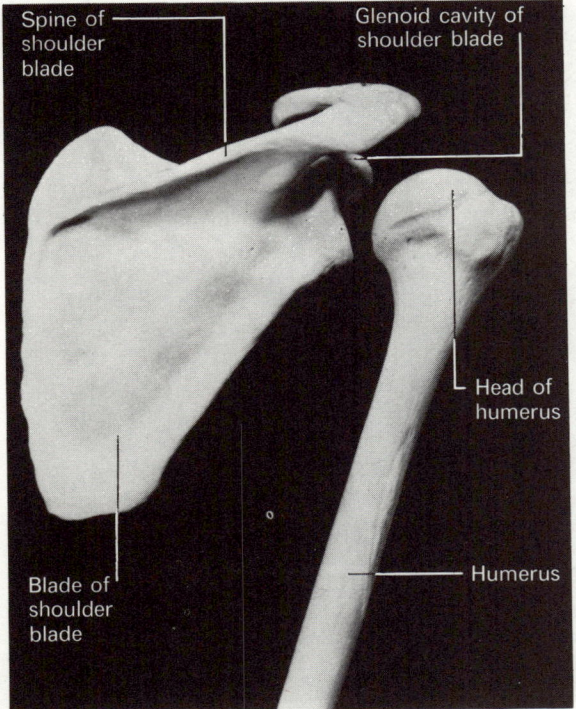

Fig. 17.15 Human right shoulder blade and top of the humerus.

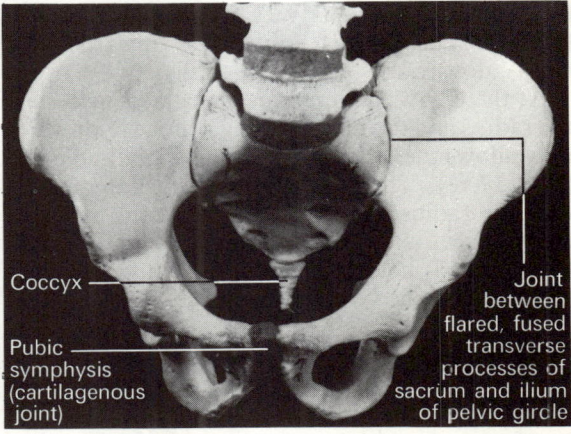

Fig. 17.16 Human pelvic girdle and sacrum, seen from the front.

205

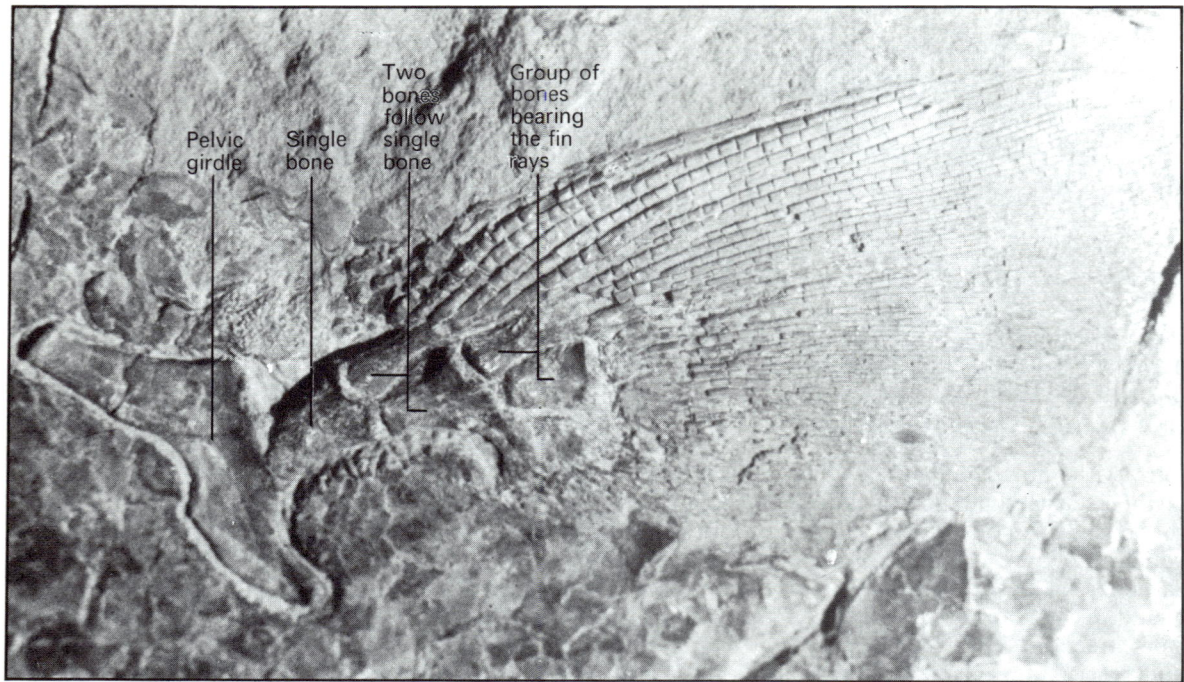

Fig. 17.17 Fossil cast of the hind fin of *Eusthenopteron* (a Devonian air-breathing fish) showing the pentadactyl limb pattern.

example: flying in bats, birds, and the extinct reptile pterodactyls; swimming in whales and porpoises; running in horses and dogs; digging in moles; and the most precise movements in the human arms and hands.

Arm

The upper end of the humerus bears a ball which is large compared with the socket into which it fits (Fig. 17.15). Because of this it is fairly easily dislocated, but it permits very free arm movements forwards, backwards, outwards, inwards, upwards, downwards, and rotationally.

At the elbow the humerus articulates with both the radius and the ulna forming a hinge joint which only allows the forearm to be moved up and down (Fig. 17.18). The wrist bones articulate mainly with the radius, which can be rotated around the ulna (hence its name), carrying the hand with it. The radius is held in place at the elbow end by a ligament attached to the ulna (Fig. 17.19).

Leg

The knee joint between the femur and the tibia (Fig. 17.20) is also a hinge joint. It is held in place by strong cross-ligaments (Fig. 17.21). The ball and socket joint with the pelvic girdle extends over

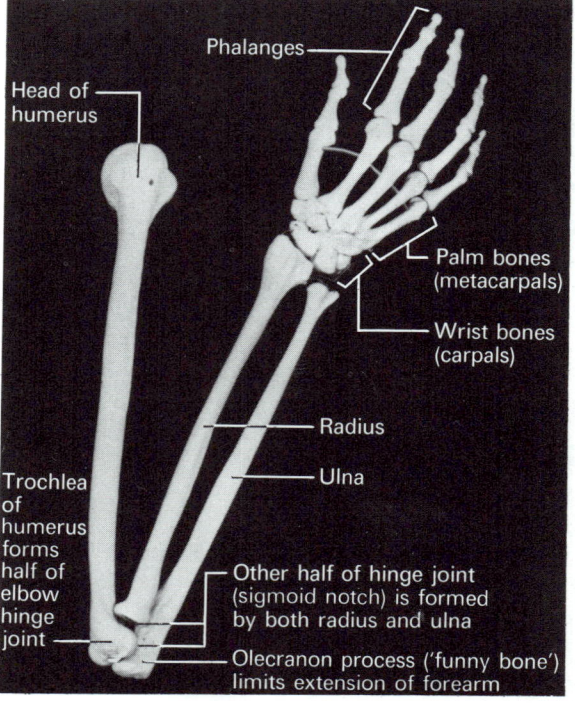

Fig. 17.18 Skeleton of a human right arm and hand from the side.

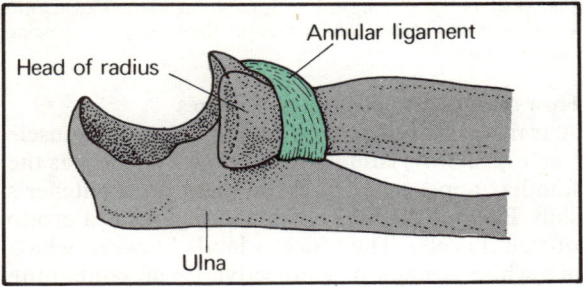

Fig. 17.19 How the annular ligament holds the radius in place and permits its rotation round the ulna.

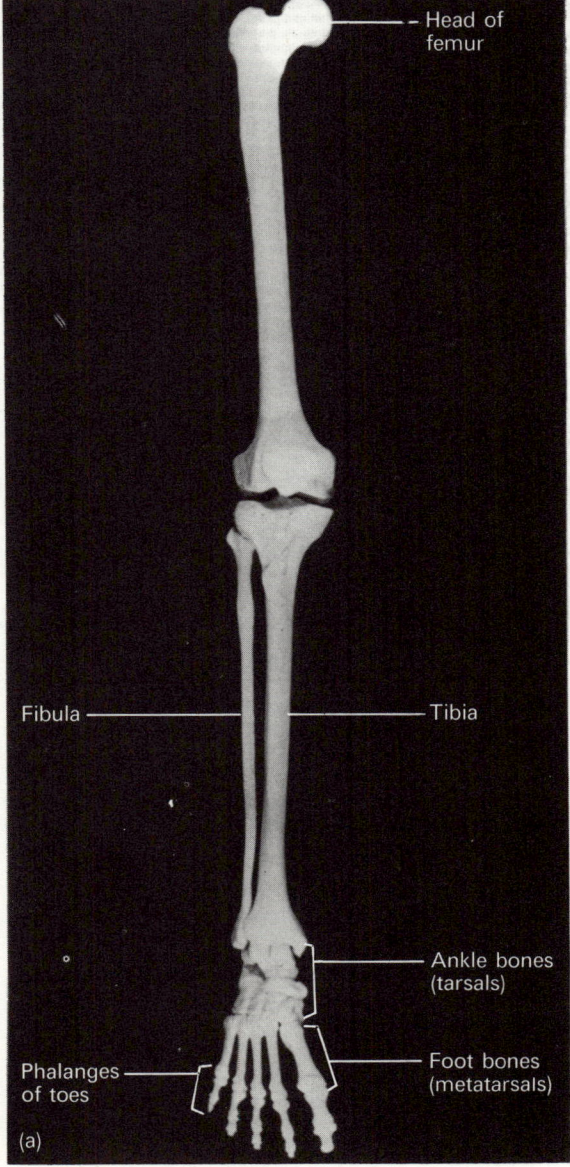

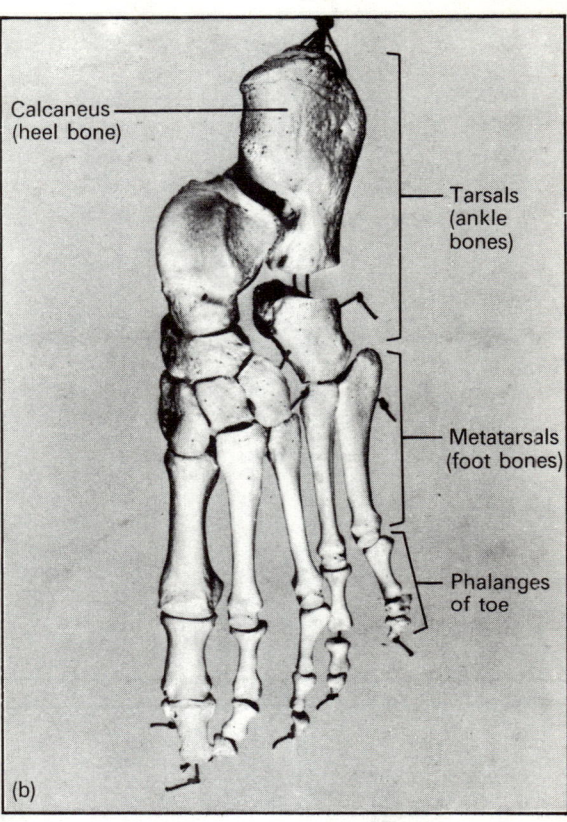

Fig. 17.20(a) Skeleton of human right leg and foot, seen from the front.
(b) Skeleton of the human foot. Is it left or right?

the area of half a sphere, and it is rarely dislocated (Fig. 17.6).

Muscles

On page 56 we encountered the long tapering muscle cells of the gut wall whose contraction brings about peristalsis (Fig. 6.16(b)). The muscles which move the bones have cells very different in appearance from these. They are called **voluntary muscles** because they can be made to contract by an effort of will, and **striped muscles** because of their microscopic appearance.

They, too, are long and thin. One voluntary muscle cell may be several centimetres long and between 10 and 100 μm wide. At their ends they taper (Fig. 17.22). Each muscle cell contains many nuclei, all situated at the edge of the cell. Figure 17.23 is a light microscope photograph of a longitudinal section through parts of three muscle cells. Inside, each is differentiated into alternate dark and light bands, giving the whole cell a striped appearance.

207

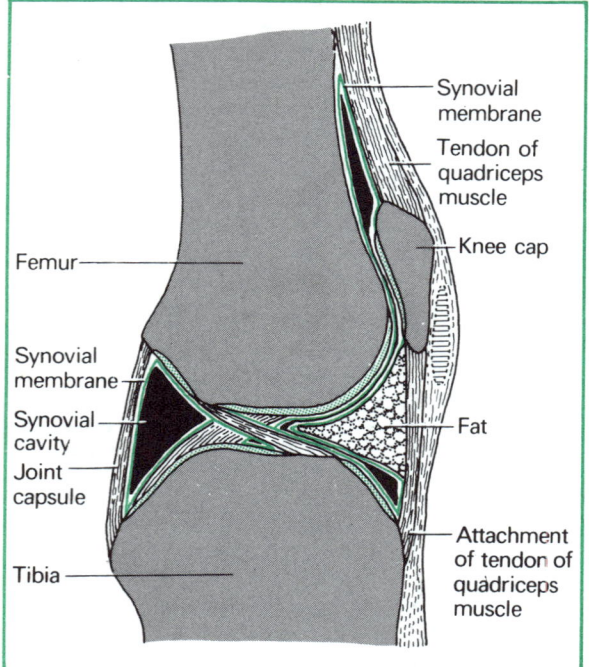

Synovial membrane

Tendon of quadriceps muscle

Femur

Knee cap

Synovial membrane

Synovial cavity

Joint capsule

Fat

Tibia

Attachment of tendon of quadriceps muscle

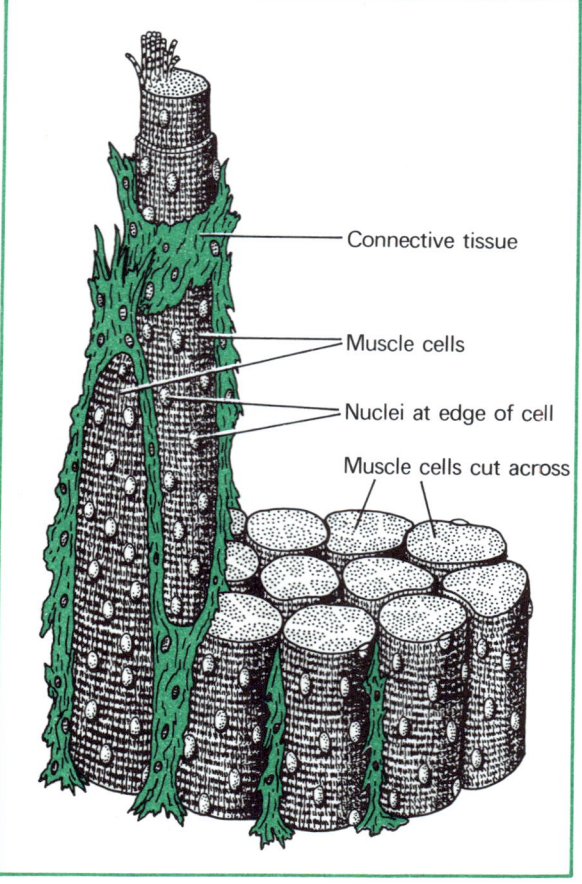

Connective tissue

Muscle cells

Nuclei at edge of cell

Muscle cells cut across

Fig. 17.22 Stereogram of a group of voluntary muscle cells. Note the tapering ends, and the fibrous connective tissue between the cells.

Fig. 17.21 Longitudinal section through the human knee joint showing the ligaments which prevent the tibia from twisting on the femur.

How muscles are connected to bones

A transverse section of a human sartorius muscle (see Fig. 2.3(a)) is shown in Fig. 17.24(a). It has the familiar appearance of sliced meat on a butcher's slab. Each of the darkly-stained regions is a group of muscle cells. They form islands between which are white strands of connective tissue containing many strong collagen fibres as well as small blood vessels and nerves.

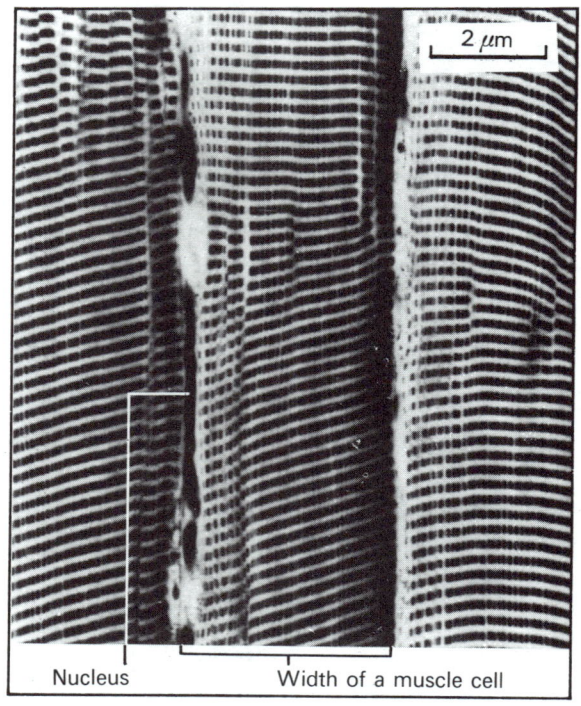

2 μm

Nucleus

Width of a muscle cell

Fig. 17.23 Longitudinal section through parts of three voluntary muscle cells, showing the banded internal structure (light microscope).

These can be seen in more detail in Fig. 17.24(b), which also reveals that the individual muscle cells are surrounded by a basketwork of collagen fibres. All these collagen fibres are inter-connected, and link with the thick sheet of connective tissue on the outside of the muscle.

Muscles are connected to bones by tendons. These thick, hard cords consist almost entirely of collagen fibres. The cells of a muscle peter out in

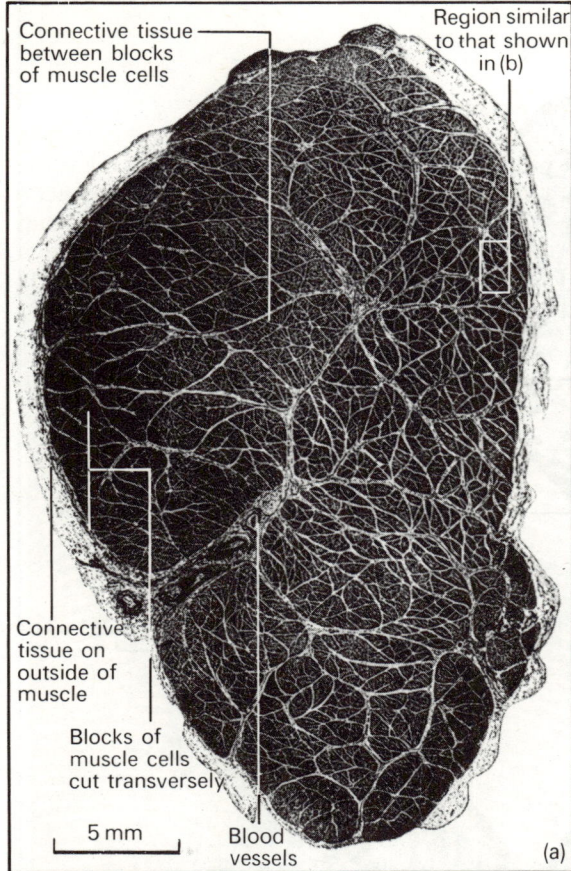

Connective tissue between blocks of muscle cells

Region similar to that shown in (b)

Connective tissue on outside of muscle

Blocks of muscle cells cut transversely

5 mm

Blood vessels

(a)

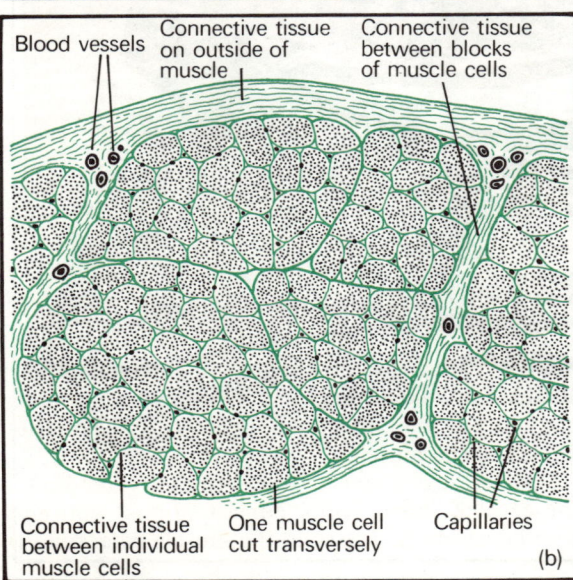

Blood vessels

Connective tissue on outside of muscle

Connective tissue between blocks of muscle cells

Connective tissue between individual muscle cells

One muscle cell cut transversely

Capillaries

(b)

Fig. 17.24(a) Transverse section of human sartorius muscle. **(b)** Detailed diagram of part of (a).

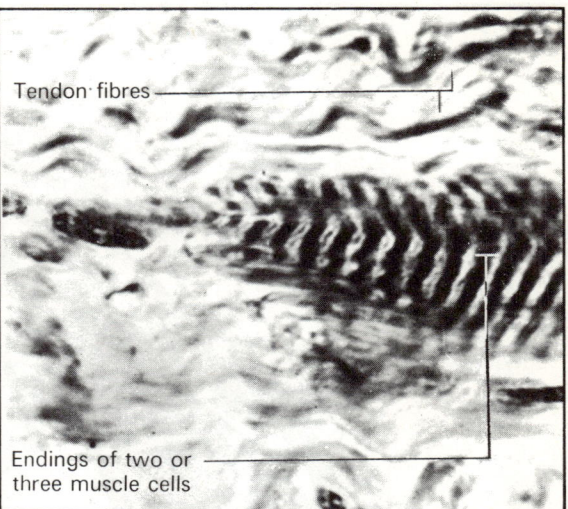

Tendon fibres

Endings of two or three muscle cells

Fig. 17.25 Ending of muscle cells in tendon.

the collagen fibres of its tendon (Fig. 17.25). Figure 17.27 shows some of the tendons which attach the biceps and triceps muscles to the shoulder and arm skeleton.

Tendons do not end blindly on the surface of a bone. Like muscle, bone is surrounded by a jacket of collagen fibres which penetrate into many of the fine holes on its surface. Collagen fibres thus extend from around the muscle cells through the tendon to form a jacket which is firmly anchored to the bone. When the muscle cells contract they pull on the collagen fibres and so on the bone.

Figure 17.26 is a longitudinal section through the finger of a six month human embryo. It shows the developing bones surrounded by their jackets of collagen fibres, together with the tendons that are attached to them.

Antagonistic muscle pairs

The only active movement that a muscle can make is to contract. To regain its former, uncontracted length, it must be stretched.

For this reason, most muscles are arranged in **antagonistic pairs**. Each member of a pair brings about a movement which is directly opposite to that brought about by the other member. An example will make this clear.

Consider two of the muscles which cause the up and down movement at the hinge joint at the elbow (Fig. 17.27). The upper end of the **biceps muscle** is attached by two tendons to the shoulder blade. Its lower end is attached by a single tendon to the upper side of the radius. If the arm is in a

209

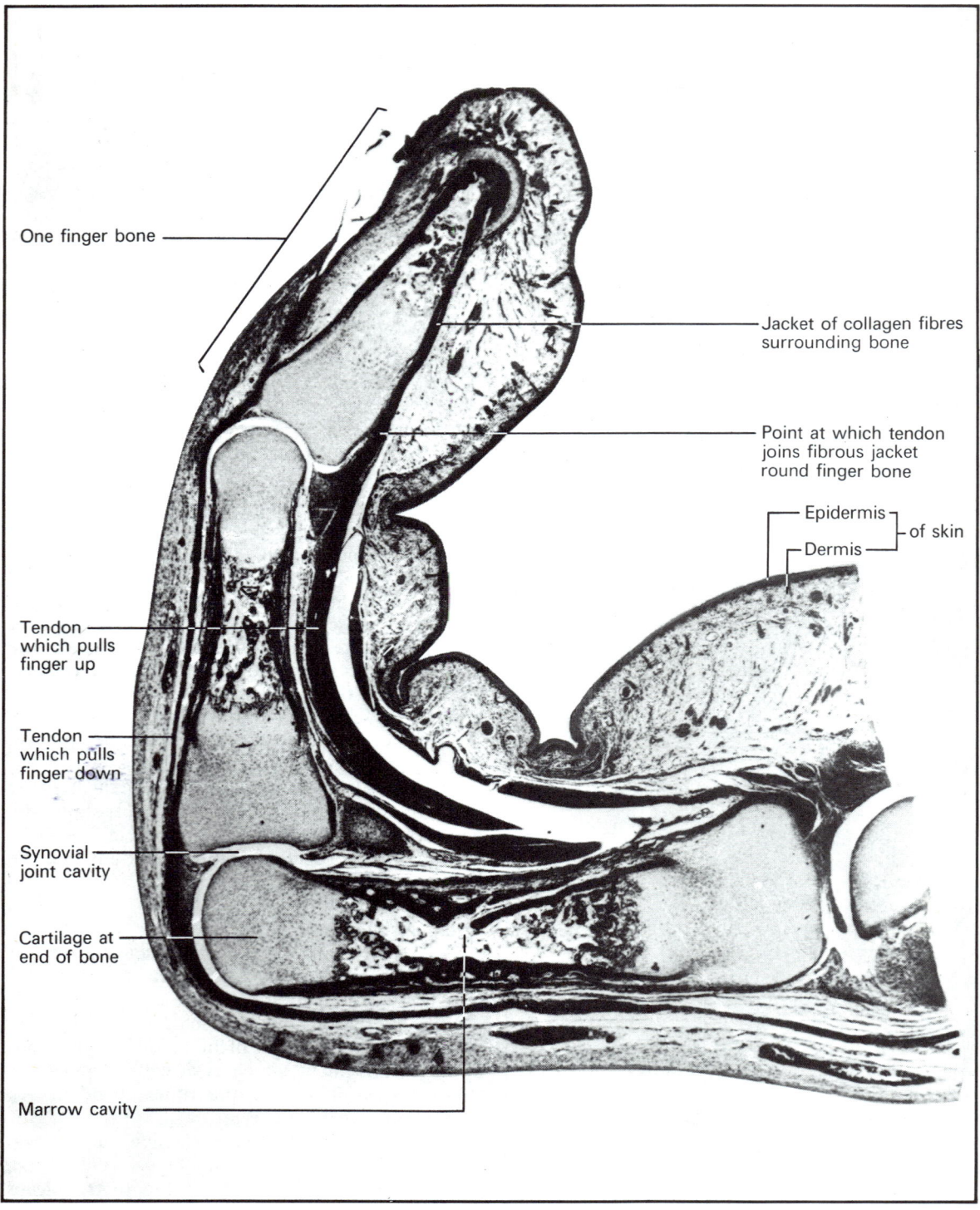

One finger bone

Jacket of collagen fibres surrounding bone

Point at which tendon joins fibrous jacket round finger bone

Epidermis ⎤
              ⎬ of skin
Dermis ⎦

Tendon which pulls finger up

Tendon which pulls finger down

Synovial joint cavity

Cartilage at end of bone

Marrow cavity

Fig. 17.26 Longitudinal section of the finger of a six month human embryo, showing the finger bones and the tendons which move them. Note the synovial joints. (Light microscope.)

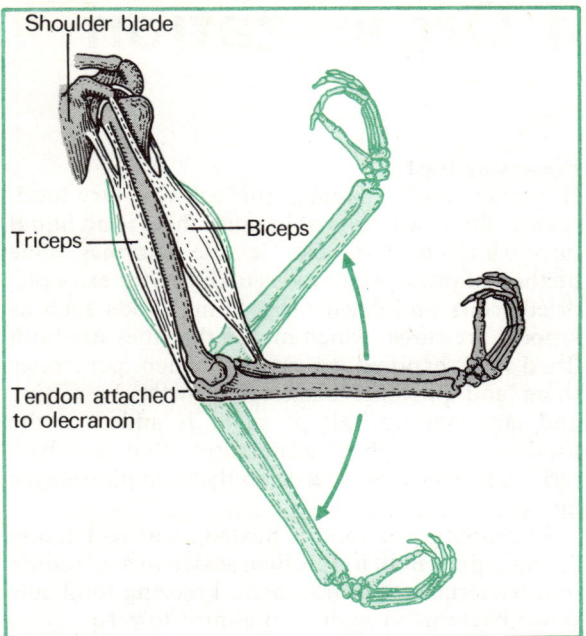

Fig. 17.27 Human biceps and triceps muscles, showing some of the tendons which attach them to the shoulder and arm skeleton.

lowered position, contraction of the biceps therefore causes the forearm to be raised.

The upper end of the **triceps muscle** is attached by two tendons to the humerus and by a third to the shoulder blade (only two of these are shown in Fig. 17.27). At its lower end it is attached by a single tendon to the end of the ulna, which lies at the rear of the hinge joint. If the arm is in a raised position (after the biceps has contracted) and its antagonist, the triceps, contracts, the rear end of the hinge joint is pulled upwards. This pulls the forearm downwards and at the same time stretches the biceps.

Conversely, starting with the arm in the lowered position (after the triceps has contracted), contraction of the biceps both raises the forearm and stretches the triceps. The biceps and triceps produce opposite movements of the forearm and the contraction of one stretches the other.

18 Other Levels of Organization

Over a million separate species of animals and 360 000 separate species of plants are known. Although, for very good reasons, this book has concentrated on the structure and function of mammals and of flowering plants, other levels of organization such as the Prokaryota, non-flowering plants, and invertebrates have been mentioned. This chapter gives details of some of the more important animal and plant groups.

Prokaryota: bacteria

Bacteria range in size from 0.2 μm to 2.0 μm. Two thousand average-sized bacteria laid side by side would fit across the head of a pin. They are the most abundant organisms in the world. One gram of soil contains millions of bacteria. Being prokaryotes, they possess no cell organelles (Fig. 1.17(a)).

Bacterial nutrition

A few species contain chlorophyll and photosynthesize. Both the chlorophyll they contain and the mechanisms of photosynthesis they use are different from those found in plants. It is probable that bacterial photosynthesis is a relic of ancient, less successful photosynthetic mechanisms in much the same way that the fossils of extinct animals and plants are relics of earlier, less successful organisms.

The vast majority of bacteria, however, are not holophytic but live on the life chemicals, such as proteins, fats, and carbohydrates, contained in the bodies of dead animals and plants. They are **saprophytes** and engage in **saprophytic** nutrition. This means that they **secrete** enzymes on to their food. The digested products diffuse back into the bacterial cell. This process rots and degrades the food substance, and is an essential feature of the living world. It returns carbon to the air as carbon dioxide and nitrogen to the soil solution as ammonium salts. If it did not take place, carbon and nitrogen, two elements which occur in almost every chemical substance in the cell, and particularly in proteins, would be locked up in dead bodies and life would come to a standstill for want of raw materials. When bacteria are prevented from rotting plant material, as in the acid conditions in peat bogs where bacteria cannot grow, plant remains do not rot but accumulate as thick layers of peat.

Preserving food

If human food is damp it too will rot. Dry food, such as flour, will not rot because there is no liquid into which bacteria can secrete enzymes. One method of preserving food is to dry it (for example, dried fruits and dried fish). Some foods such as kippers are **cured**, which means that they are both dried and exposed to smoke which penetrates them and prevents bacterial growth. In salting and jam making, salt or sugar is added to the food in such high concentrations that any bacterial cells which may settle on the food plasmolyze and die.

In canning, the food is heated so as to kill any bacteria present in it. It is then sealed in a container in a bacterial-free atmosphere. Freezing food cuts down bacterial growth on it almost to zero.

Bacteria in the air: spontaneous generation

Huge numbers of bacteria float in the air. They cannot be seen with the naked eye, but when they settle on organic material which is suitably damp they grow, causing the material to rot.

An Italian called Spallanzani attempted to demonstrate this in the eighteenth century. He took various nutrient liquids such as water in which maize or barley had been boiled and which he knew would develop large numbers of microbes if left exposed to the air. He heated the liquids in *sealed flasks* in boiling water for an hour. This, he argued, would kill any microbes that might already be present in them. Several days later he opened the flasks and found no microbes in the liquids.

His conclusion was that microbes had been present originally but that they had been killed by the heat treatment. Since the flasks were sealed, no more could settle into the liquids, so no microbes developed. This explanation was not, however, accepted by his contemporaries.

There was at the time a strongly held belief in **spontaneous generation**, the idea that microbes arise in dead food by its reacting with 'pneuma', the supposed life-spirit present in air. Since the solutions in Spallanzani's flasks had not been exposed to the air, the correct explanation of his results was that pneuma had been excluded, thus preventing the generation of new living microbes. Any pneuma left in the flasks after they had been sealed was destroyed by the heat treatment. No doubt Spallanzani found this interpretation irritating, but he was unable to disprove it.

Pasteur's experiments on spontaneous generation

About 100 years later in 1861, Pasteur carried out experiments the results of which convinced him and, eventually, all scientists, that spontaneous generation does not occur in the world today. He started with the firm belief that it did not and designed his experiments accordingly.

He had to find a means of boiling nutrient broth and then exposing it to air while at the same time preventing invisible air-borne microbes from settling into it. The apparatus he designed, his celebrated swan-necked flask (Fig. 18.1), was merely a piece of bent glass. He filled a flask whose neck had been drawn out into a fine glass tube with nutrient broth and then bent the neck into an S shape. After prolonged boiling to kill any bacteria and other microbes which might be present in the broth, he allowed the apparatus to cool slowly.

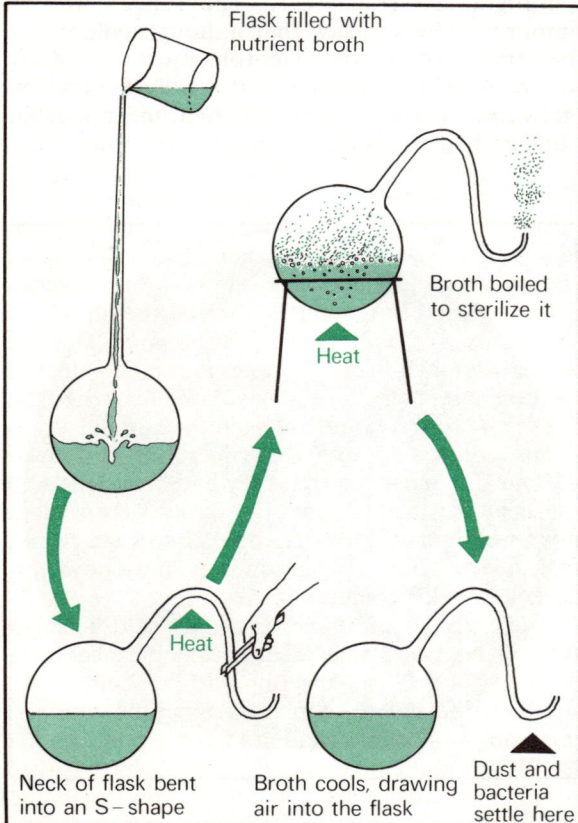

Flask filled with nutrient broth

Broth boiled to sterilize it

Heat

Neck of flask bent into an S – shape

Heat

Broth cools, drawing air into the flask

Dust and bacteria settle here

Fig. 18.1 Setting up Pasteur's swan-necked flask.

As the heated air in the flask cooled, air from outside it was drawn in. There was no question of pneuma being excluded from the flask. Pasteur found that the broth did not ferment even when kept in this way for weeks. When the flask was opened by removing the swan neck, the broth contained no bacteria or other microbes. However, it now quickly started to ferment and an abundance of microbes appeared.

Pasteur concluded that since air had been drawn into the flask slowly, any bacteria it may have contained must have settled in the dip of the swan neck. The neck acted as a bacteria trap, preventing the entry of microbes into the sterile contents of the flask.

These experiments and countless others like them since have been taken as **negative** proof that spontaneous generation does not occur. Even if it did, the task of observing it in nature would be impossible. Nobody has succeeded in showing that it does occur in a laboratory, where conditions can be controlled. Nevertheless, spontaneous generation must have happened at some time in the remote past for life to have appeared on earth at all.

Antiseptic surgery

Pasteur's work had immense significance for the welfare of mankind. If bacteria were the cause of the rotting of meat, was it not likely that they were also the cause of **sepsis**, the rotting of human flesh that commonly develops from an untreated wound? Sepsis killed nearly all the patients who underwent surgery in the eighteenth and early nineteenth centuries.

When Joseph Lister, an Edinburgh surgeon, heard of Pasteur's 1865 experiments, he became convinced that bacteria from the air are the cause of sepsis, and he spent many years developing **antiseptic** surgical techniques. Surgeons came to wear clean clothing at each operation, which was conducted in a specially cleaned room with instruments that had been boiled. Carbolic acid (antiseptic) was sprayed on to the wound to prevent bacteria from growing on it. This had limited success as carbolic acid is a corrosive substance which kills human tissues as effectively as it does bacteria. But it prompted the search for improved antiseptics and other drugs such as **antibiotics** that kill bacteria but not human cells. Today death from sepsis following a surgical operation is very rare.

Another field where antiseptic procedures had spectacular success was childbirth. This had always been extremely hazardous. A woman faced likely death from **puerperal fever** that she commonly developed after childbirth. Puerperal fever was found to be caused by a bacterial infection of the area of the womb newly exposed by the detachment of the afterbirth (Fig. 12.27). Puerperal fever is now unknown in civilized countries.

Bacteria and disease

It is worth emphasising that the majority of bacterial species are of benefit to man because they bring about the rotting processes that return carbon and nitrogen to the soil. However, a few species are parasites of living animals, causing disease. Not all diseases are caused by bacteria. But examples of diseases that are, and which may be fatal to humans, are diphtheria, tuberculosis, pneumonia, and syphilis. Death is caused partly by the rotting of human tissues and partly by the release of poisonous waste products from the bacteria into the bloodstream.

It was Pasteur who first demonstrated that a disease is caused by bacteria. In 1866 he was asked to investigate, and if possible cure, a disease of silkworms that had nearly put an end to the silk industry in France. The silk 'worm' is in fact a moth with four stages in its life cycle, zygote, larva, pupa (which occupies the silk cocoon) and imago (adult).

Examination of ground-up material from all stages of the life cycle of diseased moths showed the presence of bacteria. They were absent from the tissues of healthy moths. When bacteria from diseased moths were injected into healthy moths, they rapidly developed silk worm disease and died. Pasteur was able to develop antiseptic procedures for the raising of silk worms which saved the silk industry from ruin and made him into a public hero.

Curing diseases with drugs

In 1901 a German, Erlich, devised a drug, salvarsan, which killed the bacterium which causes syphilis but did not harm human cells. In 1937 Domagk produced a dye called prontosil which cured mice of bacterial diseases. Large numbers of derivatives of prontosil followed. They were called the sulphonamides and were the first cheap, easily-administered drugs which killed bacteria but not human cells. They formed the chief weapon against disease in the second world war.

The search for these drugs was not haphazard. As more became known about the chemical workings of cells, differences were discovered between processes in bacterial (prokaryotic) cells and mammalian (eukaryotic) cells. Attention was focussed on disrupting cell processes that were peculiar to bacteria, leaving human cells unharmed.

Penicillin: the discovery of antibiotics

The most successful attack on disease has come from drugs produced not by chemists but by bacteria and fungi themselves. These are called antibiotics. They are secreted by soil microbes. Their function in nature is to create an area round the organism in which other microbes cannot grow, thus reducing the competition for nutrients.

Over 500 antibiotics are known, but only a few are sufficiently non-poisonous to humans to be of use as drugs. Those produced by bacteria are particularly poisonous and the successful antibiotics have all been isolated from fungi. The first was penicillin, produced by the common green fungus *Penicillium spp.* which grows on jam, fruit, and other organic materials. Rubbing this fungus on to a wound had been a folk remedy to prevent sepsis since time immemorial.

The great discovery was made by Sir Alexander Fleming in 1928, working in St. Mary's Hospital, London. He was investigating bacteria which cause boils and noticed that a colony of *Penicillium* had grown on one of his experimental dishes, which he photographed (Fig. 18.2). The fungus was an impurity. Fleming saw that although colonies of bacteria were growing vigorously on the side of the dish farthest from the *Penicillium*, very few grew near it. He thought this might mean that the fungus had produced a substance which was

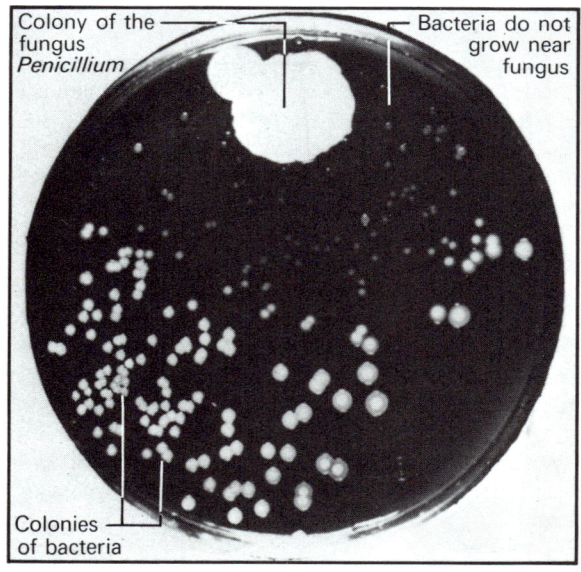

Colony of the fungus *Penicillium*

Bacteria do not grow near fungus

Colonies of bacteria

Fig. 18.2 The original culture of *Penicillium* which Fleming made in 1928.

poisonous to the bacteria; this had diffused into the agar jelly in the dish and inhibited bacterial growth.

It is often said that Fleming's discovery was made by chance, and it was chance that the fungus happened to settle on his experimental dish. But Pasteur said that 'chance favours the prepared mind'. We owe penicillin to Fleming's intuition that it was worth while putting his planned experiments aside for a while in order to follow up a clue which might have led nowhere.

Like many other scientific discoveries, the importance of this one was not recognised at the time. It was not until 1938 that Florey and Chain, working at Oxford, read Fleming's paper and decided to repeat his work. A crucial experiment performed by Chain was to take two sets of diseased mice that had been injected with bacteria and to inject one set with nutrient broth on which *Penicillium* had grown. All these recovered, while all the untreated mice died.

A great deal of work was carried out on penicillin. Industrial methods for its efficient production in large quantities were devised; it was isolated in a pure crystalline form; and eventually its complicated chemical structure was worked out. In the 1950s its mode of action was discovered. The cells of those bacteria it kills lie inside capsules made of a cellulose-like material. Without its capsule, the bacterial cell dies. Penicillin interferes with the manufacture of the capsule, and since human cells do not make this substance, penicillin is without effect on them.

Industrial fermentations

We saw on page 68 that in anaerobic respiration, glucose is not converted into carbon dioxide and water but into two 3-carbon molecules instead. When yeast respires anaerobically, the product is ethanol (alcohol), and this fact is the basis of the brewing and wine-making industries.

Some bacteria also grow in the absence of oxygen, producing alcohol and other substances, for example acetic acid (vinegar). Many organic chemicals are made industrially by bacterial fermentations.

Nature has utilised bacterial activities in the digestive tracts of many animals. For example, the stomach of a cow contains a rich bacterial flora which secretes the enzyme cellulase. This converts the otherwise useless cellulose in the cow's food into 3-carbon chemicals that form the greater part of the cow's diet (see page 63).

Eukaryota

Protista

These are minute organisms which many scientists regard as being unicellular. But each one is a complete, independent, free-living individual, unlike mammalian or higher plant cells. The Protista are therefore sometimes described as being **acellular**, that is without cells. The distinction is a man-made one and is only useful in so far as it makes us think critically about what we mean by the term cell.

Those protists that feed holozoically can be called animals. They are the **Protozoa**. Those protists that feed holophytically can be called plants, and form the **Protophyta**. Some, however, such as the group containing *Euglena* do not fit easily into either the animal or the plant category. The Euglenoids include closely-similar species, some of which ingest food and some of which do not; on the other hand similar species contain chloroplasts while others do not. Again the classification of organisms into plants and animals is man-made, and the important thing is to understand the reasons for making the distinction.

The Protista are distributed throughout the world. They are most abundant at the surface of sea. Here they form the **plankton** which, through food webs, is the ultimate source of food for all marine animals.

A few protozoans cause serious human diseases, for example amoebic dysentry (caused by *Entamoeba histolytica*) and malaria (caused by *Plasmodium sp.*).

We shall examine two protists, *Euglena* and *Amoeba*.

The Euglenoids

Most species live in fresh-water ponds, ditches, puddles and streams—and particularly in water which contains decaying vegetable matter. They cause green pools in farmyards and water butts.

Structure

Very little detail can be seen under the school microscope, where *Euglena* appears as a minute green speck in a state of continual jerky movement. High power light microscopy (Fig. 18.3(a)) shows a nucleus, a **contractile vacuole**, starch grains, mitochondria, and eye spot, a **flagellum**, and a **flagellar swelling**. This specimen had no chloroplasts. Figure 18.3(b) is a drawing of a green specimen and shows in addition the plate-like chloroplasts.

These organelles can be seen more clearly in

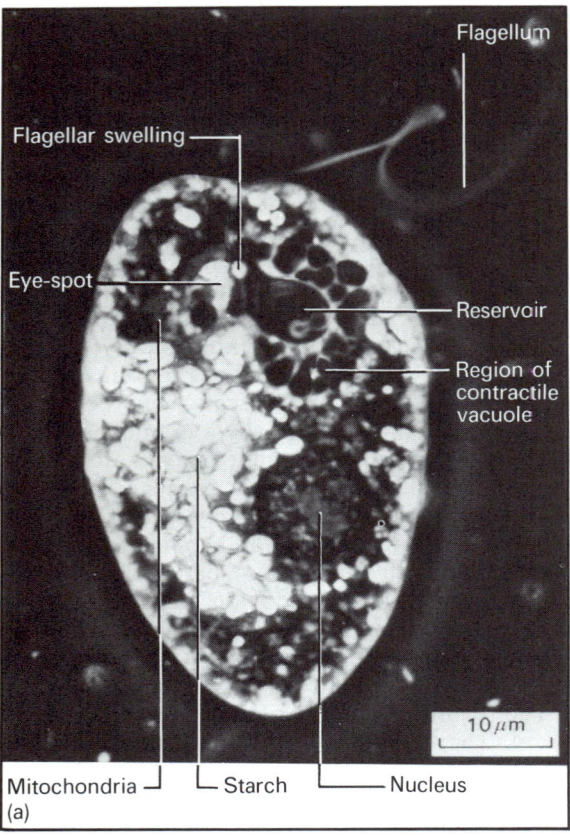

Flagellum

Flagellar swelling

Eye-spot

Reservoir

Region of contractile vacuole

10 μm

Mitochondria — ⌐ Starch — ⌐ Nucleus

(a)

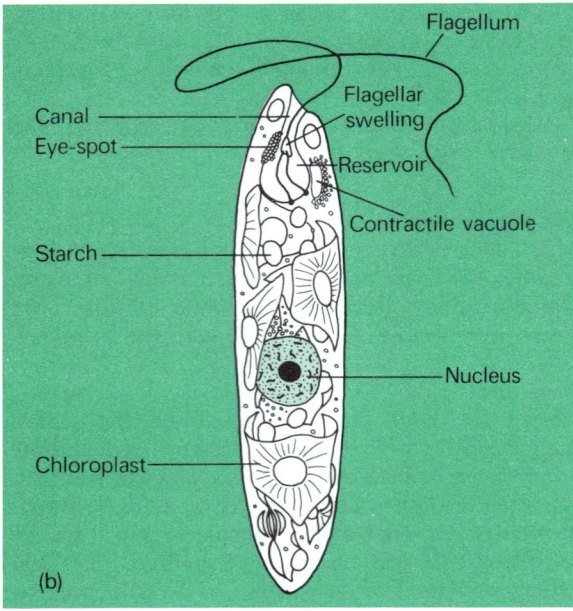

Flagellum

Canal

Eye-spot

Flagellar swelling

Reservoir

Contractile vacuole

Starch

Nucleus

Chloroplast

(b)

Fig. 18.3(a) *Euglena gracilis.* A living (unstained) cell of a strain which lacks chloroplasts. (light microscope).
(b) *Euglena gracilis* (drawing).

electron micrographs (Fig. 18.4). These reveal a highly-differentiated eukaryotic cell.

Function

Contractile vacuole Mitochondria are clustered round the contractile vacuole and probably provide the chemical energy needed for it to work. *Euglena* is like a concentrated solution separated from a weaker solution (the pond water) by its plasma membrane, which is partially permeable. It therefore draws water into itself by osmosis (see page 85), and if this were not expelled the cell would burst. To prevent this, water is secreted into the contractile vacuole which gradually swells. It then discharges into the **reservoir**. The contractile vacuole forms, discharges, re-forms, and discharges again at regular intervals of a few minutes. Figure 18.4(b) shows three Golgi apparatuses in this region and they probably also play a part in the secretion of water.

Movement The cell is pulled through the water by the lashing flagellum. At one side of the neck of the reservoir is the orange eye-spot (Fig. 18.4(a)). It lies directly opposite a light-sensitive swelling on the base of the flagellum. As the cell moves it rotates, and the eye-spot shades the light-sensitive swelling from one-sided light.

The whole cell moves in response to a one-sided light stimulus. It swims away from bright light and from darkness towards dim light. It is likely that the eye-spot and the light-sensitive swelling detect the direction and the strength of light, though how this is translated into a message that can cause appropriate movement of the flagellum is not known.

Nutrition Some species possess chloroplasts and photosynthesize. The chloroplasts are placed at the edge of the cell with their grana parallel to it. This is the optimum position for efficient photosynthesis (compare with Fig. 5.7). Some green species can grow in the dark by absorbing dissolved organic materials from the water, and this is the normal mode of nutrition for many of the colourless species.

Yet other species ingest (take into their bodies) solid food particles of dead material or living prey.

Reproduction The euglenoids reproduce only by mitosis and cell division (Fig. 18.5). All species are therefore clones. As far as is known, reduction nuclear division (meiosis), gamete formation, and fertilization, do not occur. Apart from mutation,

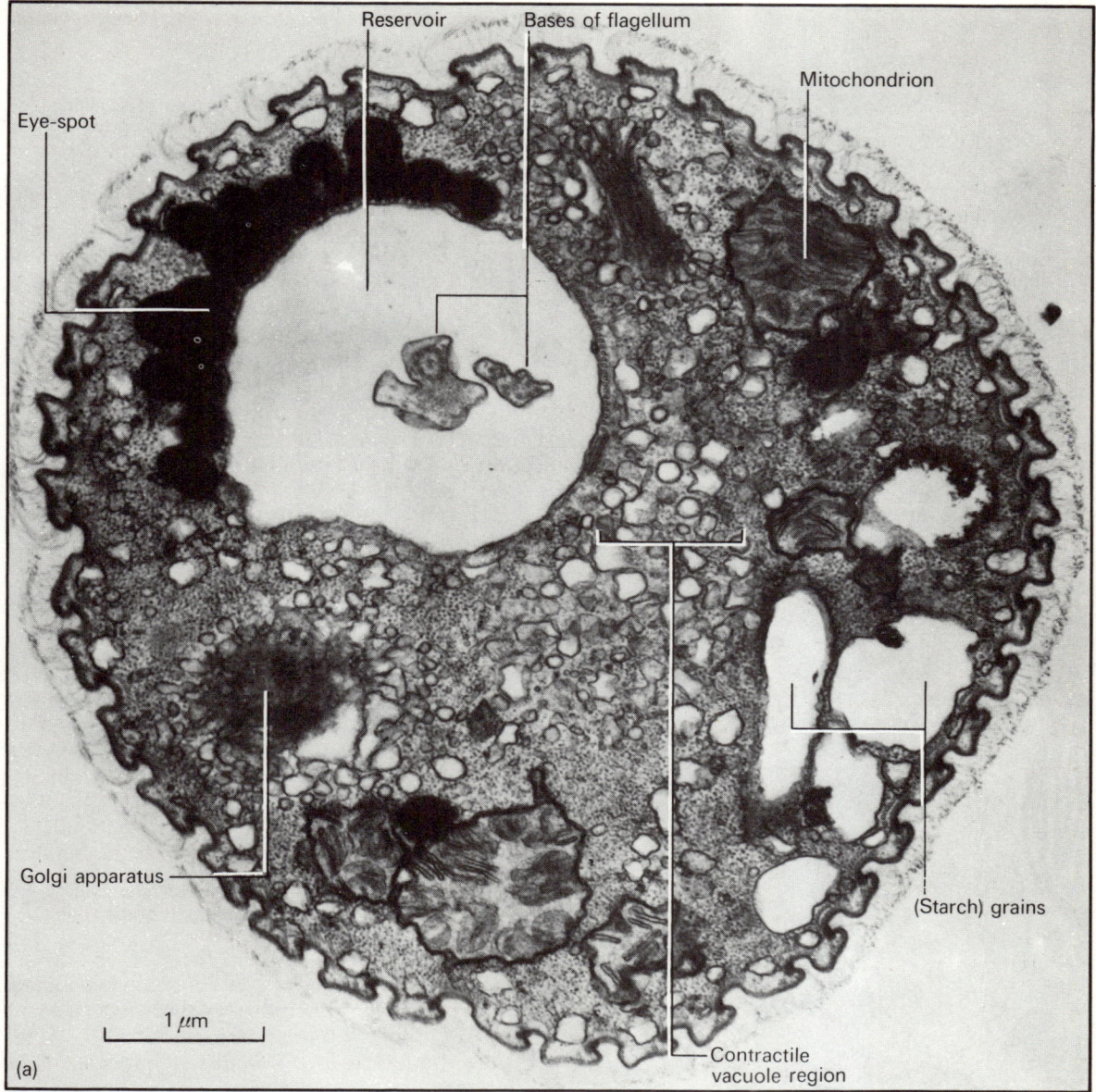

Fig. 18.4(a) Transverse section of front end of *Euglena viridis*, showing most of the features of a eukaryotic cell. The section does not pass through the nucleus or the chloroplasts. (Electron micrograph.)

therefore, the Euglenoids do not possess genetically-controlled variation.

What is the status of the Euglenoids?
1 Make a list of the features of the Euglenoids that justifies (a) the view that they are unicellular organisms; (b) the view that they are acellular organisms.
2 Make a list of the features of the Euglenoids

that justifies them being regarded as (a) animals; (b) plants; (c) neither.
3 Make a list of the features of the Euglenoids that justifies the view (a) that they are primitive organisms; (b) that they are not primitive organisms.

Amoeba
There are many species of *Amoeba*. They are cer-

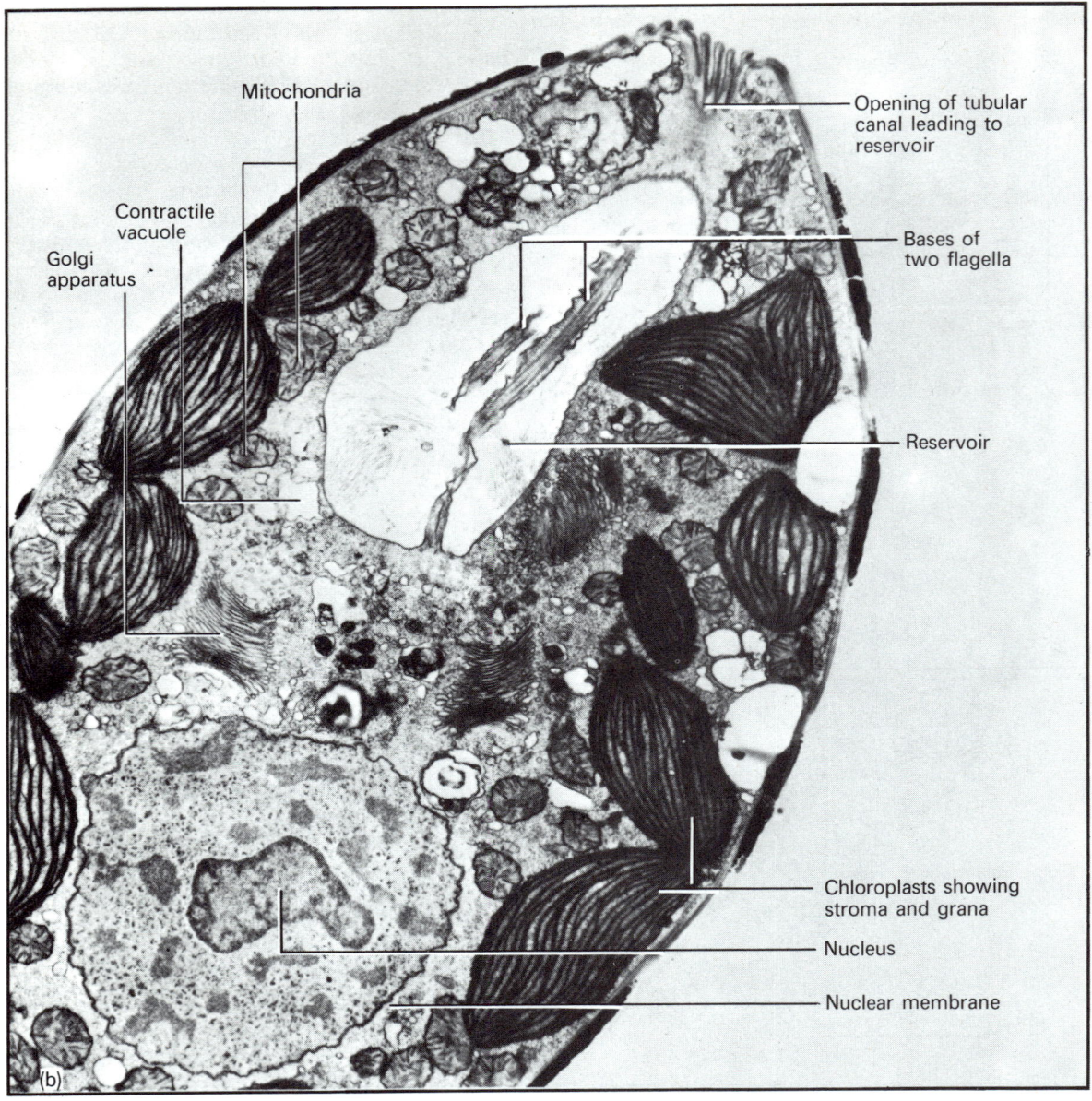

Mitochondria

Contractile
vacuole

Golgi
apparatus

Opening of tubular
canal leading to
reservoir

Bases of
two flagella

Reservoir

Chloroplasts showing
stroma and grana

Nucleus

Nuclear membrane

(b)

Fig. 18.4(b) Longitudinal section of *Lepocinclis ovum*. (Electron micrographs.)

tainly animals because they never possess chloro-plasts, and they capture prey and digest it.

They appear as colourless specks just visible to the naked eye. The largest is about 0.5 mm across. The cell has no protection against drying out and so the animal's habitat is either fresh or sea water. Fresh water species possess a contractile vacuole, like *Euglena*. Sea water species do not because their cytoplasm is not osmotically stronger than sea water so they do take up water by osmosis.

Movement An individual *Amoeba* has no precise or fixed shape (Fig. 18.6). Its cytoplasm is divided into two distinct regions. The outer **ectoplasm** is clear; the inner **endoplasm** is granular. The endo-plasm is a liquid sol; the ectoplasm is a gel. We are all familiar with the change from a sol state to a gel state when a table jelly sets.

Movement is by the formation of **pseudopodia** (Greek, false foot). In all species there appears to be a front and a back end to the animal. The back

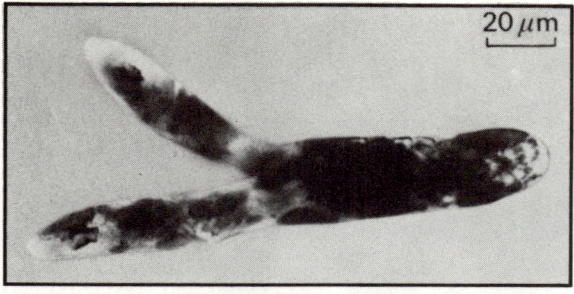

Fig. 18.5 Cell of *Euglena sp.* dividing. The two eyespots can be seen in the two cell anteriors. (Light microscope.)

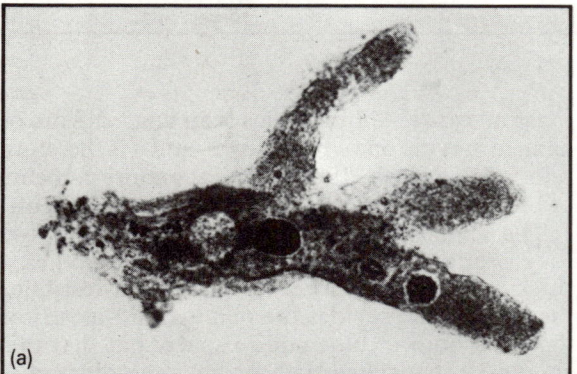

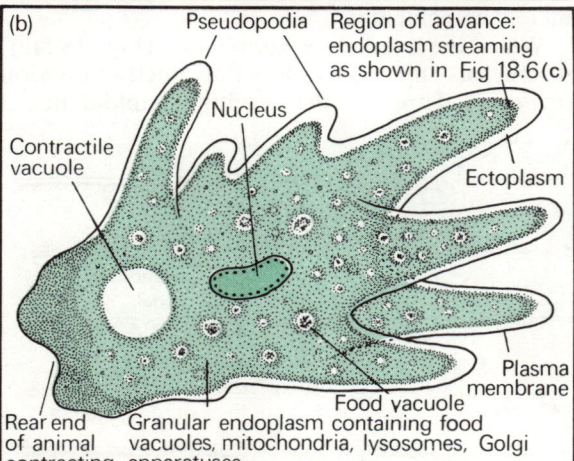

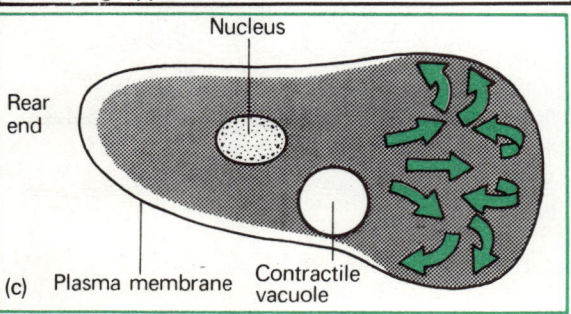

end exerts a pressure on the front, and liquid endoplasm splays out like a fountain at the front, where it is converted into ectoplasmic gel. The animal thus forms an ectoplasmic tube through which it advances (Fig. 18.6(c)).

Nutrition   *Amoeba* is able to detect the presence of food and will move towards it. When proteins contact its plasma membrane, microfilaments in the ectoplasm contract, drawing the membrane inwards and forming a food vacuole. A lysosome fuses with this, discharging digestive enzymes into it. Digested products diffuse from the food vacuole into the cytoplasm. Undigested remains are discharged at the plasma membrane (Fig. 1.15, route 1). Food vacuoles, Golgi apparatuses, lysosomes, and mitochondria, form much of the granular material that can be seen moving so freely in the endoplasm.

Respiration   **Amoeba** possesses no circulatory system. When moving, the formation of pseudopodia constantly brings endoplasm from the interior of the animal close to the surface ectoplasm.

Reproduction   Nuclear behaviour in the Protozoa has received very little attention from geneticists and is an area of great ignorance. It appears that in some species of *Amoeba* nuclear division takes place without chromosomes appearing, the dividing nucleus merely constricting into two parts. This is followed by constriction of the remainder of the cell into two parts.

So far as is known, *Amoeba* does not undergo sexual reproduction.

Algae

The algae are a large group which the fossil record shows to have been the first plants. They are found in rocks of the Cambrian period which are some 600 million years old. Land plants did not evolve until the late Silurian period, some 200 million years later.

Today the red, green, and brown seaweeds form the dominant flora of the sea, where flowering plants have never successfully competed with them. Many other species grow in fresh water. We shall examine only one, *Spirogyra*.

Fig. 18.6(a) *Amoeba sp.* (light microscope.) (Courtesy Carolina Biological Supply Co.)
(b) Drawing of *Amoeba sp.*
(c) Diagram to illustrate the fountain-like flow of endoplasm at the tip of the single pseudopodium of *Amoeba limax.*

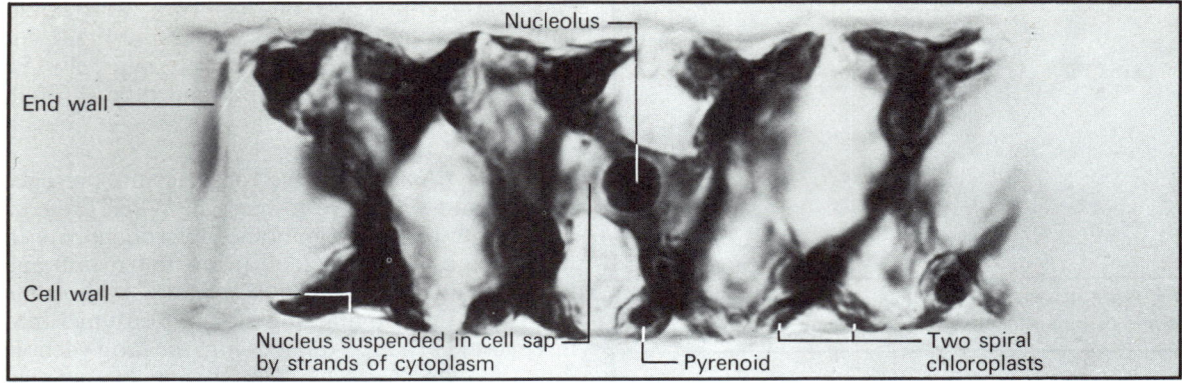

Fig. 18.7 One cell of *Spirogyra sp.* (Light microscope).

Spiroyga, a filamentous alga

This is a widespread fresh water plant growing at the margins of ponds and streams and on other water plants which it often smothers, hence its common name of blanket weed.

Structure

Each plant consists of an unbranched filament of cells which are identical except for the basal cell. This bears outgrowths which anchor the plant to the substrate.

Each cell shows the characteristic plant features: an outer cellulose wall and a central permanent vacuole filled with cell sap. The cytoplasm forms a layer pressed against the cell wall. Embedded in it are the ribbon-like chloroplasts that give the plant its name (Fig. 18.7). The chloroplasts contain pyrenoids which are regions where starches are made from sugars.

An unusual feature is the nucleus, suspended in the centre of the cell sap by cytoplasmic threads. It surrounds a dense spherical **nucleolus** which might easily be mistaken for the nucleus. The slimy feel of *Spirogyra* comes from mucilage which the cells secrete. This prevents smaller plants (diatoms) doing to *Spirogyra* what it does to water weeds: settling on it and shading it from the light.

Reproduction If the filament is broken, the detached part may become attached to a fresh substrate. This is asexual reproduction. Sexual reproduction involves some cell differentiation (Fig. 18.8).

The plant grows in dense tangled masses and filaments are always lying along side each other. Adjacent cells put out protrusions which fuse, forming a **conjugation canal**. Commonly this happens to many or all the cells in the two filaments (Fig. 18.8(b)). The entire living contents of one of the cells now rounds off and passes through the con-

jugation canal. Although this is an unusual kind of gamete, it is the one which moves, and it is therefore referred to as male. It fuses with the entire contents of the other cell, which have now also rounded off.

The filaments are haploid and when they fuse they produce a diploid zygote which becomes a thick-walled **zygospore**. Being drought-resistant, the zygospore provides the plant with a means of dispersal both in time and in space, but that this should be associated with sexual reproduction is merely coincidental.

When the zygospore germinates (Fig. 18.8(c)) its nucleus undergoes reduction nuclear division (meiosis). Three of the resulting haploid nuclei

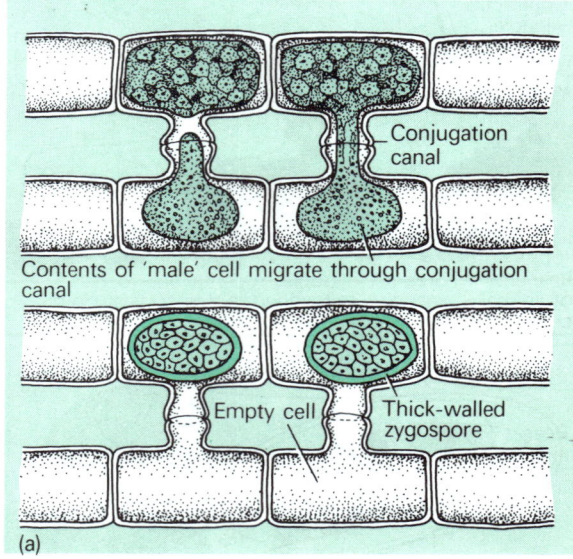

Fig. 18.8(a) Sexual reproduction in *Spirogyra*.
(b) Three stages in the sexual reproduction of *Spirogyra*. Note that all the 'male' cells are in the upper filaments. (Courtesy Carolina Biological Supply Co.).
(c) Germination of *Spirogyra* zygospore.

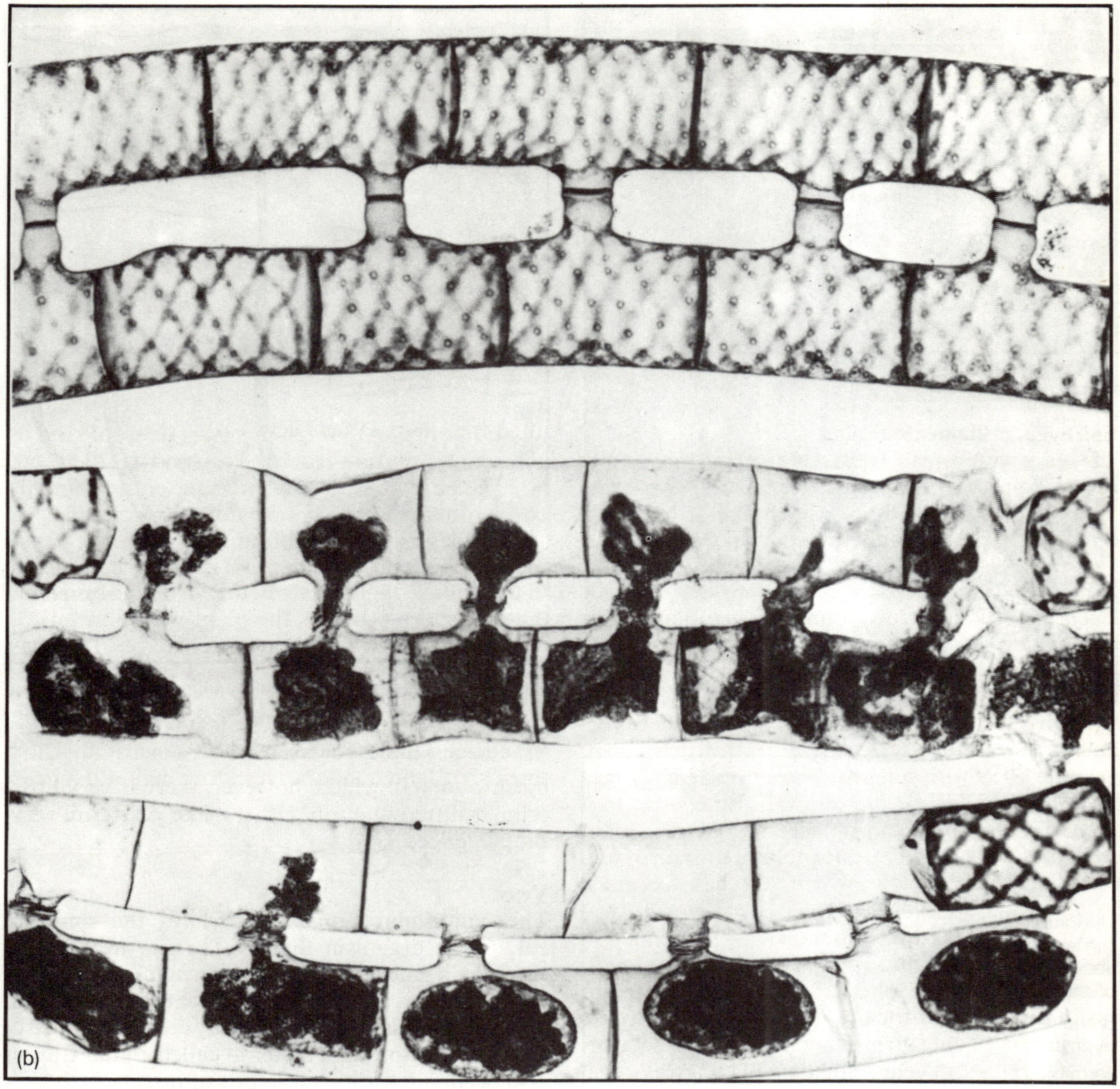

(b)

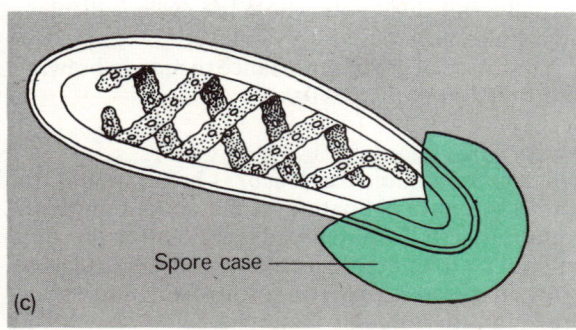

(c)

Spore case

degenerate, the fourth becoming the nucleus of the young cell.

Spirogyra is a simple plant Although *Spirogyra* has a eukaryotic cell as complex as that in any flowering plant, it is simple when judged by the criteria of differentiation and specialization. It shows no division of labour. There are no cells specialized for support, protection, photosynthesis, conduction, storage, or reproduction. Even the gametes are unspecialized cell contents borne in unspecialized cells.

221

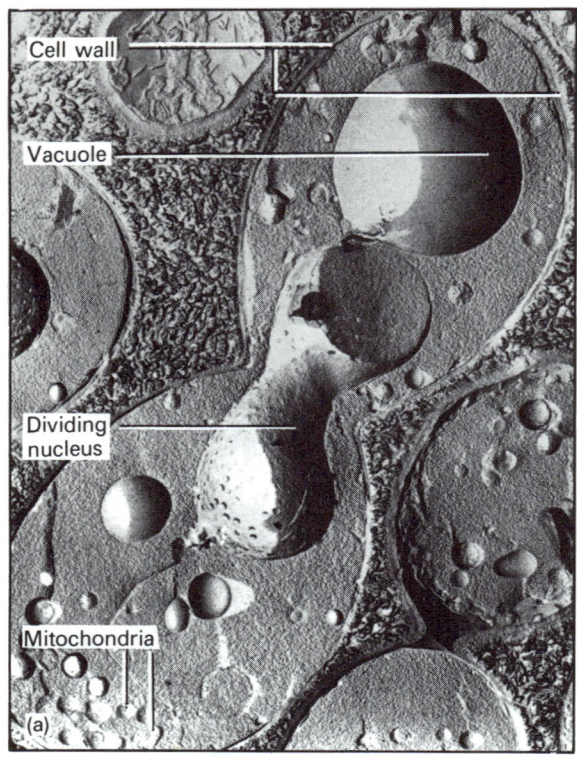

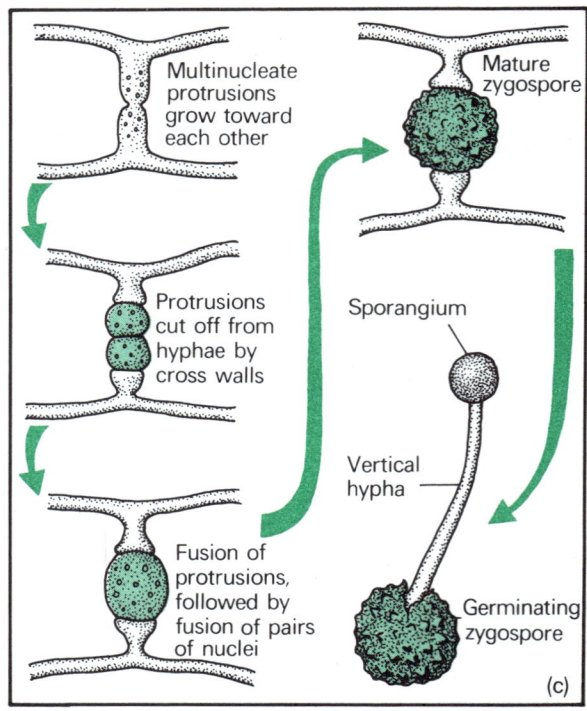

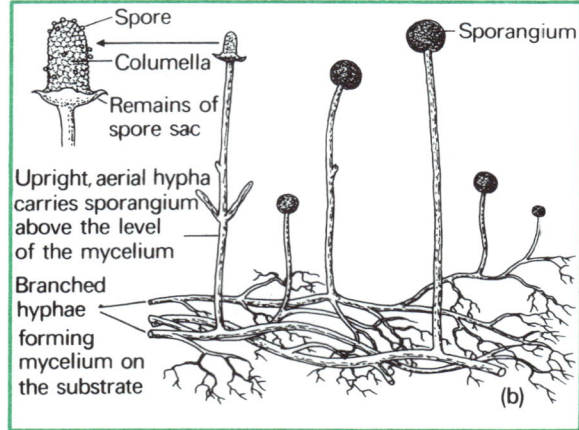

Fig. 18.9(a) Section of a yeast cell budding (Scanning electron micrograph).
(b) Hyphae of *Mucor sp.*
(c) Sexual reproduction in *Mucor sp.*

mushrooms) in which, however, there is very little cellular differentiation. Others, like yeast, are very simple indeed.

Yeast

This unicellular fungus is perhaps the simplest eukaryotic organism that exists (Fig. 18.9(a)). It contains a nucleus, a central permanent vacuole, mitochondria, and a cell wall. Like the cell walls of other fungi, this contains not only cellulose but also another cellulose-like material called **chitin**. Chitin also occurs in the exoskeletons of insects.

It reproduces asexually by **budding**, during which the nucleus undergoes mitosis. It also reproduces sexually, but during its whole life cycle it displays no cellular differentiation whatever.

Yeasts are of great importance to man in brewing and bread-making (p. 70).

Mucor

The common name for *Mucor* is bread mould. Its spores are always floating in the air of temperate countries, and if damp bread is exposed to the air it will not be long before a colony of *Mucor* starts to grow on it. It also grows on rotting fruit, horse manure, and other organic materials.

Fungi

Fungi are often placed in the plant kingdom because their cells possess walls and central permanent vacuoles filled with cell sap. However, no fungus possesses chloroplasts. They are all either parasites, living on live animals or plants, or saprophytes, living on dead organic matter. Like bacteria, they fulfil an essential role in nature by rotting other organisms. Fungi are probably best regarded as neither plants nor animals.

Some fungi have an elaborate body form (e.g.

Structure

Its body shows considerably more differentiation than yeast (Fig. 18.9(b), (c)). It is plant-like in that it possesses a wall of cellulose-chitin and a central, permanent vacuole filled with cell sap. Individual strands of the fungus are called **hyphae**, and the whole dense mat is the **mycelium**. Unlike most organisms, *Mucor* (and many other fungi) is not divided into uninucleate cells. There are no cell types, though some species show 'rooting hyphae' (Fig. 18.9(b)).

Nutrition   The hyphae secrete enzymes which digest the organic material on which they are growing. The digested products diffuse into the cell across the plasma membrane. The filamentous structure of the fungus is well suited to this saprophytic nutrition. The mycelium fits closely and intimately to the substrate so its secreted enzymes take effect immediately outside the cell. The reverse diffusion path for digested foods is very short.

Since it feeds on liquid food the fungus cannot grow in dry conditions. Like all fungi it can only grow in the presence of oxygen. Thus the mycelium never penetrates deeply into the food but always remains at the surface.

Reproduction   All saprophytes face the hazard that their food supply is bound to run out. They are therefore well endowed with efficient dispersal mechanisms so that they can find new material on which to grow.

This is certainly true of asexual reproduction in *Mucor*. Most of the mycelium grows close to the substrate, but many upright hyphae, some 5–6 cm long are also produced, and on the top of each a spore sac or **sporangium** develops. This is cut off from the upright hypha by a cross wall, the **columella** (Fig. 18.9(b)). The contents divide into some 100 multinucleate spores which are scattered when the columella swells and bursts the wall of the sporangium. Because they are situated high above the mycelium, the spores are likely to be caught up in air currents, and the effectiveness of this dispersal mechanism is shown by the ready appearance of *Mucor* on damp substrates.

In sexual reproduction two multinucleate protrusions appear in adjacent hyphae and grow towards each other (Fig. 18.9(c)). These become cut off from the rest of the hyphae by cross walls. On fusing, the enclosed nuclei fuse in pairs. The mycelium is haploid and the nuclei produced are therefore diploid.

The whole structure now develops a thick, hard wall and becomes a **zygospore**. Unlike the asexual spores this is able to resist drying out and other adverse conditions, and so provides *Mucor* with dispersal in time. However, the fact that this is associated with sexual reproduction is merely coincidental. When it eventually germinates, the zygospore immediately produces an aerial hypha bearing at its end an asexual sporangium. The diploid zygote nuclei undergo reduction nuclear division (meiosis) and the resulting spores are haploid.

Hydra: a simple animal with limited degrees of cell and organ differentiation

Hydra belongs to a group of animals called the *Coelenterata*. Marine members of the group are the

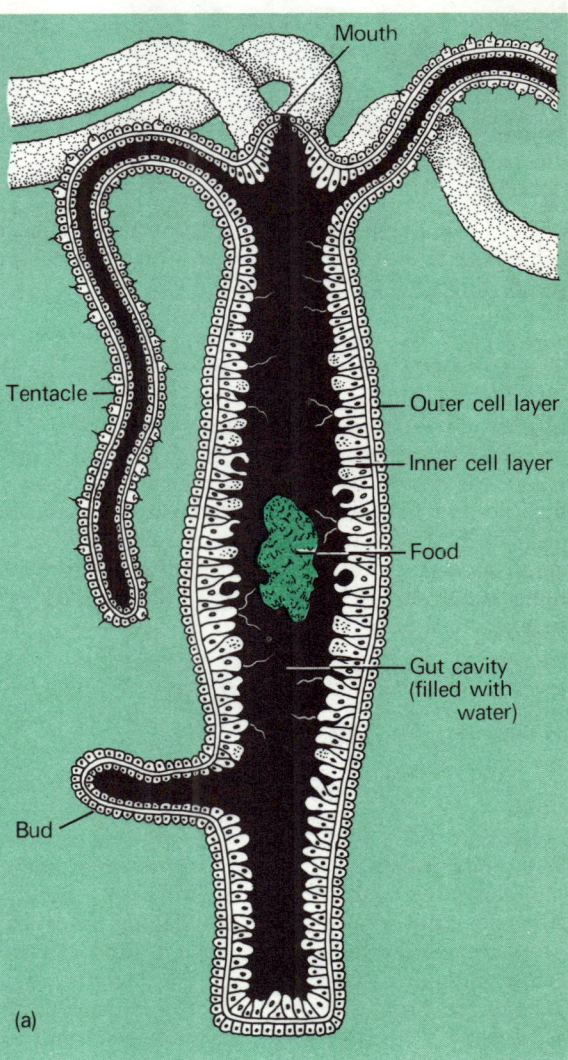

Fig. 18.10(a) Longitudinal section of hydra.

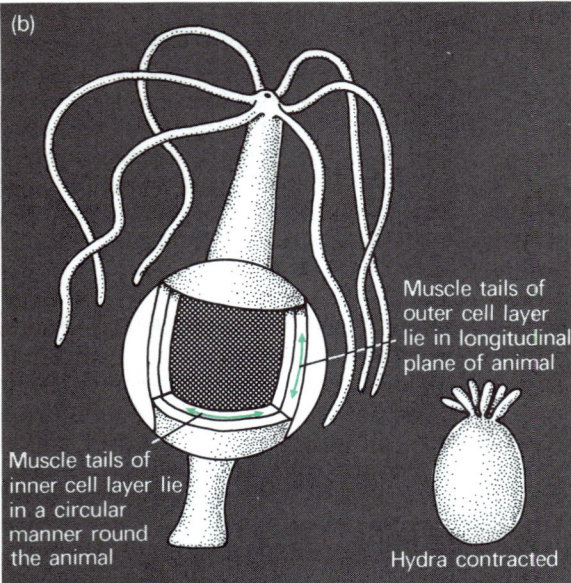

Fig. 18.10(b) Diagram to illustrate the arrangement of muscle tails in hydra.

corals, sea anemonies, jelly fish, and Portuguese men o'war; but hydra lives in fresh water—in ponds and rivers.

Life style
The body is a tube about 1 cm long (Fig. 12.3). At the top end is a small but very stretchable mouth surrounded by a group of tentacles. The animal is attached to its support by a sticky substance produced by cells at its base. Its body is flexible and mobile and if disturbed the tentacles and the body contract and the animal becomes a small blob (Fig. 18.10(b)). At other times the tentacles wave slowly in the water, and the body can bend from side to side.

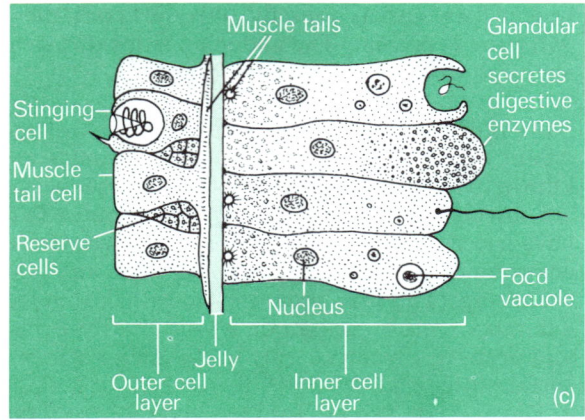

Fig. 18.10(c) Detail of the body wall of hydra.

If prey touches the tentacles, it is paralyzed by highly specialized stinging cells and drawn into the mouth which gapes to receive it. It is then stuffed inside the gut cavity (Fig. 18.10(a)). Cells lining the gut secrete enzymes which digests the food. The soluble products are absorbed by diffusion into the body cells. Eventually undigested remains are expelled from the mouth.

Specialized cells
This description reveals the need for considerable cell specialization, and for coordination. The animal must possess muscle cells to produce movement; sensory cells to detect food; secretory cells to attach it to its support; secretory cells to produce digestive juices; and nerve cells to coordinate its activities.

Sections of the animal (Figs. 18.10(a) and (c)) show that its body is made up of two layers of cells. The outer layer contains a variety of specialized cells. One type have their bases drawn out into **muscle tails** which lie in a plane parallel to the long axis of the animal (Fig. 18.10(b)). When these muscle tails shorten the animal shortens. At the same time the muscle tails round the mouth contract, sealing off the water in the animal's gut. This makes the animal assume a spherical shape.

Also in the outer layer are the **stinging cells** which contain inverted threads filled with paralyzing liquid. When the trigger is touched (Fig. 18.11) the thread turns inside-out like a pushed-in finger in a

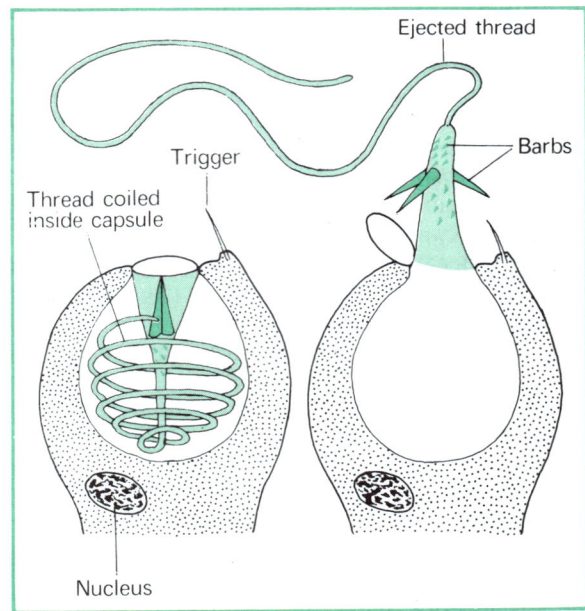

Fig. 18.11 Stinging cell of hydra before and after discharge.

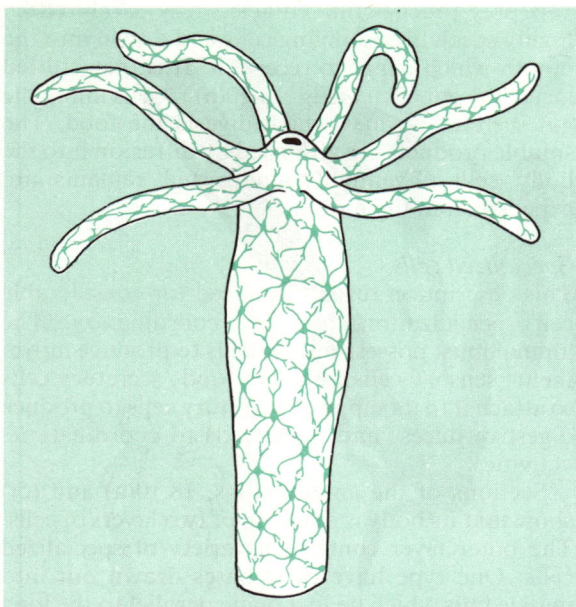

Fig. 18.12 Diagram to illustrate the nerve net of hydra. There is no concentration of nerve tissue anywhere in the body.

rubber glove suddenly poking out. These cells are constantly being used up and the **reserve cells** of the outer layer replace them. In Fig. 12.3 the tentacles can be seen bristling with stinging cells.

The inner layer contains cells which secrete digestive juices. Others produce food vacuoles like those in *Amoeba* and engulf small particles of food from the gut cavity. These cells also bear muscle tails at their bases. These, however, lie in a circular fashion round the animal. In order for a contracted hydra to lengthen, these muscle tails contract, pressing against the water-filled gut cavity and causing the cross sectional area of the animal to decrease (Fig. 18.10(b)).

A simple nervous system coordinates these activities. Nerve cells lie at the base of the outer cell layer. Each possesses many dendrites which connect in the form of a nerve net (Fig. 18.12). There is no central nervous system, and influences spread shorter or greater distances in the animal's body depending on the strength of the stimulus that produced them.

Reproduction

The animal reproduces asexually by budding (Fig. 12.3). When the bud is large enough it becomes detached and leads a separate life. In effect the parent has been split in two. Offspring and parent are genetically identical.

If hydra is chopped up in the laboratory, quite small pieces grow into new hydras. This is possible because small pieces contain all the various types of cells, and the animal has no organs except when reproducing sexually (see page 116).

Sexual reproduction　Sexual reproduction involves two organs, the testis and the ovary. The animal is hermaphrodite, but it does not fertilize itself since the testis discharges its sperms before the ovum is mature. The young embryo is a hollow ball of cells surrounded by a thick protective coat. It is drought resistant and has a resting and dispersal stage before germinating to produce a new hydra.

Insects

Insects belong to the phylum *Arthropoda*. The word means 'jointed foot'. Other members of the phylum are the *Crustacea* (crabs, lobsters, scorpions), the *Myriapoda* (centipedes, millipedes), the *Arachnida* (spiders), and the extinct *Trilobita*.

The arthropods are the largest phylum in the animal kingdom. Nearly 900 000 distinct species are known. All are characterized by a hard, jointed, **external** skeleton, (like a suit of armour) and paired, jointed limbs. The insects are distinguished from other arthropods by possessing three pairs of legs and often two pairs of wings.

Body organization

The body is divided into three regions: **head**, **thorax**, and **abdomen** (Fig. 18.14). The head bears the sensory structures (**compound eyes** and **antennae**) and the mouth parts. It is the sensory and feeding region of the body. The thorax is the locomotory region. It is divided into three **segments**, each bearing a pair of jointed legs on its underside (Fig. 18.14). Most insects possess two pairs of wings, one pair attached to the upper side of the second and the third segments of the thorax. In the two-winged flies, such as housefly, only the first of these two pairs is present. The abdomen is the digestive and reproductive region of the body.

Success of insects

About one million distinct species of animals are known to inhabit the earth. 850 000 of these species are insects. They are the only invertebrates that have successfully inhabited all types of terrestrial environment, including the most arid and unpromising. (This may be contrasted with the limited habitats of snails, nematode worms, earthworms, and protists, which can only live in damp soil and surface litter.)

The first insects were wingless and date from the Devonian period 400 million years ago. 100 million

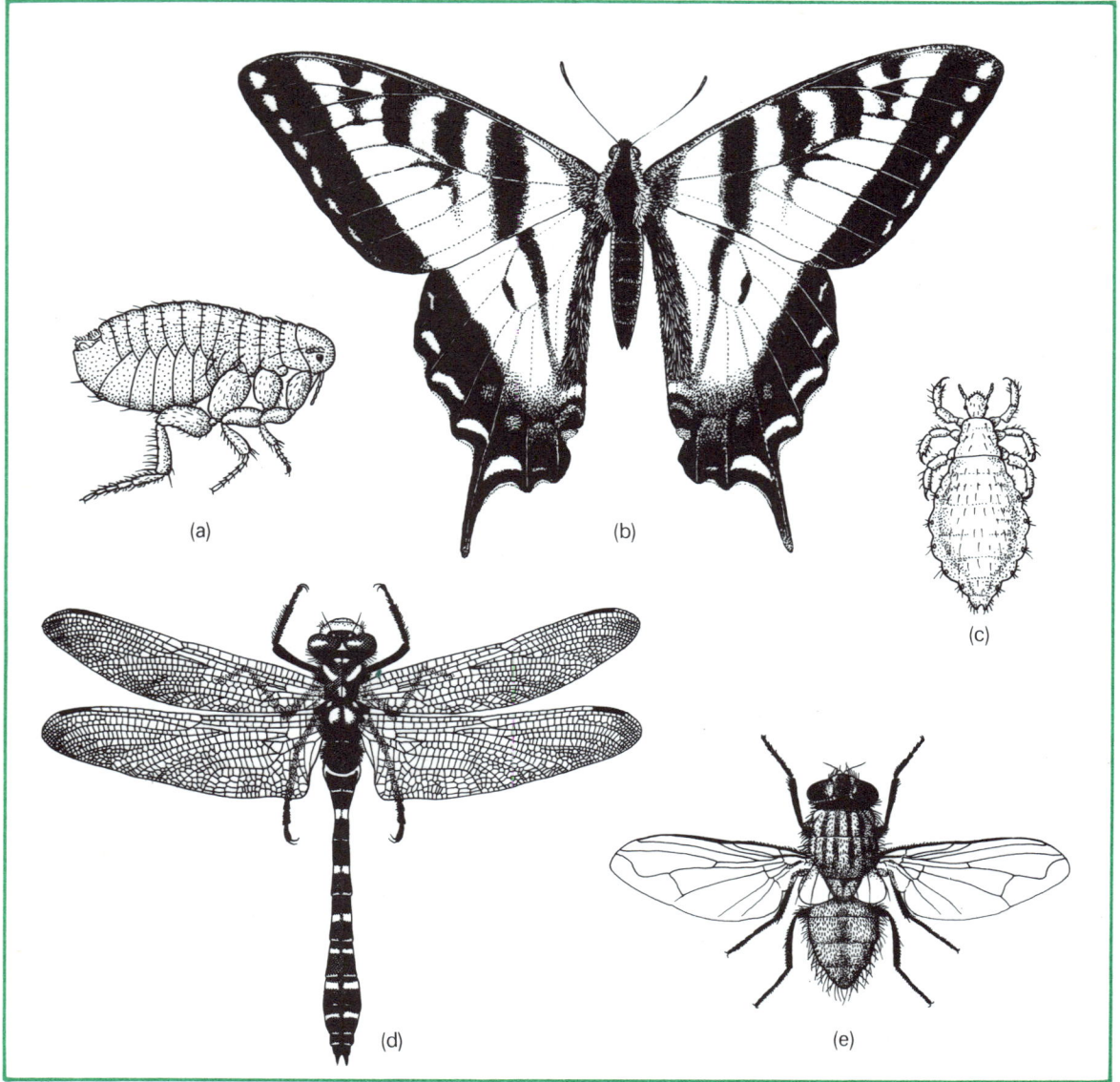

Fig. 18.13 Some representative insects. **(a)** Flea imago. **(b)** Swallow tail butterfly imago. **(c)** Sucking louse imago. **(d)** Dragon fly imago. **(e)** Housefly imago.

years later, in the Carboniferous period, some insects developed fixed projections from the thorax which enabled them to glide. From these evolved the wings which carried insects into so many new habitats. A small invertebrate cannot migrate far by walking, but small insects can easily move over long distances by flying.

In the Cretaceous period, 130 million years ago, insects underwent a major adaptive radiation which coincided with that of flowering plants. This means that both groups developed many new species at the same time, and many flowers have highly specialized pollination mechanisms that form part of a symbiotic relationship with particular species of insect (for example, white dead nettle, Fig. 13.6). Insects are of enormous importance in pollination.

Until the advent of man, insects were the most successful group of animals. Even now the annual production of insect body material in natural grasslands equals that of the herbivorous mammals. They are vitally important components of food webs (see page 247). One aspect of this is that insects of the

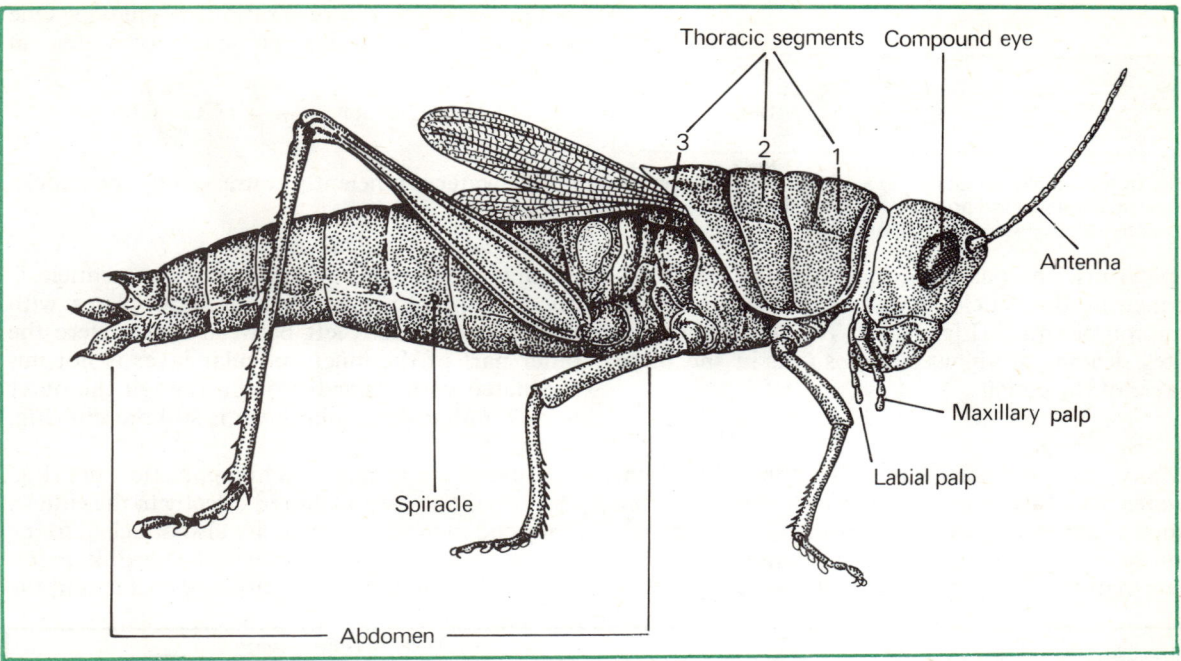

Fig. 18.14 Grasshopper, last nymphal stage, sideview. Note that the wings are attached to the second and third thoracic segments and the jointed legs one to each thoracic segment.

forest floor reduce plant material to small pieces, thus contributing to the cycling of natural materials.

Some species are harmful to man because they are parasites (e.g. lice, fleas) or are vectors (carriers) of diseases such as bubonic plague, malaria, typhus, and sleeping sickness. They also destroy crops on a large scale.

The study of insects forms a whole branch of science called **entomology**.

The insect cuticle

The insect body is bounded by the **cuticle**, shown in section in Fig. 18.15. The cuticle has played a major part in the success of the group. It is differentiated into two layers. The inner is up to 200 μm thick and consists of chitin, a cellulose-like substance. Towards the outside of the layer this is toughened by the addition of protein which is chemically hardened (tanned).

Tanned chitin is so tough that some beetles can bite through copper and lead. The insect's muscles are attached to it (Fig. 18.17), and it provides the rigidity needed for the attachment and operation of the wings. Chitin gives the animal considerable physical protection, and it is difficult for parasites to penetrate.

The outer layer is only some 1–4 μm thick. It is

covered by a thin layer of wax which renders the animal almost perfectly waterproof. Evaporation through the cuticle is normally less than 1 % of the body water content per hour. Thus the cuticle enables insects to overcome one of the major difficulties facing animals that live on the land—the conservation of water. If adult insects are rubbed gently with a fine abrasive powder, they soon dry up and die because the wax layer is destroyed and cannot be replaced quickly enough. The scratches and other damage received in normal life are repaired by wax exuded in a solvent from the cells of the

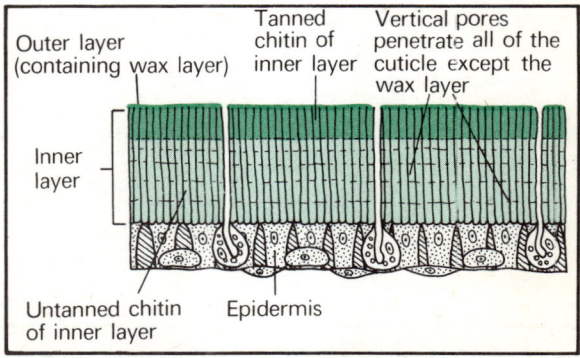

Fig. 18.15 Vertical section through insect cuticle.

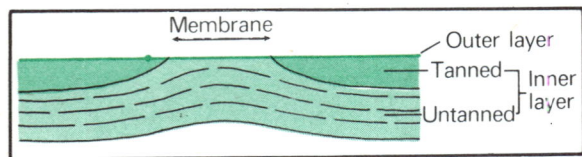

Fig. 18.16 Flexible membrane in the cuticle in vertical section: this contains no tanned proteins. Note that the outer layer, with its waterproofing wax, is continuous across the membrane.

epidermis. This passes through the fine pores which penetrate the cuticle in vast numbers—up to one million per mm² (Fig. 18.15). The solvent evaporates, leaving fresh wax in position in the outer layer of the cuticle.

Water supply
Many insects drink, but most obtain sufficient water from their damp or juicy food. Some insects, such as mealworms and flour beetles, can pass their entire life cycles in dry food whose water content is less than 1%. They make use of the water produced

in their cells by the respiration of glucose. One gram of glucose yields 0.6 grams of water on respiration:

$$\underset{\text{glucose}}{C_6H_{12}O_6} + 6O_2 \longrightarrow 6CO_2 + 6H_2O$$

The water is efficiently conserved by the cuticle.

Joints, muscles, and movement
So that movement can take place, the cuticle is divided up into a number of hard plates with flexible membranes left between them. Here the outer part of the inner cuticular layer is not impregnated with tanned protein, though the outer wax-containing cuticular layer is still present (Fig. 18.16).

Muscles are arranged in antagonistic pairs (Fig. 18.17(a)). They are anchored directly to the cuticle. They end in tendons which are also attached to the cuticle. Figure 18.17(b) shows the flexible membrane of the hinge joint between the tibia and the

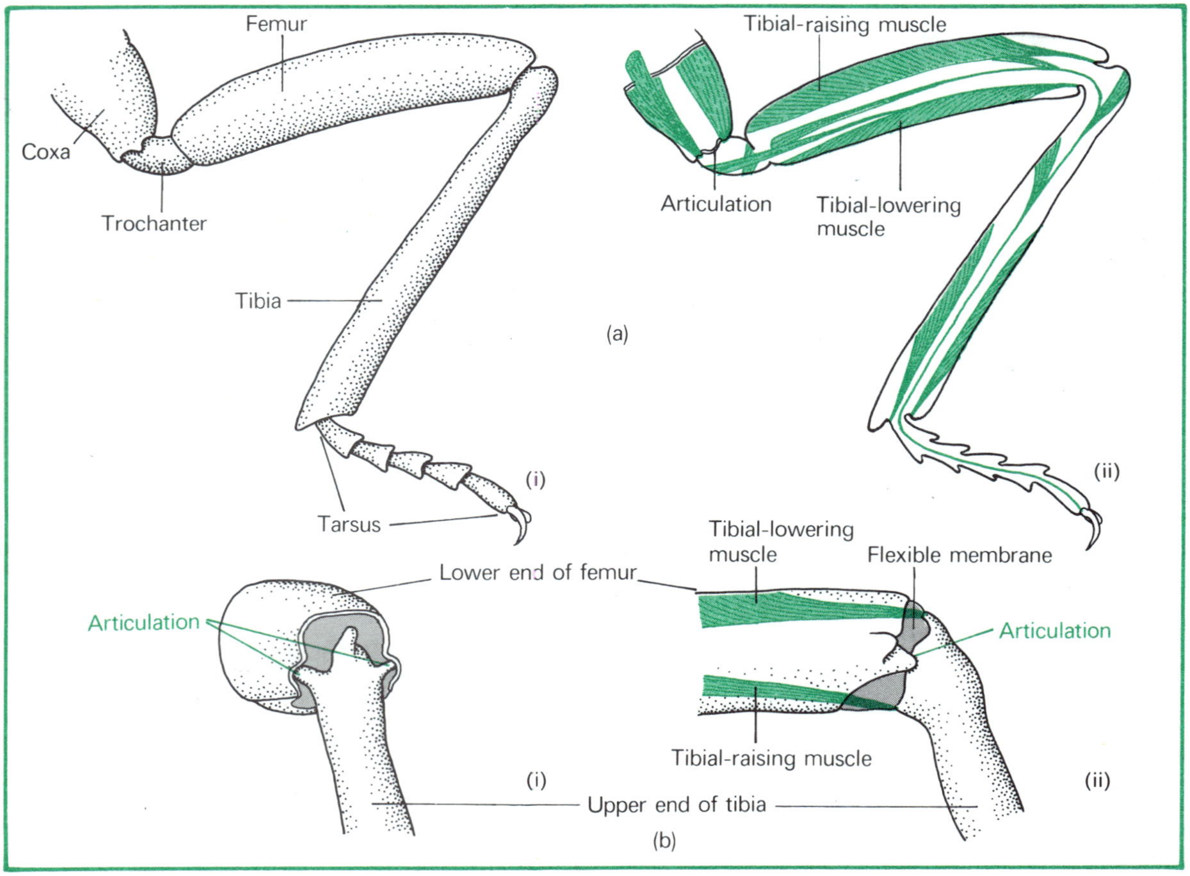

Fig. 18.17(a) Typical insect leg. (i) External view (ii) Section showing muscles and tendons. **(b)** External view of the joint between the femur and the tibia ((i)). (ii) The antagonistic muscle pair which moves the tibia.

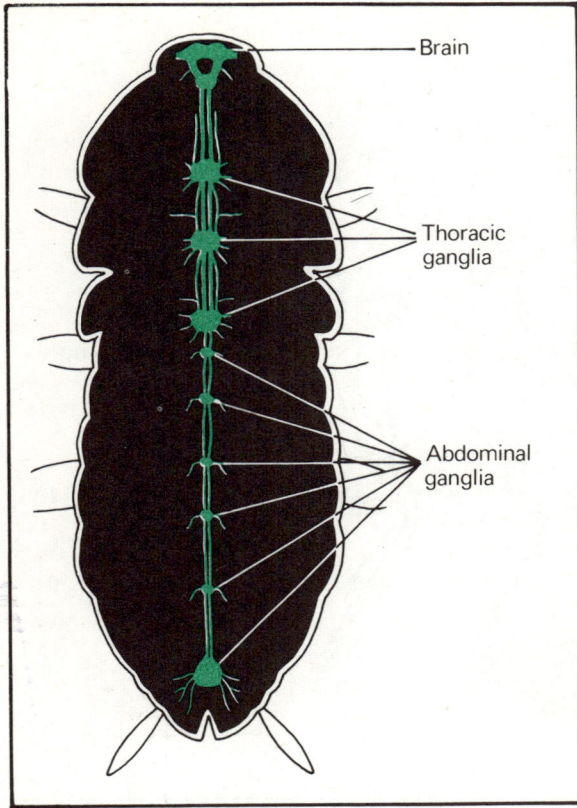

Fig. 18.18 Brain and nerve cord of a cockroach.

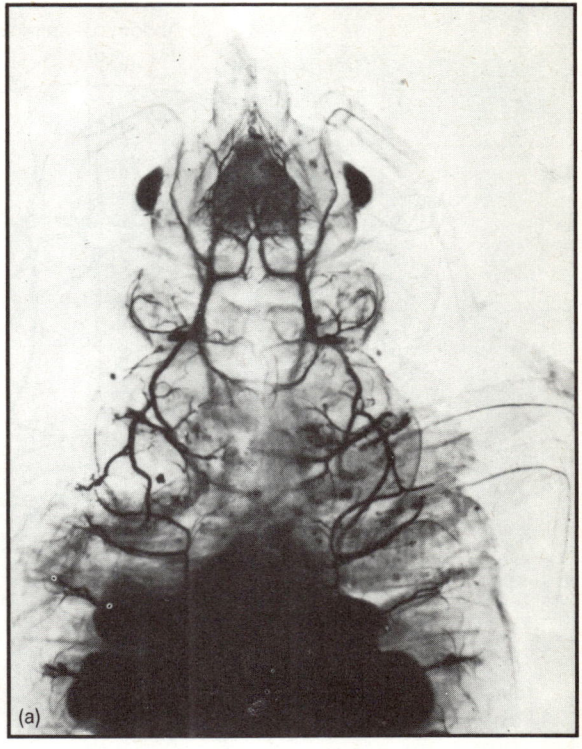

Fig. 18.19(a) Tracheal system in the bed bug, injected with opaque material. Note the rich tracheal supply to the sucking muscle between the eyes.

femur of a typical insect leg, and the antagonistic pair of muscles which produces the up-and-down movement of the tibia.

Nervous system

Whereas the vertebrate central nervous system is located on the upper side of the animal and is hollow (Fig. 2.8), that of insects is on the lower side of the body and is solid.

There is a frontal swelling, the brain, which receives sensory information from the eyes, antennae, and mouthparts. It coordinates the activity of the mouthparts. The brain is on the upper side of the body, but the twin nerve cord leading back from it immediately descends to the lower side (Fig. 18.18). Various swellings called **ganglia**, containing the cell bodies of nerve cells, are situated along its length. The three large ones in the thoracic region control movement. The abdominal region does not contain a ganglion in every segment; the final ganglion is larger than the others and serves several segments.

Respiration

Insects are very active animals with strong oxygen needs. The amount of energy required for insect flight exceeds that of any other activity in any organism. The flight muscles perform between ten and twenty times more work in proportion to their size than the muscles of a man working at maximum exertion. Insects beat their wings at anything up to 1000 times per second; but unlike vertebrate muscle, insect flight muscle does not develop an oxygen debt (see page 78). Even under these demanding conditions, the oxygen supply must be adequate to permit the complete oxidation of glucose to carbon dioxide and water.

The insect respiratory system must therefore permit two apparently contradictory things to happen: plenty of oxygen must be admitted into the body, but valuable water vapour must be prevented from leaving it.

Tracheal system   The insect body is riddled by a series of branched tubes called **tracheae** (Fig. 18.19).

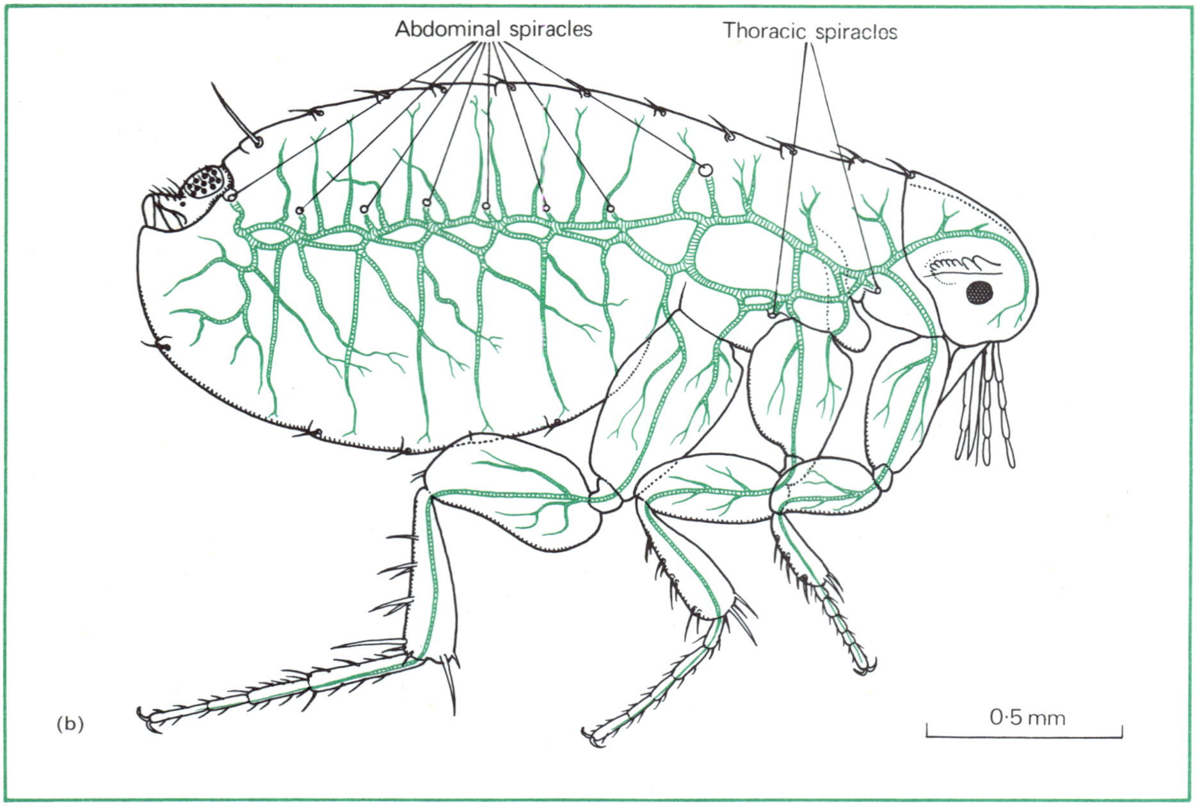

Abdominal spiracles   Thoracic spiracles

(b)   0·5 mm

Fig. 18.19(b) Half of the tracheal system of a flea, seen from the side.
(c) Part of a locust abdomen from the side, showing the spiracles.

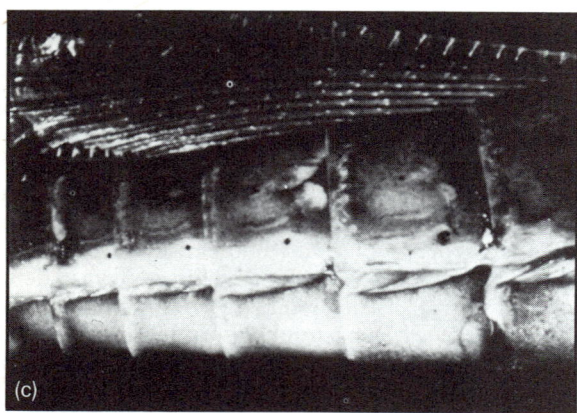

They are lined with cuticle throughout and open to the exterior at closeable holes called **spiracles**. There is normally one pair of spiracles per body segment (Fig. 18.19(c)).

As they approach the tissues the tracheae break up into bundles of fine **tracheoles** whose diameter is only 1.0–0.5 μm. (Fig. 18.20(a)). These are in fact long thin cavities inside long thin cells. They penetrate the body cells and end abruptly in them. The cavity of the tracheole is also lined with cuticle (Fig. 18.20(b)). Its blunt end is about 0.1 μm in diameter.

Thus air is able to penetrate right into the interiors of all the insect's cells, and calculations show that diffusion alone accounts for the efficient exchange of oxygen and carbon dioxide between the respiring cells and the atmosphere.

Some species possess in addition to tracheae dilated air sacs, and perform ventilaing movements with the abdomen.

Keeping water vapour in: spiracles If the tracheae were permanently open to the atmosphere the insect would soon dry up, and for most of the time the spiracles are closed. During this time oxygen in the air inside the insect is used up. When the spiracles briefly open there is therefore a strong oxygen diffusion gradient into the tracheae and a correspondingly strong carbon dioxide gradient out of

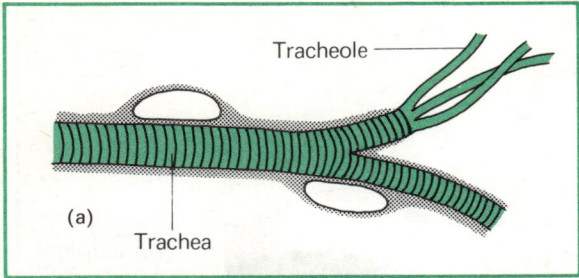

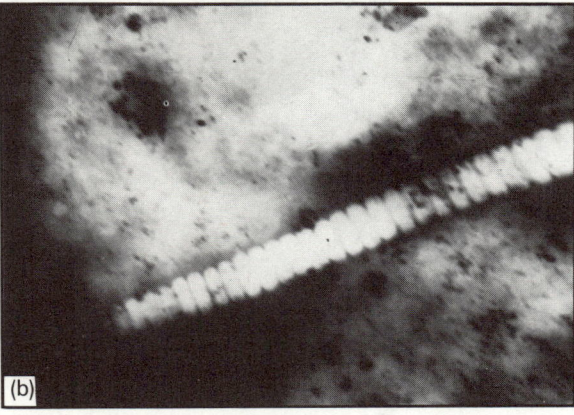

Fig. 18.20(a) Trachea dividing into a bunch of tracheoles.
(b) Electron micrograph of a tracheole within an insect cell.

them. Some water vapour is also lost during the opening period. Figure 18.21 shows how the rate of loss of water vapour shot up when meal worm larvae were exposed to an atmosphere of 5% carbon dioxide, which made them open their spiracles.

Respiration and body size Efficient though the respiratory mechanism of insects is, it places a

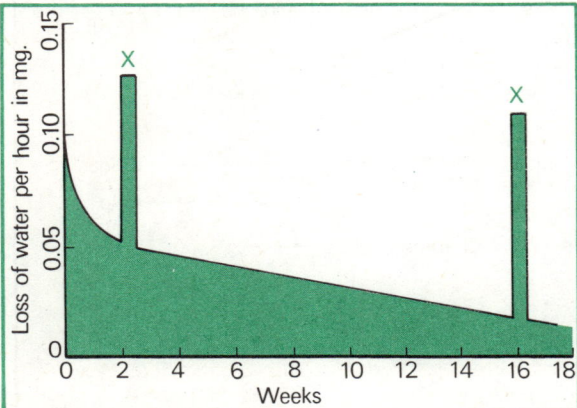

Fig. 18.21 Opening of the spiracles of mealworm larvae by exposure of the animals to carbon dioxide at the points X causes a marked increase in loss of water vapour from the body.

limit on their size. Oxygen has to pass to their cells by diffusion alone, and no cell can be situated farther from the animal's exterior than this fact will allow.

A completely different situation exists in animals in which oxygen is transported in circulating blood. Cells in the centre of an elephant, for example, are more than half a metre from the animal's hide. Few insects have achieved a size of more than a few centimetres. The largest that have ever existed are the giant dragonflies of the Carboniferous period (about 300 million years ago). These had a wing span of about 60 cm, but like present-day dragonflies their bodies were long and thin.

Feeding
Many insects feed on leaves and the structure and function of the mouthparts of the locust have been especially well studied. (Fig. 18.22).

The animal detects food at a distance by sight. This sense is supplemented by a sensitivity to the odour of leaves provided by small sensory organs situated on the antennae (Fig. 18.23).

The acceptability of food is explored by vibrating the palps on the **labia** and the **maxillae**. These are brought into repeated brief contact with the leaf surface. They bear batteries of small sensory organs (Fig. 18.24) which are able to distinguish between the very similar chemical compositions of the waxy cuticles of various leaves.

If the leaf is acceptable, the animal moves over it until it finds an edge from which it takes an exploratory bite, using the saw-edged **mandibles** and the **lacineae** of the **maxillae** (Fig. 18.22). The mandibles are hinged in such a way that they swing in and out in one plane perpendicular to the long axis of the animal. They cut like a pair of scissors. The maxillae are more mobile and as well as the cutting action of the lacineae, they help to force fragments of food upwards into the mouth.

The first bite releases fluids from the torn leaf cells. As these spread over the inner surfaces of the mouthparts, they reach sensory organs on the inside of the labrum. Many of the dissolved substances in these fluids stimulate feeding. The mouthparts of a medium sized grasshopper carry over 12 000 sense organs concerned with determining the chemical composition and hardness of the food. Thus the detection and selection of food, and the commencement and continuation of feeding, are a series of reflex actions.

Chopped-off food is stuffed into the mouth by the general activity of the mouthparts. As the foregut becomes distended, stretch-sensitive cells in its wall signal to the brain. This information

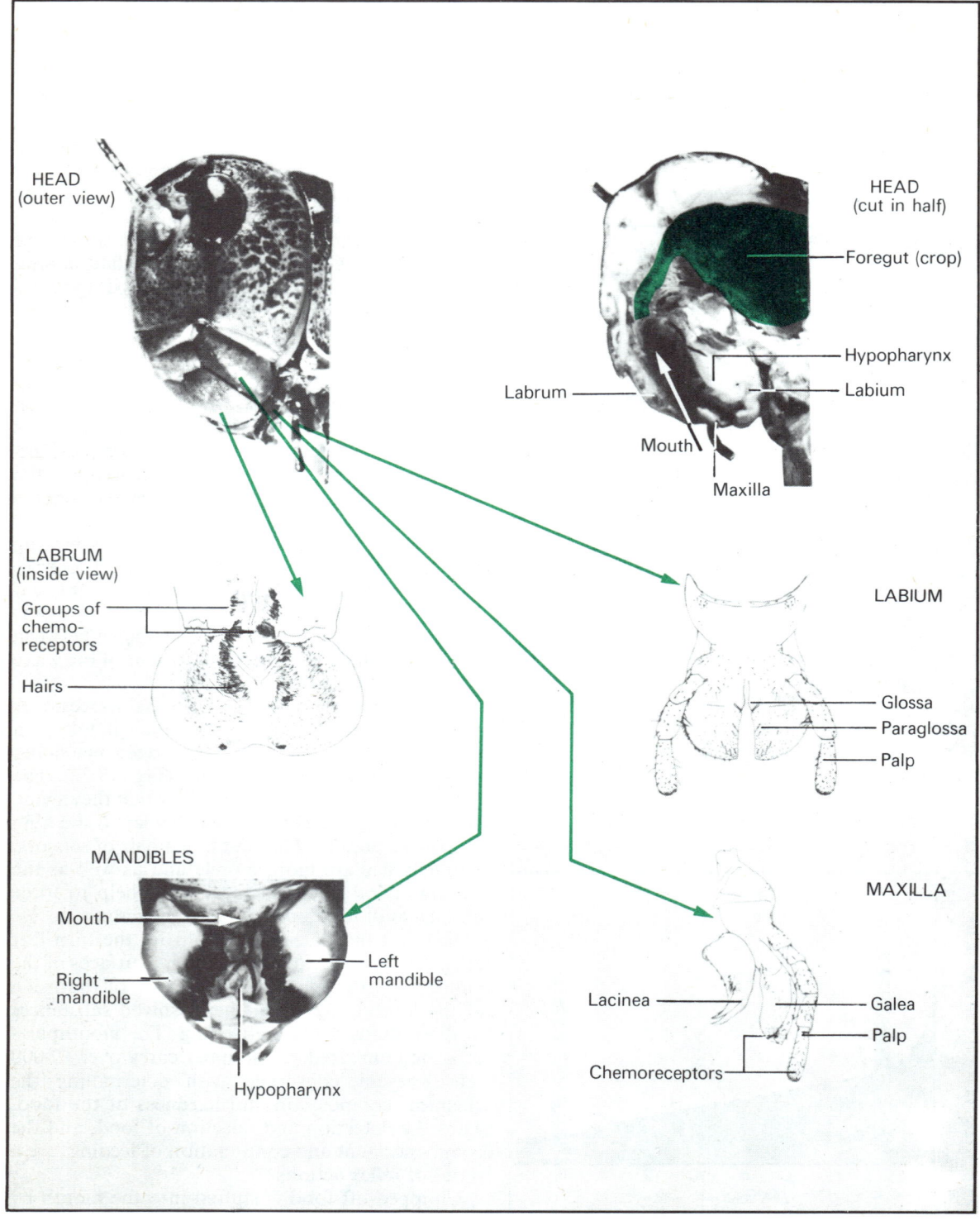

Fig. 18.22 The structure and arrangement of the mouthparts of a locust.

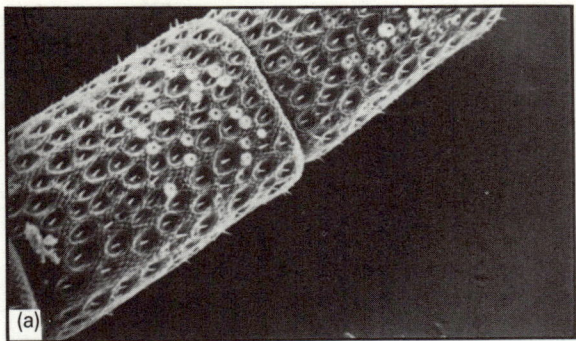

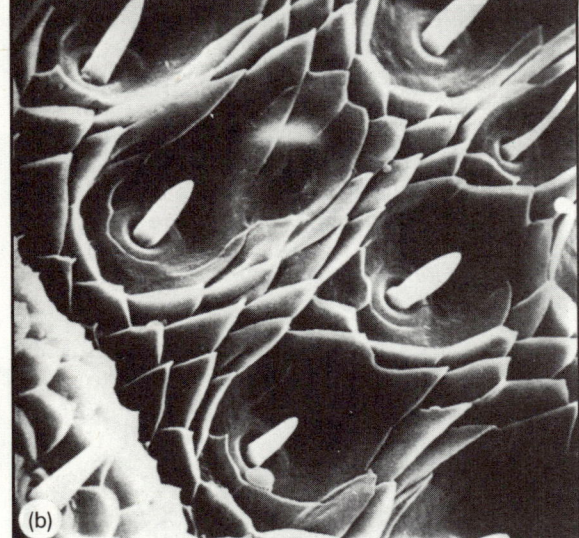

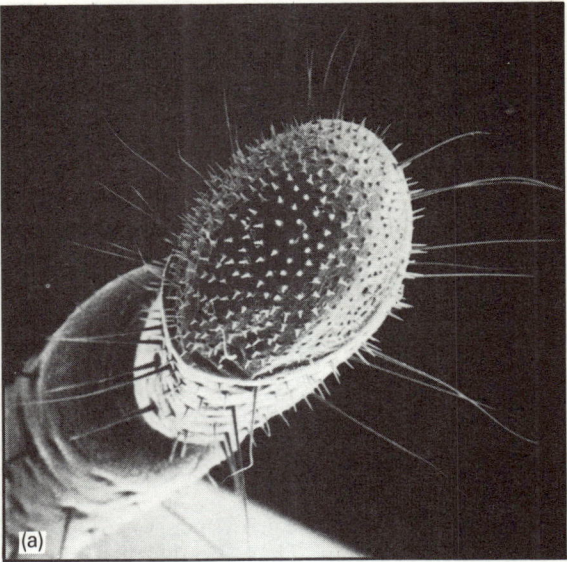

Fig. 18.24(a) Tip of the maxillary palp of a locust. (Scanning electron micrograph.) Note the hairs at the extreme tip which are sense organs for detecting the chemical compositions of epidermal waxes.

Fig. 18.23(a) Scanning electron micrograph of part of the antenna of a locust. Large numbers of sense organs are scattered over the surface.
(b) Detail of (a).

eventually causes a reflex cut-off of the nervous instructions to the muscles operating the mouthparts. The cessation of feeding is also a reflex action.

Reproduction and growth
Reproduction in insects is sexual and all species exist as males and females. The animals copulate. The male introduces sperms into the female's body; these may be stored for months and used to fertilize successive batches of eggs.

Moulting The cuticle is unstretchable, and in order to grow the young animal has to **moult**. In many species of insects the young **nymph** resembles the adult (**imago**). Growth follows each moult, but the animal retains the same detailed body plan except

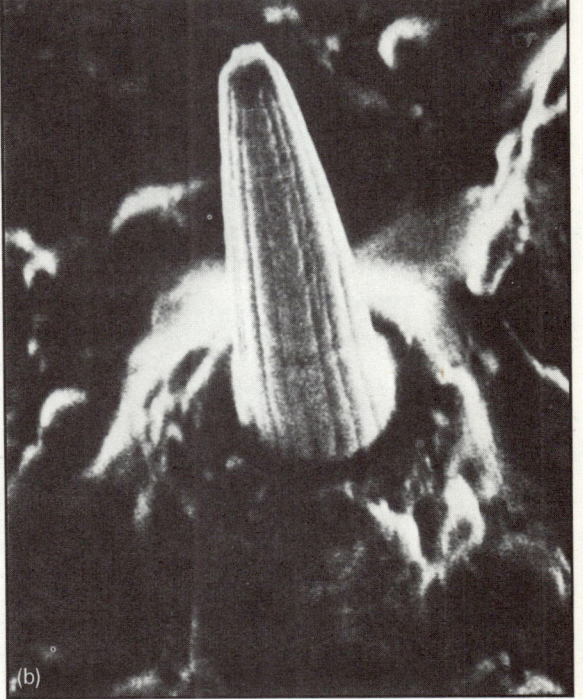

(b) Detail of (a).

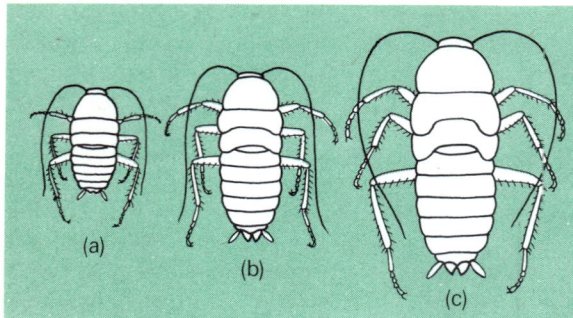

Fig. 18.25 Three nymphal stages of cockroach.

that the wing buds on the second and third thoracic segments enlarge (Fig. 18.25). After the final moult these develop into full-sized wings. The nymphs are not, however, sexually mature. Only the imagos can mate and reproduce.

Figure 18.26 shows the growth curves of a locust and a water boatman. They should be compared with the mammalian growth curve (Fig. 12.28). In the case of the locust there is a small decrease in mass at each moult caused by the loss of the

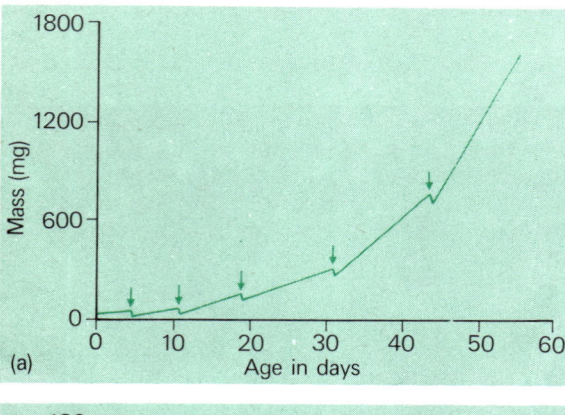

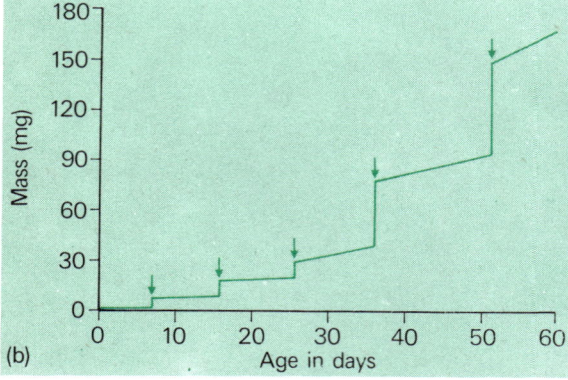

Fig. 18.26(a) Growth curve of a locust. Moulting indicated by arrows.
(b) Growth curve of a water boatman. Moulting indicated by arrows.

cuticle. This is followed by a steady increase in mass up to the next moult. The water boatman is an aquatic insect. It does not show a fall in mass at moulting but a sharp increase due to the absorption of water. This is followed by a steadier increase in mass due to growth.

Moulting is controlled by a hormone which acts particularly on the cells of the epidermis (Fig. 18.15). First these cells become detached from the old cuticle and then they divide and secrete a new cuticle. The new epidermis and the new cuticle cannot expand, for they are imprisoned by the old cuticle. They therefore become much folded.

Next the epidermal cells secrete enzymes which dissolve the old cuticle, which ruptures. By wriggling and pulling movements, the insect extricates itself from the old cuticle. It is not only the outer body layer which is shed, for the entire lining of the tracheal system is also cast off and is drawn out of the spiracles still attached to the old cuticle. The linings of the tracheoles, however, are not shed.

Insects with larvae and pupae
In other insects there is a sharp difference between

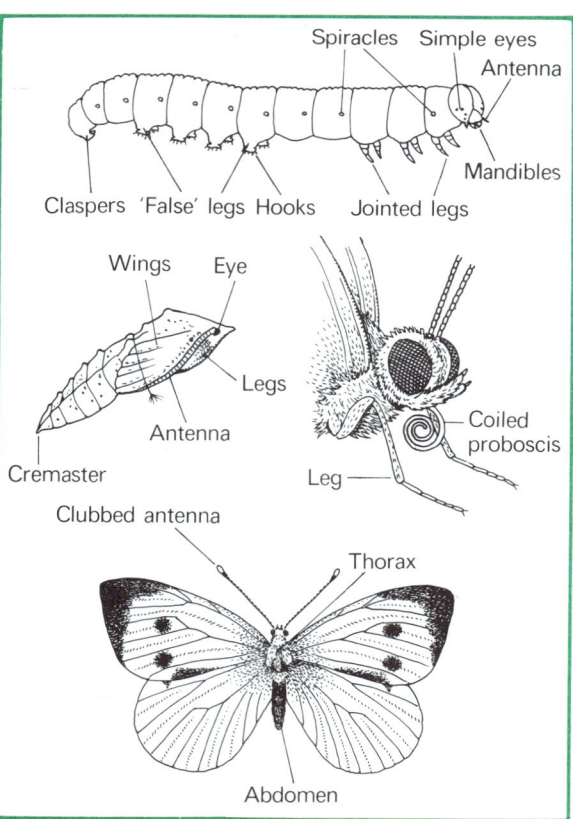

Fig. 18.27 Stages in the life cycle of the cabbage white butterfly.

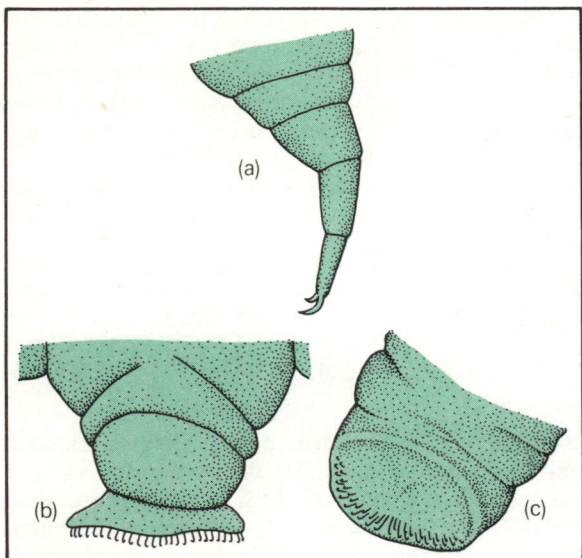

Fig. 18.28 (a) Jointed leg of caterpillar. (b) 'False' leg seen from the side and from below (c).

The **larva** (caterpillar) of the cabbage white butterfly has a head, bearing mandibles similar to those of the adult locust, antennae, simple eyes, and other sensory cells. Behind the head lie thirteen body segments. The first three correspond to the thorax and each carries a pair of jointed 'true' legs. The third to sixth abdominal segments each carry a pair of fleshy, unjointed legs which have many hooks at their ends (Fig. 18.28). Because these do not correspond to legs in the imago they are called **false legs**. They are used for clasping as the animal searches for food and, in conjunction with the true legs, for movement. The animal loops the middle of its body and draws the rear end up to the front end.

The first thoracic segment and the first eight abdominal segments each possess a pair of spiracles.

The function of the larva in the insect's life cycle is to feed. In contrast to the locust type of development, the young in this life cycle are able to exploit a different kind of habitat and different kinds of food from the imago. Larva and imago do not compete with one another. After several moults (the number varies from five to seven), the larva migrates to a dry sheltered place and spins a silk cocoon.

The pupa emerges from the last moult, and as it develops the features of the adult become visible through its skin (Fig. 18.27). In particular, the wings develop. But the whole internal body structure of the imago is so different from that of the larva that an immobile, non-feeding stage is needed so that the body can be completely broken down and rebuilt.

the imago and the earlier stages of the life cycle (Fig. 18.27). Indeed the differences are so great that it is only by studying the life history that one becomes aware that these are stages in the life of one animal and have any relationship to one another. All the cells throughout the life cycle are genetically identical. They are produced by mitoses from the zygote, so different parts of the genetic information in the nucleus are used at different times in the life history.

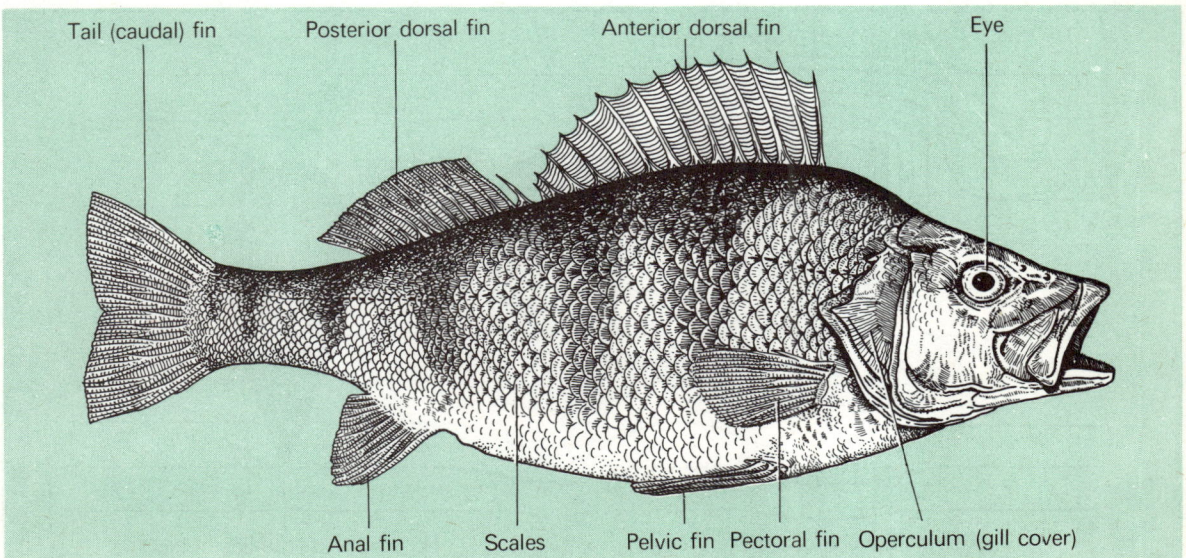

Fig. 18.29 External features of a perch.

The function of the imago is to reproduce. On mating, the life cycle is completed. The imagos of some insects do not feed at all, but the cabbage white butterfly feeds on nectar which it sucks from tubular flowers through its coiled and extensible proboscis.

Fish

The fossil record shows that fish were the first vertebrates to evolve. They first appear in the late Ordovician period about 440 million years ago and have remained the dominant fauna of the sea ever since. Today there are about 37 000 species of vertebrates in the world and 20 000 of these species are fish.

They possess the basic vertebrate body plan (Fig. 2.8). Devonian fish gave rise to the Amphibia and so to the truly land vertebrates, the reptiles, the birds, and the mammals. In this brief account we have space for only two aspects of their lives, their movement and their breathing.

Movement

All fish are streamlined (Fig. 18.29). The advantage of this shape can be appreciated by considering Fig. 18.30, which represents three objects moving through water. As the square block moves forwards it puts pressure on the water ahead of it. At its back end a region of low pressure develops, and this exerts a **drag** which resists the object's forward motion. If the front end is rounded the water is pushed aside more easily, reducing the forward area of high pressure, though the drag remains.

This is overcome by tapering the back end. The streamlined form is found in the earliest known fish (e.g. *Hemicyclaspis*, Fig. 17.1). The animal possessing it needs less energy in order to move forwards through water. This in turn means that the animal needs to find less food, giving it a keen evolutionary advantage over less streamlined forms.

Fins Figure 18.29 shows the many fins in a perch. Only two pairs, the pectoral and pelvic, correspond to the limbs of higher vertebrates. These are attached to the backbone via the pectoral and pelvic girdles (see skeleton of perch, Fig. 2.9(a)).

Pectoral fins Most fish propel themselves to a certain extent by means of their pectoral fins, and some, such as *Polypterus* (Fig. 18.31(a)) rely almost entirely on this method. Its efficiency is limited because of the weak musculature involved in moving the pectoral fins compared with propulsion by means of the tail.

Most fish also use the pectoral fins as brakes, keeping them folded against the body when moving and extending them in order to slow down (Fig. 18.31(b)). They are also used to correct **pitching** movements. These are up and down movements along the long axis of the body, see Fig. 18.32. This is especially effective when the pectoral fins are attached to the body horizontally as in sharks (Fig. 18.33) and primitive bony fish like the gar pike (Fig. 18.31(c)). Here they act like hydrofoils; but such a fish has no brakes.

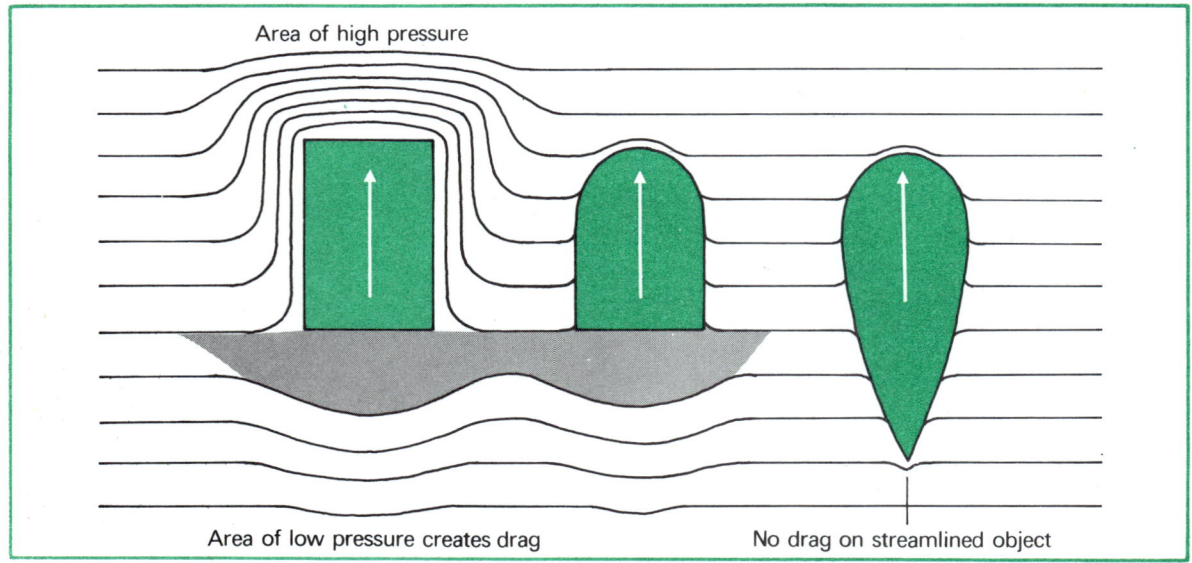

Fig. 18.30 The effects of forward pressure and drag can be reduced by streamlining.

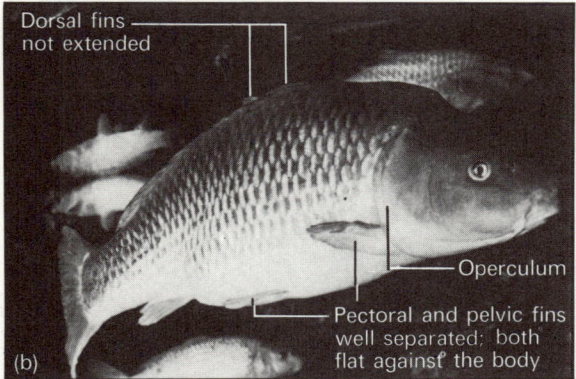

Dorsal fins not extended

Operculum

Pectoral and pelvic fins well separated; both flat against the body

Fig. 18.31(a) *Polypterus* moves by rowing with its pectoral fins. The pelvic fins are situated well back.
(b) Carp.
(c) Gar pike. Note the horizontal insertion of pectoral and pelvic fins and the very posterior dorsal and anal fins.

Pelvic fins　These are also used to correct pitching. In sharks and some bony fish (e.g. carp, Fig. 18.31(b)) the pectoral fins are placed well forwards and the pelvic fins well back, like the limbs of tetrapods. But this arrangement makes swift turning movements difficult, and in advanced bony fish such as the perch (Fig. 18.29) the pelvic fins are set forwards, almost underneath the pectorals.

Dorsal fin　Dorsal fins are sometimes fleshy but often contain skeletal rods in which case they can be

extended and retracted. Some fish, such as perch, have both an anterior movable dorsal fin and a posterior fixed one. They and the **ventral** and **anal** fins are used to correct **rolling** by extending the body surface area and so acting as stabilizers. Rolling is rotation about the long axis of the body, Fig. 18.32(b).

They also correct **yaw**. This is a sideways deflection of the front part of the body which results from the propulsive action of the tail (Fig. 18.32(c)). It is particularly strong during turning, and many fish only extend their dorsal and ventral fins when turning.

In the gar pike (Fig. 18.31(c)), yaw is corrected by a single dorsal fin set right at the back of the body and acting like a weather vane.

Caudal (tail) fin　This is like the blade of an oar. Its sideways thrust against the water by the muscular tail provides the fish's main propulsive force.

In sharks and primitive bony fish the caudal fin is asymmetrical and has a very pronounced upper lobe (Fig. 18.33). These fishes are denser than water. They are bottom dwellers and when they stop swimming they sink. Movement of the asymmetrical tail provides them with lift, and this is enhanced by the horizontally-attached pectoral and pelvic fins, which act like aeroplane wings.

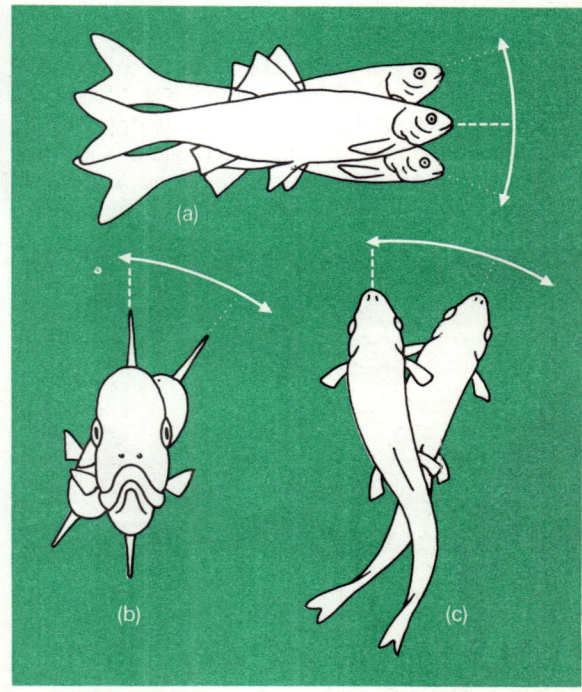

Fig. 18.32 Pitch (a), roll (b), and yaw (c).

237

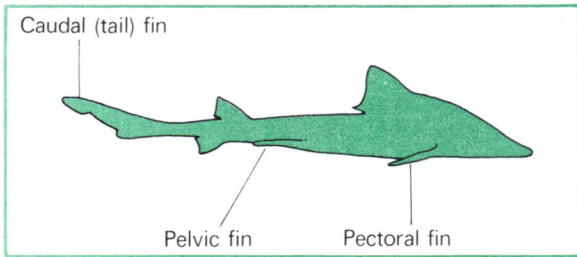

Fig. 18.33 Outline of a dogfish showing the asymetrical tail fin and the horizontal attachment of the pectoral and pelvic fins.

In order to rise from the sea bottom to pursue prey, such fish have to expend energy. In many sharks the problem of density has been solved by the development of massive livers which account for 25% of the total body volume and contain about 80% of fat whose low density makes the animal bouyant (Fig. 18.34).

In most fish bouyancy is provided in a different way. They possess a **swim bladder**. This is a large hollow sac lying above the gut. Gas can be secreted into it and withdrawn from it, adjusting the animal's buoyancy so that it can float motionless at any depth. In consequence the tail fin is symmetrical, since it is required only to provide forward thrust and not lift (Fig. 18.29). The pectoral and pelvic fins, freed from the necessity of providing lift, have become smaller, more flexible, and adapted for use as brakes.

Respiration

Under normal conditions about 2 cm³ of air dissolves in 100 cm³ of water. However, dissolved air contains about 33% of oxygen as opposed to 20% in atmospheric air because oxygen is a more soluble gas than nitrogen. Fish extract dissolved oxygen from dissolved air by means of their gills.

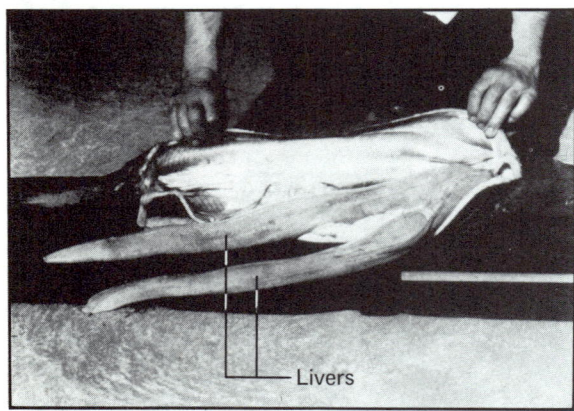

Fig. 18.34 Shark opened to display its enormous fat-filled liver.

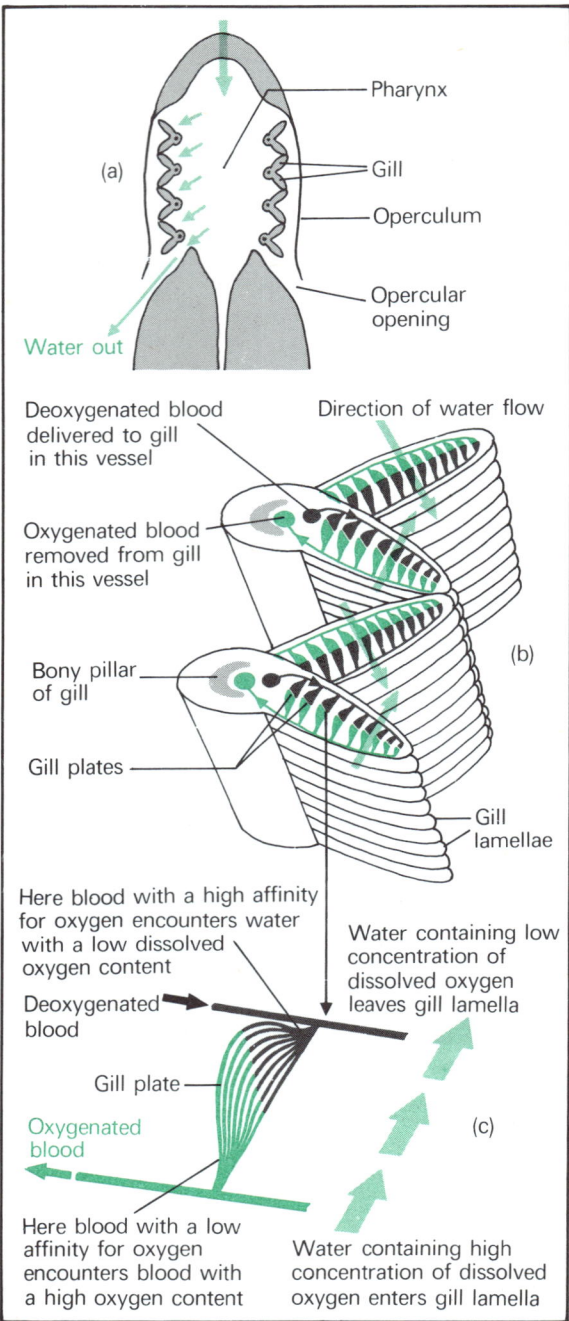

Fig. 18.35(a) Gills of a bony fish.
(b) Detailed stereogram of gill showing direction of water flow over the lamellae.
(c) Detail of one gill plate, showing how deoxygenated blood flows in the opposite direction from the water which is passing across the lamellae.

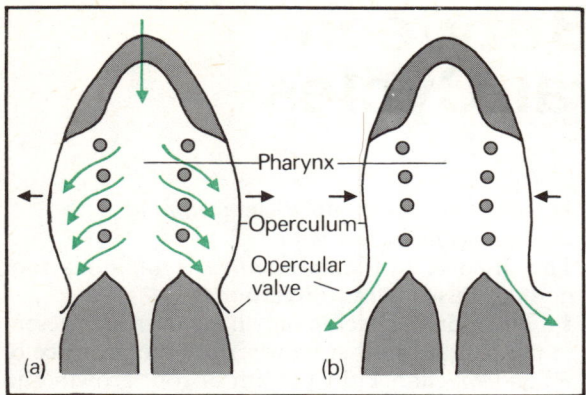

Fig. 18.36 Water is taken into the mouth cavity by the lowering of the floor of the mouth and the movement of the operculum outwards (a). Water is expelled over the gills by the reverse movements (b).

Gills Like mammalian lungs, gills are structures with a very large surface area richly supplied with blood capillaries. For example, in a mackerel there are about 10 cm^2 of gill surface for every gram of body mass.

The circulating red blood cells are brought very close to oxygenated water so that the diffusion path of oxygen from the water into them is minimal.

Figure 18.35(a) shows the structure of the mouth and **pharynx** of a bony fish in diagrammatic form. Each gill is constructed of a vertical bony pillar to which are attached two stacks of **gill lamellae**. These lie one above the other horizontally. Their large surface area is further increased by vertical projections called gill plates. Examine Fig. 18.35 and make sure that you understand the positions of the gill lamellae and the gill plates, and the direction of flow of blood through them.

Counter-current flow in the gills Oxygenated water is forced from the pharynx between the gill lamellae. It passes in exactly the opposite direction to the deoxygenated blood that is flowing through the capillaries in the gill plates.

This arrangement means that water with a high content of dissolved oxygen encounters first the blood which is leaving the lamellae. This blood has already picked up some oxygen in its journey through the gill plates, so it has a low affinity for oxygen (Fig. 18.35(c)). When the water reaches the far side of the lamellae it encounters freshly-delivered blood whose high affinity for oxygen removes the remaining dissolved oxygen from the water.

In this system oxygenated blood is flowing in the opposite direction from oxygenated water. It is therefore called a counter-current system. It removes about 80% of the dissolved oxygen from water passing over the gills. Its efficiency is shown by an experiment in which water was made to flow over the gill lamellae in the opposite direction from normal. Only one fifth as much oxygen was removed from it.

Mechanics of respiration Water is taken into the mouth and forced over the gills by the combined activities of the mouth itself, the floor of the mouth, and the **operculum**. This is a muscular flap of skin which covers and protects the gills. It can be seen in Fig. 18.31(b).

The mouth gapes. Its floor is lowered, and the operculum is moved outwards. This increases the volume of the pharynx and water is drawn into it. (Fig. 18.36). The floor of the mouth is now raised and the operculum is moved inwards. These actions squeeze the water in the pharynx and force it out over the gills. Water is prevented from leaving the mouth by the closure of two folds of skin along the upper and lower jaws.

19 Associations Between Organisms; Natural Cycles

Organisms depend on each other in various ways and live in varying degrees of intimacy with one another. In the broadest sense all animals live in association with all plants, because plants produce by photosynthesis the carbohydrate which is the source of chemical energy for all living things.

Symbiosis

An intimate association between two dissimilar organisms from which both derive benefit is called **symbiosis**, and the participants are called **symbionts**. On page 63 we saw the symbiotic association between a cow and the bacteria and protozoa in its gut. Another example is the symbiotic association between particular insect species and flowering plant species in pollination (see Fig. 13.6).

Many examples of apparent symbiosis are known and there has been no shortage of ingenious suggestions as to how the various partners benefit from their association. Only a few examples, however, have been submitted to experimental investigation.

Root nodules

Repeated over-planting of crops exhausts soils, and yields become progressively smaller. The Romans were probably the first to practice crop rotation in order to counter this effect. A depleted soil can be enriched again by sowing it with clover, vetch, or other **leguminous** crop, which is ploughed in at the end of the season. The legumes rot. Since the soil is then enriched, they presumably return to the soil more material than they took from it in growing.

The Leguminosae is the second largest family of flowering plants. Most species form swellings on their roots called **nodules** (Fig. 19.1). As these are not produced by other crop plants, it seemed likely that enrichment of the soil was associated with nodules. Towards the end of the nineteenth century, when Pasteur had established the existence of bacteria, microscopic examination of nodules showed that they contained large populations of a bacterium which was named *Rhizobium*. Since then, bacteria-containing nodules have been found on plants of many other families than the Leguminosae.

Is the association of **Rhizobium** *with the leguminous plant symbiotic?*

The discovery of *Rhizobium* in leguminous root nodules raised various questions:

1 Do nodules develop only if bacteria are present in the plant? In other words, does the presence of *Rhizobium* change the pattern of root growth and cause it to produce a nodule?
2 Can plants be raised without nodules? If they can, do they grow as well as nodule-bearing plants?
3 Suppose that the association is of benefit to the legume, then is it also of benefit to *Rhizobium*?

Some of these questions were answered by a series of experiments conducted in 1887, and Fig. 19.2 is a photograph of the results.

Plants labelled *A* are peas sown in heat-sterilized sand that had been thoroughly washed to remove any traces of nitrates or other nitrogenous substances. They were watered with a culture solution that contained no nitrate or any other form of nitrogen. The very small amount of growth they made was probably due entirely to the food reserves in their cotyledons. They did not develop root nodules.

Plants labelled *B* were grown similarly but a few cm³ of water in which soil had been shaken was added to the sand. Apart from this these plants received no added form of nitrogen. The volume of soil solution was far too small to have supplied

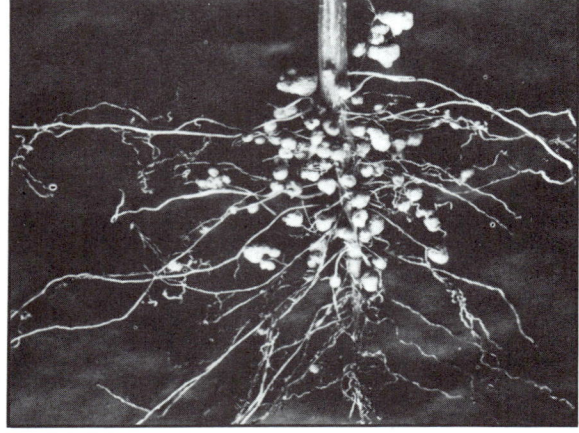

Fig. 19.1 Nodules on the root of a legume.

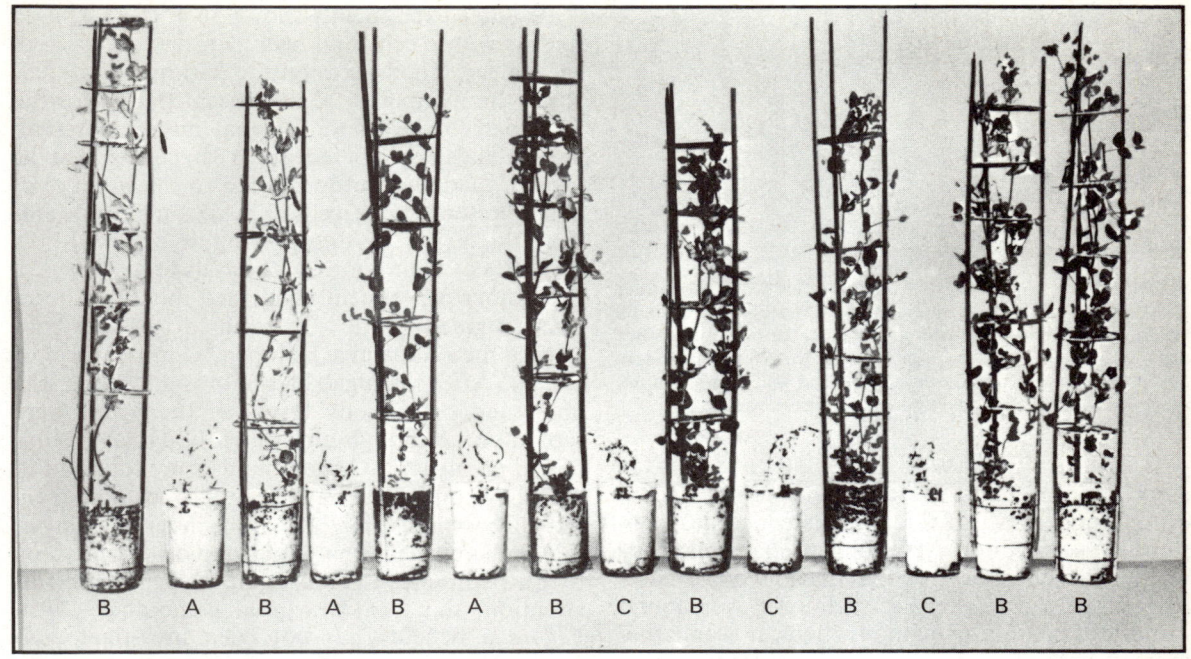

Fig. 19.2 The effects of nodule formation on the growth of peas. A. Plants grown on sterilized sand and watered with culture solution containing no nitrate or other form of nitrogen. Very little growth, and no nodules formed. B. Plants grown as A, except that a little unsterilized soil solution was added as well. Vigorous growth, nodules produced. C. Plants grown as B, except that the soil solution was sterilized. Poor growth, no nodules formed.

sufficient nitrates for growth, but these plants grew vigorously. All developed nodules containing *Rhizobium*.

Plants labelled *C* were treated in the same way as plants *B* except that the added soil solution was first sterilized. Their growth was about the same as plants *A*. They did not develop root nodules.

The conclusions drawn were as follows:

a As the nitrogen required for growth by plants *B* did not come from the soil solution it must have come from the soil air. That is to say, atmospheric nitrogen gas, a very unreactive element, had been used for growth.

b This **nitrogen fixation** occurred only in the presence of root nodules and the *Rhizobia* they contained.

c The soil solution used to water plants *B* must have contained *Rhizobium* which infected the roots and gave rise to the nodules. In the soil solution used to water plants *C*, the *Rhizobia* had been killed by sterilization.

It seemed reasonable to assume that *Rhizobium* is able to turn nitrogen gas into nitrate. This is passed on to the legume which uses it for its growth. It was only with the advent of radioactive

and other isotopes in the late 1940s, however, that this idea was proved correct. Nodule-bearing roots were exposed to soil air containing the isotope ^{15}N and it was possible to trace these atoms through the *Rhizobium* into amino acids which found their way into all parts of the legume.

Clearly the legume benefits from this association, especially in nitrate-deficient soils such as those that have been over-cropped. It is not dependent on the association, however, for it can grow well by using nitrates from soil solution. It has not been *proved* that *Rhizobium* derives any benefit from the association, though it is likely that it obtains a supply of respirable carbohydrates from the legume.

Symbiosis between algae and invertebrates

Many symbioses are known in which unicellular plants (algae) live inside the cells of invertebrate animals. The algae are rarely digested. They behave like a new organelle, photosynthesizing in the animal's cells. Experiments have shown that as much as half the carbohydrate they produce passes to the animal, which thus obtains a supplementary 'free' source of food.

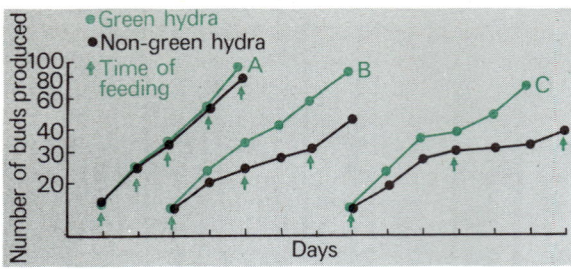

Fig. 19.3 Green and non-green hydra fed each day show little difference in growth as measured by the numbers of buds produced (A); but when partly starved by feeding every other day (B), or every third day (C), green hydra grow more vigorously than non-green and only slightly less well than when fed each day. Arrows show the times of feeding.

It is difficult to assess the extent to which the association benefits the animal in natural conditions since most of the animals concerned continue to use their normal feeding mechanisms. They will live in the laboratory if artificially deprived of their algae, but as they are seldom, if ever, found in natural conditions without them, it seems that animals with algae are more successful than those without.

The algae also produce oxygen and consume carbon dioxide during their photosynthesis, and both these processes are likely to be of benefit to the animal cells. However, these aspects have been less well studied.

Green hydra

In green hydra a population of unicellular algae, *Chlorella*, lives inside the cells of the inner body layer (Fig. 18.10(c)). Each algal cell is enclosed within a vacuole and so is not in immediate contact with the animal cytoplasm. *Chlorella* pass into the offspring of hydra either through buds or eggs. By prolonged exposure to darkness or to other experimental treatments the algae can be made to disappear and the growth of green and non-green forms of the animal can then be compared.

Figure 19.3 shows the results of such an experiment. Growth was measured in terms of the number of buds produced (Y, vertical, axis). When the animals were fed brine shrimps every day (arrows), there was no difference in the growth of green and non-green hydra (curve *A*). When fed only every other day (curve *B*) or every third day (curve *C*), the green type grew much better than the non-green and only slightly less well than those fed each day. Under these circumstances, therefore, *Chlorella* must have been supplementing hydra's food supply.

Another experiment, shown in Fig. 19.4, used hydras whose cells had been deprived of some of their algae. The figures on the X (horizontal) axis show the number of algae present per hydra cell as a percentage of the normal number present. The animals were not fed, and it can be seen that the growth made was proportional to the number of algae present in the cells. This strongly indicates that the algae were supplying hydra with the products of their photosynthesis as food.

Another experiment confirmed this hypothesis. Starved green hydra were exposed to water containing the radioactive isotope ^{14}C in the form of $^{14}CO_2$. After the algae had photosynthesized, the inner and outer body layers of the hydra were separated. Although the outer body layer contained no algae, ^{14}C compounds were found in it, and they could only have been passed there from the photosynthesizing cells of the inner body layer.

Another benefit that the symbiosis confers on hydra is the green colouration. Depending on its situation, this provides good camouflage. How *Chlorella* benefits has not been investigated. It probably obtains amino acids from the animal's digested food. It is also said that hydra 'protects' *Chlorella*, but this really means nothing.

Convoluta

This small flatworm (Fig. 19.5) is found in the intertidal zone of beaches in Brittany and the Channel Islands. It has a symbiotic association with the unicellular alga *Platymonas*, which floats in the neighbouring plankton. This is so intimate that when the creature was first described in 1910 it was referred to as a 'plant-animal'.

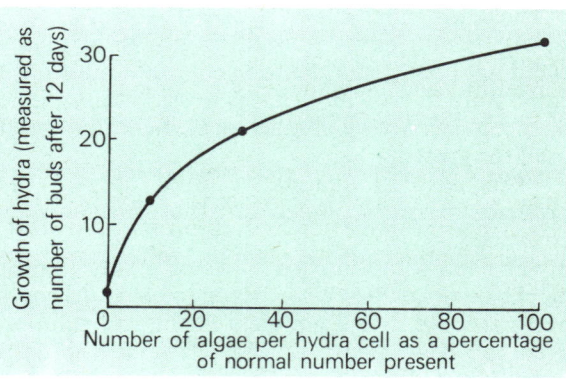

Fig. 19.4 Green hydra are deprived of some of their algae. The X (horizontal) axis shows the number of algae present per cell as a percentage of normal. The animals are starved, and their growth is proportional to the number of algae contained in their cells.

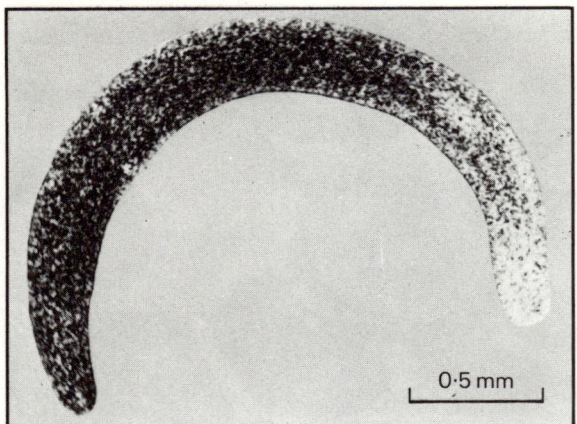

Fig. 19.5 *Convoluta*, a marine flatworm containing symbiotic unicellular algae.

The alga lives in cells in the animal's outer layer. When free-living the alga has a precise shape and structure, including an eye-spot, flagellum and cell wall, rather like *Euglena*. But when inside the animal cells, all these structures are lost. If it were not that it is enclosed within a vacuole, and so has no immediate contact with the animal cytoplasm, *Platymonas* would look very much like an integral part of the animal cell.

The animal's eggs are infected afresh at each generation, and if care is taken to prevent the alga from having access to them, white *Convoluta* develop. Although these feed, they always die. Infected green forms, however, attain maturity and then enter a prolonged phase when they cease to feed as animals and rely entirely on their algae for food. At the end of their lives, the animals digest the algae and then die.

Here, then, is an organism in which the basic distinction between animals and plants is very blurred indeed, for while it is an animal, it feeds for most of its life like a plant.

Symbiosis between chloroplasts and invertebrates

An even more remarkable symbiosis exists between certain marine molluscs and chloroplasts. (Molluscs are members of the phylum Mollusca, which includes slugs and snails.)

These animals browse on seaweeds, and a common example round British coasts is *Elysia* (Fig. 19.6). They suck out the contents of seaweed cells, but the chloroplasts thus obtained are not digested. Instead they pass into the cells lining the animal's gut, where they remain as active organelles for at least two months. Up to 40 % of their photosynthetic product has been shown to pass

to the *Elysia* cells, which use the materials in various ways.

The chloroplasts are not contained within a vacuole. They are in immediate contact with the cytoplasm of the animal cell, and behave as though they were in a plant cell. The symbiotic cell is thus very much an animal/plant hybrid.

Parasites

Many associations between organisms are parasitic. One organism of the pair benefits and the other gains nothing and is harmed, though not necessarily killed. The best arrangement from the parasite's point of view is not to kill its host, for if it does its food supply is cut off. Disease-causing bacteria living in the mammalian body are examples of parasites.

Animal parasites

Fleas and lice (Fig. 18.13) are **ectoparasites**, that is they live outside their hosts. They possess mouthparts which can pierce the skin of man and other mammals to draw the blood on which they feed.

Often such insects act as **vectors**, transmitting microorganisms which are **endoparasites**. These live inside their host, causing serious disease. Examples are bubonic plague, caused by a virus transmitted by fleas; sleeping sickness, a scourge of Africa, caused by a protozoan called a trypanozome, and transmitted by the tsetse fly (Fig. 19.7); and malaria, also caused by a protozoan, and transmitted by the *Anopheles* mosquito.

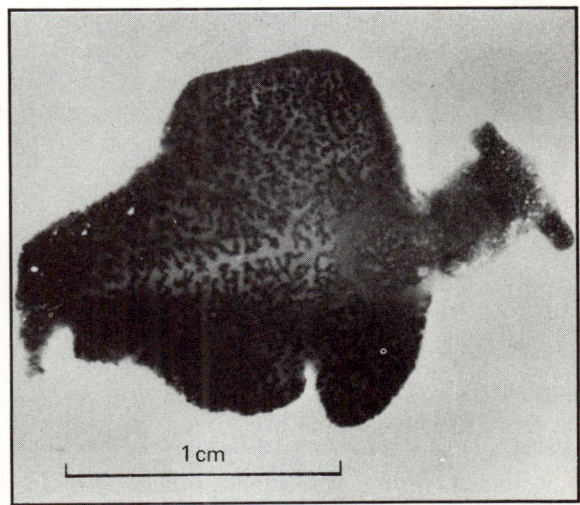

Fig. 19.6 *Elysia*, a marine mollusc containing symbiotic chloroplasts derived from the seaweeds on which it feeds.

243

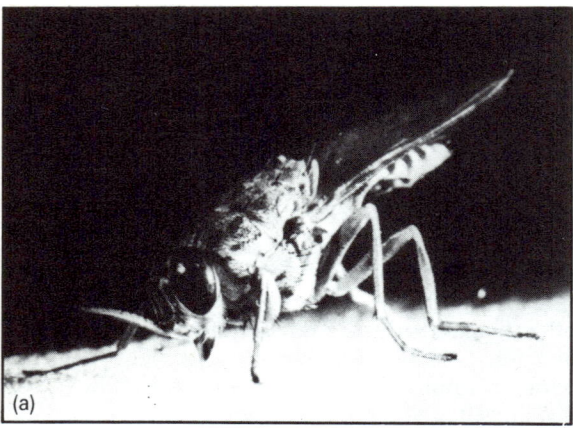

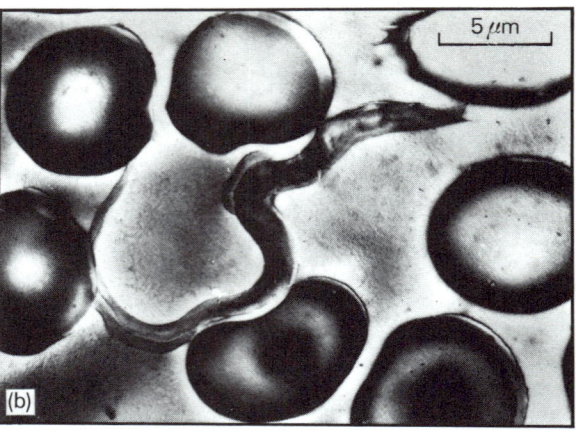

Fig. 19.7(a) Tsetse fly feeding on blood, having penetrated the skin of its host with its proboscis. The fly secretes a blood anti-coagulant which is infected with trypanozomes, thus transmitting sleeping sickness.

(b) Trypanozome in human blood smear. Trypanozomes are parasitic protozoans that cause the disease sleeping sickness because of the poisonous substances they release into the blood of the host. (Scanning electron micrograph.)

Some endoparasites are not microorganisms, for example tapeworm (Fig. 19.8), and *Schistosoma sp.*, the cause of bilharzia. This disease is spreading rapidly through Africa at the present time. Both these organisms are flatworms.

Adaptations to the parasitic mode of life A parasite has no problems of food supply, but when its host dies it must find a new one. This difficulty is often overcome by the use of a **secondary host** as a vector. For example, the *Anopheles* mosquito transfers infected blood from one person to another: the mosquito acts like a minute hypodermic syringe. Parasites often produce vast numbers of offspring because of the low chance of finding a

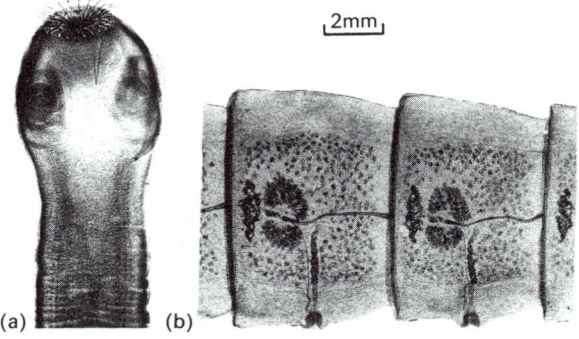

Fig. 19.8 The tape-worm of man, one of the most extreme animal parasites. It lives in the host's intestine and has no gut, sense organs, or means of locomotion. The head (a) is anchored to the intestinal wall by hooks. The rest of the body consists of many identical segments (b) which contain little but ovaries and testes. Vast numbers of fertilized eggs are passed out in the host's faeces to face the hazardous journey to the next host. (Courtesy Carolina Biological Supply Company.)

new host. Often there is a dormant, resting stage in the life cycle.

Plant parasites
These include fungi, bacteria, viruses, and nematode worms. It is worth emphasising, however, that the vast majority of bacterial and fungal species cause no harm to natural plants or to man's crops.

In 1965 the U.S. Department of Agriculture estimated that diseases of crop plants caused losses of about 3250 million dollars in the United States. The figure for world losses was about 25 000 million dollars, some 11 % of total production. A celebrated instance of a plant disease with considerable social consequences was the potato blight in Ireland in the mid 1840s. This was caused by a fungus. It eliminated the subsistence crop of the Irish, and over a million emigrated to the United States in consequence.

Damage can take many forms. The plant may be killed or only debilitated. Sometimes the disease-causing organism produces poisons which kill or weaken the plant. Many fungal parasites produce special hyphal branches called **haustoria** which penetrate the host cells and absorb nutritive materials from them.

Organisms and their environment
All organisms exist in particular environments which contain other organisms to whose life style they are therefore related. Figure 19.9 shows that about three quarters of the earth's surface is covered by the oceans. The fossil record shows

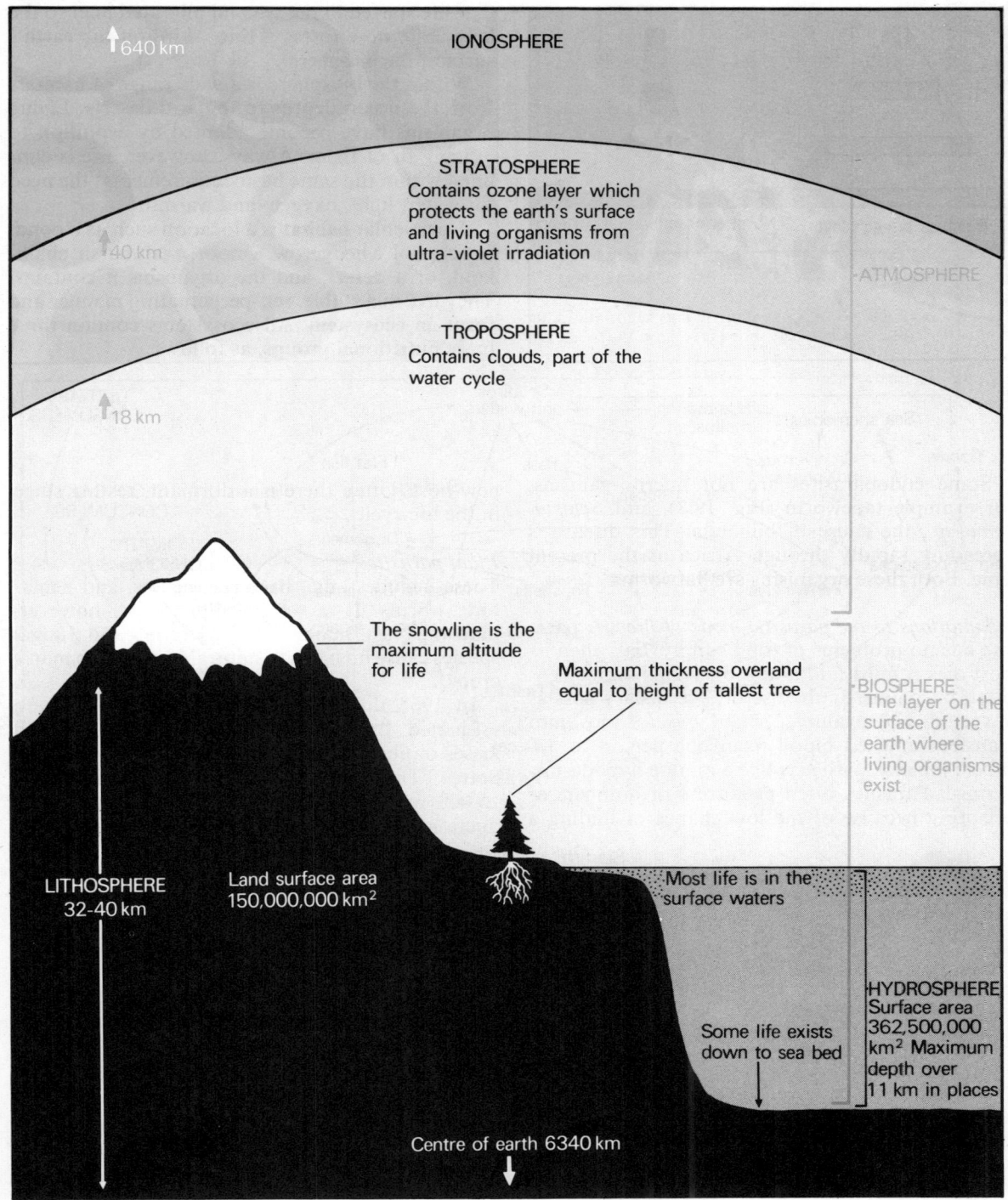

Fig. 19.9 Diagram to illustrate the biosphere and other regions in and above the earth. (After Elliott.)

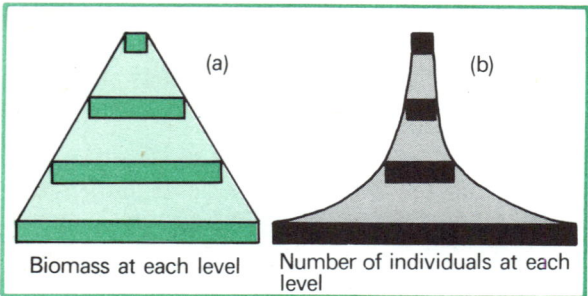

Fig. 19.10(a) Pyramid of biomass. (b) Pyramid of numbers.

Fig. 19.11(a) Food web of some shore-living organisms. (After Elliot.) (b) Pyramid of numbers showing some of the organisms in (a).

that life started in the sea and migrated later to the land. Life now forms a thin 'skin' on the earth's surface, the **biosphere**.

Within the biosphere is a wide range of habitats from the ocean depths to the arid deserts. Living organisms have become adapted by evolution to inhabit all of them. Always, however, life is confined within the same basic requirements: the need for water, light, oxygen, and warmth.

A particular **habitat** is a location such as a pond, a rockpool, a hedgerow, a moor, a saltmarsh, chalkland, or a desert, and the organisms it contains. They live in a stable, self-perpetuating manner and form an **ecosystem**. All ecosystems contain three basic nutritional groups, as follows:

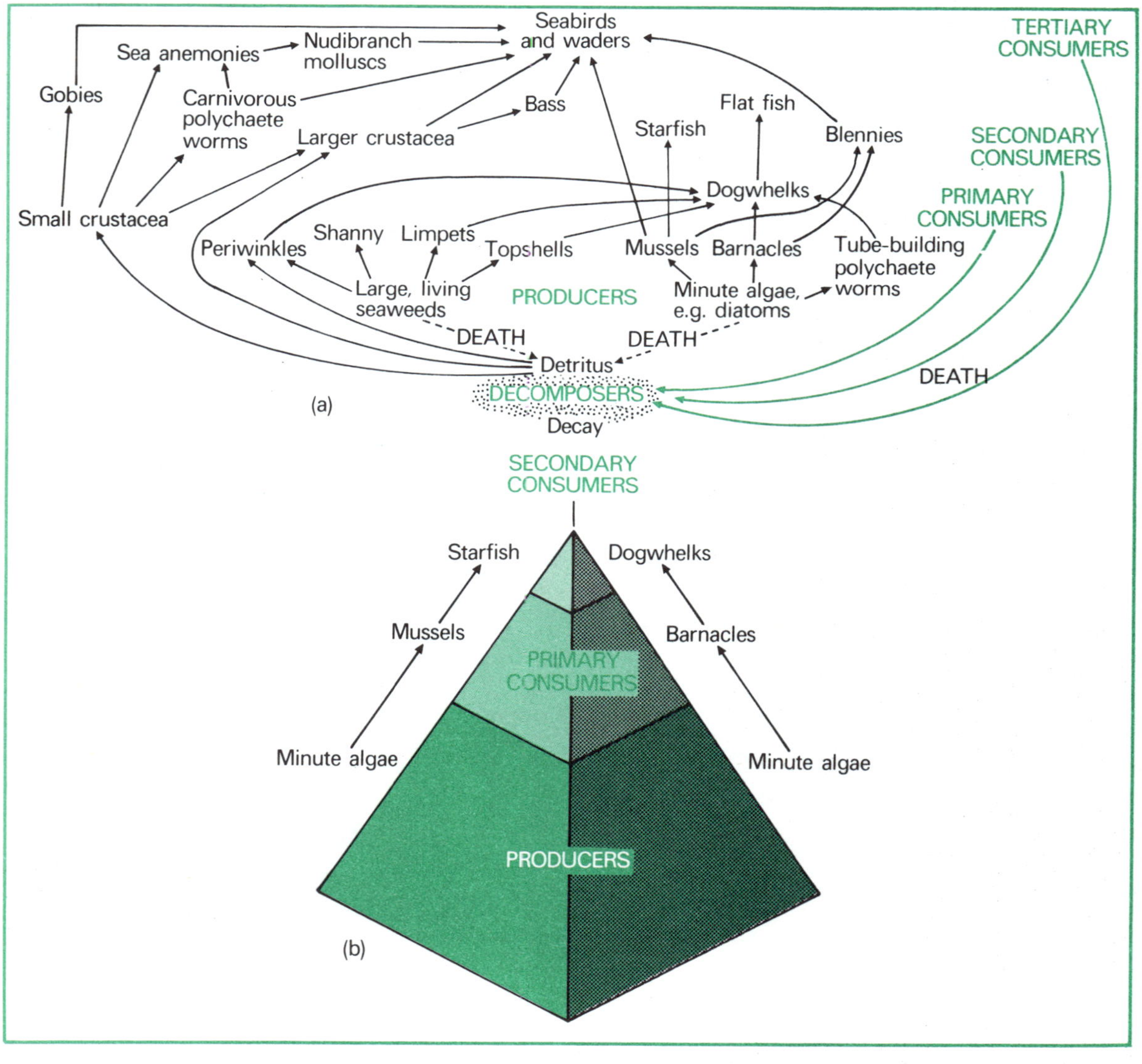

1 The **producers**. These are green plants whose photosynthetic activities produce carbohydrates and, from them, proteins.

2 The **consumers**. These are animals, and may exist in more than one 'layer', for example primary **consumers** (herbivorous animals) and **secondary consumers** (carnivorous animals).

3 The **decomposers**. These are the saprophytes, bacteria and fungi. Their secretion of digestive enzymes on to dead animals and plants returns the raw materials of life to the soil and to the air.

Food chains

A group of organisms within a habitat which rely on each other for food constitute a **food chain**. Two examples are: (i) grass (producer)→ antelope (primary consumer)→lion (secondary consumer); (ii) dandelion (producer)→snails (primary consumers)→skylark (secondary consumer).

At each stage of transfer of living material there is a considerable loss. Only 5–20 % of the chemical energy contained in plant material becomes part of the primary consumers. The rest is dissipated as heat. The same applies to transfer from the primary to the secondary consumers.

There must therefore be a large number of producers to support relatively few primary consumers, and a large number of primary consumers to support relatively few secondary consumers.

Organisms in a food chain thus form a **pyramid of numbers** (Fig. 19.10(b)). There is a corresponding **pyramid of biomass**, the total body mass of all the individuals forming a particular link in the chain (Fig. 19.10(a)).

In fact it is only in extreme habitats, such as deserts, where the range of species is limited, that simple food chains exist. In the examples quoted, antelope can feed on many types of vegetation besides grass, and lions on many types of animals besides antelope.

Food webs

The more complex inter-relationships of producers and consumers are called **food webs**. Figure 19.11(a) is a food web of sea shore organisms. Take in turn each of the organisms shown here and decide whether it is a producer or a primary, secondary, or tertiary consumer. Some of these organisms have been arranged as a pyramid of numbers in Fig. 19.11(b).

Natural cycles

The relationship

producers ⟶ consumers ⟶ decomposers

forms a continuous, dynamic cycle. Chemical elements, some of them rare, are collected together into organisms. Using light energy they are con-

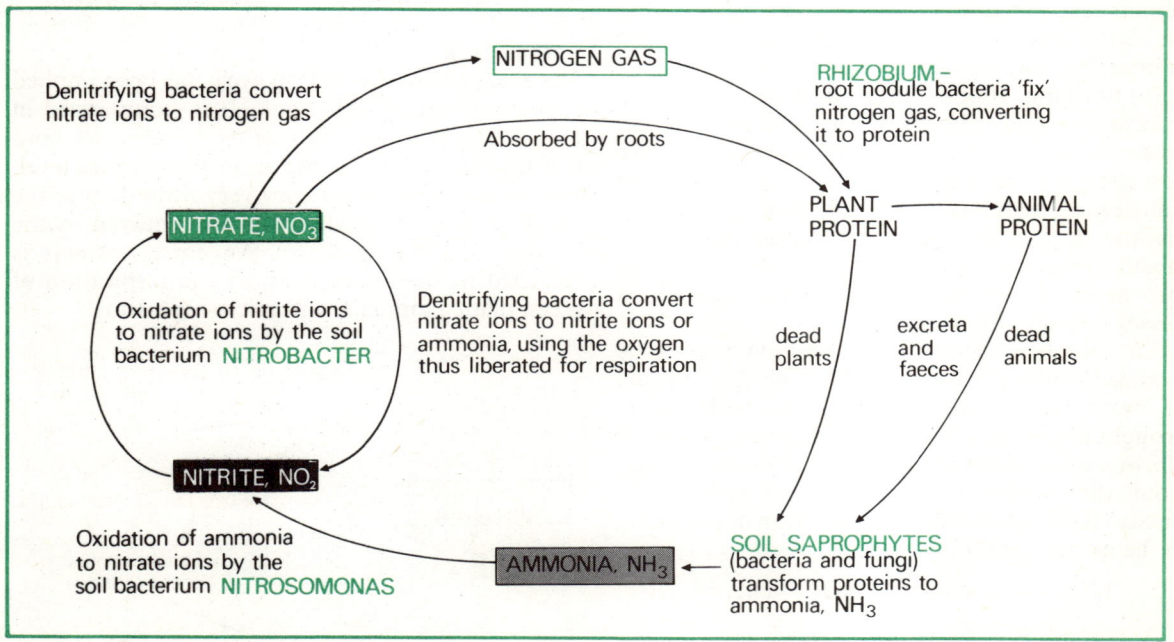

Fig. 19.12 The nitrogen cycle.

verted into cell chemicals, notably proteins, fats, and carbohydrates. When the organism dies, the chemical elements are dissipated.

Nearly all cell chemicals contain carbon, and most contain nitrogen, and there is therefore justification for considering the ways in which these elements circulate in the **nitrogen** and **carbon cycles**.

Nitrogen cycle (Fig. 19.12)

Nitrogen gas constitutes four fifths of the atmosphere. It is exceedingly unreactive. Only a few bacteria and blue-green algae possess the enzymatic machinery to use it directly.

Nitrogen enters the living world from soil solution in the form of the nitrate ion (NO_3^-) which is absorbed by plant roots. Although other nitrogen-containing ions exist in the soil solution (nitrite, NO_2^- and ammonium, NH_4^+), these are not absorbed by roots. It follows that some process must continually regenerate nitrate ions in the soil solution or they would all be absorbed and plant life would come to a halt.

Once inside the plant, nitrate ions are taken to the leaves via the xylem. Here, using carbon compounds generated in photosynthesis, they are transformed into amino acids and so into plant proteins.

Some plant material is eaten by the primary consumers, but much dies and decays. All consumers excrete and defaecate. The excreta fall on to the soil where they are rotted. Finally, the animal dies and it too decays in the soil.

In bringing about decay, bacteria and fungi secrete digestive enzymes on to the dead material. Some of the resulting soluble compounds are absorbed by the saprophytes, but as with other transfers of material between organisms, much is dissipated. The nitrogen in the dead material is mostly present in proteins. These are broken down into ammonia (NH_3), a very soluble gas which passes into the soil solution.

To complete the cycle the ammonia must be oxidized to the nitrate ion (NO_3^-) which the roots of more plants can absorb. This oxidation is brought about by soil bacteria. One species, called *Nitrosomonas*, does not derive its chemical energy from the oxidation of glucose in respiration. Instead it obtains it from the oxidation of ammonia to the nitrite ion (NO_2^-):

$$NH_3 + oxygen \longrightarrow NO_2^- + energy.$$

A second soil bacterium, *Nitrobacter* derives its chemical energy from a similar oxidation of nitrite ions to nitrate ions:

$$NO_2^- + oxygen \longrightarrow NO_3^- + energy.$$

These bacteria therefore have a symbiotic relationship with plants. Plants provide them with ammonia, while they provide plants with nitrate ions.

Adjuncts to the cycle

These are the principal chemical transformations in the nitrogen cycle, but other processes are involved. The symbiotic association of *Rhizobium* with the roots of legumes is important in nature and is extensively used in farming. A good legume crop such as clover can 'fix' up to 90 kilos of nitrogen gas per acre in one season's growth, roughly equivalent to the application of one tonne of NPK fertilizer.

NPK fertilizer contains ammonium sulphate, $(NH_4)_2SO_4$, which is produced industrially from nitrogen gas. It has a disadvantage compared with natural sources of nitrate because after the conversion of the ammonium ion to nitrate ion and its absorption by plants, the remaining sulphate ions make the soil acid and unfavourable to the growth of further crops.

Some soil bacteria cause a *loss* of soil nitrate. They convert it to nitrite ions, ammonia, or even to nitrogen gas. These **denitrifying** bacteria are particularly active in poorly aerated soils and the denitrification processes liberate oxygen which they used to oxidize sugar in their respiration.

Carbon cycle

The existence of the carbon cycle has been implied at many points in this book. It is summarized in Fig. 19.13. Carbon dioxide is a rare gas constituting only 0.03 % of the atmosphere at sea level. But its concentration varies very little despite the fact that enormous quantities are removed by the process of photosynthesis. An equal amount is returned to the atmosphere by the respiration of green plants, animals, and microorganisms.

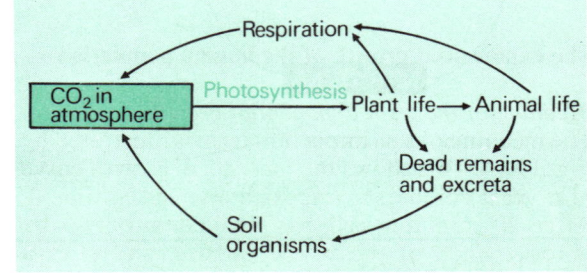

Fig. 19.13 The principal components of the carbon cycle.

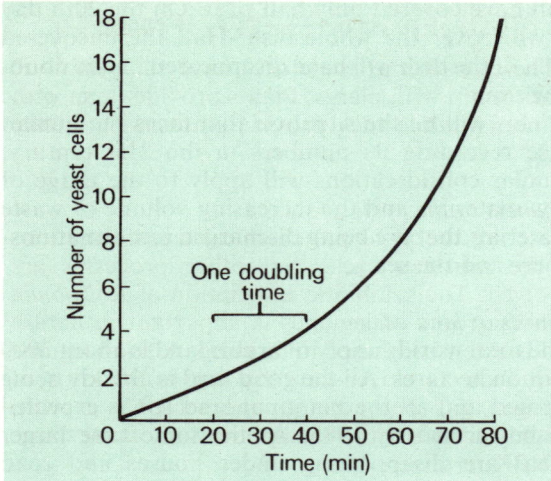

Fig. 19.14 Exponential growth curve of a yeast population.

each of these cells has divided, producing eight cells, and the last plot on the graph shows the position 80 min after the start, when 16 cells have been produced.

Such a curve, where a *quantity* (in this case the number of yeast cells) increases by a *constant percentage* (in this case 100%) in a constant time (in this case 20 min) is called an **exponential** curve. One of its important characteristics is that it displays a **doubling time**, a fixed time interval in which the variable on the Y (vertical) axis doubles in amount.

Sigmoid growth Curve

In fact population growth curves, such as that for yeast cells growing in liquid medium, are more usually of the shape shown in Fig. 19.15. This is a **sigmoid** growth curve. Initially the growth is exponential, but it later levels off for a variety of reasons. Some of these are:

1 Increasing competition between individuals for food, which may be diminishing in amount.

2 Increased competition between individuals for space.

3 Unfavourable changes in the environment. Some may be produced by the growth of the organism, such as the accumulation of waste products.

4 Death of the weaker individuals in the population, which therefore no longer reproduce.

Growth of the human population and its impact on natural cycles

The earliest known fossils of creatures that were undoubtedly human, rather than ape-men, have been found in central Africa and are between 300 000 and 600 000 years old. From an initial population of tens, man has produced a present day number of about 4000 million individuals.

The time it has taken for this to happen is a small fraction of the period since life emerged on to the land (Lower Devonian period, about 400 million years ago), or since the evolution of the first mammals (Jurassic period, about 200 million years ago).

The existence of so many people, all with needs for food, shelter, energy, transport, material goods, and all producing wastes both biological and non-biological, is in conflict with the continued existence of other species. To cite but three examples, some species of whale, tiger, and rhinoceros are almost extinct from over-hunting. It is also beginning to interfere with the whole tissue of inter-connected living organisms in the world, of which man is only one.

The exponential growth of the human population

Meaning of the term exponential growth

The meaning of this important term is illustrated by Fig. 19.14. This shows the *theoretical* growth curve of a yeast colony starting with one cell at time 0. After 20 min this cell has divided into two by mitosis. Each of these cells divides, and after a further 20 min four cells are present. 20 min later

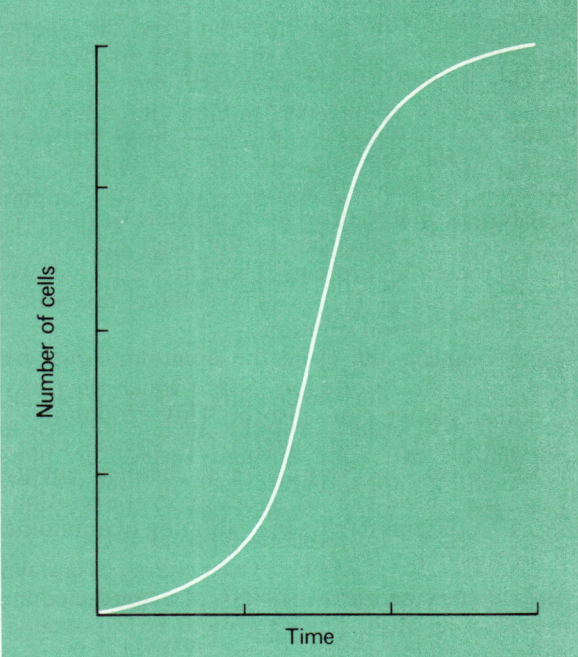

Fig. 19.15 A typical sigmoid growth curve.

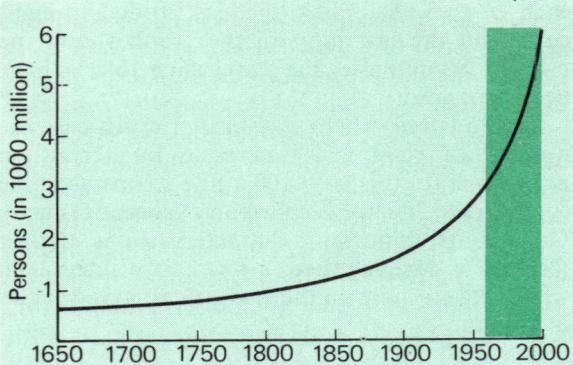

Fig. 19.16 Exponential growth curve of the human population since 1650 (figures after 1970 are estimates). Note the progressive shortening of the doubling time.

The human population is also subject to natural checks such as disease, starvation, and war. But throughout history, man has striven to reduce or even eliminate these scourges. With the spread of modern medicine and public health measures, and the increased productivity obtained from the use of improved crop plants and more efficient farming techniques, the human mortality rate has fallen. At the same time life expectancy has risen from an average of only 30 years in 1650 to an average of 53 years at the present time.

For these reasons the human population growth curve has not yet entered a levelling-off phase. It is still exponential, as Fig. 19.16 shows. This curve is based on historical data and on reliable estimates for a projection into the future. It shows no sign of levelling off before the year 2000 when there will be some 6000 million individuals on earth. An important point to note about Fig. 19.16 is that, unlike the yeast population growth curve in Fig. 19.14, the doubling time of the human population is *progressively decreasing*. Thus in the period starting 1650 the doubling time was approximately 250 years, but in the period starting 1970 it will be only 30 years.

The last doubling time

A treacherous and misleading aspect of exponential growth is that a variable such as increase in population can pass through several doubling times without appearing to reach an unmanagable size. However, after a further one or two doubling times it reaches overwhelming proportions.

Consider, for example, a yeast colony of circular outline growing on the surface of a solid medium in a circular dish. Let its growth be exponential with a constant doubling time. If it takes 20 days to cover the whole dish, then on the 19th day it

will have covered only half of it. On the 20th day it will cover the whole dish. Half the uncovered area of the dish will have disappeared in one doubling time.

This will be the situation that faces the human race regarding its numbers in the 21st century. Similar considerations will apply to the usage of raw materials and the increasing volume of waste materials that are being discharged into the atmosphere and the sea.

Limits of land usage

The total world supply of arable land is about 3000 million hectares. All the good land is already being farmed and all the marginal land left is expected to be farmed by 1985. At the same time larger areas are disappearing under houses and road schemes. Figure 19.17 shows the amount of arable land needed to support the rising human population, assuming production to remain as it is at present. (It may in fact decline from exhaustion. Some areas may have to be returned to permanent pasture.)

The curve follows the exponential path of the human population growth curve and cuts the line showing total arable land available at about the year 2000. Because the amount of arable land is decreasing (downwards dip in the arable land curve), this point may be reached somewhat earlier.

As land supplies diminish, man turns more to 'reclaiming' 'wastelands' such as estuaries and marshes, for these seem superficially to be of little value. Yet such areas are among the most biologically productive of the world's ecosystems, and estuaries are spawning grounds for fish and shellfish whose elimination would disrupt the food chains of the sea.

Limits of food production

Figure 19.18 shows the daily human protein and

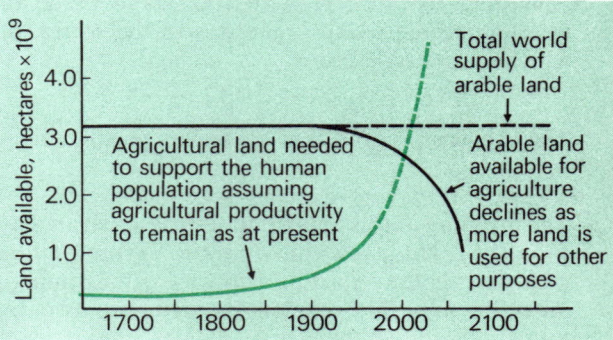

Fig. 19.17 The availability of agricultural land.

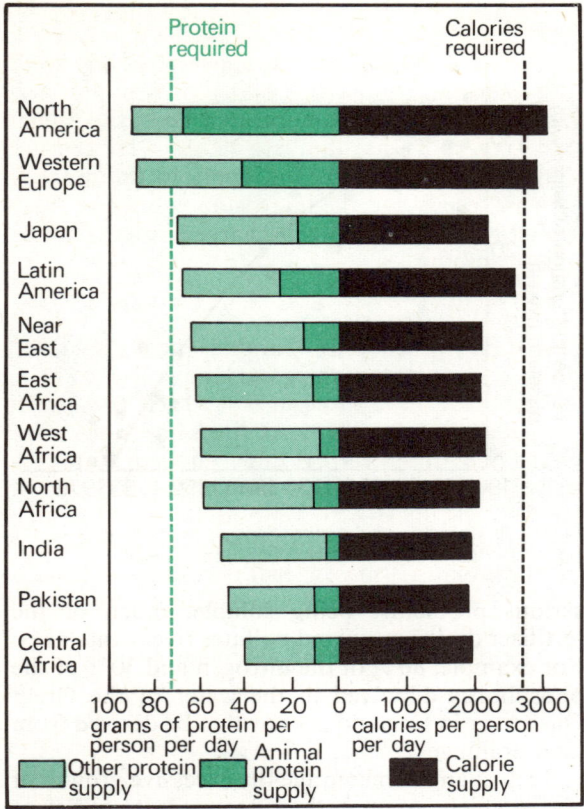

Fig. 19.18 Daily protein and calorie intakes of humans in various parts of the world. Only in North America and Western Europe does the intake equal the need.

calorie intakes in various parts of the world. (Calories are a measure of the chemical energy contained in foods, see p. 47). The vertical dotted lines show the required minimal intake of North Americans, and assuming this applies to all men, large sections of the world population—probably at least one third—are under-nourished.

For example, in Malawi, a small, poor country in central Africa, babies are weaned on to a diet consisting almost entirely of maize meal. Maize protein is deficient in two essential amino acids, and it is not surprising that the mortality rate of children under 5 in Malawi is about 25%.

The availability of food clearly depends on the availability of land on which to grow it, the amount of which is finite.

Use of pesticides and fertilizers
Between 1951 and 1966 world food production increased by 34%. To bring this about required an increase of 300% in the amount of pesticides used and of 140% in the amount of nitrogenous fertilizers.

This may be a help towards a short term solution of supporting the exponentially-increasing world population, but it brings about long term changes in ecosystems which may be irreparable. Insect and fungal pests are sexually-reproducing organisms. Their genetically variable populations contain naturally disease-resistant individuals. When pesticides are used, these individuals are favoured and give rise to resistant populations which require heavier applications of pesticide to control them. Furthermore, pesticides kill indiscriminately, eliminating predators of species which were not formerly pests because their numbers were naturally controlled. After pesticide application, such species may develop into new and equally formidable pests.

An instance of the unexpected effects resulting from the use of a particular insecticide is provided by DDT. This substance was developed during the second world war for use in jungle warfare in the far east. Later its use became widespread. Later still it was realised that DDT is not **biodegradable**, that is it decomposes slowly, if at all, in the soil. It therefore passes along food chains and becomes concentrated in the secondary consumers. When it was used to control the aquatic gnat *Chaoborus* on Clear Lake in the U.S.A., grebes (fish-eating birds) started to die five years later from DDT poisoning. The insecticide had become concentrated in their body fat from the fish on which they fed.

About 25% of the DDT that has been discharged to date is estimated to have found its way into the ocean. If it (or any other pollutant) were to kill the marine plankton, the collection of minute algae and protozoa which floats at the sea surface and which forms the basis of all marine food chains, the whole ecosystem of the ocean would be upset. This could have traumatic consequences for man's supply of fish. It would also affect the concentration of oxygen in the atmosphere, for much of this is given off by photosynthesizing algae in the plankton.

Fortunately this state has not yet been reached, and the fate of materials like DDT cannot be predicted from theoretical considerations alone. The use of DDT is now banned in Britain, the U.S.A., and Sweden, but 100 000 tonnes of it are still liberated annually. It would be unreal to expect underdeveloped countries to stop using it in malaria control programmes unless an equally effective and less harmful substitute were provided.

Chemical pollutants threaten other species
The combined effects of the many man-made

chemicals that are released into the environment, plus the clearing of land, are threatening the survival of 280 species of mammals, 350 species of birds, and 20 000 species of plants. We must remember that the genetic diversity that is represented in these and other species has taken hundreds of millions of years to evolve. To eliminate other species of life is not good sense even from a purely self-interested point of view, for the development of new hybrids of food plants and animals depends on the use of natural stocks. To drain the gene pool is to lose a resource that can never be replaced.

Damage to the ozone layer

Great concern is being expressed about the effects of chemicals called fluorocarbons that are used as propellants in aerosols such as fly killers and deodorants. Millions of tonnes of fluorocarbons have already been released into the atmosphere, and it seems likely that they may cause the break-up of the ozone layer that surrounds the earth and protects life from the effects of ultra-violet radiation from the sun (p. 265). If the ozone layer were to be removed, the only life possible on earth would be under its surface waters.

The Green Revolution

The Green Revolution is a term used to describe the development of genetic hybrids of cereals which have high-yielding capacity. It started in Mexico in 1940, and from the point of view of improving food supplies in under-developed countries has had considerable success. If the world human population were not increasing exponentially, it might result in adequate nourishment for all.

However, the new strains only produce high yields when large amounts of artificial fertilizers, up to 27 times the applications used for 'normal' strains, are employed, and this raises a new crop of

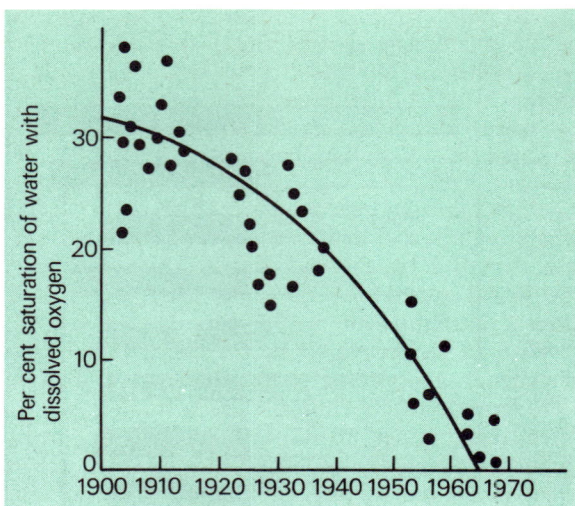

Fig. 19.20 Decrease in dissolved oxygen in the waters of the Baltic sea due to accummulation of organic wastes.

serious problems. Being soluble, much of the fertilizer drains away and pollutes rivers and lakes. For example, 80 % of the nitrogen and 30 % of the phosphorus dissolved in the water of the Great Ouse river in England is known to be derived from land drainage.

Such changes encourage the excessive growth of algae which are not then cropped in sufficient quantity by aquatic herbivores such as water snails. Algal decay produces poisonous substances which may make the water unfit for drinking and which kill fish. Decay causes a serious drop in the dissolved oxygen content of the water, also killing fish and so producing more complications by the disruption of the food web of the river or lake.

Figure 19.19 shows how the mineral content of Lake Ontario has risen exponentially since the turn of the century. Figure 19.20 shows that the dissolved oxygen content of the Baltic Sea has declined since 1900. In many parts the concentration is zero and there no form of aerobic aquatic life can exist.

Effect of fertilizers on soil structure

Excessive use of fertilizers disrupts another important ecosystem, the soil, whose saprophytic bacteria and fungi are vital components of the nitrogen and carbon cycles. Only 5 % of the nitrogen in humus is in the form of soluble ammonia/nitrite/nitrate. The rest is present as decaying organic matter which generates a good crumb structure and has important effects on the water and air-holding capacities of the soil.

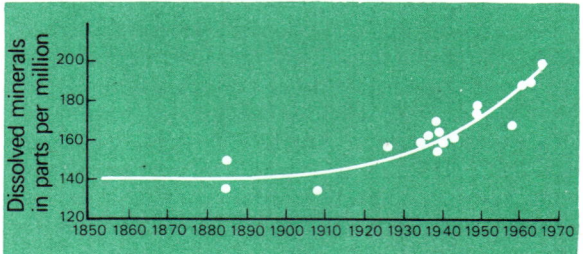

Fig. 19.19 The exponential rise of mineral wastes in Lake Ontario due to industrial, agricultural, and municipal dumping.

The heavy use of artificial fertilizers and of heavy machinery causes rapid deterioration in soil structure. This leads to bad drainage, acidity, and erosion, and between 1882 and 1952 the area of world desert has risen from 1100 million hectares to 2600 million hectares.

Changing farming practice in Britain

The population of Britain is growing exponentially at a rate of 0.5% per year. This is much less than the world average of 1.9% per year, but it still gives a doubling time of 138 years.

With more people to be fed there is pressure on farmers to become more 'efficient', and since 1950 considerable changes in farming practice have taken place. Britain is one of the most intensely farmed and agriculturally productive areas in the world, largely because of the climate and the soil structure. Its output exceeds that of Australia and New Zealand put together. Farms have traditionally been of the mixed type. A variety of crops meant that they could be rotated, and fields could be planted with legumes and also put down to grass for grazing. The mixed animal population of the farm produced much excreta, all of which was returned to the soil.

The situation is now very different. Mixed farms require much labour, and encouraged by economic necessity and Government policy, there has been a marked trend to 'monoculture', which means that each farm raises only one particular crop (say wheat) or one particular kind of animal (say cows). The eastern part of England, which is dry and sunny, and has always been 'the granary of England', now grows cereals almost exclusively, and the traditional small fields have been replaced by large ones so that farming machinery can be used easily and efficiently. This has meant bulldozing away 130 000 miles of hedges per year for twenty years. Many of these date from mediaeval times and many rare species of plants have thus been eliminated.

Perhaps of more immediate concern is the lack of nesting places for insectivorous birds which the hedges formerly provided. Insect pests now have to be controlled by spraying with insecticides. This is expensive (reducing the 'efficiency' of the whole operation). The side effects noted above have meant that resistant strains of pests have emerged which require larger applications of insecticide to produce the same effect. Animal manure has been replaced by NPK (nitrogen, phosphorus, potassium) fertilizer, but this contains no humus. The crumb structure of the soil has deteriorated badly, reducing yield and creating the possibility of erosion.

This has been made worse by the use of heavy machinery which compacts the soil.

In the west of England there has been a concentration on cow raising, but on some farms these animals are not allowed to move freely. They spend their lives in sheds, being fed on grass that has been clipped and delivered to them by tractor. Their thousands of tons of excreta, which should be returned to the soil, is treated as sewage and dumped in rivers.

In such areas there has been a rejection of the nitrogen and carbon cycles and an attempt to replace them with technology. Yield has risen by 35%, but this is the short term effect. The impact of the natural come-back has yet to be felt in full. Already yields are falling because of the damage to the soil and the spread of crop and farm animal diseases that were formerly merely a nuisance.

Combustion of fossil fuels: carbon dioxide and thermal pollution

About 97% of the energy used for industry comes from the use of fossil fuels; oil, coal, and natural gas. These deposits have been formed from once-living organisms which have not decayed completely. They have taken hundreds of millions of years to produce and man is using them at an exponential rate. Figure 19.21 shows the rise in the

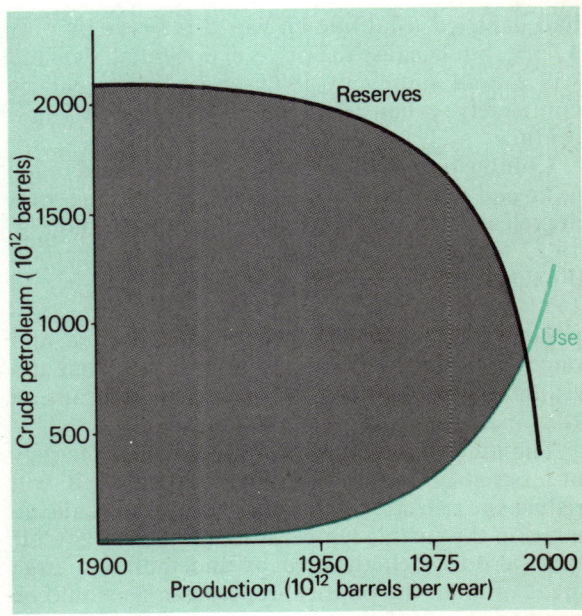

Fig. **19.21** The use of world petroleum reserves is rising exponentially, and the reserves are falling exponentially. They will be exhausted by the early twenty-first century.

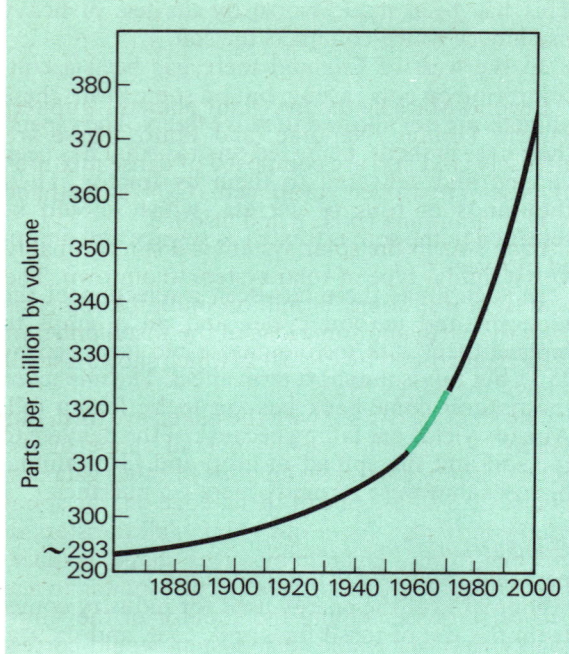

Fig. 19.22 Exponential rise in atmospheric carbon dioxide concentration in mid Pacific between 1958 and 1970 (red line). The black lines are extrapolations.

the remaining areas of natural forest in such areas as the Amazon and the Congo reduces the natural capacity to remove carbon dioxide from the atmosphere via the carbon cycle.

The energy derived from the combustion of fossil fuels ultimately enters the atmosphere and the sea as heat. For this reason also the temperature of the earth is rising.

The future

These are some of the major effects that the activities of man—or rather of so many men—are having on the natural world. Many biologists think that because of the exponential growth of all the factors involved, which all stem from the exponential human population growth rate, the human race is doomed to disaster in the first half of the twenty-first century, unless it can counter these effects in time. Incredible as it may seem, when a United Nations world population conference was held in Bucharest in 1974, many countries, notably the 'third world' and the communist countries, favoured increasing their populations.

One of the things that ordinary people can do about the problem is to be informed about it. This means understanding the nature of living organisms, and underlines the importance of biology as a part of the education of all citizens.

consumption of petroleum since 1900. By 1975 this had depleted total known world reserves by only 12.5%, but because the rise is exponential, demand will exceed supply by 1990, and reserves will be completely exhausted by the early twenty-first century.

Combustion of these fuels yields carbon dioxide as an end product. Some 20 000 million tonnes of it are released into the atmosphere each year, and Fig. 19.22 is a graph showing the atmospheric carbon dioxide concentration in mid Pacific in the period 1958–1970 (heavy lines). This has been extrapolated both backwards and forwards in time, and once again the graph is seen to be exponential and rising at a rate of 0.2% per year, which means an 18% increase by the year 2000.

The long term effects of this change are largely unforeseeable. It has been suggested that it will reduce the reflection of heat from the earth, causing a rise in the earth's temperature of about 0.5°C. If this were to melt the icecaps and inundate large areas of land, the political repercussions would be considerable. Whether the increased carbon dioxide concentration will have an adverse effect on plant life is not known. However, the trend to clearing

20 Evolution

The minuteness of the earth in space

Figure 20.1 is a diagram of part of the **solar system**. At its centre is our **star**, the **sun**, which is a globe of white-hot gas some 1.4 million kilometres in diameter. Orbiting it are the **planets**, non-luminous bodies that shine by the reflected light of the sun. Closest to the sun is Mercury, then comes Venus, Earth, and Mars.

The distance from the earth to the sun is a critical factor for the existence of life. If the earth were closer to the sun than Mercury, life could not exist because the temperature would be too high. If it were further away than Mars, life could not exist because it would be too cold. The sun exerts a gravitational pull on the earth, but because of the distance between the two bodies this pull is weak. It does not cause the oceans and the atmosphere to fly off into space. (We are all familiar with the effects of the pull of gravity on the oceans, for the moon's gravitational pull causes tides.)

Beyond Mars lie the five giant planets—Jupiter, Saturn, Uranus, Neptune, and Pluto. These are intensely cold and are composed of dense mixtures of gases—hydrogen, cyanogen (C_2N_2), methane (CH_4), ammonia (NH_3), and also ice.

Each star in the solar system is a sun, probably with a similar type of solar system to our own. The nearest star is some 38 million million kilometres from the earth and some idea of the vast number of stars is illustrated by Fig. 20.2, which shows a small part of the Milky Way.

The stars we can see with the naked eye form our **galaxy**. We cannot see the overall shape of the galaxy, but beyond it lie millions of other galaxies. Figure 20.3 shows one of them, edge on, the spiral galaxy in Virgo. As in our galaxy, all the stars in this one form a gigantic biconvex shape in space. The Milky Way in our galaxy corresponds to the belt of stars seen round the equator of the Virgo galaxy, and our whole solar system would correspond to one minute speck in it.

Fig. 20.1 Part of the solar system showing the inner planets and the distance range from the sun in which temperatures are compatible with living organisms.

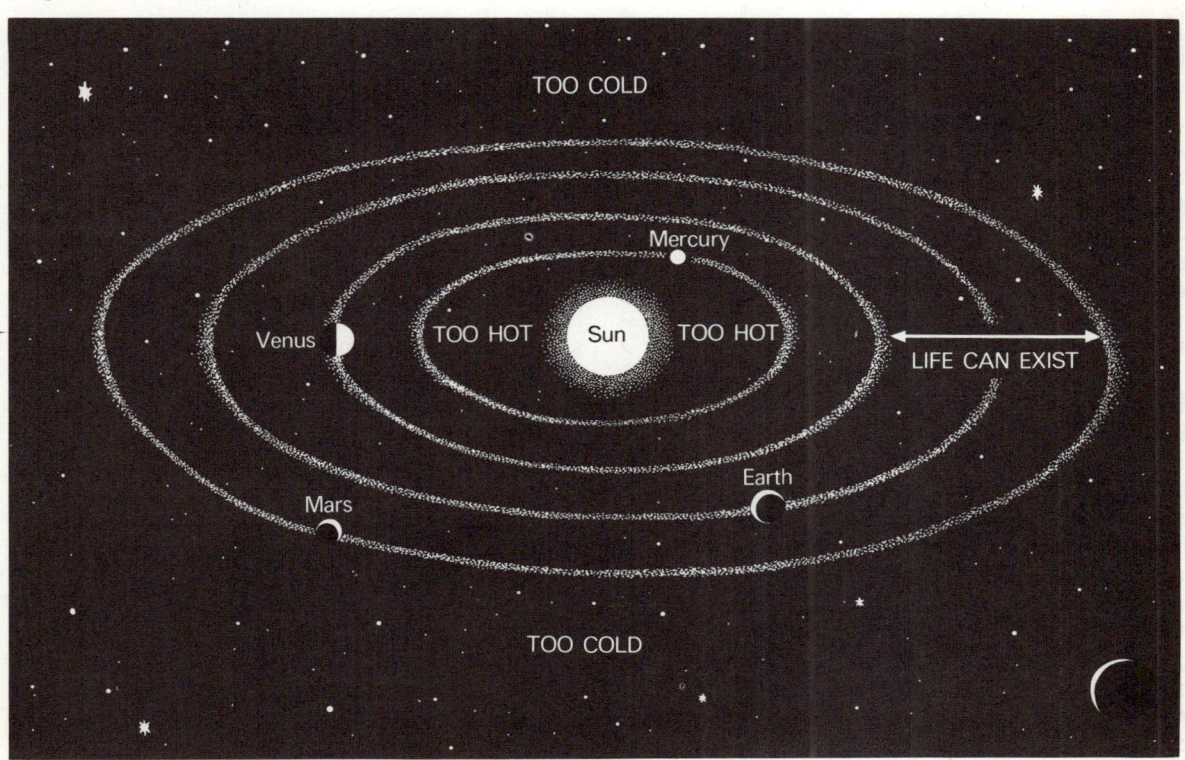

Fig. 20.2 Part of the Milky Way. Each star is a sun.

The origin of life

In our solar system the physical conditions of temperature and gravity could only be right for life on Venus, the Earth, and Mars. Space probes have shown that there is no life on Venus or Mars. However, since all the matter in all the galaxies is composed of the same chemical elements, such as oxygen, hydrogen, carbon, and nitrogen, it is very likely that whatever the combination of circumstances was that allowed life to appear and develop here, they will have been repeated in the numberless stars in the millions of galaxies in space.

Origin of the earth

It is now generally believed that the earth originated some 4500 million years ago by the agglomeration of dust particles. Vast clouds of dust still exist in parts of our galaxy. Each particle is similar in composition to rocks that are found on earth and is surrounded by a film of ice. The earth's original atmosphere is thought to have been like the present-day atmospheres of the five giant planets, with the gases hydrogen, cyanogen, methane, and ammonia condensed on the ice. Because the dust particles were small, vast areas of condensed gases were thus exposed to sunlight.

Miller's experiment

In 1953 an American scientist called Miller made an apparatus in which he placed a similar gas mixture to that of the earth's original atmosphere. He passed an electric spark through it for several days. On opening the apparatus and analyzing its contents he found traces of five amino acids. These are the molecular units from which the polymer molecules of proteins are made (see page 30). In nature they do not exist outside living cells, which make them; but in Miller's apparatus they had been synthesized from simple gases by the energy supplied in the electric discharge.

Similar experiments have shown that other essential life chemicals can be formed in this way. For example, the molecules of ATP, DNA, chlorophyll and haemoglobin, all contain a component part called a nitrogenous base, and this was also found in discharge experiments through simple gas mixtures.

The primitive soup

It seems likely that similar chemical changes took place in the gas mixtures present on the dust particles of the forming earth. Ultra-violet radiation from the sun would have supplied the necessary

Fig. 20.3 The spiral galaxy in Virgo, seen edge on.

energy. When these materials dissolved in the seas they formed a solution which has been called the primitive soup. Meteorites which fall on the earth from interplanetary space contain basic life chemicals thought to have been formed in the same way.

Minimum requirements of a living organism

Living organisms are far more complex than any mixture of chemicals. The chemical substances in cells are not themselves living, but they are organized in such a way that chemical changes take place at the right place at the right time and at the right speed. The basic essentials of a living organism are:

1 The cell must be separated from the environment by a membrane (p. 4). This allows only selected chemical substances to pass across it. Chemical processes can then take place inside the membrane that are, at least to some extent, independent of the environment. In other words, the plasma membrane creates an internal environment for the cell.

2 Proteins must exist within the cell. Enzymes are proteins, and nearly all of the hundreds of chemical reactions that take place in the cell are speeded up by enzymes.

3 A source of chemical energy must be available, together with an energy currency (see p. 67). In the primitive soup, chemical energy was probably supplied by fats. The universal energy currency in all living organisms is ATP.

4 A genetic material must exist which is able to provide the cell with information about the assembly of proteins from amino acids, and other

chemical processes. This material must be capable of being duplicated when the cell divides. The genetic material of all living organisms is made of a chemical substance called DNA. In eukaryotic cells the DNA is confined to the chromosomes in the nucleus. These duplicate in the interphase of mitosis (p. 157) and the prophase of meiosis (p. 159).

The next essential stage in the origin of life therefore involved the production of proteins and DNA from the amino acids and nitrogenous bases in the primitive soup. Many ideas have been put forward as to how this happened, and there is no doubt that life will be re-created in similar ways in the laboratory. Life is unlikely to arise spontaneously in nature today, however, for the earth is now screened from the ultra violet radiations of the sun by a layer of the gas ozone (a form of oxygen) which exists thirty kilometres up in the atmosphere. Even if life *were* newly created now, it would never be recognizable as new and would at once be broken down by saprophytes.

Mycoplasmas

The simplest organisms known to exist on the earth today are called **mycoplasmas** and they correspond almost exactly to the description just given of minimum requirements (Fig. 20.4). They possess a membrane inside which is a DNA chain and various protein particles.

Many mycoplasmas are parasites, but some are saprophytes. The first organisms, which lived on the organic material in the primitive soup, must have been saprophytes. The first known fossils are

257

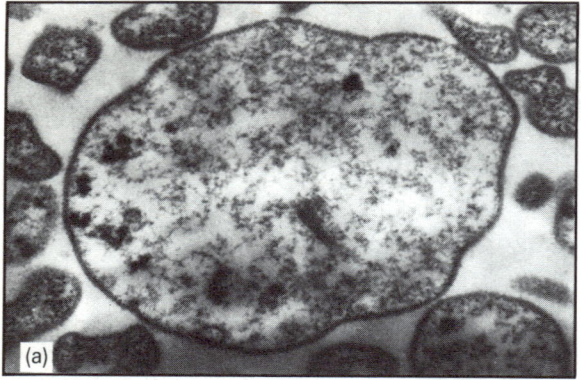

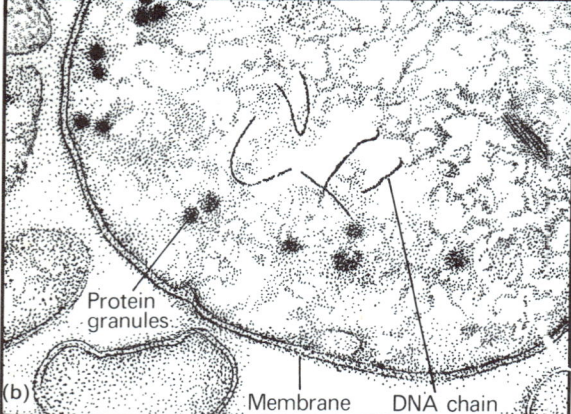

Protein granules

Membrane   DNA chain

Fig. 20.4 (a) A section of a mycoplasma showing the membrane, DNA chains, and protein particles (electron micrograph). (b) explanatory sketch.

of single-celled algae, so we are not yet in a position to say that mycoplasmas were the first living organisms to exist.

Changes in the earth's crust in the last 4500 million years

The only direct evidence that life has existed before the present time is provided by fossils. These are embedded in rocks, and as rocks change with time, fossils are destroyed.

Radioactive changes in rocks in the centre of the earth liberate heat energy, and this makes the **core** of the earth molten. Outside the liquid core is the solid **crust**. Floating on the crust are the **continental land masses** whose shapes have varied throughout geological time and which are still moving (Fig. 20.5). The oldest known rocks date from 3500 million years ago and have been found in Canada and South Africa; but the St Paul's rocks in the central Atlantic ocean, which rise directly from the molten core, are 4500 million years old (Fig. 20.11).

Weathering

Three quarters of the earth's surface is covered by the oceans, whose evaporation creates rain, snow and ice. Above the surface the atmosphere extends for some 650 km, and disturbances in it give winds of various strengths from breezes to hurricanes. The weathering changes in rocks that these agencies cause were considered on page 93.

Additional changes come from rivers, which carry stones and boulders and cut valleys. The sediment they deposit at their mouths builds up into muds, and these become converted into slates when subjected to pressure and temperature changes such as those caused by eruptions from the core. Collisions between floating land masses have caused

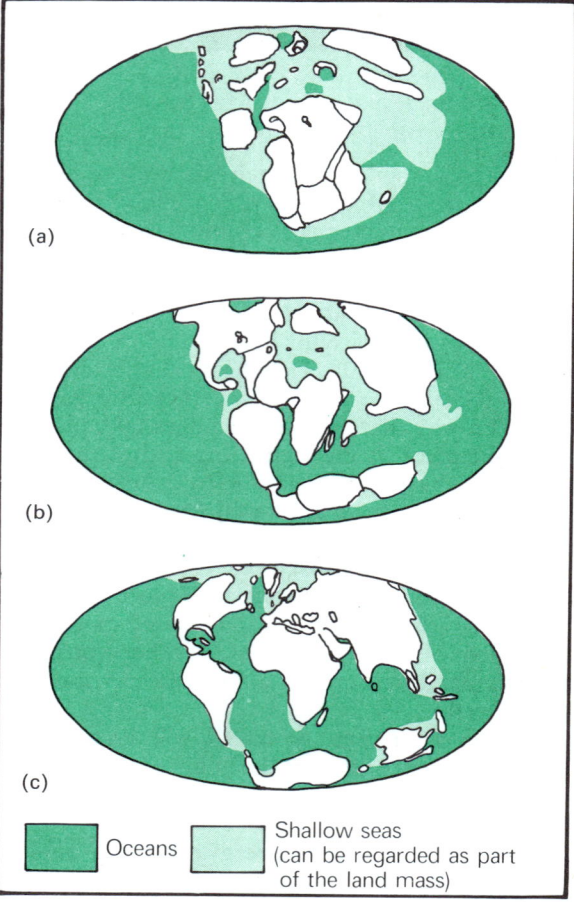

| Oceans | Shallow seas (can be regarded as part of the land mass) |

Fig. 20.5 How the continental land masses have changed their shapes over the last 300 million years, due to *continental drift*. (a) Carboniferous period, about 300 million years ago; one vast land mass. (b) Beginning of the Tertiary period, 60 million years ago; some drifting. (c) About one million years ago; present pattern established but Antarctica still very close to South America.

whole continents to be puckered and folded, giving rise to mountain ranges such as the Alps and the Himalayas.

Types of rock

From these considerations we can make a simple classification of rocks. **Igneous rocks**, such as granites and basalts, are poured out of the earth's core as hot melts which cool and solidify. They contain no fossils. **Sedimentary rocks** are formed by deposits of mud, sand, or shells, for example, the limestones. These are rich in fossils. **Metamorphic rocks** are sedimentary rocks whose primary structure has been changed by heat or pressure. Examples are slates and marbles. They may contain fossils, but often these have been destroyed by the changes caused by heat or pressure.

Fossils

Fossils are the remains of dead organisms, such as mammoths from the recent past that have been preserved in the Siberian ice. More usually they are casts of organisms in which the organic material has been replaced by minerals deposited from water. Such casts can be very detailed, and fossil casts of plant stems show all the cell walls.

The natural events following death are decay and the return of the animal and plant material to the soil and the air through the activities of saprophytes (see page 247). We have only to think how seldom it is that we see the dead body of an animal to realize that preservation after death is a rare event. Death on the land, particularly in warm climates, will be followed by rapid decay. The most likely situation for preservation is if the organism quickly becomes covered with fine mud or sand, preferably in acid waters where there are no bac-bacteria to decay the body.

The group of Carboniferous trilobites (arthropods, see p. 225) in Fig. 20.6(a) were caught in such a situation; it is likely that the group of Devonian fish in Fig. 20.6(d) were suddenly smothered by a sandstorm. *Archaeopterix* (Fig. 20.6(e) and (f)) is one of the most celebrated of all fossils because it is a bird which possesses many reptilian features. It

Fig. 20.6(a) Fossil cast of trilobites, a group of bottom-living arthropods which appeared in the Cambrian period and became extinct in the Permean period. This specimen is from the Cambrian period (600 million years old).
(b) Fossil cast of a Crinoid (sea lily). These animals were echinoderms (related to starfish). They appeared in the Ordovician period and there are several hundred surviving species today, living in deep waters. This specimen is from the Silurian period (440 million years old).

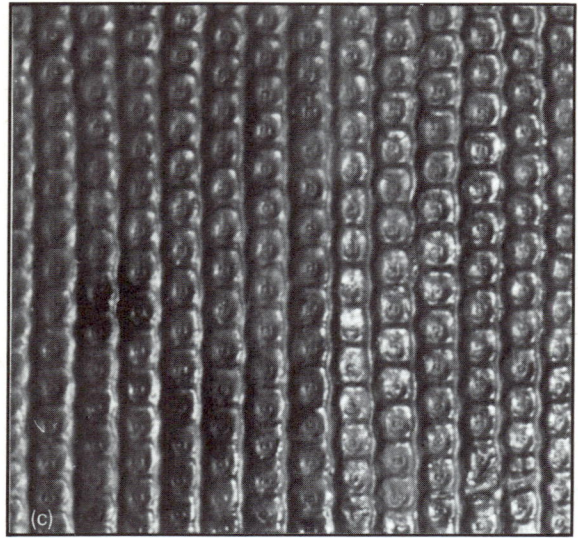

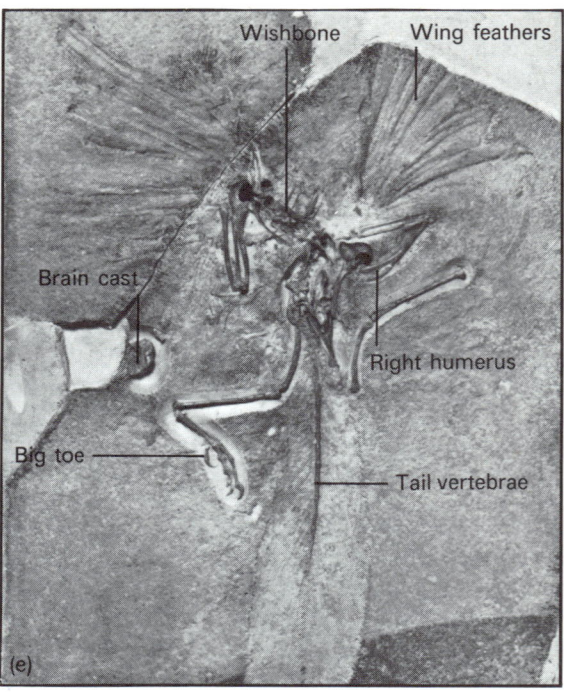

(c) Fossil cast of part of the bark of a Carboniferous tree, *Lepidodendron*, (300 million years old). At this time the northern hemisphere had a tropical climate, and forests of trees such as this gave rise to coal deposits.

(e) Fossil cast of *Archaeopterix*, in fine limestone from the Jurassic period (150 million years old). This animal shows a combination of reptilian features (long tail, clawed forelimbs, simple brain) and bird features (feathers, breastbone, grasping feet). It was about the size of a crow.

(d) Fossil cast of a group of fish from the Upper Devonian period, (350 million years old). The remarkable state of preservation may well be due to the pool in which the animals lived being swamped by a sandstorm (the rock is a sandstone). Such fish as these were the forerunners of the Amphibia.

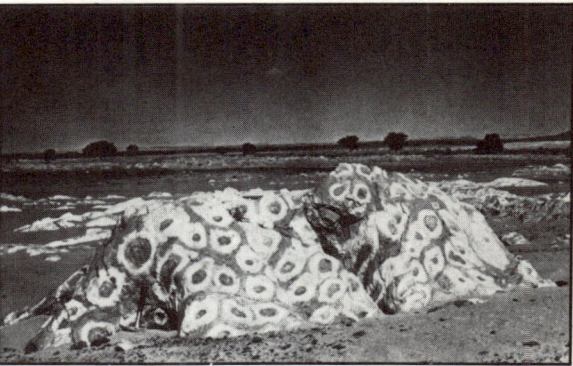

Fig. 20.7 Limestone deposits laid down by algae 2000 million year ago. They are now in the Sahara desert.

(f) Reconstruction of *Archaeopterix*.
(g) Fossil cast of a fig leaf from the Cretaceous period (135 million years old). This is when the flowering plants first appeared.

shows that birds arose from reptilian stock. This particular animal probably fell or was blown into shallow water at the edge of a lagoon, where it drowned and sank into a soft bed of tiny shells which covered it and prevented decay.

The first fossils

We can now see why, if the first organisms were mycoplasmas, it is unlikely that any will have been preserved as fossils. Mycoplasmas have no hard parts that resist decay; on death they disappear completely.

The first known fossils are 2700 million years old, half as old as the earth itself. They are not fossil organisms but layered deposits that were secreted by organisms. In the Sahara desert there are limestones deposited 2000 million years ago by algae (Fig. 20.7). The earliest known fossil cells are of blue-green algae. They have been found in Canada and are called the Gunflint fossils (Fig. 20.8). They are 1600 million years old.

The later fossil record

Figure 20.9 is a geological time scale which shows that the history of life on the earth can be divided into two principal phases. The **pre-Cambrian** phase dates from the origin of the earth to 600 million years ago; it is followed by the **post-Cambrian** phase which occupies the last 600 million years.

Pre-Cambrian life was almost entirely unicellular, and the sudden appearance of many

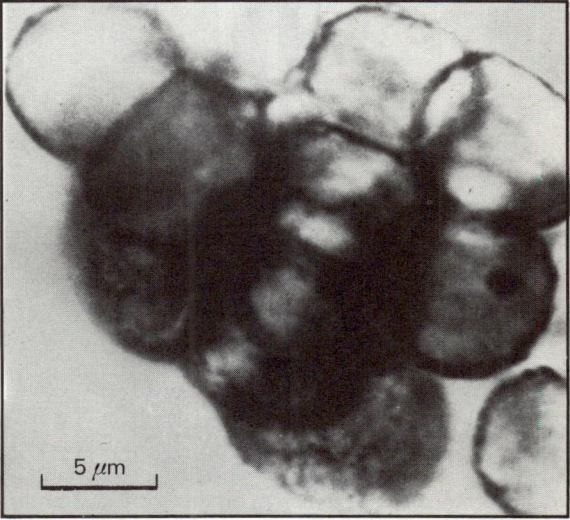

Fig. 20.8 Fossil algal cells, 1600 million years old. These are the earliest fossils we know.

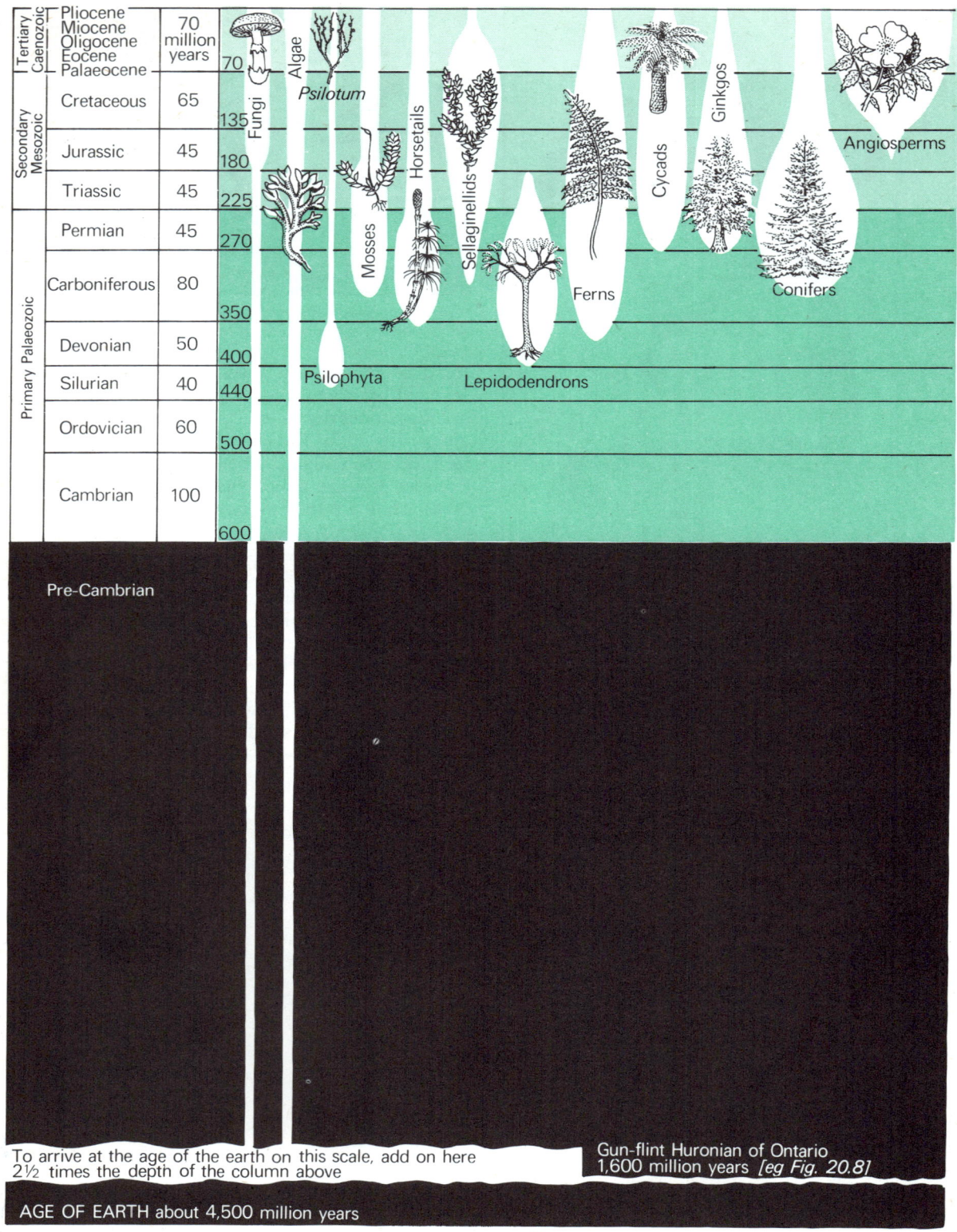

Tertiary Caenozoic	Pliocene Miocene Oligocene Eocene Palaeocene	70 million years	70
Secondary Mesozoic	Cretaceous	65	135
	Jurassic	45	180
	Triassic	45	225
Primary Palaeozoic	Permian	45	270
	Carboniferous	80	350
	Devonian	50	400
	Silurian	40	440
	Ordovician	60	500
	Cambrian	100	600

Fungi

Algae

Psilotum

Mosses

Horsetails

Sellaginellids

Cycads

Ginkgos

Angiosperms

Ferns

Conifers

Psilophyta

Lepidodendrons

Pre-Cambrian

To arrive at the age of the earth on this scale, add on here 2½ times the depth of the column above

Gun-flint Huronian of Ontario 1,600 million years *[eg Fig. 20.8]*

AGE OF EARTH about 4,500 million years

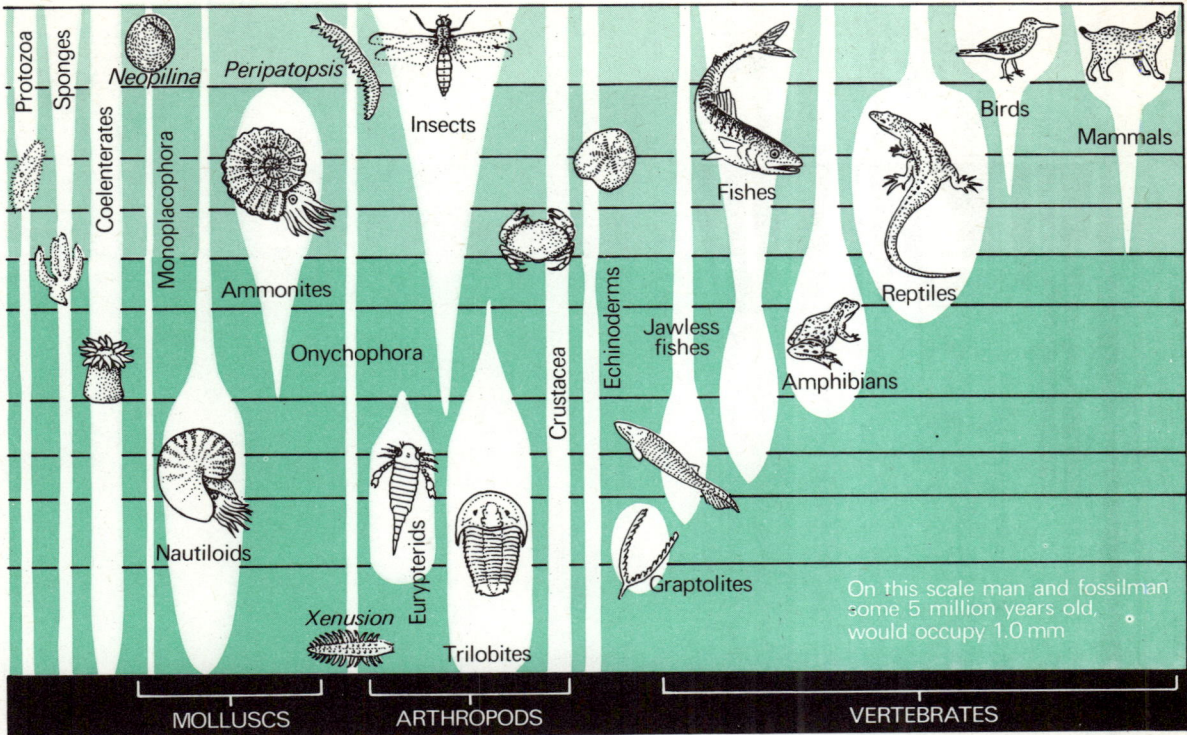

Fig. 20.9 Geological time scale, together with the appearance, disappearance, and relative abundance of some of the major plant and animal groups. Which groups have become extinct? Which are now the most flourishing? Identify the positions of the fossils shown in Fig. 20.6 on this diagram. Man evolved so recently that it is not possible to add him to this figure.

invertebrate groups at the start of the Cambrian is very striking. The thickness of the lines representing the various orders of organisms is roughly proportional to their abundance at the time. Some of the groups have come into existence and become extinct again. Note that the flowering plants are now very numerous and other forms of plant life which were once more flourishing have now diminished. Similarly mammals are very abundant today compared with reptiles; but the insects have been on the increase since they first appeared in the Devonian period.

**A major event in the history of life:
evolution of photosynthesis**

The earth's original atmosphere contained no oxygen. The first organisms must therefore have respired anaerobically using nutritional material, probably fats, available in the primitive soup. Once this had been exhausted, life would have ceased for want of raw materials. The rate of formation of life chemicals from hydrogen, cyanogen, methane, ammonia, and water could not have kept up with the needs of energy-consuming living organisms. The situation would have been broadly similar to the energy crisis today, where man is using up in decades deposits of oil, coal and natural gas that took hundreds of millions of years to form.

But the fossil cells in Fig. 20.8 are of prokaryotic blue-green algae (Fig. 1.17(b) shows a present-day blue-green algae cell). These organisms contain chlorophyll, and its use in photosynthesis is perhaps the most significant event in the history of life. Photosynthesis overcame the energy crisis facing the earth's first organisms. No longer did they have to rely on fats from the primitive soup. Their chemical energy could now be derived from sugars produced by using converted light energy, of which there has been and will be an unlimited supply until the sun, like other stars, eventually disintegrates.

**Oxygen from photosynthesis makes
possible aerobic respiration**

One molecule of glucose respired aerobically generates twenty times more ATP than when it is respired anaerobically (see page 68). An aerobically-respiring organism is therefore much more efficient than an anaerobically-respiring one.

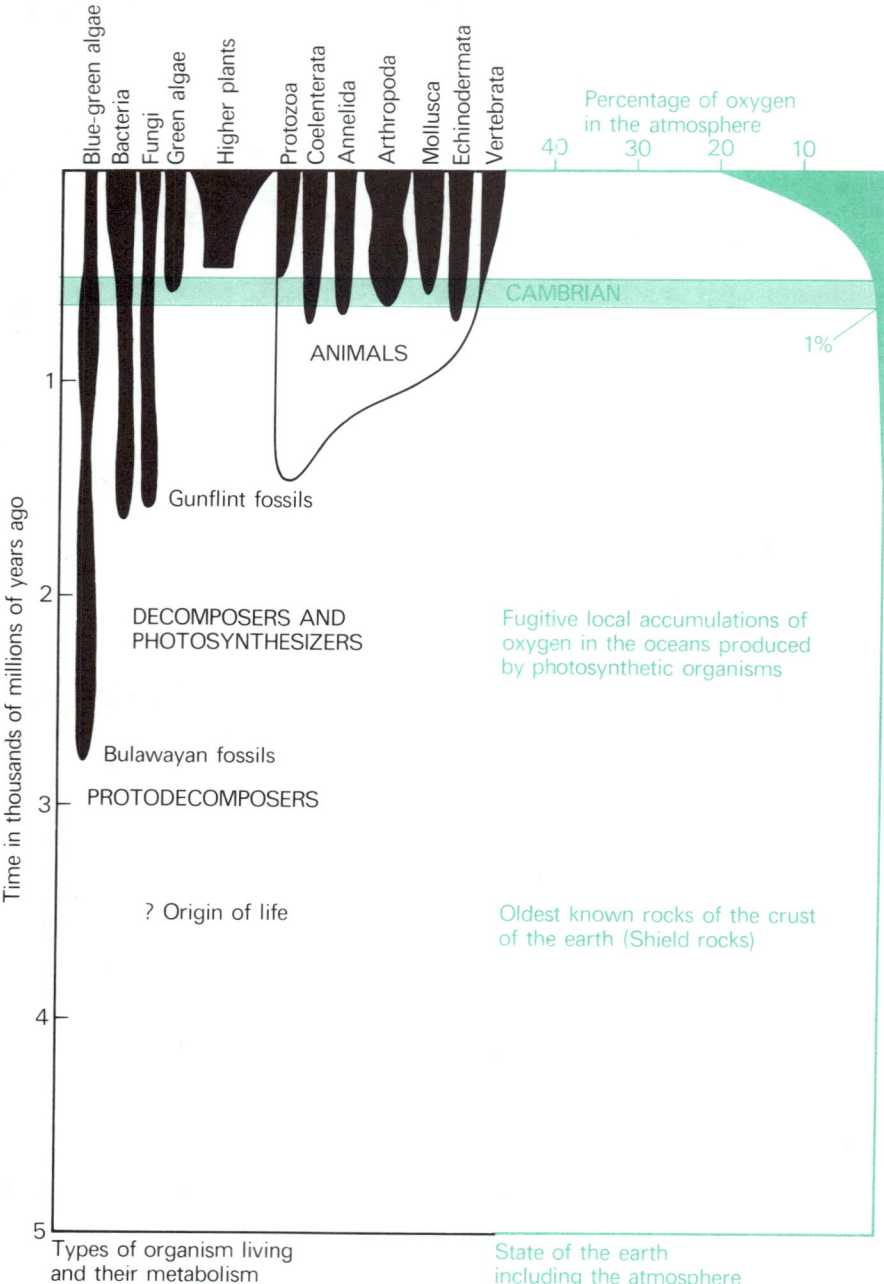

Fig. 20.10 The evolution of free oxygen in the earth's atmosphere (right side of diagram) correlated with the appearance of various forms of life over the period of geological time.

Figure 20.10 shows the percentage of oxygen in the atmosphere during the earth's history (right hand side of figure). Photosynthesis by blue-green algae began somewhere between 3000 million and 2000 million years ago, but all the oxygen liberated at first was dissolved in sea water and used to oxi-

dize chemicals in rocks. Free oxygen did not appear in the atmosphere in appreciable quantities until about 1500 million years ago. It reached a concentration of 1% of the atmosphere 600 million years ago, at the start of the Cambrian period. One percent is the oxygen concentration at which

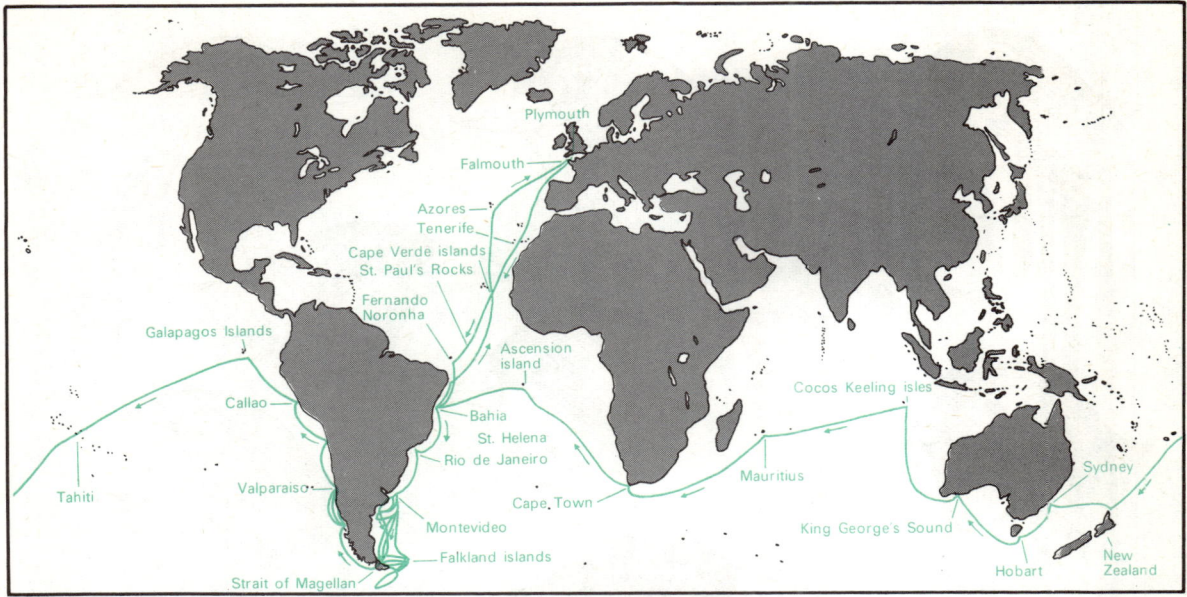

Fig. 20.11 The voyage of HMS Beagle (1831–36).

aerobic respiration becomes more efficient than anaerobic, and the sudden flourishing of life in the Cambrian period, when so many new animal groups appear in the fossil record, is probably due to this cause.

Mitochondria

Aerobic respiration only takes place inside mitochondria (see page 67). These organelles are present in all eukaryotic cells but are not found in prokaryotic cells. Blue-green algae and bacteria are the most ancient fossils, and are both prokaryotes (Fig. 1.17). How did prokaryotic cells come to acquire mitochondria and so become efficient aerobic respirers? If mitochondria existed at one time as independent organisms, it is possible that they were 'captured' in a similar way to that in which *Elysia* captures chloroplasts and uses them in its own cells as if they were its own organelles (p. 243).

The ozone layer cuts off the supply of primitive soup

The earth is surrounded by a shell of the gas ozone about 30 km up in the atmosphere (Fig. 19.9). Ozone consists of molecules in which three oxygen atoms are linked together. It is represented in symbols as O_3, and has different properties from normal oxygen, O_2. One of these is that it strongly absorbs ultra-violet light, which forms part of the sun's radiation. Very little ultra-violet light penetrates the ozone layer.

Ozone was derived from the oxygen liberated into the atmosphere by photosynthesis. Once the ozone shield was in place the chief energy source for the formation of life chemicals from simple molecules such as cyanogen and ammonia was cut off.

However, there is a beneficial side to the ozone shield. Ultra-violet light kills cells and if the earth were irradiated by it all exposed living organisms would be killed. Water also absorbs ultra-violet light, so the first organisms were protected in this way. It was not until the ozone layer was in place that life could emerge on to the land, and this occurred around Cambrian times, some 600 million years ago.

Evolution of the eukaryotic cell

Eukaryotic cells contain other membrane-bounded organelles besides mitochondria. During the course of evolution the scattered chlorophyll-bearing membranes of the blue-green algal cell developed into chloroplasts. The endoplasmic reticulum, Golgi apparatus, and lysosomes, developed from infoldings of the plasma membrane.

Prokaryotic cells have no nucleus, but in all eukaryotic cells the nucleus is bounded by a membrane and contains chromosomes. It undergoes precise changes during nuclear division that were described in Chapter 14. Mitosis provides for the duplication of the genetic material and its exact division between the two daughter cells. Reduction nuclear division (meiosis) provides for the re-

Fig. 20.12 The various species of finch that inhabit the Galapagos islands. Note the different shapes of beak.

shuffling of the genetic material. Prokaryotic cells are simple in structure and contain no organelles. Eukaryotic cells are much more complex, and most eukaryotes are multicellular organisms in which different parts have to work together. Thus much more genetic information is needed in eukaryotic cells, and it seems that the evolution of the nucleus and nuclear division is the development which enabled organisms to cope with it.

Charles Darwin and the origin of species

It is only in the last twenty years that it has been possible to do anything more than guess how life originated. The discovery of the genetic code and of

cell ultrastructure, as revealed by the electron microscope, have given important clues to the probable course of events. We have seen that the evolution of mitochondria placed the cells possessing them at a great advantage over those that did not. Accepting this acknowledges the driving force behind evolution, namely biological efficiency, and this fact was first recognized in the early nineteenth century by Charles Darwin (1809–92).

Darwin was appointed as naturalist on a round-the-world voyage of survey and mapping by HMS Beagle from 1831–36 (Fig. 20.11). At its start he was convinced, as were most biologists of his day, that species were unchangeable and had existed in their present-day forms since the creation of the earth. Amazing as it may seem 100 years later, this was accepted as having taken place in the year 4004 BC.

What is a species?

Animals and plants have been arranged in groups of like kinds by man. All such groupings are man-made, though they are usually based on sound biological reasoning. A **species** is one such category and is defined as *a group of organisms that can inter-breed but which will not breed with any other species*. The species thus perpetuates itself from generation to generation. Examples are the lion, *Panthera leo*, and the tiger, *Panthera tigris*. But if evolution is taking place new species must be coming into existence (Fig. 20.9). They can only originate from already-existing species, which cannot, therefore, be unchangeable.

Darwins's observations of island faunas

Darwin's doubts about the unchangeability of species came from his own careful observations of nature. The Beagle visited two sets of similar small islands, both of which are now known to have arisen from the ocean floor by volcanic action some 15 million years ago. The islands are the exposed tips of underwater volcanoes. They are the Gala-pagos Islands off the west coast of South America, and the Cape Verde Islands situated at a similar latitude off the west coast of Africa (Fig. 20.11). Although both groups are broadly similar in their geographical location, Darwin found that the animal life in each group was totally unlike that of the other, but was similar to but not identical with that of the adjoining mainland. In many cases a particular species was found only on one parti-cular island.

It is easy to see why this should be so if animals had migrated to the islands from the mainland and then, in different surroundings, had evolved into new species adapted to their particular modes of life.

The Galapagos finches

In the Galapagos, Darwin was particularly struck by the existence of fourteen distinct but closely related species of finch. These were similar to a range of finch species that existed on the mainland. One of the ways in which the Galapagos finches differed was in their beaks, which provide their means of gathering food (Fig. 20.12). As many as ten of these species coexist on particular islands, but because of their different beaks they do not compete with one another for food.

The explanation that Darwin offered for this situation was that in the past some related birds had migrated the difficult journey from the main-land and had found on the Galapagos an en-vironment in which there was no competition from other birds. Because of this they underwent adap-tive radiation. This means that new species evolved from the original migrant stock, each suited to a particular mode of life, including a particular food, particular seeds or particular insects.

In suggesting this, Darwin was rejecting the idea that species are unchangeable. Instead he proposed that species **evolve**. This means that *sub-groups within a species gradually accumulate differences which are inherited and passed on to subsequent generations. When these differences are sufficiently marked to prevent interbreeding between the two sub-groups, a new species has evolved. This process takes place over the course of many generations and the organisms involved are all related to each other by descent.*

Darwin's observations in South America

The Beagle called at many places in South America, and Darwin noticed that the rabbit-like rodents of the continent were all built on the same body plan. However, this differed markedly from the body plan of the rabbits in North America and the Old World. In a similar way, he found several species of ostrich-like birds (rheas) which, while resembling each other, differed considerably from the African ostrich. The reasons, he suggested, were that South American rodents were descended from a common ancestral stock which was separate and distinct from the ancestral stock which gave rise to the North American and Old World rodents. Evolution of mammals fitted to the rodent-like existence had taken separate courses in South America and North America and the Old World. (For long periods of geological time, South and North America were physically separated.) Evolution had resulted in animals which resembled one another because they were adapted to life in similar ecological circum-stances. The details of their body plans, however, were different because evolution of the two stocks had been unrelated. Similarly, the species of rhea were descended from a common ancestor which was unrelated to the ancestor of the African ostrich.

These examples were concerned with living animals, but on the pampas grasslands of Uruguay and Argentina, Darwin 'was deeply impressed' by the discovery of 'great fossil animals covered with armour like that on existing armadillos'. Figure 20.13 shows a fossil of one of them, *Glyptodon*, about 26 million years old, compared with a present-

Fig. 20.13 (a) 'A great fossil animal covered with armour like that on existing armadillos' (shown in (b)) (Darwin). (a) is *Glyptodon*, about 26 million years old.

day armadillo. Darwin argued that the animals were both descended from a common ancestral stock, and this example provided an important direct link with the past, indicating that animals now extinct have left present-day descendents.

1836–1859. The idea of natural selection

On his return to England in 1836 Darwin was convinced from his own observations of the reality of evolution. He was not the first man to have had this idea. His own grandfather had suggested in 1794 that 'all warm-blooded animals are descended from a single living filament' and had drawn attention to the profound changes brought about by man in the various breeds of horses and dogs. This had been achieved by not allowing random interbreeding to take place, but instead carefully selecting the parents that were to produce the next generation of stock. This is **artificial selection**, and it has not given rise to new species. All dogs are theoretically able to interbreed, that is dog sperm from all breeds will fertilize dog ova from all breeds. But many barriers to breeding already exist, for example the physical difficulty of an Alsation dog copulating with a Pekinese bitch, or the Pekinese bitch giving birth to Alsatian/Pekinese puppies. There are also behavioural differences and it is unlikely that a Pekinese bitch would accept an Alsatian dog as a mate.

Then in 1839 Darwin read *'An essay on population'* by an English clergyman Thomas Malthus, written in 1798. This was the first reasoned account and prediction of overpopulation, the major problem now facing the human race. Malthus pointed out that while the human population was growing exponentially, its food supply was not. If the human population was not checked, it would rapidly outstrip the supply of food.

Darwin already realized, again from his own observations, that *the reproductive capacity of animal and plant species far exceeds their survival capacity.* For example, a female cod lays something like a million eggs during her life, only a few of which, perhaps only one or none, survive to adulthood. A female frog lays something like 5000 eggs a year for three years, most of which die or are eaten by birds or other predators. A mature oak tree produces thousands of acorns in a good year, few, if any, of which will find a suitable place to germinate, or will survive to become trees if they do grow into seedlings. Such over-production of offspring seems to be a necessary aspect of the life of a species if it is to maintain *the more or less constant numbers* that Darwin knew existed in nature.

These thoughts were in Darwin's mind constantly and then they combined in a flash of inspiration (Darwin has recounted the very moment that it happened). If more individuals are produced than survive, and if all are competing with each other for food, mates, light, and water, and all are subject to predation, disease, floods, droughts, cold, and heat, is it not likely that those best suited to the environment will survive longer at the expense of others? These better-adapted individuals will grow to adulthood, and will be the parents of the next generation. Does not *nature* exert a force of selection similar to that imposed by man in artificial selection, by eliminating as parents the majority—the less well adapted—of the newly-produced population?

It must be emphasized that this idea of **natural selection** means the ability of a *group* to leave descendents. Natural selection should not be thought of in terms of the survival of individual members of the species. Neither does it imply open warfare betweeen all species. Lions, for example, are not in competition with zebras in nature, but with other lion-like carnivores, which they may never meet. On the other hand animals which use the same food source will compete with each other, though this does not imply fighting. For example many herbivores compete for the same food source—grass. The Perissodactyls are much less efficient at digesting grass and using it as a food source than are the Artiodactyls (see page 63), and this fact is reflected in the present-day evolutionary position of the two groups. The Perissodactyls are a dwindling group while the Artiodactyls are flourishing and diverse.

The keenest competition in nature arises when a new species is more efficient in a particular environment than an already-existing one. Once birds had evolved, for example, the flying reptiles (pterodactyls) stood no chance in competition with them. But birds have not affected the evolution of flying insects, with which they do not compete.

The need for variety in a population
Natural selection can only operate if the group of individuals which make up the species are genetically different from one another. There must be a *genetically variable population from which to select*. Darwin knew, once more from observation of nature, that the majority of species do possess variable phenotypes. We now understand the genetic causes of this: the independent movement of chromosomes in anaphase I of reduction nuclear division (meiosis) (page 162). This process shuffles the genetic pack at each generation. Even in the huge human population every person is genetically unique except for identical twins, triplets, etc.

But here Darwin was at a disadvantage. Like nearly all biologists of his day he believed in the idea of blending inheritance. This means that when unlike parents reproduce, their offspring will display a half-and-half mixture of their distinct characteristics. If this were true, all the variation between individual members of a species would be eliminated in about ten generations. Not only would this give natural selection no material on which to operate, it was demonstrably untrue because variation does exist within species in nature.

Unable to explain the source of variation, Darwin thought that his argument that evolution is due to the operation of natural selection had a fundamental flaw. He wrote down his ideas in 1844, eight years after his return from the voyage of the Beagle, but he made no attempt to publicize them. He showed them only to one friend.

Another eleven years passed before in 1855 another biologist, Alfred Russel Wallace, also hit on the idea of evolution operating through the force of natural selection. A joint paper from Darwin and Wallace was read to the Linnean Society (a society of biologists in London), and Darwin was

persuaded to publish his observations and ideas in *The Origin of Species*, which he did in 1856.

The irony of Mendel's work
1856 was the very year that Mendel started his researches with peas. These laid the foundations of genetics, disproved the idea of blending inheritance, and showed that the source of genetic variation between individuals is sexual reproduction. This paper, published in 1865, could have provided Darwin with the answer to a problem which vexed him for the rest of his life, but he was never aware of its existence.

It is an extraordinary thing that Darwin himself carried out experiments similar to Mendel's and observed similar results, but he made nothing of them. Crossing two pure line strains of antirrhinum differing in flower shape, he raised an F_1 generation and found that all the offspring displayed only one of the two shapes. These he self-fertilized, planting out the seeds they produced and recording that in the F_2 88 plants produced flowers of one of the two parental types, and 33 flowers of the other parental type. This is a ratio of 2.3 : 1, which, considering the sample was of only 125 individuals, is a fair approximation to a 3 : 1 ratio.

Contemporary reaction to The Origin of Species
Publication of *The Origin of Species* created an intellectual and spiritual uproar the like of which had never been seen before. It happened to coincide with a fundamentalist religious revival in England. Fundamentalists are persons who believe in the literal truth of every word of the Bible. They will not accept that any of it, such as the account of the creation of the earth in Genesis, is intended as an allegory, and they are silent about the changes of meaning that literature inevitably suffers on translation. Previous suggestions of evolution, like that of Darwin's grandfather, had been brushed aside because they were not backed up by any evidence. But Darwin could not be brushed aside, he could only be argued against, and his bewildered opponents could produce no convincing counter-arguments, only dogmas. People whose ancestors had been raised for generations to accept the description of the Creation in Genesis as literal truth therefore felt insecure and uneasy lest the whole structure of the Christian and Jewish religions were being undermined.

Chance
It was not only uneducated people who felt affronted by the idea of evolution operating through natural selection. Nearly everybody in Victorian

society had, at least at the back of his mind, an acceptance that the world and the organisms in it are the result of the working-out of a master plan, culminating in the emergence of man, the supreme animal. The adaptations that organisms possess were thought to be the result of a purpose. For example, birds have wings because they were intended to fly, or, on another level, red blood cells are biconcave discs because they were intended to have the best design for offering the maximum surface area for the absorption of oxygen.

But Darwin proposed that evolution was not the result of any design. On the contrary, it was due to chance. **Chance** means the *absence of design*, or that events *cannot be attributed to any purposeful cause*. Darwin suggested that variation in natural populations is due to chance; the raw material for evolution arises by chance. We now know the molecular basis of this, namely fortuitous and unpredictable changes in the molecular structure of DNA, which we call mutations (page 165). Adaptations arise from the gradual accumulation of beneficial changes that have occurred by chance.

The idea that evolution does not come about by the intentions of a Designer was hard to accept. Indeed it is contrary to the idea of Destiny and to the idea that events can be influenced, via a Designer, by prayer. Nevertheless, by 1880 Darwin's views had been accepted by the great majority of biologists. Today the crudeness of some of the contemporary discussion has largely evaporated, for example the Roman Catholic Church has recognized the animal ancestry of man in the Papal encyclical *Humani Generis* of 1951.

Significance of the idea of evolution through natural selection

The idea of evolution operating through natural selection is one of the greatest ideas ever formulated in the history of mankind. It has provided a deeper insight into nature than any other idea. It makes the study of biology into a coherent, logical whole. It makes it possible to understand how, starting from the simplest prokaryotic forms of life, over a million present-day species have come into existence. All these, and all extinct species, are related to one another with varying degrees of closeness.

Circumstantial evidence for evolution

Darwin's ideas stimulated biologists into a great deal of diverse work. The period 1860–1920 saw the gathering and presentation of much evidence in favour of the idea of evolution operating through natural selection. Only the merest outline of this can be given here.

The mechanisms by which natural selection operates could only be investigated when a thorough knowledge of genetics had been established, and this has been during the last forty years.

1 The fossil record

The age and history of the earth's crust is now known in considerable detail, and this includes a greater knowledge of a wider range of fossils than was available in the nineteenth century. In several instances, evolutionary lines of development have been traced. Forms intermediate between the major groups, such as the mammal-like-reptiles, have been discovered.

Fossils reveal a constant change of emphasis throughout geological time (Fig. 20.9). Thus the decline of the Lepidodendrons coincided with the rise of the **cycads**; their decline, and that of the **confers**, coincided with the rise of the **flowering plants**. This is due to competition for the same ecological niches by forms more efficient than their previous occupants. The decline of the amphibians coincides with the rise of the reptiles for this reason, their decline coinciding with the rise of the mammals.

A new pattern of life develops which has selective advantages over previous forms. Amphibians are able, to a limited extent, to exploit land life as well as aquatic life. Once they had evolved and emerged on to the land, they radiated into an abundance of species, suffering from no competition. They were the dominant land fauna until some other life pattern embodying fundamental advantages appeared and in its turn radiated into the same niches. Reptiles have scaly skins and eggs which are protected from drying out by a tough outer skin. Inside is a food supply in the form of the yolk, and a water supply in the form of the 'white'. These adaptations made reptiles able to cope with land life far better than the amphibians, whom they largely displaced.

2 Classification

Animals and plants can be arranged into a natural classification. The basis of this is similarities of structure and function which exist because of descent from common ancestors. Figure 20.14 is a classification of vertebrates. All the organisms in it are related, some closely (man, apes), some distantly (man, lampreys).

All vertebrates possess a skull, backbone, two limb girdles, a brain and spinal cord on the upper side of the body, a heart on the lower side of the body, a post-anal tail, and, at some stage in the life history, gill pouches (Fig. 2.8). A first sub-division can be made between vertebrates without jaws (the Agnatha) and those with jaws (the Gnatha). The Gnatha can be divided into those with fins (the fish) and those with five-fingered limbs, the Tetrapoda. The tetrapods divide into those not possessing an amnion (see page 124) (the Amphibia), and those with one, the Amniota.

The amniotes can be separated into the cold-blooded reptiles and the warm-blooded mammals and birds. Within the mammals we have the egg-laying monotremes of Australia (e.g. platypus) and the Ditremata which are viviparous. These separate into those without a placenta (the marsupials) and those with one, the Placentalia. Within the Placentalia there are a dozen or so orders, one of which is

the Primates. The Primates are divided into the prosimians (tarsiers, lemurs, and tree shrews) and the Anthropoidea.

The latter group contains three sub-divisions, the old world monkeys, the new world monkeys, and the apes, of which man, *Homo sapiens*, is one member.

3 Fundamental patterns of structure

As the above example shows, classification is based on—and reveals—fundamental plans of structure such as the vertebrate body plan. Another example on a more detailed level is the pentadactyl limb (Fig. 2.10). This is found in a wide range of verte-

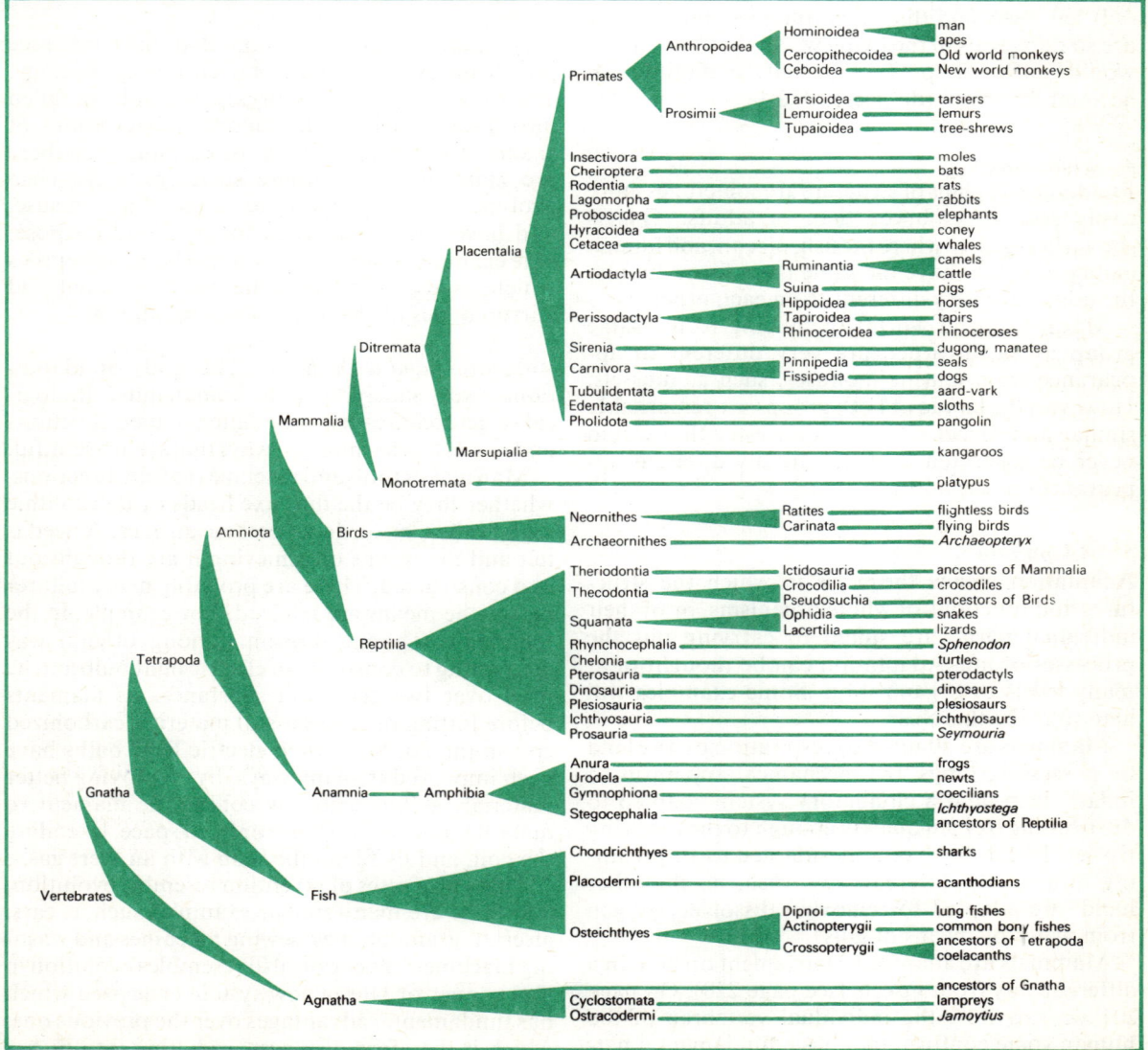

Fig. 20.14 Classification of the vertebrates.

271

brates and has become adapted for flying, swimming, running, jumping grasping, and digging. The only rational explanation of this is that all the organisms possessing these limbs are descended from a common ancestor in the remote past. We noted the appearance of the pentadactyl limb in the pelvic fin of the Devonian fish *Eusthenopteron* (Fig. 17.17). This animal preceded the first amphibians, which appeared in the late Devonian period.

Similarly, there are over a quarter of a million species of flowering plant in existence and their flowers are of the most varied kinds. Yet they all conform to the same basic pattern of form: a layer of sepals below a layer of petals, the petals below the stamens, and the stamens below the carpels.

It has been said that such fundamental patterns are so numerous in the living world that 'they alone would suffice to prove that only evolution can account for them' (de Beer, 1972).

4 Embryology
Fundamental plans of structure are sometimes more easily seen in embryos than in adults. This is shown in Fig. 2.11 where the fish, pigeon, and human embryos clearly have the same basic form, though the adults are very different from each other.

Marine annelid worms (belonging to the same group as earthworms) are very different in appearance from marine molluscs, such as mussels. However, the larvae of both groups are remarkably similar and reveal a common ancestry that would never be suspected from the totally different appearances of the adults.

5 Adaptation
Adaptation means the ways in which the structures and functions of whole organisms, or of their individual parts, are suited to carrying out the processes of life. Adaptation can be thought of at many levels of organization. Some examples will help to make this clear.

Mammals are adapted to respiration on the land by possessing lungs, the mechanical apparatus to inflate them, and a circulatory system adapted to distributing oxygen quickly enough to the respiring tissues. Insects have become adapted to air breathing in a totally different way. Fish, on the other hand, are adapted to removing dissolved oxygen from water by means of gills.

Mammals are adapted to movement on land in a different way from insects (see page 228). On page 201 we saw how the individual vertebrae of the human spine conform to a basic fundamental pattern of structure; but they are individually adapted to the special functions they perform in the different regions of the spine.

In chapter 13 we encountered a few of the many adaptations that exist for the pollination of flowering plants by insects, and for the distribution of their fruits. The structure of the flowering plant leaf is adapted in many ways to the most efficient absorption of carbon dioxide. One of these is the position and shape of the chloroplasts, and the display of chlorophyll in the grana (see page 42). On a molecular level, the fact that ATP is a *cyclic* intermediate (see page 67) is an adaptation enabling it to participate in a large number of chemical reactions though only very small quantities of it need to exist in the cell.

Adaptations are so legion that they embrace practically every structure or occurrence in biology. The only exceptions are organs which have fallen into disuse, such as the minute pelvic girdles of snakes. Even here we must be cautious, for there are many instances where structures have been evolved for one purpose, have fallen into disuse, and have become adapted for a second purpose. One example is the three jaw hinge bones of reptiles which have evolved into the hammer, anvil and stirrup bones of the mammalian middle ear.

Adaptation and a Designer   The study of adaptations is very satisfying to the human mind. Biological structures are seen to be suited to their functions and to work in harmony in ways that seem beautiful.

Man makes tools and machines to fulfil functions, whether they be the flint axe heads of Palaeolithic man or the Concorde supersonic airliner. A need is felt and the means of achieving it are thought out and constructed. There are probably many failures before the means are realized. For example, in the late 1870s Thomas Edison (among others) was attempting to construct an electric light bulb and he tried over two thousand substances as filaments before hitting on a successful material, carbonized cotton thread. Since then electric light bulbs have been improved in many ways, by employing better materials as filaments, by coiling the filament to make it longer in the same confined space, by coiling the coil, and by filling the bulb with an inert gas.

These processes of invention resemble evolution, and there are many similar examples such as cars, aircraft, gramophones, sewing machines and washing machines. Invention also resembles evolution in that sooner or later a new system is devised which has fundamental advantages over the previous one, which it therefore displaces, radiating into niches that the first system previously occupied. The filament lamp was superseded by the discharge tube. Mechanical gramophones, in which the sound was generated from the gramophone record by the physical vibration of a diaphragm, were superseded by models in which playing the record gave rise to electrical impulses which were amplified and reproduced through a loudspeaker. In turn, this system was superseded by stero records and stereo-reproduction, and this by the use of tapes instead of discs.

In these examples the desired end—electric light or recorded sound—is thought of first, and the means of achieving it follow after. This way of thinking is so habitual to man that it is very natural when examining adaptations in living organisms to think that they must have arisen in the same way. This natural thought process leads to the idea that a Designer must first have thought of the desired end and then worked out a means of achieving it. The Designer must have 'wanted' birds to fly and so have 'provided them' with wings so that they could.

This is how the book of Genesis describes the origin of the earth ('In the beginning God created the heaven and the earth . . .'). Most people now regard this passage as an allegory. In 1802 an English clergyman, William Paley, described the origin of the vertebrate eye in the following uncompromising terms: 'The marks of design are too strong to be gotten over. Design must have a designer. That designer must have been a person. That person is God'.

Natural selection and a Designer Darwin's view of evolution operating through natural selection is based on chance and rejects the explanation that adaptations arise to fulfil a predestined end. Thinking of the desired end first and the means of achieving it afterwards is so 'natural' to man, and has such obvious survival value in enabling him to solve problems, that it is very likely to have become incorporated into his genotype by the very process of evolution whose mechanism it does not explain. Anyone doubting this should reflect on whether they believe that simpler behaviour patterns, such as suckling the young in mammals, or nest building in birds, have not arisen through evolution by natural selection.

The **evolutionary mechanism** is *natural selection operating on chance variations in populations over very long periods of time*. It produces adaptation by an entirely different means from that implied by a fore-ordained plan. It is contrary to man's every-

Fig. 20.15 The Huia bird of New Zealand. The female (above) had a long thin beak adapted for reaching insect grubs; the male (below) had a shorter, stronger beak for chiseling holes in bark. Huia birds became extinct in 1907.

day experience of the world and therefore met with scepticism and resistance when first proposed.

'Unpleasant' adaptations Darwin pointed out that adherents of the idea of a wise and benevolent Designer must explain such 'unpleasant' adaptations as those of internal parasites, including disease-causing organisms. He spoke of '. . . the clumsy, wasteful, blundering, low and horribly cruel works of nature . . . which allows . . . a group of animals to have been formed to lay their eggs in the bowels and flesh of other sensitive beings'.

Over-successful adaptation: extinction The fossil record (e.g. Fig. 20.9) shows that the vast majority of animal and plant groups have come into existence, flourished for a time, and then become extinct. This is quite understandable in terms of evolution by natural selection. It is difficult to reconcile with a supreme intelligence who is, it seems, forever making serious mistakes.

Extinction is due in most cases to excess of adaptation. A species becomes so specialized to its environment that when the environment changes (as it always does, eventually) neither it nor its offspring can adapt to the new circumstances rapidly enough to permit survival. A recent example of this is the Huia bird of New Zealand (Fig. 20.15) which became extinct in 1907. It fed on beetle larvae which lived under the bark of trees and in decaying wood. In effect the species possessed two tools for obtaining these larvae. The beaks of the

Fig. 20.16 (a) The one-horned rhinoceros from India; (b) the two-horned rhinoceros from Sumatra; (c) the one-horned rhinoceros from Java; (d) the two-horned black rhinoceros from East Africa; (e) the extinct wooly rhinoceros from Siberia, of which whole specimens are known.

males were strong and stout, adapted to boring holes but not long enough to reach to the bottom of the insect-grub tunnels that they opened. The female's beak, being long and thin, was adapted to this purpose, though useless for boring holes. The birds therefore lived in pairs, neither member of which could feed without the other. Extinction was caused by a deforestation prgramme. Huia birds were far too specialized—too highly adapted—to become adapted to alternative sources of food.

6 The geographical distribution of plants and animals The similarities and differences between the fauna of the Galapagos islands and the neighbouring coast of South America and between the fauna of the Cape Verde islands and the neighbouring coast of West Africa started the doubts in Charles Darwin's mind about the unchangeability of species (see page 267).

Since 1836 the flora and fauna of the world have become much better known and recorded. In many cases the distributions of particular groups can only be explained in terms of evolution. One example must suffice here.

Similar but distinct species are often distributed **discontinuously** throughout the world. This means that they exist in different continents with no possibility of migration from one location to another. The rational explanation is that the different species originated from a common ancestor in one place. This species radiated outwards by migration, undergoing evolutionary changes at the same time and so splitting up into distinct species. Later some of the species became extinct in the intervening regions. (Herring gulls in the northern hemisphere have reached a half-way stage in this process at the present time, see page 284.)

This explains, for example, the nineteenth century distribution (now sadly depleted, for several species are nearly extinct) of the various species of rhinoceros (Fig. 20.16). In a memorable passage in *The Origin of Species* (here slightly simplified), Darwin puts his case in the following way:

'Shall we allow that the distinct species of rhinoceros which inhabit Java, Sumatra and the mainland of Malacca were created out of the inorganic materials of these countries? Without any adequate cause, shall we say that they were, merely from living near each other, created very like each other? Without any adequate cause, shall we say that they were created of the same type as the extinct wooly rhinoceros of Siberia? That without any adequate cause their short necks should contain the same number of vertebrae as that of the giraffe; that their thick legs should be built on the same plan as those of the antelope, the mouse, the monkey, the bat and the porpoise? I repeat, shall we say that a pair of each of these three species of rhinoceros were separately created out of the inorganic elements of Java, Sumatra and Malacca? Or have they descended, like our domestic races, from the same parent stock?'

Observing evolution in action

The speed of evolution

Darwin was suggesting that the five species of rhinoceros seen in Fig. 20.16 had originated from one species. This had migrated to different geographical locations, forming sub-groups which became adapted to life in the local environments. The gradually-accumulated differences between the sub-groups must have been genetically controlled for them to be handed down to successive generations. Eventually these accumulated differences were sufficiently great for interbreeding between the sub-groups to be impossible even if the animals from the various locations ever met. The sub-groups of the original species were then new species.

The speed at which such changes take place depends on the **selection pressure** exerted on the sub-group by the environment. If environmental conditions remain unchanged, there is no selection pressure, and the sub-group does not evolve. For example, in the deep sea, where temperature, light intensity, food supply, and predators change very little, the lamp shell brachiopod *Lingula* has remained unchanged for 500 million years. On the

Fig. 20.17 Whitish and melanic specimens of *Biston betularia* (a) on lichen-covered bark; (b) on soot-grimed bark.

other hand the horse has passed through eight genera, and many more species, in only 70 million years. It adapted from life in marshy ground where it ate leafy vegetation, to hard, dry, exposed, open ground, where it ate the tough leaves of grasses. This change of environment imposed a high selection pressure favouring forms that were able to move quickly, to grind hard vegetation and to digest it efficiently.

We should not expect, therefore, to observe the evolution of new species in a human lifetime. Charles Darwin believed that even the minor evolutionary changes, whose accumulation ultimately produces species, are also too slow to observe. Here he was wrong. Many examples of evolutionary changes in progress have been investigated since 1930. We will look at three of them.

1 Evolution of melanic (black) moths

In the last hundred years some eighty species of moths in Britain have evolved from populations in which the individuals are almost all light-coloured to populations in which the individuals are almost all dark-coloured, sometimes black. One species, the Peppered Moth, *Biston betularia*, has been particularly well studied.

Butterfly and moth collecting was popular in the nineteenth century, and these records show that in the first half of the century *Biston betularia* was always whitish with minute dots and pencillings. These give it such a remarkable resemblance to the lichens on the tree barks on which it rests that it is very difficult to see it (Fig. 20.17(a)). On the other hand it is very conspicuous on a soot-grimed bark (Fig. 20.17(b)).

In 1848 the first melanic or black specimen was recorded in Manchester. The phenotypic difference between the two forms is due to a single mutation. The allele producing black is completely dominant to the allele producing the normal whitish appearance. If we represent the genotype of the white form as $\frac{\mathbf{b}}{\mathbf{b}}$, that of the melanic mutant is $\frac{\mathbf{B}}{\mathbf{b}}$, then a cross between the two will produce an F_1 generation in which 50 % of the individuals are black and 50 % normal whitish (see Fig. 14.9(b)).

Selection pressure against the melanic form
This mutation is known to occur quite frequently. There can be no doubt that melanic individuals appeared before 1848, but at that time they were at a strong disadvantage because they were obvious prey for insect-eating birds. Another way of saying this is that there was a strong selection pressure

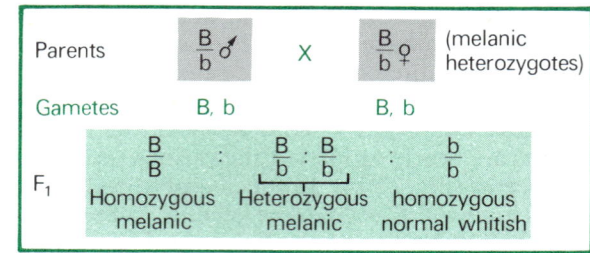

Fig. 20.18 Once heterozygotes, $\frac{\mathbf{B}}{\mathbf{b}}$, existed in the *Biston betularia* population in sizeable numbers, matings between heterozygotes became likely. These produced a 3 melanic: 1 normal whitish F_1, and the proportion of melanics in the total population thus increased.

against the melanic form, the selecting agent being predatory birds.

Experiments have been conducted to measure the strength of this selection pressure. Large numbers of melanic and of normal whitish individuals were released in unpolluted areas of England where the tree barks are covered with lichen. The numbers of each sort eaten by birds was counted by direct observation through field glasses. For every 100 melanic forms eaten, only 83 whitish individuals were taken. This means that the selection pressure acting against the melanic form was 17 %.

Selection pressure
against the normal whitish form
In the mid nineteenth century the industrial revolution was in full spate and one of its main centres was Manchester. The sulphur dioxide gas contained in industrial fumes killed all the lichens in the district. The tree barks became grimed with soot, coal dust, and other dirt. The habits of the moth and its predators did not change, but now the selective advantage lay with the melanic form. On a sooty bark it is almost invisible, whereas the normal whitish form is very conspicuous (Fig. 20.17).

Experiments have shown that a 10 % selection pressure in favour of the melanic form exists in such areas. By 1895, 98 % of the *Biston betularia* individuals in the Manchester area were melanic. This striking evolutionary change took only fifty years to accomplish.

Selection alters the proportions of the genotypes in the population
The first melanic individual must have been heterozygous, $\frac{\mathbf{B}}{\mathbf{b}}$, and must have bred with the normal

form, $\frac{b}{b}$. Both phenotypic forms belong to the same species and will accept each other as mates. But once a sizeable population of melanics existed, heterozygotes mated, and this gave rise to homozygous dominants, $\frac{B}{B}$ (Fig. 20.18). Matings between $\frac{B}{B}$ and $\frac{B}{B}$ or between $\frac{B}{B}$ and the heterozygote $\frac{B}{b}$, or between $\frac{B}{B}$ and the recessive homozygote $\frac{b}{b}$ would all give melanic offspring. The proportion of melanic individuals in the population thus rose rapidly.

However, even a population which is 98% melanic still contains the allele **b** in its heterozygous members. If the environmental conditions were again reversed, we would therefore expect natural selection to operate in favour of the normal, whitish form. This has in fact happened in the 'smokeless zones' established since the Clean Air Act in Britain (1956).

An important point to note about this evolutionary change, or adaptation, is that the melanic individuals arose by mutation. The process of mutation has been intensely studied and all experiments show that it is caused by fortuitous and unpredictable changes in the molecular structure of DNA. That is, the mutations that occur in a population are a matter of chance. In this case the mutation was at first a disadvantage. Black moths were an easier prey for birds. It was only after an equally unpredictable change in the environment resulting from the industrial revolution that natural selection favoured the melanic form and brought about its evolution.

2   Evolution in the common snail, *Cepea memoralis*.

Individuals of this species vary in colour. Some are brown, others pink, others yellow. Breeding experiments have shown that all three phenotypes are controlled by one gene which exists in *three* allelic forms. The brown allele is dominant to pink and yellow, and the pink allele is dominant to yellow. (N.B. in chapter 14 we only considered examples where a gene exists in two alleleic forms, such as R and r in Mendel's peas.)

Some *Cepea nemoralis* have plain, unbanded shells, others possess a varying number of bands. This is also due to a single genetic difference. Absence of banding is dominant to banding.

Selection pressure exerted by the song thrush

The snail is preyed on by the Song Thrush. In one

Fig. 20.19(a) *Cepea nemoralis* on leaf litter which is uniformly brown. The two shells on the left are brown, the one on the right is yellow. The fourth shell is pink. All are unbanded.
(b) Banded and unbanded shells of *Cepea nemoralis* among long grass and mixed herbage. The background is a mixture of greens. The unbanded shell is also green when the animal is inside it.
(c) Yellow and brown banded shells of *Cepea nemoralis* on short grass.

experiment a snail population of 10 000 individuals was counted near Oxford and in 17 days 863 of them (8.6%) had been eaten by Song Thrushes. This bird has the convenient habit of taking the snails to a stone or 'anvil', where it breaks them open, leaving their shells which can be counted.

Investigations have shown that when the snails are exposed against different kinds of backgrounds, the least conspicuous are the least frequently eaten, the most conspicuous the most frequently eaten. In

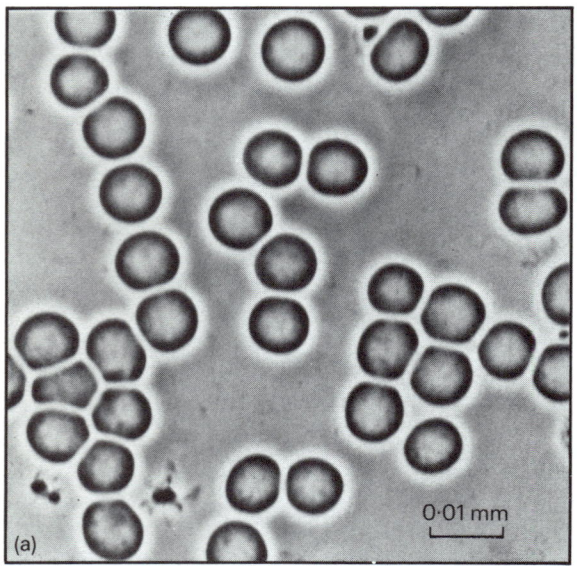

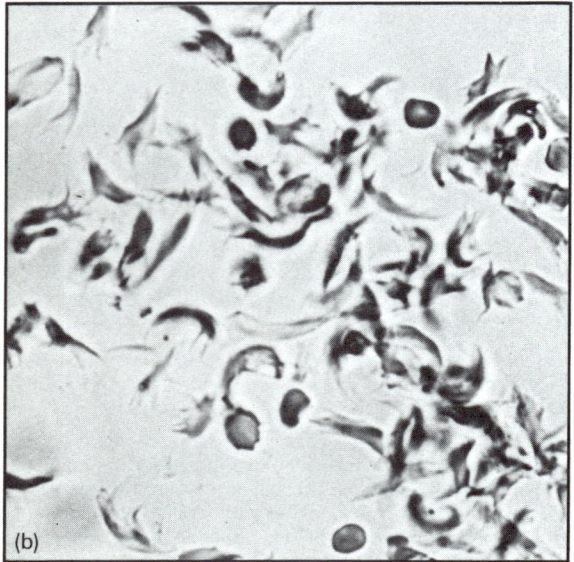

Fig. 20.20 Red blood cells of a patient who is heterozygous ($\frac{S}{s}$) for the sickle-cell gene. (a) Normal shape in the presence of oxygen. (b) Sickle-cell shape in the absence of oxygen.

other words, natural selection operates against the conspicuous forms and in favour of the inconspicuous forms.

For example, in a beech wood where the snails rested on a uniform brownish background, the yellow form was conspicuous (Fig. 20.19(a)). Only 17% or less of the snail population in this environment was yellow. Here the selection pressure acted against the yellow form, the selective agent being predatory thrushes.

This view was confirmed by comparing the numbers of yellow and brown snails found at the anvils in different environments. Against a background of mixed leaves the yellow form accounted for 85% of the population. This background is a mixture of greens, and the yellow-shelled snail is also green when inside its shell. Here selection was acting against the darker forms, and the yellow form was favoured. (Fig. 20.19(b)).

On this greenish, mixed background, snails with banded shells fared better than those with plain ones. They formed 78% of the population whereas plain-shelled individuals, which were much more conspicuous, formed only 22%. On the other hand, the banded shells are more conspicuous against a uniform background (Fig. 20.19(c)), and in this environment they formed only 40% of the population.

As the seasons change, so do the appearances of the backgrounds. The selection pressures against the various forms of shell vary accordingly. In one woodland colony the background changed from brown to green as winter turned to spring. The yellow phenotype, which is a handicap against brown leaf litter, became an advantage against a background of young green shoots, and the proportion of yellow individuals in the population rose. Here we see natural selection actively and delicately attuned to the changing environment.

3 Evolution in human populations: sickle-cell anaemia

Sickle-cell anaemia is a human disease in which the red blood cells are distorted into a sickle shape (Fig. 20.20). The haemoglobin they contain is also abnormal. Many of the patient's red blood cells are destroyed. He consequently suffers from anaemia and normally dies before the age of five.

The condition is due to a mutation in a single gene. Normal individuals are homozygous dominants for an allele concerned with haemoglobin production. Their genotype is $\frac{S}{S}$. Sickle-cell anaemics are homozygous recessives, $\frac{s}{s}$. Heterozygous persons, whose genotype is $\frac{S}{s}$, have normal-shaped red blood cells unless they are suffering from severe oxygen shortage. Then their red blood cells take on the sickle form. Their haemoglobin is a mixture of normal and abnormal types.

Selection pressure against sickle-cell anaemics

Sickle cell anaemics are $\frac{s}{s}$ homozygotes. They never reach reproductive age. In each generation they are removed from the population by early death. The selection pressure acting against them is high.

We would therefore expect the proportion of s alleles in human populations to be very low, or for s to have been eliminated altogether, except for its recurrence by mutation.

Yet in various parts of Africa the populations carry a high proportion of s alleles (Fig. 20.21(a)). For example, in parts of Tanzania 4% of the population are homozygous sickle-cell anaemics, $\frac{s}{s}$, and 32% are heterozygotes, $\frac{S}{s}$. From this it can be calculated that 20% of the alleles of this gene in this population are s and 80% S.

Selection pressure against the allele S

The reason for this situation is that the abnormal form of haemoglobin possessed by the heterozygotes $\frac{S}{s}$ confers a strong selective advantage on them: they are resistant to malaria. On the other hand, homozygotes $\frac{S}{S}$ are susceptible to malaria.

Malaria is caused by a protozoan parasite which enters normal red blood cells and feeds on them. The parasite cannot enter or feed on the red blood cells of sickle-cell homozygotes, $\frac{s}{s}$, or sickle-cell heterozygotes, $\frac{S}{s}$.

There are thus two selection pressures acting on the human population in malarial regions of Africa. One acts against the sickle-cell allele s by eliminating the homozygotes $\frac{s}{s}$. The other acts against the normal allele S by eliminating at least a proportion of the homozygotes $\frac{S}{S}$. A balance is struck where the death rate of sickle-cell anaemics, $\frac{s}{s}$, balances the death rate from malaria of normal homozygotes $\frac{S}{S}$. The higher the intensity of malaria in a particular region, the higher the frequency of the alele s (Fig. 20.21).

If environmental conditions change, we would expect the balance struck between the two selection pressures to change. This has happened in regions where malaria has been eliminated: the frequency of the allele s has declined rapidly in these areas.

As with the chance mutation that gives rise to the allele B in *Biston betularia*, the mutation from S to s is harmful in most environments. The allele s has a survival value only in malarial districts, and here natural selection operates in its favour.

Selection pressures against dominant and recessive mutations

(a) Mutation from a recessive to a dominant allele

Mutations from a recessive to a dominant allele are rare. The mutant individual is phenotypically different from the other members of the population. Since the organism already well-adjusted to the environment because of natural selection, a sudden change in phenotype is more likely to be harmful than beneficial and all such mutants will be eliminated immediately by natural selection. This was the case with the mutation from the allele b to the allele B in *Biston betularia* before the industrial revolution. We must remember that for a mutation to produce an effect on the whole organism it must take place in a gamete-forming cell, a gamete, or a zygote (see page 166).

(b) Mutation from a dominant to a recessive allele.

In an organism homozygous for a dominant allele (e.g. $\frac{R}{R}$), mutation produces a recessive allele. The

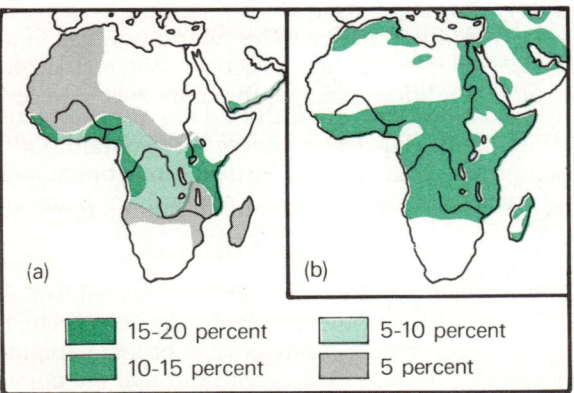

Fig. 20.21 (a) Distribution of the sickle-cell allele *s* in Africa. (b) Distribution of malaria (shaded areas) in Africa.

(a)   (b)

■ 15-20 percent	■ 5-10 percent
■ 10-15 percent	■ 5 percent

resulting individual will be heterozygous, $\frac{R}{r}$. When this individual reproduces with a normal, homozygous dominant member of its species, its offspring will be half homozygous dominant and half heterozygous (Fig. 20.22(a)). However, all the offspring will possess the dominant phenotype.

Thus a proportion of heterozygotes builds up in the population. The mutant recessive allele spreads. Hidden behind its dominant allele, it is not exposed to natural selection since it produces no phenotypic effect.

Sooner or later two heterozygotes will mate, producing $\frac{1}{4}$ homozygous recessive offspring (Fig. 20.22(b)). If the phenotype of this homozygote is less well adapted to the environment than that of the normal dominant (as it most likely will be), it will be eliminated by natural selection. This is the case with the mutant allele s which is exposed to natural selection in the homozygous sickle-cell anaemics $\frac{s}{s}$.

Natural selection does not eliminate recessive alleles

Note, however, that whereas natural selection can eliminate a dominant mutation in one generation, it eliminates only 2 out of 4 *recessive* alleles in Fig. 20.22(b). The remaining two heterozygotes

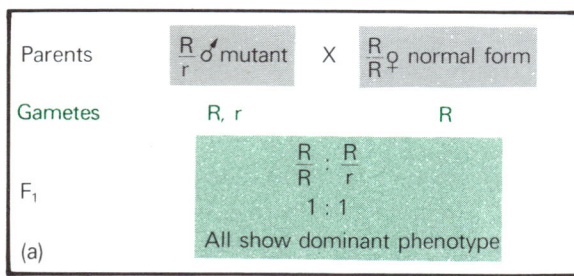

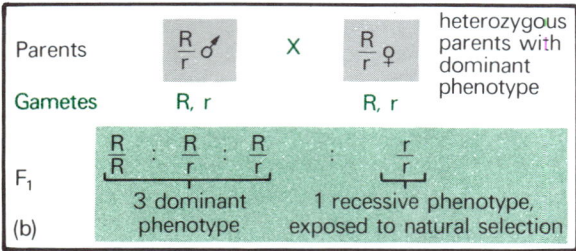

Fig. 20.22 (a) Mating of a mutant heterozygote with a normal dominant homozygote produces a 1:1 segregation of homozygotes and heterozygotes, all of which possess the dominant phenotype. (b) Mating of two heterozygotes produces a 3:1 segregation of dominant to recessive phenotypes.

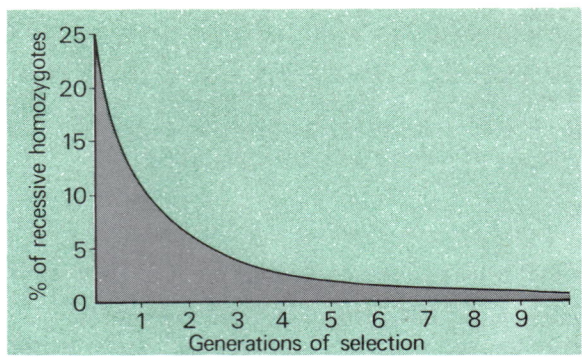

Fig. 20.23 The percentage of recessive homozygotes in the population over nine generations. The initial frequency of the recessive allele was 50%. All recessive homozygotes died and did not reproduce.

$\frac{R}{r}$ will again produce a proportion of recessive homozygotes $\frac{r}{r}$ in the next generation. Again these will be eliminated by natural selection, but even so the allele **r** will still be present in the population in the form of heterozygotes.

The effect of selection pressure against a recessive allele acting over nine generations is shown in Fig. 20.23. Starting with a population in which half the alleles of a particular gene were recessive and where all homozygous recessive offspring were eliminated in each generation by natural selection, the recessive allele was still present in the ninth generation. So, with respect to this gene, the genetic variety of the population has been maintained despite very strong selection pressure. The population still possesses both alleles, **R** and **r**.

Significance of recessive mutation

The fact that most mutations are recessive is clearly an adaptation which maintains genetic variety in a population. This adaptation has likewise evolved by natural selection. When environmental conditions change, the homozygous recessive phenotype $\frac{r}{r}$ may be more advantageous than the dominant phenotype. It will then be favoured by natural selection as the homozygous recessive $\frac{b}{b}$ has been favoured in smokeless zones.

It is also known, though the mechanism is beyond the scope of this book, that under such circumstances recessive alleles can become dominant, which hastens their spread through the population.

Natural selection operates by changing the genotype of the population

Natural selection operates on the phenotype of organisms, but it is by altering the genotype of the population that *heritable* changes arise. In each of the three examples cited above, the effect of natural selection is to alter the proportions of the various alleles in the population. In this way the phenotypic characters of the populations of descendents change.

Recombination produces genetic variation

The major source of genetic variation is the shuffling of the genetic material in reduction nuclear division, meiosis. This is produced by the independent movement of chromosomes in anaphase I of meiosis (page 162).

Shuffling can only happen if there is different genetic material to shuffle, that is, if the organism in which meiosis is taking place is heterozygous for its genes. We would therefore expect that wild populations of organisms would be heterozygous for many, or most, of their genes. This has been investigated with various organisms, all of which proved to be heterozygous for the great majority of the many genes examined.

Mutation is the ultimate source of genetic variation

Mutation is a change in a gene which produces a new allele (see p. 165). It is a very infrequent process, averaging one mutation per half million copies of the allele made. It is caused by a fortuitous and unpredictable change in the molecular structure of DNA.

Mutation cannot *by itself* account for the variety necessary for natural selection to produce evolutionary changes. It not only is but *has to be* a rare event in order to preserve the genetic stability of the individual and the population. If the mutation rate were faster than it is, it would not be possible for a favourable genotype to be selected. As soon as it was, it would start to break down because of mutation. For example, the melanic population of *Biston betularia* in industrial regions would be unstable, reverting to whitish forms by mutation. A low mutation rate has evolved by natural selection.

On the other hand it has been calculated that if mutation were to stop now, there is already enough genetic variation in organisms to allow evolution to continue as far into the future as it has gone on in the past, that is for at least 600 million years.

A low mutation rate plus the genetic reshuffling provided by reduction nuclear division allows the two apparently contradictory requirements of the genetic material, namely that it should be stable on the one hand and should provide genetic diversity on the other.

Preservation of favourable genotypes

When the environment is a changing one, as it usually is on land, natural selection produces evolutionary changes. When the environment is stable over long periods of time, as it is on the deep seabed, natural selection operates against change and prevents evolution. Thus although *Lingula* is a sexually-reproducing organism, it has not evolved in the last 500 million years.

Many organisms are capable of asexual reproduction, producing clones (see page 158). This applies to most plants and to animals that are not highly differentiated, such as hydra. When such organisms evolve a genotype which is well adjusted to the environment and the environment is temporarily stable, offspring are produced which are genetically identical with the parent. Like the parent, they are well suited to life in the same environment.

Evolutionary 'mechanisms' based on mutation

All investigations in the sciences of genetics and evolution during the twentieth century have shown that the only mechanism that can explain evolution is natural selection. However, in the nineteenth century other 'mechanisms' were put forward. None has stood up to biological or mathematical examination.

One was based on the idea that changes in the phenotype that an organism acquired during its lifetime could be inherited by its offspring. If this were so it could produce adaptation and so evolution. For example, a giraffe-like mammal might develop a long neck in its lifetime and pass on this feature to its offspring which would be born with long necks.

We have only to reflect that however darkly tanned a European may become when exposed, even for years, to strong sunlight, his or her offspring are always born white skinned, to see that this suggestion is not likely to be correct. To take another example, Jews have practiced circumcision on their boy babies for at least five thousand years, yet Jewish boys are always born with foreskins.

This could not be otherwise, because such alterations in the phenotype cannot produce the

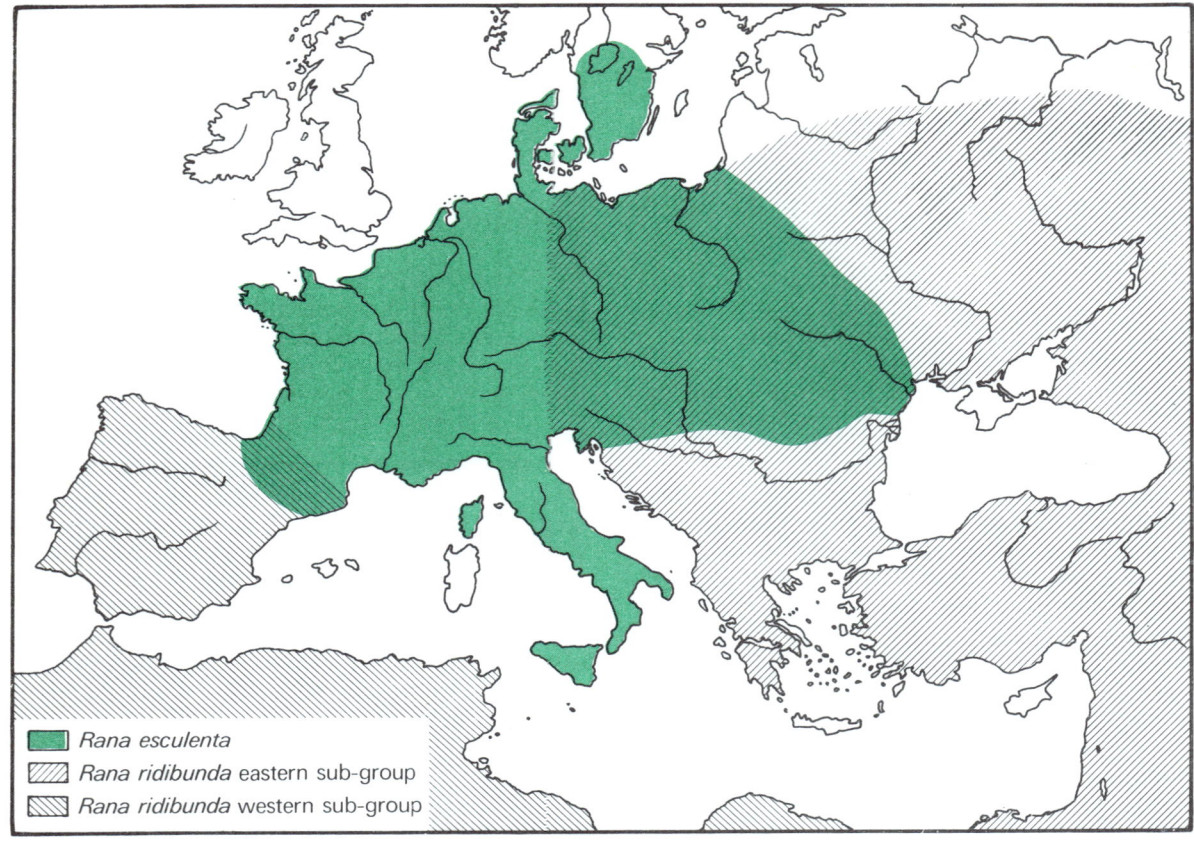

Fig. 20.24 Distributions of *Rana esculenta* and *Rana ridibuada*. Although the species overlap they rarely interbreed because their breeding seasons are different.

genetic change in the gametes that is necessary to give rise to an inherited change in the offspring. Furthermore we have just seen that mutation must be a rare event in order to preserve the genetic stability of the individual and the population. If changes in the individual's phenotype *were* to produce mutations in his gametes, the genetic material would be too unstable to preserve the organism's identity.

Perhaps this appears too obvious to need saying, yet as recently as the late 1940s belief in the genetic inheritance of characteristics acquired in the individual's lifetime was enforced by law in the Soviet Union. This line was favoured by Stalin, who liked to imagine that human behavioural changes could be inherited in this way.

He backed up phoney experiments on tomatoes by an agriculturalist called Lysenko which appeared to establish that such changes could be inherited. All orthodox geneticists in the Soviet Union were liquidated. 'Western' genetics text-

books which mentioned Mendelian genetics were all destroyed. It was declared a criminal offence to be caught reading one of them.

The origin of species
Because evolution is taking place, species are continually changing. For example, two species of fruitfly are known, *Drosophila pseudoobscura* and *Drosophila persimilis*. If these are really separate species, they must not interbreed, and they normally do not. But at low temperatures matings do take place which produce live, fertile offspring.

When the species have diverged further so that (for example) their courtship behaviour differs, or their genitalia are no longer physically matched, they will be completely separate species.

Factors that prevent interbreeding
As a species enlarges and radiates from its place of origin, its members meet with various environmental conditions. Natural selection causes them

to evolve into sub-populations which are called **races** or **sub-species**. We have seen how closely natural selection attunes populations of *Cepea nemoralis* to its various environments.

Eventually the accumulated differences are great enough to prevent interbreeding between the two sub-groups. This can occur for a variety of reasons. Some of these are:

1 The breeding seasons of the two sub-groups may become different. In mammals, the time of coming on heat may be different.

2 The female of one sub-group may refuse to accept the male of another sub-group.

3 Anatomical differences may make copulation impossible. (We noted on page 268 that Alsatian and Pekinese dogs would find it difficult to interbreed even if the female would accept the male.)

4 The sperms of one sub-group may be unable to penetrate and so fertilize the ova produced by another sub-group. Penetration depends, amongst other things, on the exact chemical composition of the wall of the ovum and (in mammals) on the chemical composition of the fluid in the Fallopian tubes.

5 If the partner chromosomes differ in many genes they will not pair with one another in the prophase of reduction nuclear division. This means that the whole meiotic process is upset and functional gametes are not formed. Thus even if a zygote is formed which matures to an adult, the

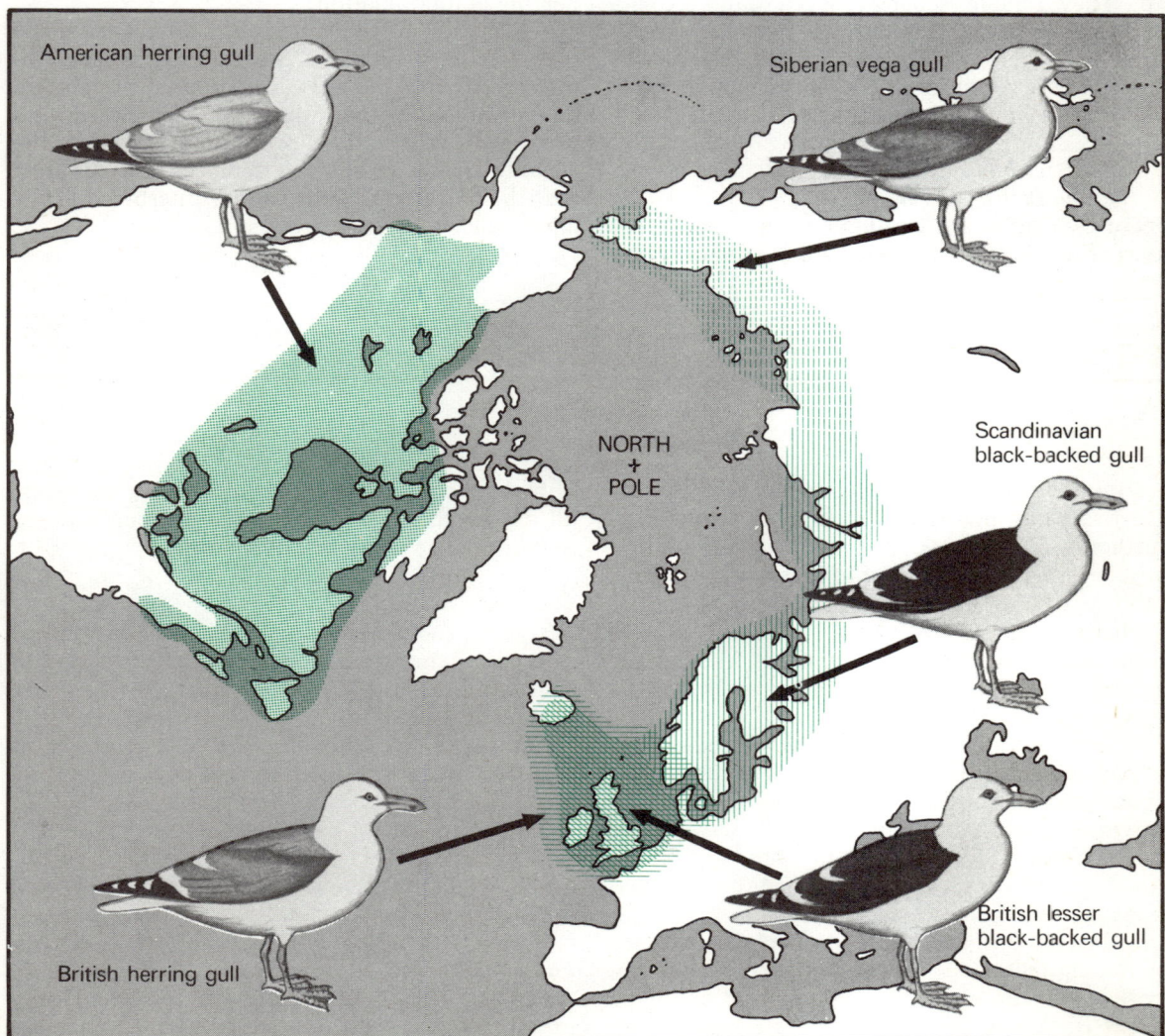

Fig. 20.25 Distribution of gull populations round the North Pole. For explanation see text.

adult may be sterile. This is so in the case of the mule, a sterile hybrid formed between a he-ass and a mare, and the tigron, a sterile hybrid formed between a lion and a tiger.

Two examples of populations on the point of forming two species

1 European edible frogs

There are two species of European edible frog. *Rana esculenta* is distributed throughout central Europe. Two sub-groups of *Rana ridibunda* adjoin it (Fig. 20.24), one in Spain and North Africa, the other in Eastern Europe.

The geographical ranges of the two species overlap. They remain separate species and do not interbreed because *R. ridibunda* normally breeds three weeks before *R. esculenta*. Very occasionally local climatic conditions delay the breeding season of *R. ridibunda* and the two species then interbreed.

2 The lesser black-backed gull and the British herring gull

These two gulls are common round the British coast (Fig. 20.25). They are distinct species (*Larus fuscus graelsii* and *Larus argentatus argentatus* respectively) and do not interbreed under any conditions. They differ in appearance (feather colour and leg colour) and in habits. The lesser black-backed gull breeds inland and migrates in winter, while the British herring gull nests on cliffs and does not migrate.

As Fig. 20.25 shows, different geographical races of these gulls occupy a ring-shaped range round the North Pole. The lesser black-backed gull grades into a sub-group, the Scandinavian black-backed gull, with which it interbreeds. Further east still the Siberian vega gull population grades into another subgroup, the American herring gull. This in turn grades into the British herring gull population.

While each of the sub-groups differs from the others in its phenotypic characteristics, each will interbreed with the next. An original population has diverged in two directions, eastward and westward. Where they overlap (in Britain), the accumulated hereditary differences are so great that the two populations cannot interbreed. They have evolved into two species.

Glossary

Absorption Taking up of dissolved material from the environment into the organism, usually into living cells. Absorption can be by diffusion (e.g. taking CO_2 into palisade cells of the leaf), when no chemical energy is expended. Absorption is often *active*. Here dissolved materials are absorbed *against* the diffusion gradient, and this requires the expenditure of chemical energy, for example in the active absorption of certain ions by root hair cells from soil solution.

Selective absorption Absorption of certain dissolved substances from the environment in preference to others which may not be absorbed at all. Selective absorption can only be active.

Accommodation (Fig. 16.7). Change in the focus of the lens in the eye. Brought about in man by changing the shape of the lens from thin to fat. Accommodation is an automatic, reflex action.

Adaptation Any feature of a living organism which improves its chances of survival. Lungs are an adaptation in land-living vertebrates that enable them to survive by using oxygen gas from the atmosphere rather than the dissolved oxygen in water that aquatic animals use. The production of vast numbers of offspring by tapeworms is an adaptation which helps them to find a new host. Evolution consists of the continual refinement of adaptations by the natural selection of those which enable their possessors to survive longer and so leave more descendents.

Adaptive radiation (Fig. 20.14). The evolution from one ancestral stock of a range of organisms each adapted to a distinct mode of life in a particular environment. For example, when the dinosaurs became extinct about 60 million years ago, the mammals underwent adaptive radiation. Till then they had been small animals with little variety of form. Now they diversified into the niches in nature formerly occupied by the dinosaurs, and evolved swimming, flying, running, and burrowing forms. Some evolved into herbivores, others into carnivores, and some of the resulting adaptations are described in chapter 6.

ADP, ATP Adenosine diphosphate and adenosine triphosphate. ATP is able to enter into very many chemical reactions in the cell in which it passes on one of its phosphate groups (P) *plus* chemical energy. Thus substance X may react with ATP to become $X \sim P$, while ATP becomes ADP. $X \sim P$ contains more chemical energy than X. Because of this it will now react with other chemicals with which X by itself could not react. ATP thus passes on chemical energy for cell reactions. It is the 'energy currency' of the cell. ATP is constantly consumed (being converted into ADP). It is replenished by the chemical reactions of tissue respiration. One molecule of glucose when respired aerobically regenerates 40 molecules of ATP from 40 molecules of ADP and 40 'molecules' of phosphate, P.

Afterbirth (Fig. 12.27). The placenta and umbilical cord which are expelled from the womb shortly after the birth of the baby.

Alveolus (Fig. 7.18) Minute air-filled sac in the lung. Alveoli are very numerous, very thin-walled, and are intimately bound-up with blood capillaries. Exchange of oxygen and carbon dioxide takes place across the alveolar wall cells.

Amino acid Chemical substance which forms one 'building block' of a protein. About 20 different amino acids occur in proteins; one protein molecule may contain thousands of amino acids. Each amino acid contains an 'amino-group' $-NH_2$. Amino acids in the animal body that are not needed for making proteins are 'de-aminated' in the liver.

Essential amino acids are those out of the twenty different sorts which the animal needs and which *cannot* be made by an animal from other amino acids that have been obtained from proteins in the food. It is therefore essential for the animal to eat foods that contain these amino acids.

Amnion (Fig. 12.21). Fluid-filled sac surrounding the developing embryo in the womb. The fluid protects the embryo from the pressure of the mother's organs and from external knocks, and acts as a heat insulator helping to keep the embryo at an even temperature.

Antibody Chemical substance formed from the protein globulin in the blood plasma. Antibodies combine with and neutralize foreign bodies, such as bacteria, which may have invaded the body.

Antibiotic Chemical substance produced by bacteria and fungi which diffuses out of the cell and kills other bacteria and fungi. Some antibiotics (e.g. penicillin, streptomycin) are very good disease-curing drugs because they kill bacterial cells but not human cells.

Antiseptic Chemical substance that kills bacteria.

Antitoxin Antibody which combines with and neutralizes poison (toxin) released into the body by bacteria.

Areolar tissue A kind of connective tissue. Surrounds nerves, blood vessels, and other structures and is widely distributed in the animal body. Forms an important 'no-man's land' across which nutrients, oxygen, and water, are exchanged between blood capillaries and body cells.

Articulation Part of the skeleton where two bones form a movable joint, e.g. the articulation of the head of the femur with the hip girdle.

Artificial selection Change in the genotype of a population produced by man by encouraging certain matings and preventing others. Many varieties of dogs, horses, cattle, sheep, wheat, and maize (to cite only a few) have been produced in this way.

Asexual reproduction Reproduction without the formation of gametes. Can take many forms, viz vegetative reproduction in plants (e.g. bulbs, corms, runners, rhizomes, rooting leaves). Uncommon in animals. The offspring are genetically identical with the parents and can form a clone.

Auxin A plant hormone which affects the elongation of plant cells. Local differences in auxin concentration in stem and root tips are thought to be the cause of phototropic and geotropic growth movements.

Biosphere The region of the earth's surface and the earth's atmosphere that contains living organisms.

Blastocyst (Fig. 12.14) Hollow ball of cells produced from the mammalian zygote and still surrounded by the wall of the ovum.

Blending inheritance Nineteenth century notion about the nature of inheritance, namely that if two organisms differing in a particular feature mated, their offspring would show an intermediate condition of the feature. Darwin believed this, and so could not understand how natural populations possessed variety, for it 'should' all be bred out in a few generations. Blending inheritance was proved to be a false idea byMendel. For example, mating a true-breeding pea plant which possessed yellow seeds with another which possessed green seeds did not produce offspring with yellowish-green seeds. Instead all the seeds were yellow.

Bronchus One of the two main branches of the trachea (windpipe).

Bulb (Fig. 3.16) A modified, swollen bud. Contains a very short stem which bears crowded, fleshy storage leaves. Bulbs provide a means of perennation (persistence from year to year) and of vegetative (asexual) reproduction. A clone could be raised from bulbs, for example a bed of tulips of a particular variety.

Cambium (Fig. 10.2) A plant tissue. Its cells are small and brick-shaped. They are living and have thin cellulose walls, but unlike most plant cells, they have no central permanent vacuole. Cambial cells undergo cell divisions and so give rise in growing stems and roots to additional phloem and xylem. Phloem cells and xylem cells differentiate from the cells produced by the cambium.

Carbohydrates Chemical substances containing carbon, hydrogen, and oxygen only, these being present in each molecule in the ratio $1C:2H:1O$, i.e. CH_2O. Sugars are carbohydrates, e.g. glucose, $C_6H_{12}O_6$. Starches and cellulose are also carbohydrates. Sugars and starches are important sources of chemical energy for the cell.

Carpel (Figs. 13.1, 13.4) The female reproductive organ of flowering plants. Consists of a receptive surface for pollen grains (the stigma) and a chamber, the ovary, which contains one or more ovules. Each ovule contains one haploid female gamete produced by reduction nuclear division (meiosis). In some species there is a stalk, the style, between the stigma and the ovary. After fertilization the zygote develops into the embryo, the wall of the ovule develops into the testa of the seed, and the wall of the ovary develops into the fruit.

Catalyst A substance which, when present in minute amounts, causes the rate of chemical change in a particular chemical reaction to increase. Catalysts are not used up during the reaction. They have the same mass and chemical composition at the beginning and at the end of the reaction. Most biological catalysts are enzymes.

Cell membranes (Figs. 1.8, 1.11) These are thin sheets which separate the cell from the world outside it (plasma membrane) or separate parts of the cytoplasm from one another (e.g. endoplasmic reticulum, nuclear membrane). They all have a similar basic structure which is like a sandwich of two outer layers made of proteins surrounding a middle layer made of fats. One of their most important properties is that they all allow water to pass freely across them, but may or may not allow substances dissolved in water to pass. They thus help to prevent the many dissolved chemicals in the cell from becoming mixed up.

Cell wall (Fig. 5.6) A feature of plant cells only. The cell wall is not part of the living cytoplasm, though it is made by it. In parenchyma and phloem it consists of cellulose (a polymer of glucose). In xylem vessels and cork cells, other chemicals permeate the cellulose wall making wood and cork respectively.

Cellulose Chemical substance which forms the cell wall of plant cells (Fig. 1.16). Cellulose is a polymer which, like starch, is made of thousands of molecules of glucose linked together. The chemical link between the glucose molecules is different in the two substances, however, and cellulose is not attacked by starch-digesting enzymes or by any enzymes found in the guts of animals.

Centromere (Fig. 14.13) Region of the chromosome specialized for attachment to the spindle.

Cerebellum (Figs. 15.21, 15.22) Roof of the hind brain. In mammals it has an outer layer of grey matter, the cerebellar cortex. The cerebellum is responsible for the coordination of complex muscular movements such as posture, walking, and manipulation.

Cerebrum (Figs. 15.21, 15.22) Roof of the fore-brain. Prominently developed in mammals, especially man, where its outer part consists of grey matter and is called the cerebral cortex. Responsible for high-level mental activity such as learning, memory, and thought.

Chlorophyll Green pigment found in some plant cells. An essential component of photosynthesis: it is chlorophyll which converts light energy into chemical energy. In plant cells other than those of blue-green algae, chlorophyll is confined within chloroplasts.

Chloroplast (Figs 5.7, 5.8) An organelle found in those plant cells which carry out photosynthesis. Photosynthesis takes place within the chloroplasts. Chlorophyll is confined to the piles of sacs (grana) and single sacs which are embedded in the stroma. Chloroplasts, mitochondria, and the nucleus, are the most elaborate organelles in eukaryotic cells.

Chromatid (Figs 14.11, 14.14) One of two identical strands making up the chromosome. Chromatids are seen (a) during the prophase and metaphase of mitosis (b) during the whole of meiosis up to metaphase II. At anaphase of mitosis and anaphase II of meiosis, the separating sister chromatids are called chromosomes.

Chromosomes (Fig. 14.11, 14.14) Cell organelles contained within the nucleus of eukaryotic cells. They consist largely of DNA (deoxyribonucleic acid) which is a chemical code in the form of a long molecule whose shape is a double helix. Distinct regions of the DNA form the code which can 'tell' the cytoplasm how to make a particular protein. These distinct regions are called genes. A chromosome resembles a string of beads in which each bead is one gene.

Chromosomes are only visible when the nucleus is dividing. At other times (that is, during interphase) they are very long, thin, and invisible in electron micrographs.

All the cells of animals and flowering plants *except the gametes* contain chromosomes in sets of identical pairs. In nuclei of the body cells of man, for example, there are 46 chromosomes consisting of 23 identical pairs of *partner chromosomes*. The sex cells or gametes (sperm and ova) contain only one of each pair of partner chromosomes. The sperm of a man and the ova of a woman therefore contain 23 chromosomes each, each one being one member of one pair of partner chromosomes.

Clone The descendents produced from a single 'parent' by asexual reproduction. Members of a clone are genetically identical.

Clotting Conversion of liquid blood to a jelly. Occurs when blood vessels are injured and is an important wound-healing reaction, preventing loss of blood.

Cork A plant tissue composed of dead cork cells. These 'cells' consist of cellulose cell walls in which fatty material has been deposited. Cork forms an outer layer of woody stems and roots. It protects the inner parts from physical blows and provides an insulating wrapping which prevents stem tissues from freezing in winter. The first 'cells' to be drawn (by Robert Hooke in 1665) were cork 'cells'.

Cotyledon See seed.

Cross A mating.

Cuticle (*plants*, Fig. 5.4) Thin varnish-like layer secreted by epidermal cells of stems and leaves (not roots) and covering the entire plant surface above ground except where there is bark. Prevents loss of water vapour and entry of disease-causing microorganisms. (*insects*, Fig. 18.15) Non-cellular layer secreted by the epidermis and forming the exoskeleton. On its outer surface is a very thin layer of wax which effectively prevents evaporation of water from the body.

Cytoplasm (Fig. 1.7) All the material in a living cell except the nucleus. The cytoplasm is bounded on the outside by the plasma membrane. It contains various organelles such as the endoplasmic reticulum, Golgi apparatus, lysosomes, mitochondria, and vacuoles, embedded in the ground substance. This appears to have no structure in electron micrographs. In plant cells the cytoplasm contains chloroplasts and a large central permanent vacuole. The non-living cellulose cell wall of plant cells is not part of the cytoplasm.

Deamination See amino acid.

Deficiency disease Nutritional disease caused by *lack* or *absence* of a particular vitamin from the diet, e.g. scurvy.

Diastole Phase of the heart beat when the heart muscle relaxes, hence auricular diastole, ventricular diastole.

Dicots A group of flowering plants, sometimes called the broad-leaved plants. Most species of flowering plants are dicots. 'Dicot' means 'two cotyledons' (two seed-leaves). Dicots also have net-veined leaves and flowers with sepals, petals, stamens, and carpels in fours, and fives. Many dicots are trees.

Diffusion Movement of freely-moving molecules or ions from a region where they are concentrated to a region where they are scanty. Diffusion applies to gases, vapours, and to dissolved substances. If a substance diffuses from region A to region B, a state will be reached when its concentration is uniform throughout A and B. Molecules of the substance will then move from A to B and from B to A with equal frequency.

In many biological situations chemical substances are being produced or consumed, thus giving rise to diffusion situations. Carbon dioxide diffuses into the leaf because it is consumed in the chloroplasts; oxygen and dissolved foods diffuse from the plasma through capillary walls into connective tissue because they are being consumed by the tissues.

Diffusion gradient The diffusion pathway from the region of high concentration of a gas, vapour, or dissolved substance, to the region of low concentration.

Differentiation The division of the parts of a cell, of cells, of tissues, or of organs, into specialized parts. Thus a eukaryotic cell is differentiated into the nucleus, mitochondria, etc.; an organism such as man is differentiated into muscle cells, nerve cells etc.; and into muscle tissue and nerve tissue etc.; and into a brain, a spinal cord, various muscles, the various bones of the skeleton, etc..

Diploid (Fig. 12.1) A nucleus containing two sets of chromosomes, i.e. two of each pair of partner chromosomes. Denoted by 2n. Mammalian and flowering plant zygotes (for example) are diploid, being formed from the union of two haploid gametes. All the cells derived from these zygotes are diploid, except the sex cells. Organisms whose body cells possess diploid nuclei are referred to as diploid organisms.

DNA See chromosomes.

Dominant allele Allele which produces its effects in

the heterozygote to the exclusion of the recessive allele.

Dormancy Period during which seeds will not germinate even though the conditions of water supply, oxygen supply, and temperature, are favourable.

Dormant A dormant organism is in a resting condition. Its metabolism is inactive. In plants it is often associated with the dispersal of seeds and spores. In animals it is often associated with the avoidance of adverse climatic conditions.

Effector cell Cell which produces an action, e.g. muscle cells which contract, or glandular cells which secrete.

Endocrine gland See hormone.

Endodermis (Figs. 1.3, 10.7) The innermost cell layer of the cortex. In the root its cells are brick-shaped with no spaces between them. Because of a fatty strip running round the upper, lower, and side walls of each endodermal cell, water (and dissolved substances) cannot soak through the endodermal walls to get to the xylem, but has to pass through the living contents of the endodermal cells.

Endometrium See womb.

Enzyme A biological catalyst composed of protein. Most of the hundreds of chemical reactions in the cell proceed very slowly at room temperature. Enzymes catalyse nearly all cell reactions. Much of the cell protein consists of enzymes, and like other proteins, the coded information 'telling' the cytoplasm how to make them is contained in DNA in the nucleus. Most enzymes catalyse reactions within the cell (such as the steps in the oxidation of glucose). However, some are secreted and do their work outside the cell, for example salivary amylase in saliva and pepsin in gastric juice.

Epidermis The outermost layer of cells of a plant or animal.

Eukaryota (Greek, proper cells). All organisms except bacteria and the blue-green algae. Their cells are very varied but differ basically from the Prokaryota in four ways. (1) The genetic material, or hereditary instructions, is organized into organelles called chromosomes which are situated in the nucleus and separated from the cytoplasm by the nuclear membrane. (2) Their cytoplasm is permeated by the endoplasmic reticulum, which forms conducting channels for chemical substances. (3) Photosynthesis takes place not throughout the cytoplasm as in blue-green algae, but in specialized organelles called chloroplasts. (4) Chemical energy (ATP) is produced not throughout the whole of the cytoplasm as in prokaryotes, but mainly in specialized organelles called mitochondria.

Evolution To define evolution we have to embrace several ideas:

1 Cumulative changes take place in populations of sexually-reproducing organisms.
2 These changes are due to changes in the genotypes of the populations and are therefore inherited.
3 The process is slow, taking place over many generations.

We may therefore say that evolution is the gradual cumulative change in the characteristics of populations of organisms related to each other by descent. It takes place over long periods of time, and therefore over many generations.

Excretion Getting rid of waste chemical products, e.g. the excretion of carbon dioxide produced by the respiration of all cells.

Exponential growth Growth by compound interest, that is where a given quantity increases by a constant percentage in a constant time.

Fat Chemical substance made by the union of a molecule of glycerol and three molecules of fatty acids. Fats are all insoluble in water, but they themselves can dissolve many of the substances present in cells. They form the middle layer of the 'sandwich' of cell membranes. Like carbohydrates, they are important reserves of chemical energy. (Their molecules, too, contain only C, H, and O, but the ratio of these atoms is not $1:2:1$ as in carbohydrates.) Fats also provide a heat-insulating layer in animals, for example in the dermis of the vertebrate skin.

Fertilization The union of the haploid male and female gametes to form a single cell, the diploid zygote. The essential feature of sexual reproduction.

Fibrinogen A protein dissolved in the blood plasma which plays a vital part in wound healing. It is converted into minute insoluble threads of fibrin which form a meshwork over the wound.

Fruit The ripened ovary of a flower containing the seeds. Fruits may be dry (e.g. Fig. 13.12) or succulent (e.g. Fig. 13.16). Fruits play an important role in dispersing seeds in a wide variety of ways.

Gamete Sex cell. See sperm and ovum.

Gene A small part of the genetic material concerned with the control of a particular developmental process. Consists of a particular length of the DNA double helix. See also chromosomes.

Gene mutation Sudden change in the chemical structure of a gene causing it to change from one allele to another, e.g. $T \xrightarrow{\text{mutation}} t$. The new allele reproduces itself over many generations. Eventually some of the copies will mutate in the reverse direction (e.g. $t \xrightarrow{\text{mutation}} T$).

The most important mutations are those occurring in the gamete-forming cells, the gametes themselves, or the zygote, for then the mutant allele will be present in every cell of the developing organism, and eventually in half of its gametes (Fig. 14.18). Mutation is *and has to be* very infrequent (see p. 281). It provides the ultimate source of genetic variation between organisms, though the re-shuffling of the genetic material in reduction nuclear division (meiosis) provides most of this genetic variation.

Genetic material The chromosomes and the 'chemical messengers' that pass from them to the cytoplasm.

Genotype The genetic constitution of an organism.

Geotropism Growth movement in plants in which the organ grows towards or away from the one-sided effect of gravity. Primary roots are *positively* geotropic (grow downwards in the plane of gravity). Primary stems are usually *negatively* geotropic. Secondary roots and stems are usually not affected.

Germination Commencement of growth of dormant structures which will develop into mature organisms, e.g. by seeds and spores.

Globulin See plasma.

Glycogen A starch which occurs in animal cells, notably in muscle cells and liver cells (Fig. 6.1). It is a polymer made of thousands of glucose molecules linked together. These can be unlinked by enzymes in the cell, providing it with a fuel (glucose), which is the source of chemical energy (ATP).

Grey matter (Figs. 15.15, 15.22) Nerve tissue in the central region of the spinal cord and forming an outer layer (cortex) of the cerebrum and cerebellum in mammals. Contains the cell bodies of nerve cells and the end-branches of other nerve cells which form synapses with them. It is surrounded by, or surrounds, the *white matter* which consists largely of nerve fibres.

Granum See chloroplast.

Ground substance (*of cells*) The ground substance of the cytoplasm fills in the spaces between the organelles. Dissolved chemicals move freely across it by diffusion.

(*of connective tissues*). Structureless material filling in the space between cells and fibres of the connective tissue. Tissue fluid is constantly passing into and out of the ground substance.

Guard cell See stoma.

Habitat A particular place in the biosphere and the organisms that it contains.

Haemoglobin Red pigment in solution in the red blood cells of vertebrates. Haemoglobin is a protein. It has a strong affinity for oxygen, combining with it to form oxyhaemoglobin. This reaction is peculiar in that its reverse takes place very easily. Haemoglobin is thus able to transport oxygen in the body. If the surroundings contain much oxygen (as in the capillaries of the lungs), haemoglobin combines with it to form oxyhaemoglobin. If the surroundings contain little oxygen, as in the respiring tissues, oxyhaemoglobin breaks up into haemoglobin and oxygen.

Haploid (Fig. 12.1) A nucleus containing one set of chromosomes, i.e. one of each pair of partner chromosomes. Denoted as n. Sperms and ova of animals are haploid, as are pollen grains and ova of flowering plants. See also chromosome.

Hermaphrodite See unisexual.

Heterozygous Carrying different alleles of a particular gene on each partner chromosome. Thus with respect to the alleles **T** and **t** in pea plants. **Tt** individuals are heterozygous. The gametes from such an individual are of two kinds with respect to this gene, and Mendel assumed that both kinds (**T** and **t**) would be produced in equal numbers both by the male (pollen grains) and by the female (ova).

Holophytic nutrition Plant method of nutrition by making life chemicals such as proteins, fats, and carbohydrates, from inorganic materials such as carbon dioxide, water, and ions absorbed from soil solution. Photosynthesis, the making of carbohydrates from carbon dioxide and water using light energy, is the start of all these holophytic reactions. Holophytic nutrition is the hallmark of plants.

Holozoic nutrition Animal method of nutrition by taking into the body other animals, plants, or their products. These are then digested and provide the animal with the life chemicals that it cannot make for itself, mainly proteins, fats, and carbohydrates. Holozoic nutrition is the hallmark of animals.

Homozygous Carrying the same allele of a particular gene on each partner chromosome. Thus with respect to the alleles **T** and **t** in pea plants, **TT** and **tt** individuals are homozygous. The gametes from such an individual are all alike with respect to this gene.

Hormone Chemical substance produced in minute quantities in one part of an organism and transported to another (the 'target organ') where it produces a profound effect. Animal hormones are secretions produced in *endocrine*, or *ductless*, *glands* (Fig. 15.28). These are organs with this special function, for example thyroxine produced in the thyroid gland, adrenaline produced in the adrenal gland. Sometimes the 'gland' is mixed up with other, quite distinct tissue, for example the Islets of Langerhans in the pancreas which secrete insulin. Sometimes the hormone-secreting cells are very scattered, for example the secretin-secreting cells of the duodenum. In general, hormones produce long-term effects on their target cells. Since they are distributed in the blood and tissue-fluid to all the cells of the body and affect only certain cells, the target cells must be able to 'recognise' the hormone. This is a property of their plasma membranes. In plants, hormones produce many effects on growth, on cell division, cell enlargement, and cell differentiation.

Imago Adult insect which is sexually mature.

Implantation (Fig. 12.15) Attachment of the mammalian blastocyst to the endometrium and its embedding within the endometrium.

Inbreeding Mating of closely related individuals of a species. Inbreeding over several generations leads to loss of genetic variety. The stock becomes a pure line, for example Pekinese dogs.

Inflorescence (Figs. 3.2, 13.7(a), 13.8). A shoot bearing several flowers.

Ingestion Taking food into the body.

Interphase Time during which the nucleus is not dividing. Chromosomes cannot be seen during interphase as they are then very long and thin. During interphase the nucleus is performing its task of supplying the cytoplasm with coded instructions as to how to make proteins.

Invertebrates All animals other than the vertebrates, e.g. Protozoa, corals, starfish, insects, spiders, snails.

Isotope Some chemical elements exist in more than one atomic form. The different forms are called isotopes. For example, oxygen exists in two forms, denoted ^{16}O (the common form) and ^{18}O, a rare form. The difference between the atomic forms is due to different numbers of neutrons in the atomic nuclei, and all isotopic forms of any particular element behave in similar ways in their reactions with other chemicals. Some isotopes are radioactive. Their presence is easily detected with a Geiger counter. If a radioactive substance A reacts with substance B to produce C and D, it is possible to find whether the radioactive part of A ends up in C or in D.

Kimetochore see centriole

Larva Young, sexually immature, form of an animal having a distinctly different appearance from the adult. Caterpillars are larvae of some insects.

Ligament (Fig. 17.21) Tough, unstretchable band of tissue joining two bones together. Made of bundles of collagen fibres.

Meiosis (reduction nuclear division) (Fig. 14.14) The two nuclear divisions which bring about the halving of the diploid chromosome number. In animals and flowering plants meiosis results in the production of haploid gametes from diploid body cells. Partner chromosomes are separated in the first division (meiosis I), halving the chromosome number. The independent movement of chromosome pairs during anaphase I of meiosis causes the resulting gametes to be genetically different from one another (Fig. 14.17). This is a fact of fundamental biological importance, for it is the source of genetic variation among the offspring of sexually-reproducing organisms that provides the material for natural selection.

Menopause The age in women (about 45) at which the menstrual cycles cease.

Mesophyll (Fig. 5.4) The internal tissue of a leaf enclosed within the upper and lower epidermises (these are not part of the mesophyll). In many species the mesophyll is differentiated into an upper palisade layer, on which incident light falls, and a lower spongy layer.

Menstruation The monthly period in women in which the lining of the womb becomes detached and is passed out of the body through the vagina.

Metabolism The chemical processes taking place within an organism, or within a part of an organism.
Basal metabolism The rate at which energy is used by a human being who is physically and mentally relaxed. This is the energy needed to maintain breathing, heartbeat, and other essential processes.

Mitochondrion (Fig. 1.10) An organelle found probably in all eukaryotic cells, but mitochondria are most abundant in cells which need a lot of chemical energy in order to carry out their work, e.g. muscle cells and kidney tubule cells. Mitochondria have been called 'the power-house of the cell' because the chemical changes which generate 38 molecules of ATP per glucose molecule consumed in tissue respiration, occur inside them. They have an elaborate internal structure.

Mitosis Nuclear division in which there is first duplication of the chromosomes and then a separation of chromatids which migrate via the spindle to opposite poles of the cell. The two daughter nuclei are genetically identical.

Monocots A group of flowering plants. Their leaves are long and thin and have parallel veins. Their fllowers bear sepals, petals, stamens, and carpels in threes. 'Monocot' means 'one cotyledon' (one seed leaf). Only a few monocots are trees (e.g. palm). Many of the grasses, all of which are monocots, are important food plants, either providing grain for man or fodder for his domestic animals.

Motor end plate (Fig. 15.12) Structure at the end of a nerve fibre where it contacts a muscle fibre.

Motor nerve cell Nerve cell which conveys an impulse ('message') from the brain or spinal cord to an effector cell (a muscle cell or a gland cell).

Natural selection The mechanism which brings about evolution. Sexually-reproducing organisms produce genetically-variable offspring. Those possessing certain inherited characteristics will (perhaps by surviving longer) contribute more offspring to the next generation than will others. Such organisms have been 'selected' by nature. For example, in areas of Britain polluted by industrial smoke, nature selects melanic forms of *Biston betularia*: these forms leave more descendents than do the whitish forms, and the genotype of the population is thus gradually changed—that is, evolution takes place.

Nephron (Fig. 11.5) A unit in the kidney which filters blood and transforms the filtrate into urine. Consists of a renal corpuscle and a uriniferous tubule.

Nerve (Figs. 15.8, 15.9) A bundle of nerve fibres, some sensory, some motor, bound together by connective tissue.

Nerve cell (Fig. 15.4) Cell consisting of a cell body, drawn out into dendrites, a nerve axon, often surrounded by a fatty sheath, and branches which end either in synaptic bulbs or motor end plates.

Nerve fibre (Fig. 15.5) The axon of a nerve cell. Often surrounded by a fatty sheath.

Nitrogen fixation Conversion of nitrogen gas into compounds such as amino acids that can be used by living cells. Very few organisms possess the enzymes necessary to bring about nitrogen fixation. The bacterium *Rhizobium*, living symbiotically in nodules on the roots of leguminous plants, is one that does, and it forms an important link in the circulation of nitrogen in nature.

Node Place on a plant stem to which a leaf or leaf stalk is attached.

Nymph Young stage of certain insects (e.g. cockroach, locust) which resembles the adult (imago) in appearance except that it has no wings, or small wings. Nymphs are sexually immature.

Oogamy Sexual reproduction in which a large, immobile, ovum is fertilized by a small, usually free-swimming, sperm. Most sexually-reproducing organisms are oogamous.

Organelle A particular structure in the cytoplasm specialized to perform a certain task. Organelles mentioned in this book are the nucleus, the nuclear membrane, chromosomes, centrioles, the spindle, the plasma membrane, the endoplasmic reticulum, the Golgi apparatus, lysosomes, mitochondria, chloroplasts, food vacuoles (these are formed in association with lysosomes), and flagella. Note that cell membranes form important parts of nearly all these organelles.

Osmosis The movement of water across a partially-permeable membrane into a solution. Osmosis can be regarded as a special instance of diffusion where the diffusing substance is water. Since plasma membranes are all partially-permeable, osmosis plays an important part in the entry of water into cells and the loss of water from cells.

Osmotic pressure The pressure needed to prevent osmosis from taking place. Osmotic pressure depends on the *number* of dissolved particles in a given amount of water. Equal masses of (say) sugar and starch dissolved in the same volume of water therefore develop very different osmotic pressures. The mass of glucose contains very many more molecules than the starch, and so will develop a much higher osmotic pressure when dissolved. It is therefore not surprising that carbohydrates are usually stored in cells in the form of starches.

Outbreeding Mating among unrelated individuals of a species. Outbreeding results in genetic variety in the offspring, and the great majority of organisms are outbreeding.

Ovule See carpel.

Ovum (Fig. 12.12) Female sex cell. Its nucleus contains the haploid (n) number of chromosomes. Usually large, immobile, and containing food reserves for the developing embryo; some ova (e.g. those of birds) are very large. Mammalian ova are small and the mammalian embryo is fed from the mother's circulation via the placenta.

Oxygen debt Amount of oxygen consumed by animal tissue in excess of normal after a period of exercise during which oxygen supply has been inadequate. Voluntary muscle commonly incurs an oxygen debt.

Oxygenated blood Blood whose red cells contain oxyhaemoglobin rather than haemoglobin.

Parasite An organism which lives in or on another living organism from which it obtains its food.

Parenchyma (Fig. 1.3(a)) A plant tissue. Its cells are roughly spherical (that is, they are not elongated). They are living and have cellulose cell walls. Parenchyma is the 'packing tissue'. It forms the central pith of stems, and fills out the spaces between veins. It forms the bulk of fruits such as apples and plums.

Partially-permeable membrane. A membrane permeable to water but not to dissolved substances. The plasma membrane and other cell membranes are partially permeable to many dissolved substances, and cells therefore form osmotic systems.

Pectoral girdle (Fig. 2.4(b)) Bony support to which vertebrate fore-limbs are attached. In man, the shoulder blades.

Pelvic girdle (Fig. 2.4(b)) Bony support to which the hind limbs of vertebrates are attached. The backbone is fused to the pelvic girdle by the sacrum (Fig. 17.16).

Pentadactyl limb (Fig. 2.10) The kind of limb found in amphibians, reptiles, birds, and mammals. First known appearance is in Devonian air-breathing fish (Fig. 17.17). Basic pattern consists of an upper limb bone, two lower limb bones, a set of wrist or ankle bones, a set of palm or foot bones, and a set of finger or toe bones. This pattern has undergone considerable adaptive radiation and is used in various vertebrates for swimming, flying, running, burrowing, grasping, and manipulation.

Peristalsis A wave of muscular contraction passing along a tubular organ such as the intestine or the ureter. Used to mix and propel the contents of the tube.

Pharynx (Fig. 18.35(a)) The region of the vertebrate gut between the mouth and the oesophagus.

Phenotype Any observable feature of a living organism discovered by using any possible test. Individuals may possess different genotypes and yet have the same phenotype (e.g. **TT** and **Tt**). The phenotype is the product of the genotype and the environment, For example, a genotypically tall pea plant (**TT** or **Tt**) has the genetic potential to grow into a tall plant, but it may grow no taller than a genotypically short pea plant (**tt**) if it does not receive enough water.

Phloem (Figs. 10.15, 10.16) A plant tissue. It contains two kinds of cells which are involved in the movement of sugars, amino acids, and other chemicals made in the plant. Most of these are made in the leaves and the phloem transports them to growing tips of shoots and roots, and to storage organs such as developing bulbs and tubers. Phloem contains sieve tubes and companion cells, which are known to do the transporting, but *how* transport takes place has not yet been discovered.

Photosynthesis Synthesis by green plants of carbohydrates from carbon dioxide and water using light energy. Oxygen is liberated as a waste product. The process may be summarized as:

$$CO_2 + 2H_2O \xrightarrow[\text{chlorophyll}]{\text{light energy}} \underset{\text{carbohydrate}}{CH_2O} + O_2 + H_2O$$

Carbohydrates are the source of chemical energy for all cells, and photosynthesis is therefore a supremely important process for life. It changes light energy, which cannot be used to drive chemical reactions in the cell (other than some of those involved in photosynthesis itself) into chemical energy in the form of ATP, which can be so used. Photosynthesis has also produced the oxygen in the earth's atmosphere without which aerobic respiration would not be possible.

Phototropism (positive) Growth movement shown by stems and sometimes leaves in which the organ grows towards the light source.

Pituitary gland (Fig. 15.21) Small endocrine gland at the base of the vertebrate brain which produces many hormones. Some of these affect body cells directly (e.g. growth hormone), others affect other endocrine glands and cause them to secrete their hormones.

Placenta Part of the wall of the womb. The embryo's blood is delivered to the placenta where it passes into capillary tufts which lie in pools of the mother's blood (Fig. 12.22). Oxygen, glucose, and amino acids pass by diffusion from the mother's blood into the capillary tufts, and the embryo's waste substances pass by diffusion from the tufts into the mother's blood.

Plankton Collection of small, mostly minute, animals and plants floating at or near the surface of the sea and lakes.

Plasma Liquid part of the blood in which the blood cells float. The plasma proteins, albumin, globulin, and fibrinogen, are dissolved in the plasma, as are soluble foods, wastes, hormones, and other substances. Plasma gives rise to tissue fluid, and is the main transport medium of the body.

Plasmolysis (Fig. 10.8) Shrinking away of the living cell contents from the cellulose wall when a plant cell is placed in a solution osmotically stronger than its cell sap. Water passes from the central permanent vacuole into the surrounding solution by osmosis, causing the vacuole, and thus all the material inside the cell membrane, to shrink.

Plumule See seed.

Pollen grain Small spore produced by reduction nuclear division (meiosis) in the anthers of flowering plants. They contain two *haploid* nuclei, one of which is the male gamete. This is brought very close to the female gamete (ovum) in the ovary of the carpel by the growth of the pollen tube. Flowering plants are oogamous, but the male gametes do not swim to the female gametes as sperms do. Dispersal of the male gamete in pollen grains has been one of the major features contributing to the success of the flowering plant group.

Pollen tube (Fig. 13.4) Outgrowth of the pollen grain which grows down the style, into the ovary wall, and into the chamber of the ovary. It penetrates the ovule through a small hole, the micropyle. The two haploid nuclei from the pollen grain are then deposited next to the haploid female gamete, the ovum. One of these fuses with the ovum, thus bringing about fertilization.

Pollution Release by man of substances (e.g. CO_2, SO_2) or of energy (e.g. heat) into the environment in quantities that damage health or resources.

Polymer A chemical substance with a large molecule formed by the linking together of many smaller molecules of a particular type. For example, proteins are polymers whose molecules are made from molecules of amino acids linked together; starches are polymers made from molecules of glucose linked together.

Polymorph (Fig. 8.1) The commonest of the vertebrate white blood cells. The name refers to the many-lobed nucleus. They are scavengers, consuming bacteria and other particles which they digest with the aid of their many lysosomes (Fig. 1.15).

Prokaryota (Greek, first cells) The bacteria and the blue-green algae. They are simple organisms in the sense that their cytoplasm contains no nucleus, endoplasmic reticulum, Golgi apparatus, lysosomes, mitochondria, or chloroplasts. The earliest known fossils are of blue-green algae (Fig. 20.8). These are 1600 million years old, one third the age of the earth. Eukaryotes undoubtedly arose by evolution from the prokaryotes.

Protein Complex chemical substance made by the linking together of hundreds or thousands of molecules of amino acids. All enzymes are proteins. Proteins form part of all cell membranes, and so are structural parts of all organelles. Some dissolve in water (e.g. salivary amylase, pepsin); some are insoluble (e.g. the proteins which form the bulk of muscle cells, the proteins of hairs, nails, and the dead layers of the skin). The numbers of ways of linking 20 different amino acids in various sequences is infinite. Each eukaryotic cell contains coded instructions in the form of DNA in the chromosomes. These 'tell' the cytoplasm the correct amino acid sequence for the assembly of each particular cell protein.
First class proteins Proteins containing all the twenty different kinds of amino acid.

Puberty The age at which the sex organs mature and become functional.

Pulmonary circulation (Fig. 8.10) The 'section' of the mammalian circulation to and from the lungs, viz the right ventricle, pulmonary artery and its branches, capillaries of the lung, pulmonary vein and its branches, left auricle.

Pupa Stage between the larva and imago in insects. Immobile and does not feed, but very large changes in body organization take place at this stage.

Pure line A strain of organisms which is homozygous for many genes because of repeated inbreeding. Pure lines lack the genetic variety characteristic of

sexually-reproducing organisms because the source of this variety, genetic differences between the parents, is absent.

Radicle See seed.

Random fertilization A condition during fertilization whereby any male gamete stands the same chance of penetrating and fertilizing the ovum as does any other male gamete.

Recessive allele Allele which does not produce its effects in the heterozygote. Its effects can only be seen in the recessive homozygote of diploid organisms.

Recessive backcross Mating the heterozygous offspring with the homozygous recessive parent, e.g. **Tt × tt**.

Reflex action A simple, automatic piece of behaviour in which a predictable response is produced by a particular stimulus, for example blinking if solid matter falls on the cornea of the eye, coughing if a food particle enters the larynx (voice box). All reflex actions involve the brain or spinal cord, and the simplest *reflex arc* or reflex nerve circuit consists of a sensory cell, a sensory nerve, a motor nerve, and a motor cell (Fig. 15.17).

Renal corpuscle (Fig. 11.6) Part of the nephron in which blood is filtered under pressure. Plasma containing both dissolved foods and dissolved wastes (notably urea) passes from the renal corpuscle into the uriniferous tubule.

Respiration This word has several biological meanings:
a Breathing, pumping air into and out of the lungs, or water over gills.
b Taking in oxygen and giving out carbon dioxide.
c *Tissue respiration* The chemical reactions that liberate chemical energy from chemical fuel such as glucose.
d *Aerobic respiration* Tissue respiration that takes place in the presence of oxygen and consumes oxygen, for example:

$$C_6H_{12}O_6 + 6O_2 \longrightarrow 6CO_2 + 6H_2O + energy$$

e *Anaerobic respiration* Tissue respiration that takes place in the absence of oxygen and which does not consume oxygen. Yeast respires glucose in the absence of oxygen, producing ethanol and carbon dioxide:

$$C_6H_{12}O_6 \longrightarrow 2CO_2 + 2C_2H_5OH \text{ (ethanol)}$$

Rhizome (Fig. 3.12) Horizontal underground stem. Rhizomes provide a means of perennation (persistence from year to year) and of vegetative (asexual) reproduction.

Ribosomes (Fig. 1.11(b)) Particles coating the outside surfaces of many of the sacs and tubes of the endoplasmic reticulum. The DNA in the nucleus makes 'blue-prints' of 'how to assemble amino acids into particular proteins'. These 'blue-prints' are chemicals called messenger RNA. Messenger RNA molecules arrive at the ribosomes which carry out the instructions and make the appropriate protein.

Root hair An outgrowth of an epidermal cell of the root. Root hairs are living and are confined to a definite area a short distance behind the root tip. Their function is to absorb water and ions from the soil solution.

Saprophyte Organism which obtains complex food materials such as amino acids from other dead organisms. The saprophyte secretes enzymes on to the food (which must be wet or damp); these digest the food and the digested products diffuse into the saprophyte. Saprophytes are an essential link in the circulation of carbon, nitrogen, and other elements in nature.

Secretion Passing out of the cell of chemicals which have been made by it. For example, the secretion of hormones or of enzymes. Such materials are often 'packaged' into membrane-bound drops by the Golgi apparatus (Fig. 15.27 shows 'packages' of insulin being secreted from an Islet cell of the pancreas of a mouse).

Seed (Fig. 13.17) A dormant embryo plant that can be dispersed from its sedentary parent. Contains an embryo shoot (plumule), embryo root (radicle) and embryo leaves (cotyledons) which often contain food reserves and are the largest part of the seed. (In some seeds a dead tissue, the endosperm, forms the food reserve.) The seed is surrounded by the seed coat (testa) and is enclosed within a fruit. Seeds provide plants with a means of dispersing themselves both in space and in time.

Selection pressure A measure of the effectiveness of natural selection in producing a change in the genetic composition of a population of organisms.

Sensory cell Cell which detects a change in its environment. This change may be outside or inside the organism's body.

Sensory nerve cell Nerve cell which conveys an impulse ('message') from a sensory cell to the central nervous system, the brain or spinal cord.

Sexual reproduction Reproduction which involves the fusion of two haploid gametes to form a diploid zygote. A supremely important process because the formation of haploid gametes by reduction nuclear division (meiosis) causes the shuffling of the genetic material. Offspring produced by sexual reproduction are not genetically identical with their parents. They form the material on which natural selection acts, producing evolution.

Species A group of organisms which can interbreed but which cannot breed with any other species.

Sperm (Fig. 12.11) Male sex cell. Its nucleus contains the haploid (n) number of chromosomes. Sperms are produced in large numbers. They move in liquid by lashing their tails.

Spindle (Fig. 14.13) System of fibres (actually micro-tubules) formed during nuclear division. Chromosomes become attached to them by their centromeres and are drawn by them to the poles of the cell.

Stamen (Figs. 13.1, 13.2) Male reproductive organ of flowering plants. It is part of a flower and always contains pollen sacs (usually four) in which the haploid pollen grains develop by reduction nuclear division (meiosis). In most species the pollen sacs are supported by a stalk.

Starch Complex carbohydrate. Each molecule is a polymer of hundreds of glucose molecules linked together. Starches are very insoluble and provide the cell with a means of storing glucose without raising its osmotic pressure. Starches are common as food (energy) reserves in seeds, tubers, woody stems, and other parts of plants. In animals, glycogen (animal starch) is stored in the muscles and in the liver.

Stigma See carpel.

Stoma (plural, stomata) (Figs. 5.11, 5.12, 5.13) A pore (hole) in the epidermis of the leaf. Gases (e.g. CO_2) and water vapour pass through the stomata, which are very numerous. Each stoma is surrounded by two specialized epidermal cells, the guard cells, whose movements open and close the stomatal pore.

Stroma See chloroplast.

Symbiosis Intimate association between two dissimilar organisms from which both are presumed to benefit. Each member of the pair is a symbiont.

Synapse (Fig. 15.10) Place where a synaptic bulb at the end of a nerve fibre contacts the plasma membrane of another nerve cell body. Each synapse forms a valve, allowing the nerve impulse to pass in one direction only, namely onwards.

Synaptic bulb (Fig. 15.11) Structure at the end of a nerve fibre which makes contact with the plasma membrane of another nerve cell, forming a synapse.

Synaptic vesicle Minute package of a hormone present in the synaptic bulb. Discharge of this hormone causes the impulse to pass across the synapse.

Synovial joint (Fig. 17.21) Movable joint in the vertebrate skeleton. The articulating bones end in a pad of cartilage (gristle). The bones are held together by the joint capsule and the joint is lubricated by synovial fluid.

Synthesis A chemical reaction in which an elaborate molecule is made from simpler molecules. Energy must be supplied to make the synthetic reaction take place.

Systole Phase of the heart beat when the heart muscle contracts; thus auricular systole, ventricular systole. During systole the blood in the heart is squeezed.

Systemic circulation (Fig. 8.10) The 'section' of the mammalian circulation other than that to and from the lungs, viz the left ventricle, the aorta and its branches, the capillaries and veins leading back to the right auricle, the right auricle.

Tendon (Fig. 17.27) Tough, hard, unstretchable cords which connect muscles to bones. Made of bundles of collagen fibres.

Tissue fluid (Fig. 8.4) Liquid exuded from capillaries into the areolar connective tissue. It carries dissolved glucose, amino acids, and oxygen to the cells and dissolved wastes back from them. It is re-absorbed into the capillaries. It is filtered blood plasma that does not contain the plasma proteins, albumin, globulin, and fibrinogen.

Trachea *(land-living vertebrates)* (Fig. 7.15) The wind-pipe, leading from the back of the throat through the neck to the thorax, where it branches to form the two bronchi.

(insects) (Fig. 18.19) Breathing tubes leading from holes on the body surface (spiracles) directly to the tissues. Branch to form tracheoles which penetrate individual body cells.

Transpiration The loss of water vapour from the surface of a plant. Most water vapour is lost through the stomata, though some passes through the cuticle, depending on how thick this is.

True-breeding stock A stock of sexually-reproducing organisms that possess constant phenotypic characteristics from generation to generation. A pure line. True-breeding stocks are homozygous for many or most of their genes. No segregation is therefore possible, and the offspring resemble their parents, e.g. the fourteen stocks of peas used by Mendel, pedigree dogs such as West Highland terriers.

Tropism A growth curvature in a part of a plant caused by a stimulus acting from one side.

Turgor (Turgidity) A state in which a plant cell or a plant tissue can be. A turgid cell has its living contents pressed hard against its cellulose cell wall, which is slightly stretched. Turgor is responsible for the functional shape of many plant organs, for example those of leaves which wilt when they lose water and lose turgidity.

Umbilical cord (Fig. 12.22) Cord connecting the developing mammalian embryo to the placenta. Contains two arteries and a vein.

Unisexual Producing either male or female gametes but not both. Most animals are unisexual (e.g. man and woman). However, some produce both male and female gametes (e.g. hydra, earthworm, snail) and are said to be hermaphrodite. Many flowering plants are hermaphrodite.

Ureter (Fig. 11.2) Muscular tube leading from the kidney to the bladder.

297

Urethra (Figs. 12.4, 12.9(b)) Muscular tube leading from the mammalian bladder to the exterior. In man it is joined by the two vasa defferentia from the testes, and therefore conducts both semen and urine through the penis. In woman it is distinct and separate from the vagina, and conveys only urine to the exterior.

Uriniferous tubule (Fig. 11.5) Tubular portion of the nephron. Most of the water and all of the dissolved foods in the filtrate from the renal corpuscle are absorbed in the uriniferous tubule and returned to the blood stream.

Vacuole A liquid-filled space in the cytoplasm bounded by a membrane. Food vacuoles are used to digest materials inside the cell; contractile vacuoles are used to expel excess water from the cell; a large central permanent vacuole is present in all living plant cells except cambial cells.

Vector An animal which transmits parasites from one host to another.

Vertebra (Fig. 17.8) One of a chain of bones which together form the backbone or vertebral column of vertebrates. Each vertebra contains a hole, the neural canal (Fig. 17.10(b)). The soft spinal cord lies in the neural canals and so is protected by the vertebrae (Fig. 15.13).

Vertebrates Animals which possess a skull and backbone. There are five vertebrate groups, the fish, amphibia, reptiles, birds, and mammals.

Villus (Fig. 6.21) Finger-like or leaf-like projection. Villi increase the surface area across which exchange of materials can take place, e.g. intestinal villi increasing the surface area for the absorption of dissolved foods.
Microvilli Finger-like projections on the surface of individual cells, visible clearly only with the electron microscope. Examples are the microvilli of the columnar cells of the intestinal villi (Fig. 6.21(e)) and the microvilli of the cells lining the uriniferous tubules in the kidney (Figs. 11.7, 11.8).

Vitamin Complex chemical substance which an animal must obtain in its food, though it is needed only in minute amounts. *Lack* of a particular vitamin in the diet may lead to the development of a deficiency disease.

White matter See grey matter.

Womb (Fig. 12.9) A sac in the lower abdomen of a female mammal. The vagina leads from it to the exterior. Its lining (endometrium) receives the blastocyst which develops here into the baby. A special part of the womb concerned with nourishing the embryo is called the placenta.

Xylem (Fig. 10.1) A plant tissue. Most of its cells are dead. They are woody tubes up which water flows from the roots to the leaves. Dissolved in this water are ions absorbed from the soil solution by the root hairs. Xylem forms the bulk of the stems of trees, but in herbaceous stems and in leaves, xylem is only present in the veins.

Zygote A diploid cell formed when the haploid sperm penetrates (fertilizes) the haploid ovum. In flowering plants one of the haploid nuclei from the pollen grain fertilizes the haploid ovum to produce the zygote. The zygote nucleus contains all the genetic information necessary for the development of the adult organism.

Index